NEW HORIZONS

IN

CRIMINOLOGY

PRENTICE-HALL SOCIOLOGY SERIES

HERBERT BLUMER, *Editor*

NEW HORIZONS

IN

CRIMINOLOGY

Harry Elmer Barnes

Negley K. Teeters

THIRD EDITION

Englewood Cliffs, N. J.

PRENTICE-HALL, INC.

LIBRARY OF CONGRESS
CATALOG CARD NO.: 59-5873

First printing . . . January, 1959
Second printing . . . August, 1959
Third printing . . . May, 1960
Fourth Printing . . . August, 1961

PRINTED IN THE UNITED STATES OF AMERICA

6 1 4 4 9 – C

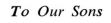

To Our Sons

Preface

Since the first edition of *New Horizons in Criminology* appeared in 1943, considerable emphasis has been placed on training in the fields of corrections and crime prevention. More college students are considering careers in these fields. Courses in criminology and corrections are increasing in universities and colleges, and correctional curricula are being developed in many institutions of higher learning.

Although this third edition may well serve as a basic text in such curricula, it is primarily written for the citizens of tomorrow—for those who will vote and who will eventually serve on committees, commissions, as members of social agencies, and as citizens who accept their share of community responsibility. It is also written for future journalists, social workers, teachers, jurists, lawyers, and members of legislatures.

The second edition of *New Horizons in Criminology* was published in 1951. One of its purposes was to appraise such new trends in crime and corrections during the war decade as the shifts in the character of organized crime and racketeering, the development of syndicated crime and its infiltration of and influence over the political scene, the growth of rural and juvenile crime, and the evolution of more humane and progressive philosophies of readjustment and rehabilitation of those sent to prisons for their crimes. Signal progress has been made in prison management, correctional personnel, and treatment procedures since the publication of the first edition.

Although a text must be authoritative, it should be readable and even interesting. This can certainly be realized without jeopardizing scholarship or academic discipline. In this third edition, the authors have attempted to achieve this objective.

This revision has been consummated with the advice of several teachers and correctional personnel across the country who have graciously made valuable suggestions regarding subject matter, illustrative material, photographing, and deletions that seemed expedient. The authors have warmly appreciated these friendly suggestions, and the table of contents reflects this advice. Many diverse features have been added. The authors have come to grips with the extent of crime in this country as compared to other countries. Some of the most glaring fallacies concerning crime, criminals, and prisons are appraised. The trend away from "penology" and toward "corrections" may be seen throughout this third edition. We have repudiated the word "convict" which is derogatory and merely engenders bitterness among those who seriously try to rehabilitate. Reforms in court procedures, law-enforcement techniques, prison programming, and parole and probation philosophy, although slow, are pointed out throughout. And, of course, statistics and information have been brought up to date.

We are more and more impressed by the number of instances in which average people, wronged by criminals, ask for understanding of the offender rather than for vengeance. This trend is a hopeful one and bears watching by us all. Another significant and gratifying trend is the insistence of jurists on pre-sentence investigation for the disposition of cases where guilt has been established. More and more judges are calling for professional help.

It has also been gratifying to the authors to see the wide use of the previous editions of this text among criminologists and correctional people in foreign countries. This is ample proof of the professional tie between American and foreign correctional authorities and personnel. The use of the text in in-service courses of state correctional institutions as well as those affiliated with the armed forces is also encouraging.

HARRY ELMER BARNES
NEGLEY K. TEETERS

Contents

Part II: Theories and Factors in Causation of Criminal Behavior

Book Two: PENAL AND CORRECTIONAL PROCEDURES

Part I: Ancient and Medieval Concepts of Treating the Offender

Part II: The Era of Reform: The Emergence of the Concept of Imprisonment

Part IV: The Resocialization of the Offender in the Community

Book One

CRIME AND
THE CRIMINAL

Crime is eternal—as eternal as society. So far as we know, human fallibility has manifested itself in all types and forms of human organization. Everywhere some human beings have fallen outside the pattern of permitted conduct. It is best to face the fact that crime cannot be abolished except in a non-existent utopia. Weakness, anger, greed, jealousy—some form of human aberration—has come to the surface everywhere, and human sanctions have vainly beaten against the irrational, the misguided, impulsive, and ill-conditioned. For reasons too subtle and too complex to understand, the ordinary pressures and expectancies that pattern the individual's conduct into conformity, break down in given instances. They have always done so; they always will. No way of drawing the scheme of the good life has yet been discovered which will fulfill the needs of all human beings at all times. (From "Foreword" to New Horizons in Criminology, First Edition, *Englewood Cliffs, N.J.: Prentice-Hall, Inc., 1943.)*

FRANK TANNENBAUM

Part I

AN ANALYSIS OF CRIME IN AMERICA

1

Apathy and Fallacy

Offsetting the many admirable qualities and achievements of American civilization are certain general facts of which the student of present social conditions must take cognizance. . . . [It] has created the largest body of laws and the most complex system of government now in existence as restraints and controls upon individual and social conduct, but every stage in [its] development has been characterized by a large and ever increasing degree of lawlessness and crime. No candid investigation can ignore [this] fact, or the conclusions which [it] naturally suggests. (From the Minority Report *of the National Commission on Law Observance and Enforcement, Vol. I, Washington, D.C.: U.S. Government Printing Office, 1931.)*

HENRY W. ANDERSON

The Folklore of Crime

Crime in America is a heavy burden upon society as well as an indictment of our culture. Its gravity and its prevalence are matched only by the apathy and ignorance they encounter, even in intellectual circles. There is no doubt that we are seriously troubled but such an acknowledgement does not seem to result in much constructive action. At all levels of thought the truth is hopelessly mixed with a kind of folklore, even romanticism, that attaches to crime, criminals, and prisons.

With perhaps no other area of social problems so permeated with legend, fiction, and downright error, it is small wonder that those charged with combating crime find their work so difficult. It will not be less so until citizens and voters in substantial number become more familiar with their problems than they can be through a quick reading of the day's headlines. The present study assumes that the reader is willing to seek and to face the truth, so far as it is known.

Many people are prone to regard the criminal as a social pariah who molests *other people,* not them. Nor do they have the remotest idea that they will ever commit a crime or be arrested. Such attitudes are far from reality.

Substantial citizens and high-placed public officials are arrested for embezzlement, income-tax evasion, conspiracy, fraud, or even assault and battery or manslaughter. There have been instances in which professional, nationally-known people have "stumbled" into crime, much to their dismay. Sons and daughters of prominent people, businessmen, or college professors, have committed offenses and brought grief to their families. College students, mild-mannered and presumably inoffensive, have run afoul of the law. Perhaps "it can't happen here," but it does.

We live close to crime, detached though we may be, because it is so effectively publicized. Not a day passes that we do not read of some crime that holds our attention, and scarcely a week goes by without screaming headlines announcing some nationally notorious trial. These stories spellbind millions of Americans who like to read of crime for its vicarious thrill. The amount of space devoted to crime news has greatly increased in recent times. The modern machinery of news distribution—gigantic presses, leased wires, specialist reporters, amazing improvements in the technology of communication—makes it possible for millions to learn at once of a bizarre crime and to follow a sensational trial "play by play."

Some years ago, Charles Merz of the *New York Times* commented on this. In one notorious and sordid case, he noted that twelve million words were telegraphed from the scene of the trial in 24 days, "Words enough, if put into book form, to make a shelf of books twenty-two feet long. This is the literature of the nation . . . because it does not wait for its patrons on bookstore shelves or gather dust in libraries, but is sold out, read and realistically debated within two hours after it comes smoking from the press. It needs no pushing, needs no advertising, needs no criticism."[1]

The avid interest in crime may be attributed to an ambivalence common to us all: we fear the criminal, but we also secretly identify with him, and are thus willing victims of romanticism. Crimes that shock our sense of decency may not fall into such a category, but the crimes of bravado, such as robberies of banks, give us a vicarious thrill that we secretly enjoy. Interest in criminals who have committed reprehensible crimes, such as kidnaping or murder, may be associated with the primitive urge for vengeance for our own grievances, real or imagined.

---

[1] Charles Merz, "Bigger and Better Murders," in *Great American Band-Wagon* (New York: Harper & Bros., 1928), p. 81. The trial referred to was the Hall-Mills case in New Jersey. This was prior to the Bruno Hauptmann trial for the kidnaping and murder of the son of Charles A. Lindbergh in 1932. These trials took place years before the development of television which can bring the graphic details into the living-room. The Shepherd trial in Cleveland in 1955 had a much wider coverage than the trials mentioned by Mr. Merz.

Also symptomatic of functional attitudes is the popularity of the mystery stories that come off the presses by the hundreds and are portrayed nightly by popular television and radio performers. The unfolding of a real crime story, reported day by day in the press, comes too slowly; we can absorb crime "double quick" by reading the mysteries or looking at television thrillers.

The list of popular fallacies that follows is by no means exhaustive, but as we proceed through this book we will find that many others have been exposed by our study.

1. *That the United States is the most criminal country in the world:* we shall defer our discussion of this until later (see pp. 10–12).

2. *That most of our crimes are serious:* the best data we have regarding criminality come from the *Uniform Crime Reports for the United States,* issued semi-annually by the Federal Bureau of Investigation, U.S. Department of Justice. These data are divided into (1) serious offenses, or felonies; (2) less serious offenses, or misdemeanors (referred to as Parts I and II). In previous years traffic violations amounted to about 85 per cent of all offenses reported, about 12 per cent were misdemeanors, and about 3 per cent, felonies. However, if violations of city ordinances are not reported, the percentage of all serious offenses tends to rise. For instance, serious offenses (Part I) amounted to 11 per cent of all reported and recorded crimes in 1957. If drunkenness, prostitution, and other vices had not been recorded, the percentage of serious crimes would rise still more. (For further comment on this see pp. 78–79.)

3. *That most of those who violate the law are caught and convicted:* we shall take one year's data to refute this fallacy. In 1956, we note that of 558,561 serious offenses reported, 151,561 were cleared by arrest. Of the number held for prosecution, 106,709, only 64,411 were found guilty as charged and 9,308 pleaded guilty to lesser charges. This represents 69.1 per cent of those held for prosecution. But of the number of offenses known to the police, it means that only 13.2 per cent ended in conviction, and of the number of persons arrested, only 50 per cent were found guilty.[2]

4. *That most of our criminals are locked up in prisons and reformatories:* less than 200,000 persons are incarcerated in the 200 prisons and reformatories of our country at any one time. Each year, approximately as many men and women (convicted of crimes) are released from prison as enter. They may have served out their sentences or have been released on parole or conditionally on so-called "good time" provisions as operating in most

---

[2] *Uniform Crime Reports,* Vol. 28, No. 1 (September, 1957), Table 20, p. 61. Virgil W. Peterson, operational director of the Chicago Crime Commission, maintains that of over 1,000,000 burglaries committed in some 2,200 American cities between 1947 and 1951, more than 800,000 remain unsolved. Furthermore, of the 184,388 robberies committed during the same period, 108,413 were not solved through any arrest. See "Crime Does Pay," *Atlantic Monthly,* Vol. 191, No. 2 (February, 1953) pp. 38–42.

states. Add to this number an unknown but quite sizeable group who have committed crimes but who have escaped detection by the police or who were not brought to trial because of insufficient evidence. It has been estimated by competent students that vastly more criminals are free at any moment than are behind prison bars.

5. *That more severe penalties and longer prison terms are deterrent:* the history of corrections is ample proof that corporal punishments and long prison terms do not deter those who commit offenses. We cannot scientifically state that the crime rate, through the years, has gone up or down, due primarily to the lack of precise data. We do know that crime has continued at a rate that is alarming and considered high during each generation. Periodically we are told by law-enforcement officials that crime is on the increase.[3] Frankly we have no way of knowing. Severe penalties do, no doubt, deter some—perhaps most—people from committing crimes. But those who commit crimes do not give evidence of being disturbed by penalties, no matter how severe they are. This is especially true of crimes calling for the death penalty. The fallacy of long sentences deterring crime is cogently pointed out by James V. Bennett, Director of the Federal Bureau of Prisons, in his annual report for 1957. He states:

> If the purpose of long sentences is to deter other potential offenders, it cannot be demonstrated that they have any substantial effect on the reduction of the crime rates. If, on the other hand, the purpose of a long sentence is to increase the severity of the penalty for the sake of punishment alone, history and experience tell us how self-defeating this can be. Recognizing that certain dangerous criminals must be kept out of circulation for the protection of society, the long and mandatory sentence nevertheless rejects the basic correctional concept that an offender should be given the opportunity to retake his place in society when there are reasonable grounds to expect that he can do so without again violating the law.[4]

6. *That we have periodic "crime waves":* it is popular for the police and newspapers to report the existence of a "crime wave." There is no evidence that such a phenomenon exists although statistics might reflect a slight increase in specific crimes for certain periods of time. It is a truism in criminology that crime is relative to time and place. With an increase in laws there is a natural reflection in crime rates. But these level off when they are viewed through the years and corrected for increases in population. It is human for the public to be alarmed when a repetition of news stories indicates that a certain type of offense is being committed locally with more

---

[3] For a critical analysis of this position, see "Crime in the U.S.," *Life,* Vol. 43, No. 11 (September 9, 1957). This is the first of a series of articles on the crime problem, ending with the issue of October 14, 1957. In the first installment Professor Thorsten Sellin, an authority on crime statistics, discusses the difficulty of appraising the annual crime rate.

[4] *Federal Prisons—1957,* A Report of the Work of the Federal Bureau of Prisons, United States Department of Justice (Washington, D.C.: U.S. Government Printing Office, 1958), p. 3.

than average frequency. Hence the cry of "crime wave" is heard. An example of this was the rash of "bombings" in and around New York City during 1956 until a mild-mannered but disgruntled man was arrested. He had planted some twenty bombs in the New York vicinity over a sixteen-year period. His reason was that the company for which he worked had "not done right by him" when he had been incapacitated during his employment. While the police were going "all out" to find the culprit other cities gave much attention to reported bombings or fake bombings. By the time the perpetrator was finally apprehended, the "bomb scare" had spread to several cities, with little or no basis in fact.

7. *That poverty is the chief cause of crime:* we select this so-called cause since it is so frequently blamed for crime. Few crimes are committed because of *need.* The economic determinists have exploited this point of view for generations, and while many crimes may be traced to economic causes, most such offenses are committed through *greed,* not *need.* We shall discuss this more in detail in Chapter 9.

8. *That criminals are below the average in intelligence:* it is quite possible that the reverse is true. Studies made by clinical psychologists of prison populations (inmates) demonstrate that those behind bars compare favorably with the general population in intelligence. Since we seldom arrest and convict criminals except the poor, inept, and friendless, we can know very little of the intelligence of the bulk of the criminal world. It is quite possible that it is, by and large, superior.

9. *That most of our criminals are insane or at least suffering from mental disorders:* most of our mentally affected persons do not commit crimes. There are many such people in our communities. On the other hand the bulk of our criminals are not emotionally disturbed. In general it would seem reasonable to assume that a cross section of the criminal world would show no more emotional imbalance than could be found in the general population.

10. *That crime costs are fantastically high:* our annual crime bill is estimated by J. Edgar Hoover, Director of the F.B.I., at about $20 billion, ten times as much as our annual donations to churches, and about $500 for each family in the United States.[5] While crime is known to be expensive, however, it is difficult to estimate even its approximate cost.

Probably the most comprehensive and careful study on this subject was published in 1931 by the National Commission on Law Observance and Enforcement, better known as the Wickersham Commission. After an exhaustive study which embraces over 650 pages of material and data, this report states:

> This report does not attempt any lump-sum estimate of the total cost of crime. . . . Any attempt at such an estimate would . . . be a mere guess. . . . We are of the opinion that no such aggregate figure can be worked

---

[5] *Time,* Vol. 63, No. 7 (February 15, 1954), p. 21. Speaking before the convention of the American Bar Ass'n. in Los Angeles on August 25, 1958, Mr. Hoover estimated the annual cost at $22 billions.

out with even an approximate accuracy, and we are unwilling to indulge in vague estimates which could, at least, be no more than a guess.[6]

While the Commission did not minimize the economic cost of crime, it did point out the difficulties in establishing just what costs should be included. For example,

> If a thief steals $100 from X, X is $100 poorer, but it by no means follows that the community as a whole is economically worse off to the extent of $100. Indeed, to suggest an extreme instance, it may well be that the success of Robin Hood's mythical attempt at a redistribution of wealth by robbing the rich to give to the poor would have been an economic benefit to the community as a whole in twelfth-century England. This extreme example merely emphasizes the impossibility of regarding the total amount of such losses to individuals due to crime as a net loss to the community; and the same considerations obviously apply to expenditures for protection against crime, whether public or private. They are part of the *immediate* cost of crime, but not necessarily part of the *ultimate* cost.[7]

A list of some of the items to be considered in the cost of crime was submitted and explained by the Commission. The items add up to less than a billion dollars annually—that, of course, was in 1931. The items are summarized here:

(1) The administration of criminal justice and law-enforcement. Under this heading we find the cost of maintaining thousands of police forces throughout the country and the heavy expense of maintaining the various courts. Part of this item, however, should be charged against the cost of social services rendered by the police, such as traffic control, overseeing public events, supervising fires, and the like; another part is chargeable to the costs of civil litigation in our courts.

(2) The cost of correctional institutions in addition to the money spent on probation and parole services.

(3) Private expenditures for protection against the criminal element— insurance, armored cars, and private police protection (this expense, however, cannot be or is not charged to taxpayers but rather to the higher costs of doing business reflected in costs to the citizen).

(4) Losses due to commercialized fraud, extortion, embezzlement and other predatory crimes. But here, again, we must question the actual costs to the taxpayer rather than to individuals or insurance companies.

(5) The indirect loss of the productive labor of prisoners and law-enforcement officers. This item is likewise difficult to compute. If prison inmates were productively worked the cost of maintaining prisons (mentioned in item two) would be reduced to some degree.

---

[6] National Commission on Law Observance and Enforcement, "Report on the Cost of Crime," *Report No. 12* (Washington, D.C.: U.S. Government Printing Office, 1931), pp. 11, 67-70.

[7] *Ibid.,* pp. 65-66.

There have been no recent over-all studies of crime costs. The Special Crime Study Commissions of California issued a pamphlet in 1949, entitled "Sources for the Study of the Administration of Justice,"[8] in which several pages are devoted to crime costs but do not go beyond the Wickersham Commission's conclusion that no aggregate figure can be arrived at with even approximate accuracy. The only national figures are Bureau of the Census estimates of the total general expenditure for corrections: 107 million dollars in 1947. The pamphlet then lists costs in certain categories by states.

But financial losses do not tell the whole story. Intangible costs are much more serious to contemplate; nor can they be measured or estimated. The wholesale tragedy to victims that results from such crimes of violence as assault and battery, mayhem, kidnaping, rape, manslaughter, criminal negligence, and murder is almost inconceivable. Thousands are thus affected, and their physical and mental suffering is very real. In addition, the innocent members of the families of delinquents and criminals suffer embarrassment due to social stigma. Each year thousands of families must attempt to make new adjustments as a result of criminal acts. Quite often these adjustments cannot be made, and various demoralizations result that exact a heavy toll from the individuals concerned. Society, in the long run, must pay this intangible cost. We see the results of this maladjustment in bloated relief rolls, in increased institutional treatment of all types, and in the vagrant and migratory population.

11. *That crime does not pay:* it is assumed in this injunction that crime does not pay the criminal. This is seriously open to doubt. We might counter with, "Crime does not pay whom?" As we mentioned above, crime does not pay society so far as its costs are concerned; nor does it pay its victims, either financially or psychologically. But does it pay the criminal who engages in it?

As early as 1877, Richard Dugdale, prison investigator who made the classic study of the Juke family, warned against taking for granted the common notion that crime does not pay. He wrote:

> We must . . . dispossess ourselves of the idea that crime does not pay. In reality there are three classes with whom it does: 1st. The experts, who commit crimes which are difficult to detect or who can buy themselves off. These are the aristocrats of the profession. 2nd. The incompetent, who are too lazy to work and too proud to beg, or too young for the poor-house. 3rd. The pauper, who steals because prison fare and prison companionship offer higher inducements than poor-house fare or poor-house society. This stock amounts to 22.31 percent of all criminals, as seen by our study.[9]

---

[8] Prepared by Dorothy Campbell Tomkins (Sacramento: California State Board of Corrections), see pp. 25-28.

[9] *The Jukes* (New York: G. P. Putnam's Sons, 1910), p. 100. See Peterson, "Crime Does Pay," *supra,* f.n. 2.

The Italian criminologist, Napoleone Colijanni, insisted in 1887 that crime pays and that the honest man earns less than the thief and, in addition, is more likely to be injured or killed than the criminal.

Of course, crime does not pay all who engage in it, for many of them do not understand the intricacies of the game they are playing. Most of the criminals who are caught, convicted, and sentenced to a prison term are the inept or those who do not have the means to employ adequate counsel. But those who are professionals in the business of crime, and many white-collar criminals, find it a fairly safe and exceedingly lucrative line of activity. The well-trained criminal who has good biological equipment and who studies what he is doing in the same serious manner as the trained executive in a large manufacturing company, has little cause to worry about being deprived of a livelihood. No business or activity in which risks are involved is 100 per cent safe. It is primarily the moralist who still believes that crime does not pay.

It is time that we dispense with the quaint, even absurd, idea that we can deter American youth from crime by frightening them into obeying the law. Radio and television programs portray stories from F.B.I. and police records, and in most of them, usually in a deep bass voice, the commentator booms out: "Crime does not pay!" Also, in some of our large cities, patriotic organizations, cooperating with police departments, display electric chairs and other punitive equipment in special trucks. A lecturer demonstrates how the electric chair works, permits children to sit in it, and pontificates on the shop-worn subject: "Crime does not pay."

It is not the old-fashioned morality, but the old-fashioned technique of attempting to instill it into modern youth that is at fault. No one answer to the crime problem is adequate—least of all the slogan that crime does not pay.

The above represent only a few of the fallacies that are widespread concerning crime and criminals. We could mention others, such as: that there is such a thing as a "born" criminal; that minority groups are inherently more criminal than native-born whites; that *most* probationers and parolees commit additional crimes; that the jury system is a guardian of democracy.

Just How Lawless is the United States?

It has been almost universally accepted that we are the most lawless people in the world. Not only foreigners, but we ourselves, have pointed up the high homicide rate, the seeming increase in both crimes of violence and against property, and in addition the high incidence of graft and failures of public trust. It is alleged there is an annual increase in sex crimes. Rarely do we find a rebuttal to these serious charges and nowhere do we find an attempt to explain the amount of crime we do have. Frequently this country is compared with England as to the prevalence of crime and the comparison is always invidious. One Englishman, Patrick Pringle, in his delightful book dealing with the origins of the London Metropolitan Police, contends that our antiquated police system, based on the "parish-constable" system of eighteenth century England, encourages disregard for law, and thus a high crime rate.[10] While this observation has merit, as do others criticizing our courts, penal codes, and parole and probation systems, all of which we shall appraise in later chapters, such criticism and the areas it covers represent only a small part of the American criminal activity.

Certainly what is needed is more and still more research in this field. Lacking this the best we can do is generalize. We do not know, for example, how much crime we can normally expect in a land of 175,000,000 people— diverse, highly mobile, heterogeneous, jealously espousing freedom of thought and action, all engaged in a highly competitive struggle for economic status and prestige. We may well explore this question since it haunts us time and again and no answer is readily forthcoming.

1. *Our Vast Area and Our Highly Decentralized Governmental Units.* In our country 175,000,000 people are scattered unevenly over an area of some three million square miles in forty-eight states, each with its own government and laws. Counties, townships, and incorporated towns and villages are further political subdivisions. Each has its own local government and ordinances. Then there are the urban areas, some compact and centralized, others with many governments over one social complex. There are states larger than many European countries. Some of them are heavily populated, while others have only a few persons per square mile. Our economy varies from agriculture to highly mechanized manufacturing, with grazing, mining, logging, and many other methods of earning a livelihood. The foreign visitor is usually bewildered at our many governmental and law-enforcing units but if he is astute he will notice the ease with which most Americans manage to take this complexity in stride. The late Margery Fry, prison reformer of England and genuine friend of this country, wrote of this as follows:

> In writing of the United States it seems equally difficult to exaggerate and to generalize. In almost everything, from the primitive to the sophisticated, from poverty to wealth, from the "toughest" to the most sensitive, the extremes are so unbelievably distanced, the mean so hard to ascertain. This is most bafflingly true of those social institutions which depend upon the administrations, Federal, State, and local, whose triple network covers the country. Not only do they vary widely from State to State, but within the States there are often startling discrepancies. In the matter of penal administration this is particularly true. Everywhere four sets of authorities are responsible for different kinds of places for the housing of prisoners; everywhere two sets of laws are working, the State law and the Federal

---

[10] *Hue and Cry* (New York: William Morrow and Company, 1956), p. 12.

law, each with its own courts, police, probation officers, etc., and, to a certain extent, prisoners.[11]

Certainly this involved situation cannot be overlooked when we appraise our crime rate.

2. *Mobility.* Vast though our country is, we are eternally on the move. Millions think nothing of traveling four or five hundred miles over a week-end "just for a visit." Farmers often drive two hundred miles after a day's work to see a ball game or prize-fight in the nearest metropolis, returning the same evening. Entire families cruise through the countryside, stopping overnight at motels, until they reach their destination, two or three thousand miles from home. Our highways are choked with automobiles day and night; people on the move. Air travel has grown tremendously.

This high degree of mobility, even of the masses, is almost unknown in most other countries. A murder may be committed on the East Coast in an evening and the criminal be on the West Coast the next morning, perhaps before the police have knowledge of the crime. A jewel thief may make his haul and arrive in another part of the country within a few hours. Thus, high mobility of Americans must also be considered in appraising our crime rate.

3. *Heterogeneity.* We are an extremely heterogeneous people. There may be an American *type* but we are ethnically and racially different—an amalgam of peoples. Our people come from many countries and relatively few have American ancestry prior to 1850. The cultures of many ethnic peoples have merged into a nebulous American type but there are many differences in folkways, attitudes and values.

4. *Love of Freedom and Individualism.* Americans are reared on democratic principles, including an annual eulogy of the Bill of Rights. They espouse freedom early and with enthusiasm, but responsibility is inculcated at a slower tempo. Our tradition encourages the belief that each person is as "good" as any other. The average citizen resents autocratic control, close supervision, and condescending advice or moralizing. This is a part of the American heritage and it too must be evaluated in scrutinizing the crime rate.

5. *The Acquisitive Urge.* In an earlier edition of this book the authors dealt with an attitude they believe to be prevalent in the United States, one which may explain much of the crime with which we are familiar. We referred to it as the "something-for-nothing" psychology. There was some criticism that this thesis oversimplified the problem. The causes of crime cannot be oversimplified. We are convinced that this attitude, which we

---

[11] *Spectator* (London), August 6, 1943, pp. 122-123. See also "Notes from an American Diary," *Howard Journal* (London, 1943), pp. 162-170. This and other aspects of the crime picture in our country are ably presented by Hermann Mannheim in his *Group Problems in Crime and Punishment* (London: Routledge and Kegan Paul, 1955), Chapters 7 and 8.

shall refer to here as the *acquisitive urge,* throws considerable light on much of the crime in the nation.[12]

Organized crime, racketeering, political graft and corruption, and "white-collar" crime certainly represent the modern manifestation of the acquisitive urge that has colored our culture since the era of colonization. In our next two chapters we describe these types of crimes and refer to them as the "upper world" of criminal activity. They are distinct from the conventional or "traditional" crimes with which each generation is more familiar; such crimes as homicide, rape, robbery, assault and battery, and the like. We discuss these crimes in Chapter 4.

A large share of the settlers who came to these shores lived in the hope that great wealth could be made in this land of opportunity. Some were dedicated to hard work, but many believed fortunes could be obtained by short-cut methods. What follows is not to be taken as an indictment of the vast majority of the industrious and frugal American settlers whose virtues have been extolled in our history books; rather is it a discussion of a pattern that gradually evolved in a way not matched in any other country.

One early instance of this pattern is both bizarre and revealing. The early settlers of Absecon Island—now a part of Atlantic City, New Jersey —lured boats onto the dangerous shoals of the vicinity by means of lanterns. As the helpless ships foundered on the rocks the villagers would loot them. Meanwhile, the women and children of these land pirates would collect on the beach and pray that the ruse would work.[13]

Nothing but sheer starvation, as at early Jamestown, could discourage the hope of gaining riches with little effort. Gold and silver were sought, but as these failed to materialize, a sort of substitute was found in lucrative fur trade with the Indians. The "El Dorado" complex was then expressed in land speculations and the rise of lotteries. It persisted in the building, first of canals, and later of the railroads.

Following this era, gold was discovered in California. The rush for easy money was further stimulated. This westward trek encouraged outlaws to mingle among the miners and other early settlers so that the *acquisitive urge* was easily satiated through openly lawless means rather than through cunning.

Stock jobbing in connection with railroad expansion came next. Many of the early railroads were considered by their owners and directors as agencies for successful stock gambling rather than as a means of transportation. This represented an era of financial chicanery we have never quite equalled since. But the pattern became set. It was conspicuously prominent in the development of public utilities and financial houses up to and in-

---

[12] Havelock Ellis once remarked that crime was an index of social progress. See "The Measurement of Civilization," *Forum,* Vol. 67, No. 4 (April, 1922), pp. 278-289.

[13] *New York Times* (October 9, 1949), p. 47.

cluding the Great Depression of the 1930's, when federal legislation attempted to curb it.

In time the *acquisitive urge* seeped down into the lives and attitudes of the masses. Many of these were descendants of immigrants who had earned an honest living by hard and honest toil. The younger generation, however, eagerly embraced the American tradition of gaining success by developing shrewdness, sagacity, and sophistication rather than by hard work. Family discipline and community controls began to be upset by technological advances and rapid communication. New machines threw many young men out of work, for the most part temporarily. Fewer of our youth cared to learn a trade, finding employment in the white-collar ranks easier and more pleasant. The blue shirt became a stigma rather than a badge.

Perhaps the greatest social upheaval in the history of the country was the advent of Prohibition. Here for the first time a law was passed by a bare majority that was regarded by many as a rank encroachment on the freedom of a minority composed of many millions. The provision of illicit liquor for this minority furnished a perfect testing ground for the ideas and techniques of organized criminals. "Bootlegging" became a remunerative criminal activity that was especially difficult to control by ethical inhibitions or the force of public opinion, since violation of the Volstead Act was regarded in many quarters as respectable, smart, and even indispensable in the service of national need.

For every "bootlegger" who supplied liquor, there were dozens of "respectable" law-abiding citizens who condoned the flouting of the law by serving as customers. This breakdown of ethical and legal norms was conjoined with the economic and political plunder of the Harding era and, in turn, with remarkably widespread corruption in municipal affairs. The more recent refined term "conflict of interest" was unknown at that time. The *acquisitive urge* was gradually sharpened so that men, in high and low places, were sorely tempted to "get theirs" unethically or even illegally.

Much of this national pilfering is petty. For instance, the New York *Journal of Commerce* (January 13, 1956) estimated that approximately four million dollars in produce is pilfered on the New York waterfront annually. How much of this was taken by longshoremen is not recorded. But what is to stand in the way of a stevedore's "swiping" a ham, a bunch of bananas, or a crate of oranges from a shipment?

Each job has its traditional perquisites. Managers of department stores, super-markets, and other chains know that some of their clerks walk off occasionally with articles, sometimes in large quantities—especially food. College professors make personal 'phone calls from their offices and use stationery for research papers; some office clerks use the company's postage stamps for their personal letters. Salesmen have their expense accounts— "swindle sheets"—all costs on which are, of course, charged up to the costs of production. It is a terrific strain for an American to live a life of scrupulous moral rectitude, never deviating one iota from a rigid code of ethics. Millions of persons are a bit shady in their morals but only in little things. Each has his limit of immorality or unethical dealings and each can rationalize his conduct. This philosophy can easily be carried over into slight deviations from the legal code. It is found most frequently in the violation of traffic regulations. Others press in on areas of more serious illegality. Limits are stretched often to the breaking point. This is the American pattern, although we dislike admitting it.[14]

Conspicuous spending and consumption, so caustically described many years ago by Thorstein Veblen, has become commonplace in the United States. This is understandable because of the emphasis placed on the production of material gadgets that make work easier, afford more time for leisure, and enhance one's social status among the members of his in-group. The media of mass communication whet the appetites of the millions who are thus a veritable captive audience.[15] If ready cash is not available, the potential customer is encouraged to plunge into installment purchases. The evil day of payment is postponed to a dim future. Banks and loan companies compete through advertising how "painless" it is to borrow money for material gadgets.

So the emphasis in our culture is not on self-control but rather on greed, comfort, and even cupidity. A culture that accepts, even implicitly, the acquisitive propensity not only as justifiable, but as an approved way to ease and affluence, is creating in its midst very serious potentialities for crime.[16]

A mild amount of much-needed criticism of "giveaway" television programs has been noticeable in recent years. In 1957 over 4 million dollars was given away through television programs. One may ponder the effect upon the youth of this country when it sees people, some of them children, acquiring large sums of money merely by answering a few questions. A child of twelve wins $164,000 for answering a number of questions dealing with his hobby, the stock market; a stock-clerk wins $200,000 for being a "walking encyclopedia." The "easy money" attitude is always rewarded on these programs, yet they are viewed in large part by young people whose parents must struggle hard for the support of these children.

We have only recently begun to consider that our carefree, materialistic culture may be in mortal danger. Foreigners have for years been unable to understand the casualness with which we regard this *modus vivendi*. While we know that millions of our citizens are decent, law-abiding, highly ethical, and serious in their daily activities and in their philosophy of life, we are

---

[14] For a discussion of this tendency to deviate, see E. S. Atiyah, *The Thin Line* (New York: Harper, 1952).

[15] See: A. K. MacDougall, "Conspicuous Consumption: 1957," *The Nation*, Vol. 184, No. 9 (March 2, 1957), pp. 185 ff; Vance Packard, *The Hidden Persuaders* (New York: McKay, Company, 1957).

[16] See Donald Taft, *Criminology*, 3rd edition (New York: Macmillan, 1956), Chapter 18, for an elaboration of this point.

forced to admit that there are many opportunities for easy money through unethical or illegal manipulation. It is, or at least has been, the American *ethos* that must also be appraised if we are to understand our crime rate.[17]

One more fallacy that has only recently been exploded; that Chicago leads the country in crime. Now *Time Magazine,* using F.B.I. *Uniform Crime Reports* for 1957,[18] shows that Chicago ranks as the second most law-abiding of the 22 most populous cities in the nation, judged by its felony rate. With rates per 1,000 in population for murder, non-negligent manslaughter, aggravated assault, robbery, burglary, larceny, and auto theft, we find the following, starting with Los Angeles as the most lawless:

Los Angeles	51.0	Indianapolis	26.5
Atlanta	44.7	Cleveland	23.0
St. Louis	43.8	Minneapolis	21.2
Denver	39.3	Boston	21.0
Seattle	39.3	Pittsburgh	20.0
Newark	37.4	New York City	17.7
Houston	35.3	Philadelphia	16.9
Dallas	35.2	Cincinnati	16.0
San Francisco	34.8	Kansas City	13.3
New Orleans	29.2	Chicago	12.9
Detroit	28.0	Buffalo	8.5

---

17 Cf. Louis Finkelstein, "The Business Man's Moral Failure," *Fortune Magazine,* Vol. 58, No. 3 (September, 1958), p. 116 ff.

18 See *Time Magazine,* June 30, 1958, p. 18.

2

The Overlords of Crime

The great and truly dangerous criminals of the present are the directing heads of the syndicates in control of bookmaking, slot machines, organized gambling, prostitution, narcotics, the loan shark racket, swindling schemes, organized murder, and the host of extortionist rackets playing upon legitimate business and labor in many different fields—men who are almost unknown to the public and whose names never appear currently on any police blotters. Criminals of this latter type are always difficult to identify. An inevitable effect of organization is to clothe the activities and participation of the leaders in secrecy. The directors and master-minds of organized crime are seldom apprehended or arrested. They live peacefully and luxuriantly, enjoying a full sense of security, and with complete confidence that they will not be disturbed in their criminal activities. Their only real fear is concerned with the ambitions and competition of rival mobsters. (California Crime Study Commission on Organized Crime, Second Progress Report, March 7, 1949.)

The Extent of the Problem

During 1950 the televised hearings of the Special Committee to Investigate Crime in Interstate Commerce provided one of the most absorbing public spectacles ever offered. The meetings of this Kefauver Committee, so named for the Senator who presided, kept millions of viewers close to their sets to see and hear a discomfiting revelation of criminal activity.[1] Perhaps to many it was just a "good scandal"; to others, an

---

[1] A special committee of the United States Senate, it was appointed May 3, 1950 (81st Congress, Second Session). The committee's final report was filed May 1, 1951. For an abridged account, see Estes Kefauver, *Crime in America* (New York: Doubleday, 1951).

17

exciting drama of the courtroom variety. But there is no doubt that to many more it was a shocking disclosure of the extent to which the really clever criminals have established themselves in our society.

One by one, witnesses from this overworld of crime—syndicate executives in the gambling or narcotics "businesses," and politicians with whom they have more than casual association—gave through their often reluctant testimony a refresher course for the American public on what big-time crime means in their land. If, nearly a decade later, it is premature to say that the hearings have marked the end of the overlords, it is still probable that the average citizen no longer thinks in narrow terms regarding the crime menace that confronts us. The secrecy cited by the California Study Commission has been at least partially lifted.

Eight years after the Kefauver hearings, the government evidently felt that dramatic new measures against the overlords were necessary. The Justice Department announced the compilation of a list of 100 public enemies on whom the highest-priority investigative efforts will be concentrated. Attorney General William P. Rogers made it clear that the targets were not conventional "public enemies" of the Dillinger variety, but that "many of them have invaded the legitimate fields of business and labor unions."[2] The McClellan Committee, during 1957–1958, has verified this.

In addition to the secret list, the following steps were announced: (1) strengthening of the Organized Crime and Racketeering Section of the Justice Department's Criminal Division, (2) setting up of a special group of former prosecutors in New York to spearhead the drive, (3) assignment to the F.B.I. as a principal concern the investigation of statutes on racketeering and extortion, obstruction of justice, and interstate fraud by wire.

The Kefauver Committee gave out no information on "white-collar crime," another logical outgrowth of the acquisitive urge. Since the conditions for its nurture seem to have much in common with organized overworld crime, however, this type of borderline criminality will also be discussed in this and the next chapters. Only after we have placed these neglected areas of antisocial activity in the foreground of our discussion, where they belong, will we be able to consider the contributions of the traditional, old-fashioned criminal in the right perspective.

The lone-wolf crook, the masked burglar so beloved of cartoonists, did not always play second fiddle to the gangster, but his decline began during World War I. A second phase of criminal activity began thereafter, when Prohibition called forth unprecedented organized lawbreaking, and the gang as we know it came to dominate the underworld economy, in response to the desire of millions of citizens for a commodity then declared illegal. The third phase commences with the end of World War II, and is characterized by regional, even national, syndicates of criminal gangs. Phases Two

---

[2] As quoted indirectly in the *New York Times* (April 11, 1958), p. 1.

and Three parallel the growth of "respectable" white-collar crimes in business, government, and the professions.

Although we must continue to focus our attention on the underworld of crime, since our traditions and institutions are primarily geared to this phase of the problem, we must recognize the vastly more sinister overworld of crime, whether it is composed of gangsters and racketeers, or allegedly respectable business or professional men, or public servants. They are all engaged in predatory, antisocial activity.

An example of ostensibly bona fide business venture characteristic of the 1930's was known as the holding-company. This device, a super-corporation that obtained control of a number of business interests through buying out the controlling stock, made possible legalized looting of the legitimate companies. Ambitious promoters organized these holding companies, not to produce anything tangible for consumption, but merely to siphon off millions of dollars of the investors' money.

We do not intend to imply that the equivalents of racketeering, political graft, and white-collar crime did not exist in our earlier history. Exploitation of society as a whole, as opposed to the exploitation of specific individuals (the province of traditional crime) is at least as old as history. But it was often tolerated in our own country's past, if not actually condoned, on the ground that it was a necessary adjunct of rugged individualism or free enterprise in exploiting natural resources and in settling a new land.

Conventional crime, irritating and even serious though it is during each generation, actually pales into insignificance when its inroads on society are compared with those of syndicated crime and the machinations of dishonest businessmen and public officials. Traditional crime is admittedly more violent, but with an adequate police force it can be dealt with and controlled. The reader must not be misled into the beliefs that the 200,000 persons in our prisons are a representative cross section of this country's criminals or especially that those in prison represent the ones most dangerous to society. Most of this group are merely our traditional criminals. Few of the other classes ever end up in custody.

The Racket as a Part of the Crime Picture

While the word "racket" is often applied to any business or activity clouded by suspicion, it is defined here as an operation in which intimidation or extortion—the threat of violence or reprisal, is present. In its less sinister forms it can be found in the activities of "car watchers" at ball parks, municipal zoos, and other places where there is usually free parking. There is usually an implied threat of reprisal if the victim does not "pay up" for amounts ranging from twenty-five cents to a dollar, or even more.

It is generally agreed that Alphonse "Scarface" Capone was the first to bring racketeering to public attention and to exploit its potential at the level of "big business." Born in Italy in 1897, Capone was brought to this country by his parents at an early age and grew up in Brooklyn. He developed such a reputation as a resourceful and ruthless thug that his fame soon spread to Chicago. In 1919, one of the leaders of the Chicago underworld, "Diamond Jim" Colosimo, brought him to the Windy City to serve as his bodyguard. But despite this precaution, Colosimo was killed the following year.

In due time Capone took over control. He became supreme in the illicit liquor field during the Prohibition Era. The first five years after Capone's ascendancy saw innumerable battles among the liquor elements in Chicago. Other rivals attempted to "muscle in" on the lucrative field. Capone demonstrated his superiority through the notorious Valentine's Day massacre in 1929 when his gunmen, disguised as police officers, herded members of the "Bugs" Moran opposition gang into a garage and mowed them down with machine guns.

It was estimated that at the height of his power Capone controlled bootlegging in four states and his gross income from illicit liquor control was

The grisly result of gang warfare in Chicago: the "St. Valentine's Day Massacre" of 1929. (Courtesy Wide World Photos.)

six million dollars per week. But Capone could never have been the power he was unless there were thousands of respectable people who were willing to evade the law by purchasing illicit liquor. Organized crime would be seriously handicapped if its many patrons would refuse to break the letter or spirit of our laws. This is currently the case in connection with gambling and has always been true of prostitution.

During the Capone era there were over eighty gang killings in Chicago, including one assistant district attorney, without a single conviction. Capone finally fell from power, not because of his long list of violent crimes, but because of federal income tax evasion. He was convicted on this charge by the federal authorities, sent to Alcatraz, and eventually released. He died in 1947 in his elaborate Miami, Florida home.[3]

But Capone was a novice compared to his successors, who capitalized on the racketeering technique initially developed by the Brooklyn thug. While it has been used by one gang of criminals against another, it was employed to greater advantage primarily during the 1930's in intimidating legitimate businessmen. In the 1950's it was extended by corrupt labor leaders.

To illustrate how the racket worked during this period we may take a typical form as applied to the laundry business. A representative of a gang

Racketeers slugged the driver of this laundry truck and tossed him inside, where he was burned alive among the gasoline-drenched clothes.

[3] For details of Capone's life, see Fred D. Pasley, *Al Capone: The Biography of a Self-Made Man* (New York: Ives, Washburn, 1930); for more recent books dealing with the pre-Capone and post-Capone story of crime in Chicago, see: Virgil W. Peterson, *Barbarians In Our Midst* (Boston: Little, Brown, 1952); Alston Smith, *Syndicate City* (Chicago: Regnery, 1954). For the story of the efforts of government officials to smash the Capone grip, see Eliot Ness and Oscar Fraley, *The Untouchables* (New York: Messner, 1957).

visits the proprietor of an urban laundry and suggests that he contribute a sum of money for "protection." The proprietor may protest that nobody has ever harmed him and he sees no reason why he should contribute for protection against imaginary danger. The representative of the racket assures him that if he has not needed protection in the past, he will surely need it in the future. The argument is often graphic and concrete enough to convince the owner that protection is worth the fee—from $50 to $500 per month, depending on the volume of business. If, however, the victim is stubborn, skeptical, or threatens to call the police, he may later find one of his trucks overturned and the contents destroyed. The owner may call the police, but in some of the cities where such methods have been used, this action has little or no effect since the police have been "taken care of" in one way or another. After a few such demonstrations the laundryman decides it will be cheaper to pay the sums demanded. He is then obliged to pass on this extorted tribute to his customers. Thus, in the long run, the public pays for the racketeering although actually unaware of its existence.[4]

During the racketeering era hundreds of extortion practices were in operation covering every line of legitimate business. J. C. R. MacDonald, in *Crime Is a Business,* listed over a hundred in the swindling category alone.[5] A. R. Lindesmith, in a cogent article on the racket, listed the following men alleged to have been connected with the late Jimmie Hines, a Tammany leader in New York City who was sent to prison in connection with graft:

> Owen Madden—beer and rum, race-tracks, night clubs, coal and laundry; J. T. Diamond—alcohol, narcotics, night clubs; Louis "Lepke" Buchalter and Jake (Gurrah Jake) Shapiro—garments, furs, movies, flour, poultry, labor unions and narcotics; "Lucky" Luciano—narcotics, liquor, Italian lottery, prostitution, and receiving stolen goods; Philip Kastel—Montreal night club owner, badger game, bucket shop, rum, slot machines; Frank Costello—slot machines; Meyer Lansky and "Bugsy" Siegel—execution; Larry Fay—milk, taxicabs.[6]

Some of these criminals have since been "rubbed out," executed, or deported. Siegel was killed under mysterious conditions; Buchalter was executed in Sing Sing prison; Luciano was deported to Italy. But some are still operating illegal businesses or have taken over legitimate concerns.

---

[4] To get an idea of what the public pays for the tribute forced from businessmen, see Courtenay Terrett, *Only Saps Work* (New York: Vanguard, 1930).

[5] Palo Alto: Stanford University, 1940. See also, John McConaughy's *From Cain to Capone: Racketeering Down the Ages* (New York: Brentano, 1931).

[6] "Organized Crime," *The Annals of the American Academy of Political and Social Science* (hereafter be referred to in the footnotes as *"The Annals"*), Vol. 217 (September, 1941), p. 126. For additional reading on the earlier racket era, see: Martin Mooney, *Crime Incorporated* (New York: McGraw-Hill, 1935); Courtney Ryley Cooper, *Here's to Crime* (Boston: Little, Brown, 1936); Sid Feder and Joachim Joesten, *Luciano Story* (New York: McKay, 1955). For story of how the racket operates in prize-fighting, see *Sports Illustrated,* August 4, 1958, "Paging Mr. Grey," pp. 25 f.; also *Newsweek,* August 4, 1958. See also, *Big Bankroll: The Life and Times of Arnold Rothstein,* by Leo Katcher (New York: Harper, 1959). Rothstein is sometimes referred to as the "father" of organized crime.

Much of the fraud, violence and other forms of illegal activity in labor unions follows the racket pattern—that is, of extortion. Unscrupulous and arrogant labor leaders have betrayed the rank-and-file by filching large sums of union money, often to support ostentatious living. In devious ways they have also terrorized and victimized management and the public in order to consolidate their control.

The A.F.L.-C.I.O. under its leader, George Meaney, has been attempting to clean up racketeering and corruption in labor unions. The only penalty the parent body can impose on unions that are dominated by corrupt leaders has been expulsion. A few years ago the longshoremen's union was expelled but thus far little has been done directly to clean up this octopus that controls New York City's waterfront. In 1957 the A.F.L.-C.I.O. expelled the giant teamsters' union, once headed by Dave Beck and succeeded by James Hoffa, and the laundry and bakers' unions.

The Senate subcommittee probing racketeering in both labor unions and management, had before it during 1957 and 1958 hundreds of labor bosses great and small. It unearthed testimony that described corruption, fraud, "goon" tactics, dynamiting, larceny, and kidnaping, as well as much unethical and immoral activity on the part of both management and labor. In 1955 the labor columnist, Victor Riesel, was blinded by acid thrown by small-time thugs, allegedly in the hire of labor bosses. A few of the minor culprits were brought to trial and convicted in the conspiracy but the "higher-up" who ordered it has never been convicted.

These hearings, headed by Senator John McClellan, went into the affairs of the International Teamsters' Union very carefully, notably the manifestation of Dave Beck's arrogant use of his power as president to mishandle large portions of the union's funds. Later Beck was convicted of the theft of union funds with which he bought an expensive automobile. His son, Dave, Jr., was also convicted of peculation of funds from the same union.[7] The affairs of Jimmie Hoffa, his successor, were also carefully scrutinized.

Racketeering known as the "shakedown" is engaged in by the police of many cities. For instance, many tavern keepers are expected to make regular payoffs to the police. In some districts tavern keepers, fruit vendors, storekeepers who have juke-boxes in their establishments, and other small merchants successfully violate many city ordinances regulating their businesses merely because they have paid off the police officer in their districts. This type of blackmail is well known to students of urban law enforcement. Some police rookies come on to the force steeped in the "payoff" tradition, and regard petty graft as a perquisite of the career. Sincere and honest police chiefs are constantly confronted with this practice. If a police chief

---

[7] For details of this investigation, see the reports of the *Senate Select Committee on Improper Activities in the Labor or Management Field*, 1957, 1958, Govt. Ptg. Office. See also, *Life Magazine*, August 18, 1958, "Dark and Strange Doings In Teamsters," pp. 15 ff.

resorts to firing a corrupt officer he may become enmeshed in trouble with civil service commissions, often composed of corrupt politicians.

In another racket, the member of a crime ring enters a restaurant, for example, and states that from now on he is a full-fledged partner in the business. The victim dares not protest because of potential damage to his business or retribution through official sources. Such businessmen know full well the close tie-up between criminal elements and the politicians. Owners of legitimate businesses, irked by next-door nuisances, may protest to city inspectors. Instead of getting relief they are suddenly besieged by various inspectors who advise them to make extensive repairs in plumbing, put in new floors, or install a new wiring system. In short, the most successful rackets are dovetailed with political power organizations whose support is expressed in various ways.

Crime Syndicates

It has long been debated whether crime syndicates actually operate on a nationwide basis. The Kefauver Committee in 1950 admitted that it had ". . . no direct evidence of a national syndicate of crime, but contended that the circumstantial data were characterized by too much co-incidental happenstance to dismiss the probability of such a combine."[8] Certainly the California Crime Study Commission (1949-1950) expressed little doubt as to the existence of a nationwide gambling syndicate. The journalistic fraternity also shares this opinion, if we may accept as factual the books and articles published.[9]

A crime syndicate is an aggregate of organized criminals, national in scope, which through strong-arm methods controls any vice or business it may choose to enter. It is a perverted epitome of the large scale business concern. A legal business syndicate is composed of capitalists or financiers who have pooled their resources to pursue some operation that requires large capital investment. A criminal syndicate patterns its activities along these same lines with the purpose of extracting from the public vast sums of money by devious and illegal methods. Perhaps the best statement concerning the existence of a crime syndicate comes from Sidney Lens:

> The crime syndicate looms as something hazy, perhaps only a mirage invented by sensational newspapers. Until some undercover agent can reach the top of its pyramid—as FBI agents, for instance, have penetrated close to the top of the Communist Party—the inner operations of this potent force in our society remain nebulous. Yet in the absence of exact knowledge, a few strong hints have been dropped over the years which do lend some insight. The syndicate works pretty much like a political party or a giant corporation. It is a loose federation, highly centralized in some re-

---

[8] *New York Times* (December 24, 1950), p. 8 E.

[9] For example, see Joseph F. Dineen, *Underworld U.S.A.* (New York: Farrar, 1956).

spects (such as dealing out "justice" to its traitors), but decentralized in execution of business ventures. It recruits carefully the individual who shows promise but who is outside its ranks; and it infiltrates into key organizations which it needs to round out its power. . . . The syndicate is departmentalized, like any good business or political machine. It has an official in charge of gambling, another in charge of prostitution, a third in charge of legitimate business, another in charge of relations with police and politicians, another who handles night clubs and sports, and still another in charge of labor activity. Each of these branches is expected to help the other. . . . The syndicate is usually thought of as a muscled monster; but most of its activity requires tact and finesse rather than muscle. The syndicate is always trying to do something for important people, building up its great store of friendships. . . . Though it has its inner circle of muscle-men and leading departmental figures, it is certainly not a membership organization. It is more of a loosely-knit force with tens of thousands of "fellow-travellers."[10]

The first inkling that such an illegal combine existed came from a book written by Martin Mooney entitled *Crime Incorporated,* published in 1935. He stated:

The modern racketeer is a suave business man. His activities are adroitly camouflaged by hundreds of obviously legitimate corporations. Modern crime has gone big business. And modern crime's bulging kitty is adequately protected by an efficient chamber of commerce.

While the combines in each section of the country are groups that work independently of others, all of them take counsel from a national association whose slogan is "The public be pleased." Miracles can be accomplished when vile medicine is given that candy taste. That's what is happening in America today. . . . The boys find it profitable to operate their lucrative rackets today under the cloak of big business rather than behind the barrel of an automatic.[11]

Contemporary crime syndicates are not managed by ignorant thugs; they are controlled by alert and sagacious businessmen. Mooney describes their operations graphically, comparing them to giant corporations composed of many regional and geographic subsidiaries. He alleges their boards of directors meet in swanky office suites in our large cities. Under a central board are special committees that maintain and extend the activities of organized crime. A "new projects" committee investigates new fields and techniques; a "secret service" committee is designed to gather information concerning persons in private and public life, which may be referred to when needed; another committee deals with new contacts, its function being to get in touch with persons who can be "shaken down" for blackmail and extortion; and finally there is the lobby committee that maintains relations with Congress, state legislatures, city councils, and public officials.

---

[10] Sidney Lens, "Labor Rackets, Inc.," *The Nation,* Vol. 184, No. 9 (March 2, 1957), pp. 179-183.

[11] By permission from: Martin Mooney, *Crime Incorporated* (New York: McGraw-Hill Book Co., Copyright 1935), pp. 5-6.

The top operatives of syndicated crime do not resort to violence unless absolutely necessary. They have gone far beyond the days of the Capones and the Morans. Most of what violence they resort to is disciplinary in the competitive struggle between the parent organization and disgruntled off-shoots that attempt to take over or "muscle in" on their operations or encroach on pre-empted territory. The well-oiled combine has on its payroll professional gunmen whose major, if sporadic, occupation is to "rub out" those who dare challenge the authoritarian rule of the top-flight moguls.

When Martin Mooney's book appeared it was received with great skepticism and doubt. But we know that in 1940 an organization of big-time criminals, nationwide in scope, known as "Murder, Incorporated," was discovered and partially liquidated by the then District Attorney of Brooklyn, William O'Dwyer. Joseph Freeman described the activities and methods of this gang in *The Nation*.

He stated that the district attorney received much of his information about this combine from an informer, Abe (Kid Twist) Reles, who turned state's evidence. This gangster, a product of the Brooklyn slums, confessed to at least eighteen murders during his brief career. Reles compared the combination to a "tree with all its branches branched out" and to "an airplane trust." He added that there were hundreds of thousands in the nationwide monopoly, and in the five boroughs of New York City, several thousands. As he is quoted by Freeman, Reles states that:

> . . . the old rule was, I'll do you and you'll do me . . . I looked to kill you and you looked to kill me. . . . There was no sense in that. So the leaders of the mob said, why not stop this crazy competition and go out and make money instead? So the leaders got together and said, 'Boys, what's the use of fighting each other? Let's put our heads together, all of us, so that there can't be a meeting without one another.' That's how they all got together, to make no fighting.

Freeman himself observes:

> The crime trust, Reles insists, never commits murders out of passion, excitement, jealousy, personal revenge, or any of the usual motives which prompt private, unorganized murder. It kills impersonally, and solely for business considerations. Even business rivalry, he adds, is not the usual motive, unless "somebody gets too balky and somebody steps right on top of you." No gangster may kill on his own initiative; every murder must be ordered by the leaders at the top, and it must serve the welfare of the organization. . . . Any member of the mob who would dare to kill on his own initiative or for his own profit would be executed. . . . The crime trust insists that murder must be a business matter, organized by the chiefs in conference and carried out in a disciplined way.[12]

This informer, Reles, was killed in November, 1941, under suspicious circumstances, when he tried to escape from his detention quarters. The rope

---

[12] Joseph Freeman, "Murder Monopoly," *The Nation*, Vol. 150, No. 21 (May 25, 1940), pp. 645 ff.

he had fashioned out of bedding gave way and he plunged to his death. Reles' fate, which has never been satisfactorily explained, aroused attention and speculation as recently as 1951, during the Kefauver hearings at New York.

All the data summarized here regarding this murder syndicate have been verified.[13] Several of its murderous henchmen, including its ruthless leader, Louis "Lepke" Buchalter, were convicted of murder and received the death penalty. Following are the types of "craftsmen" employed, and their "salary scale." Murderers' apprentices or "punks" started with "piddling" chores at a salary of $50 per week (this was in the late 1930's). Their jobs included stealing cars for the murders, changing license plates, and crowding convoy cars to confuse witnesses to the crime and any police officers who might accidentally be nearby. The curriculum, according to Meyer Berger, writing in *Life* magazine, included courses in "schlamming" (severe beating), "skulling" (assault short of murder), and the various techniques in committing murder with such weapons as meat cleavers and ice picks. A "trooper" was the finished killer. His salary started at $100 a week, and experts who could be depended on at any hour of the day or night could make as high as $200 to $250 a week. Many of these killers might commit ten or twenty murders. The combination was known to have been responsible for at least 63 murders in nine years; how many more were actually committed will never be known. Most of those killed were "squealers" or "double-crossers." Many of the killers were forced to kill their friends, but usually, when they learned that these men were double-crossers, they did not hesitate.

The organization of contemporary large scale crime syndicates illustrates how far crime has developed beyond the stage of traditional criminality. To the beginner in the field, these commonplace facts about organized crime are likely to appear dubious but they have been attested to in many crime commission reports and by reputable writers.[14]

The operations of nationwide crime syndicates, and their actual control at any given time, are always obscure. Even the police of our large cities are confused as to the identity of their leaders. For many years the name of Frank Costello of New York was frequently heard as that of the overlord of crime in the country. An attack was made on his life in 1956 but it failed. For years Albert Anastasia, another New York thug, has been referred to as the "lord high executioner" of "Murder, Inc." Some thirty murders had

---

[13] Meyer Berger, *Life,* Vol. 9, No. 14 (September 30, 1940), and Vol. 16, No. 9 (February 28, 1944). Also Burton B. Turkis and Sid Feder, *Murder, Inc., The Story of "The Syndicate"* (New York: Farrar, 1951). Mr. Turkis assisted in the prosecution of some of the members of the "syndicate."

[14] The reader is referred to the various brochures published by the Chicago Crime Commission or by the California Crime Study Commission; also see Morris Ploscowe (ed.), for the American Bar Association, *Organized Crime and Law Enforcement,* 2 vols. (New York: Grosby Press, 1952-1953).

been attributed to him without a conviction. In October 1957 he was shot to death as he sat in the barber shop of a fashionable hotel in New York City. The following month a bizarre "convention" of some sixty known criminals met in a rustic community near Binghamton, New York. They came from all parts of the country and from Puerto Rico and Cuba. The state constabulary had apparently been tipped off and swooped down on the mob as they were preparing a barbecue at the palatial home of one of their kind. As they were not engaged in illegal activity and none was wanted by the police, they were released after questioning. The authorities suspected that the gathering was either of a syndicate, meeting to carve up the empire of Anastasia, or a convention of the Mafia, a terroristic Italian organization. Since all had obvious Italian names, the latter conjecture may have been correct. Later, many were jailed in New York City for their refusal to "talk."

The dreaded Mafia has for a hundred years or more been closely linked in this country with extortion, mayhem, kidnaping, and other forms of violent crime. It had its origin in ancient times in Sicily and spread to this country during the days of large scale Italian immigration. It is governed by a drastic code of strict silence and a vow never to deal with legal authorities in any matter involving crime. The antisocial organization is often known as the Black-Hand.[15]

Nor have syndicates kept out of legitimate business. As far back as 1943 the federal government convicted the leaders of the Capone mob of conspiring to extort more than a million dollars from the motion picture industry. The principals were sentenced to ten years' imprisonment and fined $20,000 each. In 1950 the Kefauver Committee stated that big-time gamblers bought up defunct breweries or concerns that brewed inferior beer and forced taproom owners to buy their products or suffer the consequences. Individual members of crime syndicates have invested their millions in legitimate enterprises such as hotels, especially resort hotels, race tracks, dance halls, and sporting arenas.

Local crime syndicates may be operating out of sight in many communities. Each town and city has its quota of shrewd criminal operators, many of them known to the police, but all too frequently seeming to lead charmed lives. They participate in the operation and control of vices such as prostitution, the numbers and lottery business, other gambling, and the sale of narcotics. Not infrequently these small time criminals are frozen out when representatives of the national syndicates decide to move in.

Another type of local syndicate, having no connection with the overlords of crime, combines to exploit the unsuspecting public. Confidence games, bogus stock sales, and business frauds of various descriptions are included in these operations. The Better Business Bureaus in our large cities are constantly on the alert for these frauds. The principle involved is always the

---

[15] For details see Edward Reid, *The Mafia* (New York: Random House, 1952). See also, *Time Magazine,* July 21, p. 15, 1958; and July 28, p. 15, 1958.

same—a promise to the victim of large financial returns for a small invest-ment. The yearning of many people for a cheap, quick financial turnover with huge profits has made this type of crime very successful.[16]

Fly-by-night concerns that advertise a "complete renovation" of the home, or aluminum screens "installed," or some other major home need at ridiculously low prices do a land office business in our large cities. Al-though they are regularly exposed by the Better Business Bureaus and dis-trict attorneys, there are always more enterprises to start up where others have been routed. Such outfits as these might be considered white-collar crime rather than types of crime syndicates, but the line between the two is often blurred, and in these operations we seldom find well-known, re-spectable people as we do in the white-collar crime.[17]

The Syndicate and Its Control of Gambling

The emergence and development during the 1940-50's of gambling syndi-cates controlling bookmaking, slot machines, baseball and prize-fight pools, and the like, is recognized by criminologists as the most prominent, the most threatening, and the least understood feature of our national crime picture.

These syndicates have exploited the traditional urge possessed by mil-lions to gamble occasionally, frequently, or habitually without condoning criminality in principle. The motto of the crime moguls might well be "Give the people what they want." It is the ambivalence in people's wants that contributes to the success of the syndicates.

This situation is attested to by the careful reports assembled by several state and city crime commissions. It is further documented by a constant stream of articles on gambling in the large slick-paper magazines. *The Annals of the American Academy of Political and Social Science,* a profes-sional journal, devoted its entire issue of May, 1950 to the ramifications of organized gambling in 1950.

The first significant report on this amazing phenomenon was published by the Chicago Crime Commission in 1945. This was followed by the report of the California Crime Study Commission (1949-50) and by a citizens' survey in Massachusetts in 1949. These reports contended that the over-lord of nationwide gambling was at that time Frank Costello of New York, for some years an alleged entrepreneur of big-time crime and a mysterious and sinister figure. These reports also showed, through documentary proof, that Costello and his lieutenants were able to thrive in various cities and states because of their alliances with public officials. This was verified by the

---

[16] For a series of articles on confidence games, see William J. Slocum, "The Postal Inspectors: Nobody Beats the Law," *Collier's,* Vol. 125, Nos. 2-6 (January 14-Febru-ary 11, 1950).

[17] See "Beware of Home Repair Racketeers," Stanley Frank, *Saturday Evening Post,* Vol. 229, No. 3 (July 21, 1956), p. 17 ff.

Kefauver Committee during 1950-51. In recent years the aging Costello seems to have been supplanted by more vigorous thugs, as yet not precisely identified.

Frank Costello (second from right, at table) testifying before the Senate Crime Investigating Committee in New York (1950). (Courtesy International News.)

Gambling, like social drinking, is widespread. It afflicts both rich and poor. Here is one estimate of the amount of money spent annually on gambling.[18]

Illegal bookmaking	$ 8,000,000,000
Numbers (policies, lotteries)	6,000,000,000
Slot machines	3,000,000,000
Pari mutuel bets (at the tracks)	1,600,000,000
Football and baseball pools	1,000,000,000
Legally bet in Nevada	1,000,000,000
Total	$20,600,000,000

As to the number of persons who gamble, the noted authority on gambling, Ernest E. Blanche, makes the following estimates: 26 million persons play bingo, buy lottery tickets, take part in raffles, or attempt to beat the football and baseball pools; about 22 million persons play dice and cards for money; 19 million bet on athletic and political events; about 15 million play punchboards; 14 million are addicted to slot machines of vari-

---

[18] Estimated by Murray Teigh Bloom in "Gambling: America's Ugly Child," *Pageant* (April, 1950), pp. 12 ff. Mr. Bloom states that the above does not include such gambling as bingo, beano, car raffles, state fair, carnival gambling, or local betting on fights and ball games. See also, Kenneth Rudeen, "The Global Gamble," *Sports Illustrated,* August 4, 1958, pp. 10 f.

ous kinds; 8 million engage in the numbers game; and 8 million play the horses. These figures, of course, do not give a good estimate of the total number of persons participating in all forms of gambling, since many people participate actively in more than one form.[19]

The lottery was the first American gambling pastime. But as corruption in its operation began to spread it lost support and was gradually outlawed.[20] With the waning of the lottery, professional gamblers turned to another lucrative activity known as "policy" or the "numbers" game. This universal form of gambling was popular a century ago in the large cities but it did not become a serious menace until Arthur Flegenheimer, better known as "Dutch Schultz," took over and developed control of its operations during the 1920's and 1930's. It is actually a lottery with a high return to the one who successfully guesses certain selected numbers, such as the total volume of daily bank clearings, the statement of the United States treasury balance, some sequence of horse race betting payoff figures, or some similar public announcement. But the game itself and its results can be thoroughly rigged by its promoters.

That the numbers game is deeply intrenched in our larger cities is demonstrated by the openness with which it operates. The numbers jargon is the vernacular of housewives, servants, small storekeepers, and even school children. It flourishes in many urban neighborhoods as well as in hundreds of small-town poolrooms and taprooms. Investigations have often shown that in many of our large cities hundreds of persons on relief supplement their incomes by writing numbers tickets. Other vestigial remains of the old-time lottery are athletic pools, operating seasonally almost anywhere.[21]

Off-track bookmaking, identified with the popular sport of horseracing, is an area of gambling that transcends any other single form of illegal gambling activity. It is estimated that between three to fifteen million people wager on horseracing away from the track.[22] It is further estimated that some 20,000 "bookies," or agents, take bets in and around New York City alone. As an indication as to how high up in "respectable" circles illegal bookmaking does go, some years ago eight officials of the Guarantee Finance Company of Los Angeles were convicted on charges of criminal conspiracy. While these persons were not publicly identified with the underworld, they were involved in a multi-million dollar bookmaking organiza-

---

[19] See Ernest E. Blanche, "Gambling Odds Are Gimmicked!" *The Annals,* Vol. 269 (May, 1950), pp. 77-80.

[20] See Ernest E. Blanche, "Lotteries, Yesterday, Today, and Tomorrow," *The Annals,* Vol. 269 (May, 1950), pp. 71-77.

[21] For more information on these types of gambling, see John Drzazga, "Gambling and the Law—Policy," *Journal of Criminal Law, Criminology, and Police Science* (hereafter referred to as *"J. Crim. Law"*), Vol. 44, No. 5 (January-February, 1954), pp. 665-670.

[22] For an analysis of how the various types of horserace gambling operate, see Louis A. Lawrence, "Bookmaking," and John I. Day, "Horse Racing and Pari-Mutuel," *The Annals,* Vol. 269 (May, 1950), pp. 46-54.

tion. Two members of the Los Angeles sheriff's organization were involved.[23]

Today criminal syndicates have taken over bookmaking and slot machines, the gambling media of the masses, and have made them big business of the first magnitude. The bookmaking field in particular has been aggressively organized. Monopolistic control of bookmaking is achieved through the leasing of "wire service," without which bookies cannot operate. Control of this service means the power, in any and every community in the country, to select those who can make book and those who cannot. The California Crime Study Commission (1949) has the following to say about this wire service:

> Although the "wire service" is a very juicy financial plum indeed, it is the aspect of monopolistic control blanketing the Nation that has made the "wire service" such a prize for racketeers, and such a threat to our institutions. Bookmaking is illegal practically everywhere and is always accompanied by the "fixing" and corruption of public officers. It has long been apparent to underworld leaders that the personnel and organization of the "wire service" and the bookmaking racket could, in the proper hands, be as effectively utilized in the operation of the rackets other than gambling. According to James R. Ragen, who for years headed the "wire service," the long and bloody struggle of the Capone Syndicate of Chicago to obtain control of the "wire service" was due as much to their ambition and design to use the organization of the bookmaking racket in the operation of unrelated rackets such as prostitution and narcotics, as it was to their desire to secure for themselves the immense profits of the "wire service." The great danger to the public lies in this same fact, that the "wire service" can and does serve as a framework reaching every sizable community in the country upon which a whole series of criminal rackets can be organized and operated.[24]

The bookie operations function through spotters equipped with binoculars who are located outside the race track. As soon as the race is run these spotters phone the results, as well as the odds from the pari-mutuel boards, to bookies all over the country. Speed is the essence of the operation; thus it is essential to the success of the betting that telephone wires be under control of the syndicate.

The slot machine syndicate, according to the California Crime Commission and other reliable reports of the gambling problem, was and may still be controlled on a national scale by New York's Frank Costello. This form of gambling operates more openly than any other type. It was estimated by the California Crime Commission that some 20 billions of dollars are involved annually in this allegedly harmless pastime. Although slot machine gambling is illegal in all but a few states, these machines are in operation in nearly every hamlet, village, town, and city of the nation.

---

[23] *The New York Times* (December 17, 1950).

[24] *Second Progress Report* (Sacramento: March 7, 1949), p. 15. See also the *Second Interim Report* of the Kefauver Committee (Washington, D.C.: U.S. Government Printing Office, 1951).

Estimates place the total number of pinball machines in operation in this country between 250,000 and 300,000, although the United States Treasury receives taxes from a mere handful of the places where the machines are operated. Fifty per cent of the gross earnings of each machine are usually paid to the establishment in which the machine is placed, thus making it a valuable source of revenue. While the public regards the playing of pinball machines by millions with little more than casual interest and practically no concern, the concessions to operate them in a community represent a widely sought privilege. The McClellan Senate Committee found in 1957 that large sums of money were paid out in Oregon to control these concessions.

Manufacturers of pinball machines produce about 4,000 per week.[25] The turnover is high since the novelty appeal soon declines and a succession of new gadgets must be provided for the moronic public that delights in playing them.

"Variety is the spice of life" to those who play the machines; but "play the sucker for all he's worth," is the motto of the operators who have succeeded in fleecing hundreds of millions (as a minimum estimate) and billions (as a maximum estimate) from the little, intellectually dull people throughout the nation. In short, our annual losses in the gambling realm are probably larger than ever before in history. Also, the element of chance has been largely eliminated. True gambling has been replaced by the much more corrupt principle of criminal deception, both in the numbers game and the slot and pinball machines.

In December, 1950, Congress passed a law making it illegal to transport slot machines across state lines, and in 1958 the Attorney General urged broadening of this law, in addition to his support of two more bills that would prohibit the interstate transmission of gambling information—belated legislative recognition of the role of the "wire service."[26] But the presence of machines in thousands of cities today indicates that there are ingenious ways of thwarting the law. A move to tighten up on the syndicate through legislation merely prompts the adroit operator to invent a new technique to carry on his criminal activity.

Public Control of Gambling

It has been often suggested that the only way to deal with the gambling syndicates and their cohorts, the crooked politicians, is to legalize gambling and thus "make it respectable." It is worth our while to examine this possibility.

---

[25] California Crime Commission Report, 1949, p. 125. For other interesting information, see: "Slot Machines and Pinball Games," *The Annals,* Vol. 269 (May, 1950), pp. 62-70; John Drzazga, "Gambling and the Law—Slot Machines," *J. Crim. Law,* Vol. 43, No. 1 (May-June, 1952), pp. 114-123; D. D. Allen, *The Nature of Gambling* (New York: Coward-McCann, Inc., 1952).

[26] *New York Times* (April 11, 1958), p. 16.

Big-time gambling is heavily weighted against the participant. This is equally true of all games of chance, mechanical or otherwise. The gullible public, in general, knows that the odds are against them in any gambling venture. They know the games of chance are rigged by the promoters. Ernest E. Blanche points out that gambling of one kind or another motivates the daily lives of over 50,000,000 Americans. He lists reasons why these persons can seldom win in a gambling game: every system of betting breaks down or fails sooner or later; the mathematical probabilities are always against the bettor; so-called games of skill are really games of chance that even the most skilled players cannot beat. He states that most of the Irish Sweepstakes tickets sold in this country are counterfeit, and that a person has only one chance in 2,000 of getting any money in a chain letter scheme or a Pyramid Club participation.[27]

In addition, it has been cogently pointed out by Virgil W. Peterson of the Chicago Crime Commission that practically all of the manufacturers of gambling equipment make their money in crooked machines. These include crooked dice, disguised playing cards, controlled roulette wheels, faked slot machines, and the like.[28]

The subject of gambling, in itself, cannot be approached from a moral standpoint. Those who are alarmed at the present state of affairs are not concerned with individual gambling between persons, placing bets on card games at home, or betting on ball games or even horse races. The disturbing factor is the business of gambling and the corrupt alliances between politicians and crime syndicates and the debauchery resulting when persons in places of trust succumb to the blandishments of professional gamblers. Bank employees sometimes fall victim to the gambling mania and embezzle large sums of money. This latter phase of the menace is admirably presented, again by Peterson, in his article, "Why Honest People Steal."[29]

Exclusive of permitting wagers at race tracks, Nevada is the only state that has legalized gambling and permits it to operate under licensing.[30] Other states have attempted licensing of various types of gambling but their experience has been generally unsatisfactory. In recent elections the voters of California, Montana, and Arizona turned down legalized gambling and the electorate of Massachusetts repudiated an attempt to legalize lotteries.

> Any unbiased study of the history of the gambling business makes it impossible for us to ignore certain salient facts. As a business gambling is entirely parasitic. It is completely unproductive. It creates no new wealth and performs no useful service. One factor common to every legitimate

---

[27] Ernest E. Blanche, *You Can't Win: Facts and Fallacies About Gambling* (Washington, D.C.: Public Affairs Press, 1949).

[28] "Gambling—Should It Be Legalized?" pp. 266-267.

[29] *J. Crim. Law*, Vol. 36, No. 2 (July-August, 1947), pp. 94-103.

[30] See Joseph F. McDonald's "Gambling in Nevada," *The Annals*, Vol. 269 (May, 1950), pp. 30-34.

commercial enterprise or profession is that it can exist only because there is an opportunity of a mutual advantage to the operator of the business and the patrons as a class. This is true whether the business is engaged in the manufacture, distribution or sale of a product or a service. Without this element of mutual advantage the business cannot exist. In the business of gambling, even when fraud and manipulation are absent, it still operates on a one-sided percentage basis that makes it impossible for the patrons as a class to derive any benefit. . . .

Licensing proposals are based primarily on the contention that because of the human desire to gamble it cannot be suppressed and, therefore, it should be licensed and legalized with the people sharing the profits instead of allowing the hoodlum element to reap the gains. . . . Based on past experience and keeping in mind the true nature of the gambling business it is evident that the usual proposals to license gambling would ultimately lead to complete failure. It should be clear that there is absolutely no justification for a license setup that would legally place the control of gambling in the joint hands of any local or state political machine and the parasitic professional gambling interests.[31]

The arguments against legalized gambling are plausible. But what are the arguments in favor of legalized gambling? Whether it is legalized or not, we would be obliged to rely on honest officials, including police officers, to see that city ordinances and state laws were scrupulously enforced. This is a big order. Yet it is the duty of decent citizens everywhere to demand that honest officials be elected to places of trust and law-enforcement. We shall get nowhere in any phase of crime control until we have honest politicians and public officials.[32]

If gambling were legalized, there would be far less temptation for public officials to take graft or to identify themselves with the big-time criminal element. We found that driving the legal liquor trade underground through prohibition laws failed to work; in fact, this procedure brought out the gangsters and criminals in the first big lawless era the country had ever known. It is asserted by exponents of legalized gambling that the underworld can be controlled by permitting certain forms of gambling. There are certain social customs—referred to by many as evils—participated in by millions. Social drinking is one of these—which may lead to alcoholism in some cases; prostitution is another; and gambling is still another. Individual vices have been with us for centuries. Social customs cannot be stamped out by legislative fiat. What is needed is some form of social control.

There are many thoroughly conscientious Americans who cannot comprehend how anybody can advocate legalized gambling. They feel that this would mean the complete debasement of our country's morality; it would be a veritable affront to our Christian civilization.

---

[31] See Mr. Peterson's article "The Myth of the Wide Open Town," *J. Crim. Law*, Vol. 39, No. 3 (September-October, 1948), pp. 288-297.
[32] Peterson, "Gambling—Should It Be Legalized?" pp. 321-322.

Yet Great Britain is a Christian land and is blessed with little organized crime. Its record for a low rate of criminal offenses committed and a high rate for arrests and convictions is far better than ours. Great Britain has permitted gambling, and today legalizes the main types that are outlawed here and that were so severely criticized by the Kefauver Committee: (1) betting with bookmakers on credit by mail, telephone, or telegraph; (2) betting in person at dog or horseracing tracks with bookmakers or totalizer machines; and (3) wagering small sums by postal money order on weekly sporting pools. And this sane philosophy has not increased big-time crime nor resulted in crime syndicates controlling such activities.

A British Royal Commission was appointed in 1949 to examine the whole problem of gambling in the United Kingdom and to recommend any needed changes in the laws. It carried on its work for two years, held 25 meetings, and listened to about 150 witnesses, running all the way from jurists and police officials to the leading gamblers of the British Isles. The Commission presented a report that came precisely to the opposite conclusions and recommendations of the Kefauver Committee.

This British Commission on gambling held that there is nothing inherently wrong or sinful in gambling; that it does not contribute significantly to poverty; that legalized gambling does not produce much juvenile delinquency or increase the crime rate; and that it has no serious effect on general social behavior. The report further states, "We can find no support for the belief that gambling, provided that it is kept within reasonable bounds, does serious harm either to the character of those who take part . . . their family circle [or] the community generally." Such a conclusion is, of course, diametrically the opposite of the attitude held toward gambling in the United States for the past century.

Amazing as it may be to the average American reader, the Royal Commission, following the gambling debate, actually recommended that gambling be made easier, so that the little man with only a few pennies of ready cash in his pocket could take part in gambling as well as his more fortunate superiors in the economic scale.

As matters stand now in the United States, those who have time and money can gamble legally at the tracks. Yet it is illegal for others to place bets away from the tracks. If off-track betting were legalized much of the control of gambling would be taken from the criminal element. Certainly the answer is not to expect the impossible from public officials. They, too, are a part of the American scene and are under pressure from the millions of people who want to gamble and demand that right.

We find many types of petty gambling not now under control of the criminal element—church raffles, bingo games, drawings by women's clubs, private pools on baseball and football, and the like. Many civic organizations make their programs attractive through private lotteries. With legalized gambling, punch boards, slot machines, pinball, and other such gadgets could be controlled. Off-track betting, baseball pools, and the like, could then be placed under the control of city ordinances.

In any event, laws must be revised to plug up tax loopholes and prohibit the wire service that gives the underworld a monopoly of gambling. In the last analysis, however, what is needed are honest public officials and an efficient, incorruptible police force. This is essential whether gambling is legalized or not. Regardless of whether or not laws are passed to control gambling, there will always be millions of suckers who are willing to lose their money through games of chance. The problem then, is not to suppress gambling but to eliminate the underworld criminal in its control.

3

The Upper World of Crime

Power tends to corrupt and absolute power tends to corrupt absolutely. (From Freedom and Power.)

LORD ACTON

The Public Official and His Contribution to the Crime Picture

The upper world of crime is that area in which men and women who are traditionally held in respect by the public either aid and abet criminal activity or engage in it directly. These people cloak their criminal behavior in the implied trust of the large body of citizens, who are usually unaware of their predatory operations. Integrity is expected of public servants and businessmen despite the fact that here and there, in all eras, it has not been forthcoming. In this chapter we wish to discuss crime, corruption, and fraud in public places and in business. It must be emphasized, however, that it is a minority of people in these fields who engage in criminal activity or violate their trust.

To anyone familiar with urban life, the terms "machine" and "organization" refer to the political party in power at the moment. We hold no brief for any party; as an eminent professor at the University of Chicago once put it: "The Republican and Democrat parties are but two wings of the same bird of prey." The machine dominates city and state politics and often exerts a powerful influence over national politics. The machine manipulates the majority by dispensing jobs, charity, and certain types of petty immunity.[1] By using common psychological devices of the party system, such

---

[1] For a revealing book dealing with the philosophy of a political boss, see Edwin O'Connor, *The Last Hurrah* (Boston: Little, Brown, 1956). In thinly disguised fiction, it tells the story of James Michael Curley of Boston. For an interesting apologia from "Boss" Curley, see *Life* (September 10, 1956), Vol. 41, No. 11, pp. 118 ff. The University of New Hampshire invited "Boss" Curley to speak on the book, which he did in rare good humor. In 1957 Curley published his own story: *I'd Do It Again: A Record of All My Uproarious Years* (Englewood Cliffs, N.J.: Prentice-Hall).

as party names, shibboleths, catchwords, and symbols, the politicians very frequently acquire control of the city government. They remain in power by continuing the same techniques. Control of a city depends on the creation of a powerful voting machine, and this machine is maintained by rewarding the workers with political jobs. In return for their jobs the workers can usually expect to be "maced"—they are expected to contribute to campaign funds, attend "$100 dinners" around election time, and remain loyal to the party. In order to hold his job at City Hall, the worker must not only vote the straight ticket but be able to control the votes of the members of his family and his friends. Thus the voting potential of office holders plus that of the groups who have been aided by the organization represents a sizeable handicap to the opposition party in waging a campaign to "throw the rascals out."

In reaching out for more and more jobs, the politician frequently resorts to unethical practices, then to graft, corruption, and exploitation. Greater opportunities for graft in the metropolitan areas make politicians determined to hang on to their sources of spoils. Control over millions of dollars' worth of municipal contracts enables the machine to enlist the influence of respectable business concerns. This pattern of municipal control is universal, not restricted only to the large and lucrative cities.[2]

The cost of the political graft that flows from this system far exceeds the losses sustained by conventional petty crimes. The history of graft in American cities is one of enormous siphoning of taxpayers' money into the pockets of the bosses and their henchmen. Often this graft turns up in national politics. Congressmen have frequently been linked with corruption and other illegal activity. Some of our presidential administrations are remembered mostly for the graft and corruption that characterized them. This is notably true of the Grant and Harding eras. During World War II, a congressman was sent to prison for graft in connection with war contracts; two went to prison for demanding "kickbacks" of paychecks from their employees; still another served time in prison for income tax evasion and, strangely enough, was re-elected by his constituents at the next election.

We were familiar during the Truman administration with the "five percenters" and the "Friendship Racket" that existed between the higher-ups close to the President and contractors who craved war contracts. Bruce Catton's *The War Lords of Washington*[3] best reveals the story of the callousness of some businessmen during World War II. It indicates the indifference, greed, ineptness, and arrogance of many of those responsible for conducting a war to preserve our democratic ideals—ideals that were

---

[2] Political bosses are still with us. See Charles R. Allen, Jr., "McClure of Pennsylvania: Bosses Are Still in Business," *The Nation,* Vol. 183, No. 4 (July 28, 1956), pp. 71-73.

[3] (New York: Harcourt, 1948); see also Jules Abels, *The Truman Scandals* (Chicago: Regnery, 1956).

shelved by these men in dispatching their duties. The fine line between criminal activity and immorality either in business or in government is often difficult to discern. The "Friendship Racket" of the Truman administration had its counterpart in the Eisenhower administration. This was referred to as "conflict of interest." Neither activity is criminal, but both represent a "public be damned" attitude by persons in high government posts.

Big financial and business interests looking for lush contracts or desiring to secure freedom from public regulation or to receive tax reductions or rebates align themselves with friendly politicians. They contribute heavily to the campaign funds of the machines and candidates. This is not done through any loyalty to the party or the politician but rather for the financial favors that have consistently flowed from the halls of political control, national or local.

All too frequently the criminal element "muscles in" by buying up public officials. This is done by paying "protection" money, stuffing ballot boxes, intimidating independent voters, by discouraging political reformers with threats and violence, and otherwise aiding the political machines, but always for a price. For their services, racketeers want a free hand in operating whatever vice seems lucrative.

Anyone familiar with Chicago during the reign of "Big Bill" Thompson and the Kelly-Nash machine, with New York City under Mayor "Jimmie" Walker, with Philadelphia under the Vare machine, with Kansas City, Missouri, under the control of the Pendergast machine, with Atlantic City dominated by Enoch L. (Nucky) Johnson, with Memphis controlled by the Crump organization, with Jersey City under Frank (I Am the Law) Hague —to mention only a few of the more notorious—knows just how difficult it is to wrest control from a ruthless, graft-ridden political machine. Many years ago it was Lincoln Steffens in his *The Shame of the Cities* who acquainted the public with the machinations of corrupt politicians, but there is no dearth of current clinical evidence on the subject. The story is the same wherever citizens of a community become apathetic concerning the day-by-day workings of their city government. Graft, corruption, and the alliance between public servants and the criminal world are commonplace. However, it should be noted that in recent years there have been fewer exposures of wholesale graft and corruption than in the period between 1920 and 1950. Aside from large-scale embezzlement by the auditor of the state of Illinois, who in 1956 was sent to prison for his crime, there has been no glaring incident of malfeasance in public office.

Not that it does not happen; we can only know where graft and corruption flourish when they are exposed. Apparently the muckraking era is past or the muckrakers are lying dormant. But we must not forget that the activities we have been describing are part of a deeply-rooted pattern in American public life. Many intelligent men and women are an integral part of the political machine. They are loyal to the men higher up. Technically trained accountants, engineers, and others are willing to play the game in order to hold their jobs. Wealthy financiers who deal in city bond issues, and persons who seek franchises or new concessions, must support the political organization. The criminal element also wants concessions from the machine. All work together to retain power. As one New Jersey politician put it:

> I want . . . elected because I want power and patronage. I want to see my friends get jobs. I want to be able to name judges and deal out patronage. We get nothing if there is a Republican victory. But when you're in power it's really wonderful. When you're in you have the patronage and power and everything that goes with it.[4]

Seldom is such a frank statement of motive so blatantly expressed.[5]

White-Collar Crime in Our Society

The analysis of white-collar crime in our culture is usually associated with the criminologist Edwin H. Sutherland. It was he whose penetrating work in this area of crime focused our attention on its grave effect on the total crime picture.

But Professor Sutherland was preceded by others who were aware of the dangers to society from the upper socio-economic groups who exploited the accepted economic system to the detriment of the masses. Professor Albert Morris calls attention to a paper entitled "Criminal Capitalists" read by Edwin C. Hill before the International Congress on the Prevention and Repression of Crime at London in 1872. In this paper the writer noted the "growing significance of crime as an organized business requiring the co-operation of real estate owners, investors and manufacturers . . . and other 'honest' people."[6] As early as 1934 Professor Morris called sharp attention to the necessity of a change in emphasis regarding crime.[7]

Sutherland regarded white-collar crimes as acts committed by persons in the upper socio-economic level of American society in accordance with their normal business practices. If a broker shoots his wife's lover, that is not a white-collar crime, but if he violates the law and is convicted in connection with his business he is a white-collar criminal.[8]

---

[4] From the Asbury Park *Press* (October 22, 1949).

[5] For a discussion of the corruption and graft in some of our cities see: Virgil W. Peterson, *Barbarians In Our Midst* (Boston: Little, Brown, 1952); Ed Reid, *The Shame of New York* (New York: Random House, 1953).

[6] Albert Morris, "Changing Concepts of Crime," *Encyclopedia of Crime,* edited by V. C. Branham and S. B. Kutash (New York: Philosophical Library, 1949), p. 51.

[7] See his *Criminology* (New York: Longmans, 1934), especially Chapter 3, "Moral and Social Costliness."

[8] Sutherland first called attention to this type of criminality in a paper presented before the American Sociological Society in December, 1939 and entitled "White Collar Criminality." It was published in the *American Sociological Review* (hereafter, *"Amer. Soc. Rev."*), Vol. 5, No. 1 (February, 1940), pp. 1-12. Some years later he wrote *White-Collar Crime* (New York: Dryden, 1949).

Since Sutherland's initial research into this heretofore relatively neglected area there has been a sizeable literature accumulated which deals with various aspects of upper-level business dealing. Distinctions between crimes and violations of trust, unethical business practices, violations of the "spirit" of the law rather than its strict letter, of convictions or just mere indictments, are finely sifted. Definitions of "legal" criminals and "sociological" criminals have also been made.[9]

We wish to differentiate between crimes recognized as the white-collar variety, and crimes committed by criminal syndicates. It would seem that this distinction could be based on the aura of *presumed* respectability that is part of the former's stock in trade. For example, a person who possesses some degree of a good reputation and sells shoddy goods is committing a white-collar crime. But if a group of persons, unknown to their victims, sell the same type of shoddy goods, that would not be white-collar crime. In this latter category we might place bogus stock operations by criminal syndicates, fly-by-night operators of get-rich schemes, and the innumerable frauds described by Better Business Bureaus. These criminal activities are sometimes carried on by lone operators, but often they are in the hands of calculating criminal gangs. Quite often they operate within the strict letter of the law and exploit the credulity of their victims.

Sutherland carefully examined the depredations of 70 large corporations. He found that the courts had declared a total of 980 legal decisions against these giant combines. The charges included contracts, combinations, or conspiracies in restraint of trade, misrepresentation in advertising, infringements against copyrights and trade marks, "unfair labor practices," rebates, financial frauds and violations of trust, violations of war regulations, and other miscellaneous offenses. Although only 158, or 16 per cent, of the decisions listed against the 70 corporations were made in criminal courts, 60 per cent of the corporations had an average of four criminal convictions and could therefore be classed as "habitual criminals" in many states.

The public knows very little about the trickery of big-business criminals. One reason is that it is not generally considered a crime when some big national combine runs afoul of the Sherman Antitrust Act or other federal

---

[9] The reader is referred to: Paul W. Tappan, "Who Is the Criminal?" *Amer. Soc. Rev.,* Vol. 12, No. 1 (February, 1947), pp. 96-102; Marshall Clinard, "Criminological Theories of Violation of Wartime Regulations," *Amer. Soc. Rev.,* Vol. 11, No. 3 (June, 1946), pp. 258-270; Frank Hartung, "White Collar Offenses in the Wholesale Meat Industry in Detroit," *American Journal of Sociology* (hereafter, *"Amer. J. Soc."*), Vol. 56, No. 1 (July, 1950), pp. 25-34; Hartung, "White Collar Crime: Its Significance for Theory and Practice," *Federal Probation,* Vol. 17, No. 2 (June, 1953); and to the following books: Sutherland, *White-Collar Crime;* Marshall Clinard, *The Black Market* (New York: Rinehart, 1952); Donald R. Cressey, *Other People's Money* (Glencoe, Ill.: Free Press, 1953). For an excellent summary, see Robert C. Caldwell, *Criminology* (New York: Ronald, 1956), pp. 67-70. Also, Robert C. Caldwell, "A Reexamination of the Concept of White-Collar Crime," *Federal Probation,* Vol. 22, No. 1 (March, 1958), pp. 30-36.

law. For example, in one large eastern city a few years ago, the four largest department stores were indicted, convicted, and fined for price-fixing. Their only apology was that they "didn't know they were violating the law"; yet all four had expert legal advice.

While the public pays little attention to this type of crime and actually regards it as not too important, the attorney general of the United States, in 1950, said that antitrust violations are "the most pressing domestic problem of our time next to unemployment."[10] Another reason the public is apathetic is that the legal battles involved are dragged out for years in the courts, with the result that charges are forgotten long before they are settled. Then too, the decisions are not published in media accessible to the average reader. Few editorial writers crusade against big-business crimes because in most instances they are so remote and so impersonal.

Under federal law the maximum fine that can be levied in antitrust suits is $5,000 on each count. Prison sentences are also authorized, but in more than fifty years of such prosecution not a single businessman has ever been sent to jail. The Department of Justice asked Congress to increase the fines to a $50,000 maximum for criminal antitrust violations and also to strengthen the existing law, but thus far these requests have been ignored. Actually the Department is so hamstrung financially that only a bare fraction of complaints can be prosecuted. Its annual appropriation for this work is only $3,750,000—less than some cities spend on their police departments.

The story is the same with the Federal Trade Commission—inability to prosecute. Occasionally one reads that this watch-dog of fair advertising and trade practices has issued a warning to a business house to "cease and desist" in what it considers a fradulent practice. The objectionable practice is usually stopped, but after a time other firms carry on the same exaggerated claims for their products. Advertising in newspapers and magazines, and radio and television commercials provide ample proof of the frequency of borderline frauds.

There has always been crime among businessmen. There have always been instances of violation of trust. Most of us have read of chicanery and plunder in the history books and such acts have often constituted the central theme of the fiction of earlier times. But the American people seemed to believe that anyone who betrayed a trust or who mulcted the widow's mite in a shady but legal deal, would eventually suffer—if not here, surely in the hereafter. Existing practices, however, were generally accepted as being within the canons of good business. Business, therefore, was justified in pull-

---

[10] Quoted in *The Machinist* (Washington, D.C.: International Association of Machinists, April 13, 1950), p. 1. This magazine lists the names of 263 corporations and 13 trade associations haled into court during 1949 on charges brought before the Department of Justice (see pp. 4-5). For further information regarding the offenses of business firms as well as possible reasons, see Robert E. Lane, "Why Business Men Violate the Law," *J. Crim. Law*, Vol. 44, No. 2 (July-August, 1953), pp. 151-165.

ing a shrewd deal. The victim either did not report what was done for fear of being ridiculed, or received little sympathy because he had been fleeced in a socially approved and even legal deal. *Caveat emptor*—"let the buyer beware"—expressed the prevalent attitude.

A series of federal laws passed in the 1930's, including the Securities Exchange Act, the Public Utility Holding Act, and the Investment Company Act, make many types of white-collar practice illegal. But despite these safeguards, thousands of small investors still lose their money in get-rich schemes conceived and developed by clever operators. Perhaps the most sensational in recent years was that of a substantial and well-liked bank president in a small town in New York who wrecked not only the bank, but the town as well—at least temporarily. The citizens of the village, as well as friends from other communities and the government insurance agency, rallied and saved the situation. There was little censure for the actions of the banker since there was no evidence that his laxity in granting business loans was anything more than a loyal enthusiasm for his home town. Yet he "played" with other people's money and violated the law governing banking in his almost zany transactions.

Reports in the daily press and magazines on embezzlements by persons of trust (which amount, according to the Chicago Crime Commission, to $400,000,000 annually) tend to shake us from complacency and force us to reappraise our whole thesis of crime and criminals. Nothing could be more erroneous than to continue to regard criminals as products of slums, broken homes and of the lower intellectual or social classes. Notwithstanding the many volumes and articles written to discredit this outmoded and naïve outlook, the American people have become so habituated to a distorted picture of crime and the criminal that it is almost impossible for them to think of the problem realistically.

The Bond Salesman and Investment Broker

The epitome of white-collar sophistication is the stock salesman, who may represent old-line conservative bond houses or dubious promotion schemes. Prior to the lush 1920's this breed had become almost a parasite on society. Although the public has been more adequately protected in recent years through federal legislation governing the issuance and sale of stock, there is still much room for fraud and malicious misrepresentation concerning stock issues.

The fact that thousands of unwary people are still being fleeced of hundreds of millions of dollars annually demonstrates how difficult it is for legislation to keep abreast with the promoters of bogus stock. In recent years with the development of atomic power thousands of investors have been mulcted of millions of dollars in dubious or worthless cobalt and uranium mines.

Patent Medicines, Food Adulteration, and Fraudulent Advertising

A particularly antisocial group of white-collar business firms operates in the patent-medicine, cosmetic, and food fields. They are aided by newspapers, magazines, and radio and television stations that are willing to take their misleading and even fraudulent advertising. Over two hundred million dollars are spent annually on patent medicines, and that sum is exceeded in the realm of other medical cures. The field is most fruitful for those who deliberately flout the law or skirt it precariously, thus taking advantage of human misery. Many patent medicines are not only worthless but harmful.

In this area, together with the cosmetic field, we find two groups: one stays within the law and seeks to prevent more stringent legislation from being passed through powerful lobbies; the other violates the existing laws until caught and convicted. The first group may not break the letter of the law but, by violating its spirit, constitutes an unsocial economic element. Patent medicine manufacturers maintained such a powerful pressure group in Congress that it took more than thirty years to amend the almost innocuous Pure Food and Drug Act passed in 1906.

Deliberate attempts have frequently been made to adulterate and even poison the food sold to the unsuspecting public. A former auto swindler dissolved radium salts in water and sold the poison to gullible males to cure a wide variety of ailments. This poison sent at least two men to horrible deaths, and many others became seriously ill from drinking it. A manufacturer seeking a cheap adulterant for Jamaica ginger came upon a certain kind of phosphate. Ginger mixed with this chemical was sold in drugstores throughout the nation, and caused from fifteen to twenty thousand cases of deformity and paralysis. Many of the victims died.[11]

It has been estimated that from eight to fifteen per cent of all food consumed in this country is adulterated or contaminated. Each year over three million people are made seriously ill as a result of eating tainted or doctored food. Hundreds die of ptomaine poisoning and other associated ills. A few years ago a news story told of a packing house in Denver that was accused of putting horse meat in sausage and selling it to the public. The Better Business Bureau of Los Angeles prosecuted a butcher for coloring ground beef with sodium sulphide. These are a few examples of the relatively widespread practice of adulteration of food products.

Fake and misleading advertising is another area in which the white-collar criminal operates. The only protection the public has in this area is the Federal Trade Commission which cannot possibly keep up with all of the

---

[11] For an enumeration of such tragedies, see: Arthur Kallett and F. J. Schlink, *100,000,000 Guinea Pigs* (New York: Vanguard, 1932); Ruth de Forest Lamb, *The American Chamber of Horrors* (New York: Farrar, 1936).

unfair and technically illegal claims so frequently made in advertising. Not only is much of this misleading to the public but unfair in competition.

White-Collar Crime in the Medical Profession

The vast majority of medical doctors are decent, law-abiding citizens, proud of their profession and fully aware of their responsibility to patients and the public—but that can also be said of most bankers; yet we read occasionally of a banker violating the law and ethics of his calling.

We have all heard of the grossly unethical practice of fee-splitting which, though not illegal, is definitely frowned on by the profession. We have often heard and read of illegal operations committed by physicians. We might also add the close tie-up in certain instances of a physician and a criminal gang. A gangster with a bullet injury knows where he can have a wound dressed without fear of being reported. The physician also has relatively easy access to narcotics and can be of great service to addicts. At the same time he can add many dollars to his income by faulty income tax returns. Quite a few physicians have been heavily fined or sent to prison for this oversight. Then, too, some physicians have accepted money from cigarette companies for endorsing the bogus claim that the cigarette in question does not irritate the throat. Once again we find confusion between unethical practice and actual illegal acts.

Regardless of what we may think of the practice of abortion, the commission of such an operation by a physician is illegal, unless performed to save the life or health of the mother. How many physicians perform illegal abortions rarely, occasionally, or habitually, we have no way of knowing. Careful students have estimated that seven hundred thousand to two million abortions are performed annually in this country, although many of these are performed by the women themselves. In 1945 it was estimated that the number of criminal abortions in San Francisco exceeded the number of live births.[12]

Occasionally law-enforcement officers unearth an "abortion mill" in some city. Personnel include nurses, matrons, and "runners," the latter being used to draw in the clients who are usually desperate for an operation.[13]

It is estimated that a minimum of ten thousand deaths result each year from abortions and that many other thousands of women are made invalids for life. Only the relatively wealthy can afford to have abortions performed under medically safe and psychologically satisfactory conditions, albeit in violation of the law on the part of the physician. The poor have recourse only to charlatans and marginal physicians.

Since it is entirely evident that no amount of social, religious, or moral

---

[12] *Life,* Vol. 27, No. 26 (December 26, 1949), p. 63.
[13] See Jerome E. Bates, "The Abortion Mill: An Institutional Study," *J. Crim. Law,* Vol. 45, No. 2 (July-August, 1954), pp. 157-169.

pressure can do away with the prevalence of abortions, either self-imposed or by unscrupulous physicians, some sociologists and medical authorities recommend the complete legalization of abortion under proper safeguards. They argue that if we are going to have them anyway, we might better make them both legal and safe. Further, our current efforts to punish those who perform them fail almost completely. It is almost impossible to secure the necessary witnesses to effect a conviction. In short, the abortionist may ply his illegal trade with impunity, and the most unscrupulous of the group extract from their victims all that the traffic will bear.

Sutherland, in his study of white-collar criminality, makes these statements concerning the medical profession and criminal activity:

> In the medical profession, . . . because it is probably less criminalistic than some other professions, are found illegal sale of alcohol and narcotics, abortion, illegal services to underworld criminals, fraudulent reports and testimony in accident cases, extreme cases of unnecessary treatment, fake specialists, restriction of competition, and fee-splitting. Fee-splitting is a violation of a specific law in many states and a violation of the conditions of admission to the practice of medicine in all. The physician who participates in fee-splitting tends to send his patients to the surgeon who will give him the largest fee rather than to the surgeon who will do the best work. It has been reported that two thirds of the surgeons in New York City split fees, and that more than half of the physicians in a central western city who answered a questionnaire on this point favored fee-splitting.[14]

Crime in the Legal Profession

In writing of white-collar crime in the legal profession we ordinarily think only of shyster lawyers in negligence cases and of those who advise organized criminals. But that overlooks those lawyers, a minority certainly, who play a leading part in promoting and facilitating white-collar criminality. For lawyers—often of the highest standing at the bar—who specialize in corporation and constitutional law have suggested or guided the criminal or quasi-criminal activities of corporations under the sway of finance capitalism, especially the wholesale white-collar crimes of public utility in the past.

The lesser luminaries among unscrupulous lawyers include the "alley" lawyer who spends his time hanging around the courts looking for a poor man's case and sometimes engages in such unethical or illegal practices as the "liability claim" operation, better known as "ambulance chasing." It is the lawyers of this group that the public believes to be the real danger in unethical or illegal practice of the law.

Questionable practices in one profession tend to contaminate others. A prominent surgeon, addressing a national meeting of the American College of Surgeons in 1958, stated that the methods of lawyers in liability cases

---

[14] Sutherland, "White-Collar Criminality," *Amer. Soc. Rev.,* p. 2.

made it impossible for expert medical witnesses to tell the truth, twisting conscientious testimony to give a false picture, or confounding one doctor with other presumptive "experts." No doubt a conscientious lawyer could present an equally cogent case for the damage initiated by unscrupulous medical witnesses.

The tactics of shyster lawyers are familiar to many people, since they have been exposed many times. One of the practices in which they specialize is "fake claims" or liability in bogus automobile accidents. Lawyers of this breed line themselves up with other equally unscrupulous individuals—in some cases, "lone wolves," and in others, powerful crime syndicates who specialize in these "squeeze plays"—and profit handsomely.

The lawyer at the center of this chicanery may have a whole staff at his beck and call—other lawyers, runners, doctors, hospital assistants, X-ray technicians, and even professional perjurers. Insurance executives have claimed that the take in these illegal activities amounts to some fourteen million dollars annually.[15]

Then there are the lawyers who are retained by high-powered members of the overworld of crime. These lawyers know that they are dealing with the moguls of the criminal world and they sell their knowledge and help for a sizeable consideration. This activity comes within the same category as the physician who performs plastic surgery on a fugitive from justice to avoid recognition. As Frank Tannenbaum puts it:

> Thus the professional crook seeks out the professional criminal lawyer. Every large professional criminal gang has its trusted mouthpiece, the lawyer who can be depended upon to arrange things. He is on hand with the bail bond or the habeas corpus writ to "spring" the members of the gang who have had the misfortune to be arrested. If the case can be fixed, the lawyer will find the ways and means to fix it. If, for some reason, it cannot be fixed, the lawyer can be depended upon to arrange for a perjured defense or for the bribery or intimidation of jurors and witnesses. "There is a group of criminal lawyers," wrote the Chicago City Council, "whose work includes dealing with the police, furnishing professional alibis and professional witnesses, jury fixing, spiriting away of witnesses, exhaustive continuances, and all the underground activity of all around 'fixers.' "[16]

Many of these lawyers lead apparently respectable lives in their communities and mingle freely in the "better" element of society. Lawyers on the payroll of corrupt public officials are only one step removed from the gangster element. Occasionally lawyers who have been elevated to the bench accept bribes from commercial concerns or criminals. Federal Judge Martin T. Manton of New York City—the ranking judge in the nation next to

---

15 For further details, see Robert Monaghan, "Fake-Claim Rackets," *Forum,* Vol. 103, No. 2 (February, 1940), pp. 87-91.

16 Frank Tannenbaum, *Crime and the Community* (Boston: Ginn, 1938), p. 265. By permission of Ginn and Company.

those on the Supreme Court—was found guilty and convicted of accepting bribes amounting to $664,000. The six or eight commercial concerns that bribed Judge Manton were not prosecuted.[17]

A lawyer must do his duty in defending a person accused of crime. This applies equally to those from known criminal elements. There is a distinction between a criminal lawyer and a lawyer-criminal. So long as he handles his defense in a legal and ethical manner he is only dispatching his duty as an attorney. It is when he carries his defense beyond the law and the ethical standards of his profession that he may be labeled a lawyer-criminal.[18]

Demoralizing Effects of White-Collar Crime

We have passed in review a number of the better known professions which offer juicy temptations to crime as well as to unethical practices of which many are at least quasi-crimes. Some members in every profession succumb to criminal activity. There was considerable crime of this sort during the war years. We all remember the black and grey markets thrived, and illegal chiseling was not uncommon among those eager to procure defense contracts, and even among the population as a whole that jockeyed for a favorable position in securing rationed consumer goods.

We wish to emphasize again that none of these practices is particularly new. There have always been crooks and unethical men in business, the professions, and in public life. But we have neglected to denounce such persons in schools, churches, homes, and other places where the public is being trained for citizenship and character building. As a result, a large portion of the population believes that illegal activity can be carried on with impunity, immunity, and no loss of caste or prestige.[19]

White-collar crime flows from a competitive economy and a philosophy that reveres success based almost exclusively on money and material consumption. The job of the schools, courts of justice, legislators, and a regenerated public is to wipe out this insidious philosophy of the "acquisitive urge."

---

[17] For details of this case, see S. Burton Heath, *Yankee Reporter* (New York: Funk, 1940), pp. 243-287.

[18] This is further discussed in connection with defense attorneys. See page 244.

[19] For a journalistic account of businessmen and crime, see Robert Rice, *The Business of Crime* (New York: Farrar, 1956).

4

Traditional
Crimes and Criminals

The mood and temper of the public with regard to the treatment of crime and criminals is one of the most unfailing tests of the civilization of any country. (Spoken before the House of Commons in 1910, after witnessing John Galsworthy's stirring play, Justice. *Quoted by Evelyn Ruggles-Brise,* The English Prison System, *London: Macmillan, 1910.)*

WINSTON CHURCHILL

Introductory Statement

This chapter will focus on conventional crimes and criminals. It must be stated, however, that some of these also fit into criteria we established in our earlier discussion. In general we regard conventional crime as that which is described in the day-by-day news stories, crimes such as robbery, burglary, shoplifting, assault and battery, rape, and murder. Even fantastic and brutal crimes that might be considered compulsive are also included in this category. It is not that most of these offenses are not dangerous and costly, but rather that they are traditional, having occurred throughout the history of Western society. They are usually leveled against one or a few victims rather than against society as a whole.

Most of the criminals we find in our jails and prisons are of the traditional variety. We very infrequently succeed in apprehending or convicting the overlords of criminal syndicates (except occasionally for an income-tax evasion) or those belonging to the upper socio-economic groups who commit white-collar crimes. Many traditional criminals are essentially dangerous and the depredations they commit, either against persons or against property, are serious and should not be minimized. But they and their social harmfulness are insignificant compared to the costs to society of the other types.

Traditional criminals may operate singly or in groups; may be of low or high intelligence; may be occasional, habitual, or compulsive; their crimes may be petty or highly dangerous; may involve violence or not. The suave confidence man or woman, check passer or forger, is a traditional criminal but so is the petty purse-snatcher or pickpocket. Some traditional criminals are steadily employed at legitimate work, some merely casual workers, and still others rarely engage in legally gainful employment. In short, this is a diverse group of individuals.

While many traditional criminals are "lone wolves," others work in pairs or groups. They may be disciplined as in a criminal gang but they may also be "pick up" or casual acquaintances who are together merely to commit a crime. Sometimes two or more ex-prisoners meet and plan a crime. But the old-style disciplined criminal gang is worthy of some description since, in years past, and to some degree even now, it represents one of the more potent threats to peace and property.

Criminal Gangs

The criminal gang is not new in our culture. We are all familiar with stories of the old Jesse James gang, the Daltons, and others of the post-bellum West. The New York plug-uglies of the early nineteenth century have been described by Herbert Asbury,[1] and we are familiar with the Black-Hand or Mafia (Unione Siciliona) which laid tribute on Italian immigrants and otherwise terrorized hundreds of communities during the early new-immigration period. This organization is by no means dead and periodically is suspected of threatening and despoiling individuals and legitimate business.

The 1930's and '40's were conspicuous for the number of colorful if brutal gangs that roamed the country. They included the Dillinger gang, the Purple Gang operating around Detroit, the Tri-State gang that terrorized Pennsylvania, Maryland, and New Jersey, and the Barker gang headed by the notorious "Ma" Barker and her ruthless sons.

In gang criminality, violence is the most conspicuous characteristic. In recent times those thus engaged have been equipped with machine-guns, sawed-off shotguns, bullet-proof vests, tear gas, and high-powered cars (mostly stolen) for quick getaways. In some cases the armament of these gangs is stolen from state armories. The usual operations of criminal gangs in the modern era are bank robberies (following the older pattern of the James Boys), pilfering of trucks and warehouses (hijacking), and, during the 1930's, kidnaping. The notorious Brinks robbery of January 1950 in

---

[1] *The Gangs of New York* (New York: Knopf, 1928).

Boston was accomplished by a criminal gang, and all the participants may well be labeled traditional criminals.[2]

It would be difficult to estimate the actual amount of loot stolen annually by criminal gangs. Much of what they get is recovered by the police. It has been estimated that the yearly loss of trucking companies by hijacking runs to $50 million.[3] The total losses are undoubtedly high, but not so great as is generally believed.

The exploits of criminal gangs often thrill the public. They read the news stories, highly colored by clever journalists. This was evident from the radio banter and the cartoons in the papers following the Brinks robbery. Reporters also play up the fanciful names of some of these hoodlums; typical are "Pretty Boy" Floyd, "Baby Face" Nelson, "Machine Gun" Kelly. These thugs, together with John Dillinger, one of the most publicized of the 1930 decade, were small-town products who gained their notoriety initially by the news stories that helped build their egos.

Can Traditional Criminals Be Classified?

It has always been popular to attempt to classify offenders into types. As far back as the time of the eminent Italian, Cesare Lombroso (1836-1909), the practice has been common. Classifications seem to satisfy a certain interest in orderliness, a desire for neat categories; classification of criminals holds out the hope of arranging offenders tidily in distinct categories, where they may be expertly examined and analyzed. From the standpoint of convenience, classifications have an advantage. But, at best, they are merely an approximation of categorical accuracy vulnerable to all who disagree and who wish to present their own classifications.

The simplest classification is twofold: those who commit few crimes (the occasional offender) and those who commit many (the professional or habitual). Another popular type of dichotomous classification is that distinguishing between the apocryphal "born" or "instinctive" type and the person who commits a crime due to environmental factors. Many criminologists of the nineteenth century developed their own classifications, some of

---

[2] This robbery of the Brinks Armored Car Service of Boston was referred to by Herbert Brean, *Life* (September 16, 1957), p. 82, as a "classic of modern crime" representing "the biggest haul in modern history," $2.7 million dollars. *Life* continues: "The 11 men who accomplished it had spent two years watching the Brinks plant. They had broken into it more than 20 times before the robbery, removing locks, having keys made and replacing the locks before dawn. They attempted the actual robbery at least a dozen times but were scared off by unexpected lights or by people in the neighborhood. But on January 17 everything was fine and the plan worked. . . . They hid the money, led modest lives and managed to elude the police for six years until one of them, in prison for another crime, finally talked. Today, nine are in prison, two are dead." Courtesy *Life* (Copyright, 1957, *Time* Inc.).

[3] See Irwin Ross, "Hi-Jacking Is Big Business," *Minneapolis Sunday Tribune* (October 15, 1950).

which were quite elaborate.[4] The penchant for classifying criminal types is universal among psychiatrists, sociologists, psychologists, and prison administrators. However, about the only value in the practice is that it does enable investigators to make comparisons with those in other disciplines as an aid in their own work, whether this be diagnostic, therapeutic, or theoretical.[5]

We are never certain just what causes a given person to enter upon a criminal career, or even to commit a specific crime, until we have prepared a careful case history of that person. Each offender is different and must be treated as such.[6]

To the average person, a "criminal" is an inmate of a correctional institution. But prison inmates represent only a small and limited group within the whole criminal class and throw little light upon the nature of criminals in general. By and large, prisoners are either the most inept, least influential, and most inexperienced portion of the criminal population, or those who have committed a compulsive crime through force of circumstances.

Nonetheless, it is important to learn as much as possible about prisoners since they are probably in some ways typical of the large class of traditional criminals and will sooner or later be released into society. Treatment while in prison may have a bearing on their ability to lead a law-abiding life. The following pages will review some of the most useful classifications and their criminological significance.

The Situational Criminal

Thousands of men and women are in our prisons for the first time and for their first offense. Many of these may be referred to as *situational criminals*. They are persons who have committed crimes under pressure of circumstances. The offense may be a relatively minor one, such as assault and battery or petty larceny, but most frequently it is serious—murder, arson, felonious assault, armed robbery, burglary, auto theft, and kidnaping are examples. There are many young men in our federal prisons who have stolen or "borrowed" a car and unwittingly crossed a state line or who have, again unwittingly, violated the Mann Act by foolishly "transferring a female across a state line for immoral purposes." Many individuals have been caught in a situational web and are paying dearly for the results.

---

[4] For a treatment of classification of criminal types, see Maurice Parmelee, *Criminology* (New York: Macmillan, 1922), Chapter 13.

[5] For typical, though useful, classifications in sociological terms, see: Paul S. Horton and Gerald R. Leslie, *The Sociology of Social Problems* (New York: Appleton, 1955), pp. 103-113; Ruth Shonle Cavan, *Criminology,* 2nd edition (New York: Crowell, 1955), pp. 21-29; and Marshall B. Clinard, *Sociology of Deviant Behavior* (New York: Rinehart, 1957), Chapter 8, "Types of Offenders."

[6] Refer to Paul Reiwald, *Society and Its Criminals* (New York: International Universities Press, 1950), for an interesting and unique approach to the subject of criminals. See especially Chapter 6, "The Criminal of Reality."

Often the offense seems the outcome of only one moment of despair, terror, hate, or thoughtlessness, but a premeditated crime may also be committed by the situational offender. Some brood over real or fancied wrongs and commit their offenses to "get even," or to terminate in hope or desperation an impossible state of affairs. While it may be argued that the personality makeup of many of these persons is pathological, we are here concerned with that group in which no pathology is indicated.

Many murderers are in this category. They kill wives, the "other man," their best friends, or even their children. It may be a person's only offense and thus he cannot be considered "occasional," "pathological," "psychopathic," or "mentally retarded." The crime was to him the only way out of a single unbearable or confusing situation. Some situational criminals are executed.

One of the most shocking offenses committed by a situational "criminal" in recent years was that of John Gilbert Graham, a Denver youth of 24 who placed a bomb in the suitcase of his mother, Mrs. Daisy King, as she was preparing to board a plane on November 1, 1955. Graham had seen to it that his mother took out a large insurance policy at the airport. Not only did his mother die when the bomb exploded shortly after take-off but 43 other innocent persons were killed. Graham was married and had two children. He was executed in the gas chamber of the Canon City, Colorado prison on January 11, 1957. If Graham was psychotic no such diagnosis was ever made public. It may be argued that he had a deep-seated hatred for his mother but this does not invalidate his being considered a *situational* criminal. One might mention the cases of many kidnapers as examples of this type of criminal.

While we believe habitual arsonists may be pathological, there are many persons who set fire to barns, stock, or woodland merely for spite. Men discharged from their jobs commit offenses against their companies or their employers merely "to get even." There are persons, heavily in debt, who see no way out of their dilemma except to rob or steal. It is clear that many criminals can be usefully considered to be in this category, assuming once again that they have not been diagnosed psychotic or emotionally disturbed in the psychiatric sense. We believe the latter group requires another category inasmuch as the correctional problems involved are quite different.

The Chronic Offender

It is well known that some situational or "first" offenders become "occasional" or even "chronic" offenders. In many cases it is because of their inability to regain their self-respect owing to inadequacies of prison life, to guilt feelings, or to inability to make adequate adjustment after release from prison.

Chronic offenders may be placed into two sub-categories: first, those who have stumbled into criminal activity, often primarily because of personality inadequacies, even though environmental factors would not predispose them to such activity: they are usually family men steadily employed (when out of prison). They tend to gravitate toward others who are no more adequate than they, and together plan a crime. The other type is the true chronic or habitual whose antisocial attitudes have regularized as predatory crime. There are many of both types in our prisons at all times.

True chronic or habitual criminals may engage in petty crime or in the more serious felonies. They commit a wide range of offenses. Some are pickpockets, some jewel thieves, some confidence men, some forgers or check passers. Some commit crimes in which violence plays a part if deemed necessary. There are coarse, culturally low, cynical persons in this group as well as suave, intelligent, and sophisticated individuals. If arrested, convicted, and committed several times, they are our recidivists.

The attitudes of chronic criminals follow a well-defined pattern. There are money, prestige, power, and personable friends of a sort to be gained by engaging in shady deals; but work, as understood by conventional citizens, is not an element.

There is a caste society in the underworld, based on the degree of skill, daring, color, and swagger in criminal activity. In the over-all group of chronic offenders there is a wide divergence in intelligence and culture. The professional is the aristocrat of the criminal world. He has contempt for the lower breeds of pickpockets, purse snatchers, or those who resort to violence. In general he is satisfied with his lot. He gains status among his inner group by the success of his exploits and evasions of the law. He usually has only contempt for ordinary social conventions and the laws of society. There is definitely a hierarchy among chronic offenders and the professional regards himself as a person with considerable status.[7]

Wellington Scott, a "professional," writes in his autobiography: "Each act is a business proposition, considered from a business standpoint and measured only by dollars and cents and the opportunity for a clear 'get away.' "[8] Another professional, Danny Ahern, boasts: There's some theatres you can stick up, where you wind up with probably six grand. . . . You take the same risk for that as for $100,000. . . . It always pays to go out and steal big."[9] Everett Debaun, writing in *Harper's,* graphically describes the modern art of armed robbery. He cooly admits the risks in-

---

[7] For a fascinating story of the world of pickpockets, "con" men and others of this "fringe" of the criminal world, as told by an old-line "cop," see Daniel J. Campion with Myron M. Stearns, *Crooks Are Human Too* (Englewood Cliffs, N.J.: Prentice-Hall, 1957).

[8] *Seventeen Years In the Underworld* (Nashville: Abingdon, 1916), p. 18.

[9] *How To Commit a Murder* (New York: Ives Washburn, 1930), p. 107.

volved, enumerates the technological changes in the profession, and boasts of the skill of the topflight men in his craft.[10] The classic utterance of the professional is that of the notorious criminal Le Blanc, as quoted by W. A. Bonger, the Dutch criminologist:

> If I were not a thief by profession, I should become one by calculation; it is the best profession. I have computed the good and bad chances of all the others, and I am convinced by the comparisons that there is none more favorable or more independent than that of the thief, nor one that does not offer at least an equal amount of danger.[11]

These quotations may all be attacked as mere boasting to compensate for a feeling of inadequacy, yet students of the crime problem contend that the true professional shows little evidence of inferiority or shame concerning his exploits or career, even when he is apprehended. A few years ago *Life* carried the amazing story of a shrewd jewel thief who unblushingly told of exploits that netted him well over a million dollars and for which, eventually, he was sent to the Auburn, N.Y. prison for a long term. He is now reported to be a law-abiding citizen.[12] Herbert Emerson Wilson, who had been sent to San Quentin prison, was interviewed on the radio and, although claiming he had stolen sixteen million dollars, stated that crime, true to tradition, does not pay.[13]

The other groups of chronic offenders cannot be so well isolated and described. Some of them, like most professionals, are "lone wolves." Others work in pairs or in small groups; but are not to be confused with the criminal gangs described earlier. While some hold down legitimate jobs, often merely as a front, others shun toil or gainful occupation studiously. The "public opinion" of this mass of chronic offenders is oriented toward antisocial activity. Scorn may be heaped on a member who attempts to reform. He is regarded by his primary group as a contemptible backslider, a quitter, or a fool. This feeling is usually reinforced by his female associates.

It has always been a debatable question as to the loyalty of chronic criminals toward each other. The police discount this dubious virtue and prize highly their own connections with the underworld—the "informers" who aid them in clearing up crimes. But there is evidence that among the chronic offenders there is a "code" far more rigid than any found in law-abiding

---

[10] "The Heist: Theory and Practice of Armed Robbery," *Harper's*, Vol. 200, No. 1197 (February, 1950), pp. 69-77. For the exploits of Joseph R. "Yellow Kid" Weil, who is alleged to have swindled some eight million dollars during his career, see his book, *"Yellow Kid" Weil* (Chicago: Ziff-Davis, 1948).

[11] W. A. Bonger, *Criminality and Economic Conditions* (Boston: Little, Brown, 1916), pp. 585-586.

[12] "Confessions of a Master Jewel Thief," *Life*, Vol. 40, No. 11 (March 12, 1956), pp. 121 ff.

[13] N.B.C. network's "Monitor" program for Saturday, March 17, 1956. See his bizarre book, *I Stole $16,000,000* (New York: New American Library, 1956).

societies. The inured criminal will rarely reveal to the police the name of another criminal who has mortally wounded him. Criminals go to the gas chamber or the electric chair without "squealing" on another. They take great risks to aid others of their kind who have done them favors in the past. For another reason we find inmates of prisons refusing to "rat" on their fellows. While this tradition may be better ascribed to fear than to "honor," the fact remains that prison officials rarely receive information regarding irregularities from inmates.

But to the police all criminals are potential "rats" and "stool pigeons." There has been little objective research in this area, although the existence of the "code" is well known in the prison community. Both dogmas are accepted by the public with qualifications. Perhaps no doubt the truth is that some criminals are weak and under certain conditions will betray their colleagues. Drug addicts are obviously in this category, simply because they are dependent on their supply of narcotics when questioned. But it is no doubt true that professional criminals rarely inform on their fellows.

We stated earlier that there is a type of "chronic" offender who is not a part of the criminal world and whose orientation is vaguely rooted in law-abiding society. He may have been a situational offender who, released from prison, finds adjustment difficult and thus in his ambivalence "toys" with crime. He may resort to crime only occasionally. He may simply be bitter toward society. That this miscellany is numbered in thousands there can be no doubt, but little research has been done on their attitudes. We do know that much of the articulation of these offenders is pure rationalization. A bit of doggerel probably expresses the attitude of many such offenders:

> I'm walking about a prison,
> What do you think I see?
> A lot of dumb-bells doing time,
> While all the crooks go free.[14]

It is this group that is easiest to rehabilitate if the proper influences are available. Many of them are constructively helped in prison while others may find a successful adjustment in the community after their release. Their lives represent a conflict between wholesome influences and social inadequacies.

An example of this conflict was provided many years ago by the "Mark Twain Burglar." The humorist's home in Redding, Connecticut was broken into and robbed in 1908. The thieves made off with some silver plate but were easily apprehended. They were sent to the Wethersfield, Connecticut, prison for a term of years. One of the men, an immigrant boy, later wrote a story of his life in this country in a book entitled *In the Clutch of Cir-*

---

[14] Quoted by Albert Morris in "Criminals' Views on Crime Causation," *The Annals,* Vol. 217 (September, 1941), pp. 138-144.

cumstance.[15] A series of unfortunate incidents involving a bartender uncle, an avaricious employment agency, a mistaken identity, and unwarranted detention in a county jail embittered him so much that he became an amateur burglar. His inept exploits landed him in the Wisconsin state prison at Waupun where he was subjected to the old contract system of exploitation which further embittered him. Wandering back to New York he read in the paper of the lovely country home of Mark Twain and he decided to rob it. After his release from prison he was befriended by Clara Clemens Gabrilowitsch who furnished him the means to enroll in a study course. He was also encouraged by the great prison reformer, Thomas Mott Osborne. This is a story of circumstances that can make a criminal and also reform one.

There are other discernible types of offenders, to be discussed in the next chapter. These are the offenders who suffer from pathologies—the vagrants, alcoholics, drug addicts, prostitutes, sex offenders, and the so-called psychopaths. While they have pathology in common, they vary widely in their personalities and in the types of offenses they commit.

The Recidivist and the Status of Habitual Criminal Laws

One of the most serious problems confronting society is that of recidivism, the proneness of many criminals to continue a life of crime. There are no precise figures available at any one time that will offer us more than a guess regarding the number of repeaters ("recidivists") in the prison population. Most observers and students of the crime problem maintain that the bulk of the inmates of our prisons are recidivists. In a survey made by one of the writers, and colleagues, in 1952 the repeaters in a population of 85,350 inmates, representing sixty-eight adult prisons, not including reformatories, was 48,470, or 56.7 per cent. This does not include "first offenders" who had been previously incarcerated in reform schools.[16]

Even in the early days of the prison, the problem of the repeater was recognized. As early as 1817, New York attempted to meet this by passing this country's first habitual criminal law. Since then it has been as-

---

[15] Anonymous (New York: Appleton, Copyright 1922). After the robbery of his home Mark Twain posted a placard on his front door which read:

"To the Next Burglar:

"There is nothing but plated ware in this house now and henceforth.

"You will find it in that brass thing in the diningroom over in the corner by the basket of kittens.

"If you want the basket put the kittens in the brass thing. Do not make a noise—it disturbs the family.

"You will find rubbers in the front hall by that thing which has the umbrellas in it, chiffonnier, I think they call it, or pergola, or something like that.

"Please close the door when you go away." *Ibid.,* p. 178.

[16] "A Limited Survey of Some Prison Practices and Policies," *Prison World* (now known as *The American Journal of Correction*), Vol. 14, No. 3 (May-June, 1952), pp. 5 ff.

sumed by legislatures in most states that such action against the confirmed criminal, calling for long sentences (usually for life) on the third or fourth offense, was the most effective way of dealing with the chronic offender. In general such legislation sets forth increased penalties but does not make their invocation mandatory.

One of the results of such severe laws is that it is difficult to find a jury that will bring in a conviction when it realizes the decision will mean life imprisonment or an excessively long sentence for that last trivial offense, though it be the defendant's third or fourth time before the court. Then, too, such laws encourage sympathetic prosecutors to permit a recidivist to plead guilty to a misdemeanor in order to avoid the stigma of "habitual criminal."

In a nationwide survey of habitual crminal laws, Professor Paul Tappan found that they are regarded by most attorneys general and commissioners of correction with considerable disfavor. The main objections voiced were "their excessive severity, their encroachment upon the powers of the judiciary, their interference with diagnostic and clinical goals in failing to individualize treatment according to the particular requirement of the offender, and their encouragement to nullification wherefrom a loss in the deterrent efficacy of the law results."[17]

The problem of recidivism is a serious one, but drastic legislation is not the answer. It is usually brought forth in an atmosphere of hostility or hysteria rather than one of helpfulness to society. Great Britain for many years had on its statute books what was known as "preventive detention," which permitted a judge to add extra time to a sentence for a specific offense. This practice was repealed in 1948. A new provision calls for "corrective training" for all those recidivists who seem to have some potentiality for rehabilitation.[18]

Certainly meaningless long sentences are not the answer. In fact, long-term sentences are not often carried out in full, as is generally believed by the public. Life imprisonment means, on the average, only about ten years.[19] Yet many prisoners do languish in institutions for long periods of time.

---

[17] "Habitual Offender Laws in the United States," *Federal Probation,* Vol. 13, No. 1 (March, 1949), pp. 28-31. This article presents the date of each statute, the added penalty, and the conditions under which it is imposed.

[18] See: Norval Morris, *The Habitual Criminal* (London: Longmans, 1951); Charles B. Thompson, "A Psychiatric Study of Recidivists," *American Journal of Psychiatry* (hereafter, *"Amer. J. Psychiatry"*), Vol. 94, No. 3 (November, 1937), pp. 591-604; Max Grünhut, "The Treatment of Persistent Offenders," *Journal of Criminal Science* (London: hereafter, *"J. Crim. Sc."*), Vol. II (entire issue devoted to a symposium on the Criminal Justice Act of 1948, L. Radzinowicz, ed.; London: Macmillan, 1950), pp. 65-89.

[19] There are so many different types of "life" sentences that it is difficult to state just how long the average lifer remains in prison. Lifers who die shortly after imprisonment tend to bring down the average time spent. For an analysis of this question, see Alfred M. Harries, "How Long Is a Life Sentence for Murder?" *Proceedings,* American Correctional Association, 1939, pp. 513-524.

It is not unusual to find inmates in some of our prisons serving "double-life," that is, *two* life sentences; or a fantastic sentence of "life and 99 years," or ten sentences of 10 years each—some to be served *consecutively,* which means 100 years. Long sentences are a reflection of society's frustration in attempting to combat the crime problem as well as lack of realism on the part of the sentencing judges. In general the long-term complex exhibited by many jurists shows lack of knowledge of the criminal. Treatment personnel in our prisons contend that in most cases there is a psychological time to release the inmate with a good prognosis for rehabilitation. Yet he must often remain many years longer to satisfy the judge's sentence or a vengeful public. Just how long must a person who has committed a repugnant crime, such as murder or rape, remain in prison to expiate his offense? Nathan Leopold, with his companion, Richard (Dicky) Loeb, committed the sensational kidnap-murder of a young boy in 1924. After serving 32 years of a life sentence in the Illinois penitentiary where he made his imprisonment worthwhile to society in medical research and experimentation as well as in helping to build up a modern prison library, Leopold speculated over this question in 1957:

> I've spent 32 years here. Is that sufficient punishment for what I did? I don't know the answer to that because I don't know how to measure punishment. I know that I have lost those near and dear to me while I was here. My father, my aunt who was a second mother to me, my brother. I know that I have forfeited any chance of amounting to anything. I've forfeited every chance of happiness. I've forfeited a chance for a family. Whether this is sufficient, I don't know.[20]

Many inmates are released before any reformation takes place, while others, who could conceivably be released shortly after entering prison, remain for many years. Nathan Leopold was able to make a contribution to society even though incarcerated, but here is a poignant and bitter statement from a gangster-murderer who spent over 20 years in Alcatraz. He writes:

> Maybe you have asked yourself how can a man of even ordinary intelligence put up with this kind of life day in, day out, week after week, month after month, year after year. You might wonder whence I draw sufficient courage to endure it. To begin with, these five words seem written in fire on the walls of my cell—"Nothing can be worth this."—this, the kind of life I am leading. No one knows what it is like to suffer from the intellectual atrophy, the pernicious mental scurvy that comes of long privations of all the things that make life real because even the analogy of thirst cannot possibly give you an inkling of what it is like to be tortured by the absence of everything that makes life worth living—a prisoner

---

[20] Quoted in the article, "Nathan Leopold After 32 Years," *Life,* Vol. 42, No. 9 (March 4, 1957), p. 61. Courtesy *Life* (Copyright, 1957, *Time* Inc.). For further details of the Loeb-Leopold case, see pages 258-260. Leopold was finally paroled in 1958. See his *Life Plus Ninety-Nine Years* (New York: Doubleday, 1958).

in prison cannot keep from being haunted by a vision of life as it used to be when it was real and lovely. At such times I pay with a sense of delicious overwhelming melancholy—my tribute to life as it once was.[21]

Long prison sentences stifle all hope of reform. The prison poison permeates the inmate, and when he is finally released he is almost helpless to make an adequate adjustment to free society. We could not deliberately devise a more absurd system for dealing with criminals than that of imprisoning them for long periods.

The detainer device creates problems closely identified with those of the repeater. Perhaps a fifth of the inmates of our prisons are subject to detainers, under which, upon release, they will be tried in other jurisdictions on additional charges. It is not unusual to find a prison inmate being wanted in several states, often on minor charges. The detainer problem is a serious one from the points of view of the prison administrator and parole board, since no effective rehabilitation can be effected in a prisoner if he knows he must serve more time as soon as he is released. The problem has been under constant consideration among penologists, but thus far no effective solution has emerged. Possible solutions would call for interstate compacts whereby all claimants to a criminal convicted in another jurisdiction would waive their detainers under certain conditions, or turn the matter over to the Interstate Commission on the Control of Crime.[22]

Another frantic effort that failed to curb the professional criminal or recidivist resulted in laws forbidding the carrying of concealed weapons. Nearly every state has some type of act restricting the purchase and use of firearms, but only law-abiding citizens obey them. Criminals who want pistols have little trouble in securing them, although decent citizens are usually browbeaten and insulted by the police when they ask for a permit to carry a gun. In those rare instances where it is difficult for a criminal to procure a gun legally, there is always a "black market" that sells pistols. Fences and some pawnbrokers will sell firearms at fancy prices to those who must have them for their illegal work. Gangsters have been known to smuggle guns into the country and, of course, they are prepared to steal them from stores.

The Criminality of Women

We hear it stated with no little emphasis that women commit fewer crimes than men because they are inherently more moral and innocuous. Although we cannot accept this thesis without empirical evidence, statistics have consistently shown women to be less criminal, legally speaking. Yet

---

[21] Quoted with the permission of James V. Bennett, Director of the Federal Bureau of Prisons.

[22] For further details regarding the detainer and its evil effects on progressive correctional practice, see *Federal Probation,* Vol. 9, No. 3 (July-September, 1945).

it would be naïve to assume that such data are accurate. Women are protected in a male-dominated world and are traditionally relegated to the role of homemaker and bearer of children, far from the competition of the market-place. One could adduce many reasons why women appear to be less criminal than men. One that is often overlooked is that women, by and large, fear social disapproval more than men; thus they tend to cling to the prevailing mores more than do males.

Professor Otto Pollak has shown that statistics are most unreliable so far as giving an adequate index of female crime. He contends that most crimes committed by women are concealed by their "indirection and deceit," and that women engage in a wide variety of crimes incidental to the roles they play as wife, mother, nurse, and rearer of children. Among these are murder, most frequently by poisoning, adultery, abortion, infanticide, blackmail, larceny, shoplifting, and forgery—crimes that, by and large, do not call for daring or physical strength.[23]

It has been argued that poverty is the basis of most crime, but it might be contended as well that sex is at the root of crime. Men steal for women; women kill to rid themselves of unwanted mates; both men and women kill each other because of sexual jealousy. While it is dangerous to generalize, it is almost proverbial that most offenses committed by women involve men, directly or indirectly.

Aside from traditional "gang molls," we find a class of women criminals, calculating, fascinating, and intelligent, who capitalize on their charm and femininity. They tend to favor the old confidence game, usually involving large sums of money, and the equally old blackmail game. Perhaps the classic example of the former was Mrs. Cassie Chadwick, who startled the financial world during the first decade of the century. Posing as a close confidante of Andrew Carnegie, she fleeced otherwise prudent bankers of hundreds of thousands of dollars, merely by gaining their confidence. Paradoxically, she gave most of her ill-gotten loot to charity. She was finally apprehended and sent to the Ohio penitentiary. A similar case, much more recent, was that of the wife of a respectable New Jersey man who was discovered to be a shrewd, old-time swindler and embezzler who had served time in more than one prison. She had married the New Jersey businessman late in life and had been quite active in church work. At the same time she was "investing" large sums of money for the members of her church. She lulled unsuspecting prospects with little deeds of charity and then mulcted them of large sums of cash. It was estimated that a million dollars had passed through her hands during her career as a confidence woman. She had eight aliases, including such colorful names as Mildred Sidebottom and

---

[23] *The Criminality of Women* (Philadelphia: University of Pennsylvania Press, 1950). See also Walter C. Reckless, "Female Criminality," *National Probation and Parole Association Journal* (hereafter, *"N.P.P.A. Journal"*), Vol. 3, No. 1 (January, 1957), pp. 1-6.

Amelia Boniface. She told her friends she owned a farm "just dripping with oil" and a "large plum orchard." Her operations were suspected by the police only when she slipped back into the prison vernacular, thus indicating that she was a woman with a record. She was finally sentenced to eight years for her peculations.

Other swindlers who gained national notoriety were Mrs. Marie Fuller who, with several male accomplices, fleeced several retired farmers from the midwest of some $200,000 by pretending to have an "in" with the Henry Ford family,[24] and Mrs. Nora Warfield of Birmingham, Alabama, a minister's wife who in 1947 embezzled $60,000 from the community welfare fund.[25]

A spinster Sunday-School teacher in Norfolk, Virginia, startled the public a few years ago. Minnie Mangum, trusted employee of a building and loan association, had methodically embezzled over two million dollars. Much of this stupendous sum Minnie, like her prototype Cassie Chadwick, gave to charity and friends.[26]

More sinister crimes, such as murder and kidnaping, can be committed in cold blood by women. Nannie Doss of Oklahoma dispatched four husbands, in succession, and an Alabama woman fed arsenic to her mother, two of four husbands, and three of five children, for no apparent reason whatever. The most famous case in this country was that of Lizzie Borden, 32-year-old spinster of Fall River, Massachusetts, who in 1892 was tried for the murder of her aged father and her stepmother. She had hated the stepmother and for several years they had not eaten together, and rarely spoke. Lizzie was acquitted, but local opinion always assumed her guilty. The story has been celebrated in song, drama, and other popular forms; it is now almost a national legend.

Murder is more frequent among female crimes than among those committed by males. In 1957, for example, women were arrested for 373 murders and non-negligent cases of manslaughter or 0.2 per cent of all offenses leading to arrest, whereas the 1,631 cases of murder and nonnegligent manslaughter for which men were arrested were only 0.1 per cent of their total offenses. Often a woman in an "eternal triangle" case is a contributing if not a proximate cause of the resulting homicide, whether the object is to be rid of her own husband or her lover's wife. Such cases are given wide publicity by the press, but the unfortunate protagonists are obviously not professional criminals.

Many women do, however, make a career of crime. Shoplifting can be lucrative, both for the women themselves and for their "fences," who "wholesale" the stolen goods on a percentage basis. A notorious woman

---

[24] *Life,* Vol. 29, No. 15 (October 9, 1950), p. 47.
[25] *Reader's Digest,* Vol. 57, No. 342 (October, 1950), p. 26.
[26] *New York Times* (February 12, 1956).

fence was Mother Mandelbaum, who operated in New York City during the 1870's, on a scale that enabled her to retain counsel at $5,000 a year. Stores often hesitate to prosecute shoplifters because of unfavorable publicity. This is especially true when the stealing is the result of the mental aberration of kleptomania, which occurs among rich as well as poor; the culprit and her family may be among the store's best customers.

Educated women resort to forgery and other activities involving skill. With these exceptions, however, female crime follows the dull pattern of drunkenness, vagrancy, prostitution, disorderly conduct, drug addiction, and the like.[27]

The Young Adult Offender

The stark fact is that older adolescents seem to predominate in courtroom, jail, lock-up, and prison, a fact noted many years ago (1930) by the Wickersham Commission when it reported that 54.9 per cent of the inmates of adult prisons had been under 21 when convicted. The *Uniform Crime Reports* consistently demonstrate that in many categories of serious offenses youth loom large. Approximately a quarter of a million youths are arrested annually on serious charges. We read from the 1957 *Reports* as follows:

> An examination of the distribution of 2,068,677 arrests in 1,473 cities by age groups, reflects that 253,817 of these, or 12.3 per cent, were of individuals who had not yet reached their eighteenth birthday at the time of arrest, and 39.0 per cent of these were under the age of 15.
> Although youths under 18 account for only 12.3 per cent of arrests for all age groups, they make up 47.2 per cent of the arrests for the Part I crimes of murder and nonnegligent manslaughter, negligent manslaughter, rape, robbery, aggravated assault, burglary—breaking or entering, larceny-theft and auto theft. The extent of participation of youths in crime for the Part I classes is weighted by arrests for crimes against property.
> Persons under 18 represented 53.1 per cent of all arrests for the Part I crimes of robbery, burglary—breaking or entering, larceny-theft, and auto theft, but only 10.3 per cent of all arrests for the Part I crimes against the person of murder, negligent manslaughter, rape, and aggravated assault.[28]

In the individual Part I property crime classes, youths under 18 represented 26.4 per cent of those arrested for robbery, 54.8 per cent of those arrested for burglary, 51.3 per cent of those arrested for larceny, and 67.6 per cent of those arrested for auto theft.[29]

While offenses committed by this age group tend to be higher than we should reasonably expect, the public frequently receives a distorted pic-

---

[27] See table on page 80 for data on female offenses.

[28] *Uniform Crime Reports, 1957,* Vol. 28, No. 2 (April 23, 1958), p. 113. See next page for a partial explanation of this.

[29] *Ibid.,* p. 113.

ture. Virgil Peterson, operating director of the Chicago Crime Commission, cogently points out just how the F.B.I. data are misinterpreted so that youthful crime appears more serious than it is. He writes:

> A typical example is found in a leading article on juvenile delinquency in the October 1955 issue of a magazine with a national circulation. This article states with finality that FBI reports show that juvenile offenders under 18 accounted for 49 per cent of all burglaries committed in the United States last year (1954). Actually, these reports show nothing of the kind. Instead, the reports clearly state that of available records examined by the FBI of persons arrested for burglary, 49 per cent were under 18. But the same reports also show that of all burglaries known to the police in 1954, only 29.6 per cent were cleared by any arrest. Hence, out of every thousand burglaries reported to the police (in that year) no one knows who committed over seven hundred of them.
>
> Obviously, youthful amateur offenders are more easily apprehended than those who make burglary a profession. It is only natural that arrest figures will show a disproportionate number of youthful offenders. And based on such data, no one can properly state that almost half of the burglaries in the country are committed by juveniles under 18 years of age. This is particularly true since the identities of the persons responsible for over seven-tenths of burglary offenses are absolutely unknown.[30]

In other words, one reason young offenders appear to be more criminal than adults is that they are most easily caught so as to become statistics. Also, the police in many of our large cities use the F.B.I. categories in their recording of arrests, regardless of the age of the offender, rather than the blanket term "juvenile delinquent." Thus the number of rapes recorded for children under fifteen for 1956 is 126. The proportion of these offenses accompanied by violence is impossible to determine since the number includes, in the F.B.I. classification, statutory rape of the type, "no force used—victim under age of consent." In the same year 569 in this age group were arrested for drunkenness, 865 for robbery, 12,921 for burglary. One might ask just how serious these were, too. Stealing a hubcap from an automobile by a 12-year-old is, or might be, recorded as a burglary.

There are many reasons why so many young people are prone to crime. Our socio-economic structure is potent for disorganization and frustration. Old-fashioned family discipline has broken down and many parents themselves are bewildered by the changing mores. Divorce and desertion are on the increase, leaving as an aftermath many disoriented children. The complexity of our civilization, the increased contacts and stimulation of urban life, new concepts of education that develop a premature spirit of sophistication without a philosophy of self-discipline—all must be reckoned with.

Each generation is convinced that its young people are arrogant, selfish,

---

[30] Chicago Crime Commission, *A Report on Chicago Crime for 1955* (Chicago: The Commission, 79 West Monroe Street, 1956), pp. 5-6.

undisciplined, and irresponsible. The newspapers and magazines seem to affirm this. Yet if we may look backward at the meager data and stories of the youth of preceding generations it is at least tenable that the present is no worse than the past. Adults today forget the "flaming twenties," the prohibition era, the dismal decade of the Great Depression, and the war years when they excoriate the youth of today. Juvenile gangs have always existed in urban communities and probably always will, together with their senseless criminal activity. One of the earliest preserved writings is an attack on the youth of that day for their evil ways; it might have come, almost word for word, from the editorial page of today's paper. It is from the Sumerian era, approximately from 1750 B.C. The tablet upon which it is inscribed was discovered about fifty years ago but has only recently been deciphered by Dr. Samuel Noah Kramer of the University of Pennsylvania. A father bitterly rebukes his adolescent son for making him "sick unto death with his fears and inhuman behavior." He "is deeply disappointed at his son's ingratitude."[31]

Courtney Ryley Cooper reviewed the crimes of "young punks" of the 1930's in his startling book, *Designs in Scarlet*. It was widely read and just as widely criticized as being too spectacular. But it is often through journalism and a journalistic style that changes come about in the thinking of laymen; the dry, stilted phraseology of the professional journals is rarely an open challenge to complacent ignorance. Another popularization, written by a person using the pseudonym, "William Bernard," relates in a sensational manner the exploits of juveniles during the 1940's.[32] The writer is well-informed, and he paints a vivid picture of confused and frustrated youth, many of them girls and young children.

The nation is outraged periodically when a news story announces wanton killing by a teenage youth or by a young adult in his early twenties. Hundreds of such seemingly cold-blooded murders occur annually. Some years ago a carefully written series of articles appeared in the Minneapolis *Tribune* authored by Victor Cohn and entitled "Who Are the Guilty?" This writer stated that during 1946, 808 boys and girls under 21 were arrested in this country for homicide. Of these, 256 were under 18, and 69 were under 15.[33] The story is the same every year.

Few objective studies have been made of young killers to determine why they kill or what type of persons they are. Some of them should have had psychiatric treatment while they were still in the preadolescent stage. Had they been critically observed in school, there is no doubt that certain symp-

---

[31] Samuel Noah Kramer, "A Father and His Perverse Son: The First Example of Juvenile Delinquency in the Recorded History of Man," *N.P.P.A. Journal,* Vol. 3, No. 2 (April, 1957) pp. 169-173.

[32] Courtney Ryley Cooper, *Designs in Scarlet* (Boston: Little, Brown, 1939) and William Bernard, *Jailbait* (New York: Greenberg, 1949).

[33] See his articles in the Minneapolis *Tribune* (December 1948).

toms of abnormal behavior would have manifested themselves. Follow-up treatment might well have been of much social benefit.

From the results of one study of a group of youthful killers, we find that earlier symptoms of abnormalities were not present. We refer to a study by Dr. C. H. Growden at the Bureau of Juvenile Research in Columbus, Ohio. It describes 54 cases of boy and girl homicides between the ages of 9 and 19. Comparing these cases with over 8,000 unclassified delinquents the observation made was that a great similarity exists between the children regardless of the offenses. This similarity manifests itself in such items as intelligence, emotional instability or "normality," age, and physical condition. Aside from one lone homicide case, the bulk of the young killers committed their crimes without premeditation. Most of them, on being questioned as to motive, cited "mistreatment, continued irritation or dictatorial interference in their personal liberties." The study's conclusion makes this observation: "Only rarely do children who commit homicide show a tendency to manifest anti-social and non-cooperative attitudes, or a tendency that, given the same degree of constructive attention as to other delinquents . . . they are as good a social risk as those who commit any other form of serious delinquency."[34]

The following news clipping describes a gang of hoodlums that was finally brought to book and broken up but not until it had spread fear throughout the "jungle" neighborhood of that city for several years. Addresses and legal names have been deleted.

GREEN STREET COUNTS

The "Green Street Counts," organized in the spring of 1952, were as unsavory a bunch of juvenile delinquents as ever impinged on the Philadelphia scene.

With white T-shirts that bore the lettering of "The Counts" and wearing black soft hats with white bands, the young thugs were quite an addition to the night life around 20th and Green Streets. That is, until the law caught up with the leaders on a murder rap and the gay clothing was traded for the more subdued garb of prison lifers.

But before that the gang lived high, wide and handsome on the money from holdups, robberies and strong-arm jobs. Organizers, police investigators said, were "Angel," then 17, and "Maka Saki" then 18.

"Angel," specializing in car thefts, was just out of [a New Jersey reformatory] while "Maka Saki," studying burglary, had just been freed from White Hill [Pennsylvania] Industrial School.

"We need organization," Angel said, repeating a remark he had often heard from hardened criminals. "Cops can't beat organization."

---

[34] C. H. Growden, "A Group Study of Juvenile Homicide" (Columbus, O.: Bureau of Juvenile Research, State Department of Public Welfare, October, 1949). See also: Lauretta Bender and Frank J. Curran, "Children and Adolescents Who Kill," *Journal of Criminal Psychopathology* (known, with variations, as *"Journal of Clinical Psychopathology"* beginning with Vol. 6, July, 1944), Vol. 1, No. 4 (April, 1940), pp. 297-322; Ralph S. Banay, "Homicides among Children," *Federal Probation,* Vol. 11, No. 1 (January-March, 1947), pp. 11-19.

Recruiting for the gang was easy. The uniforms drew some, the flashing bankrolls drew others. Many were enlisted by the simple expedient of calling them "chicken" if they didn't.

Organization was on military lines. There was a "secretary of war" and "chief of war intelligence" among other high sounding posts. And when a rival gang (the Brewerytowners) caused a noticeable drop in the amount of loot, ["Angel"] suggested collaboration. The Brewerytowners were absorbed into the Counts after formal signing of 61 young hoodlums of a "Treaty of Peace" on Feb. 8, 1953.

Things really boomed, then. The Counts rented an apartment for a "common post," recruitment continued to grow and everything was clover until April 4, 1953 when a $48 taproom holdup went sour.

Five of the Counts staged the holdup at the cafe at Front and Richmond. Two of the "outside guards" got nervous, shots were fired, and a taproom patron was shot to death.

Five Counts drew life terms for the slaying [including "Angel" and "Maka Saki"].[35]

This represents one aspect of youthful criminality.[36] However, we see another in John Bartlow Martin's book, *Why Did They Kill?*[37] which tells of the wanton and sadistic murder of a nurse in Ann Arbor, Michigan by three boys "all from good families." More recently we find Californians concerned over a sudden and compulsive murder of a young college girl by her young neighbor, a student at Stanford University, for presumably no reason at all. Both victim and perpetrator of the baffling crime were from upper middle-class families.[38]

While much gang behavior stems from a "working-class" culture, as analyzed by Cohen,[39] who refers to it as "non-utilitarian, malicious and negativistic,"—a class protest against upper classes—the internecine warfare between gangs cannot be thus explained, nor can middle-class murders be so explained.

How much of this beserk behavior among the youth of the nation is due to drug addiction we do not know. Certainly the peddling, sale, and use of narcotics has become a major problem, especially in our larger cities. Estimates of the number of youthful addicts, not only of marijuana but of heroin as well, have varied widely. Apparently there was a decided upswing in drug addiction among youths during the early 1950's, followed in more recent years by a substantial drop. It is doubtful, however, that many young

---

[35] *Philadelphia Inquirer* (May 10, 1957).

[36] For a graphic account of the jungle life of the ruthless young adult and his nihilistic antisocial world, see Rocky Graziano, with Roland Barber, *Somebody Up There Likes Me* (New York: Simon & Schuster, 1955).

[37] (New York: Ballantine Books, 1953.)

[38] Hale Champion, "The Nice Murderer: Search for a Motive," *The Nation,* Vol. 186, No. 12 (March 22, 1958), pp. 255-257.

[39] Albert K. Cohen, *Delinquent Boys: The Culture of the Gang* (Glencoe, Ill.: The Free Press, 1955).

hooligans of predatory habits are addicted, although addicts must often resort to stealing.

Approaching the Delinquency Problem

A number of interesting angles have been taken in the sustained attack on the problem of youthful criminals. We have had statistical studies of the symptoms, or typical expressions, of juvenile delinquency. Lying, stealing, truancy, running away from home, "incorrigibility," pathological stubbornness, sex offenses, and various destructive acts are the more usual manifestations.

The effort has been made to provide intensive individual study or case histories of typical juvenile delinquents. The pioneer was Dr. William Healy, who began his work in Chicago in 1909 and later went to Boston to assume the directorship of the justly famous Judge Baker Guidance Center.

While Healy's work was primarily devoted to an analysis of the personality of the juvenile delinquent, a new attack on the problem was launched by the late Clifford R. Shaw in Chicago in 1926. He examined the areas in which delinquent city youth live, a contribution that we shall set forth in detail in Chapter 10.

Professor Frederic M. Thrasher's study, *The Gang,* was one of the first to consider this prevalent and challenging trend in the behavior of juvenile delinquents. Gangs carry on a large number of borderline acts of criminal behavior, and Thrasher performed a signal service in the field to clarify the relationship of gang life to delinquency. Out of a wealth of experience in studying the activities of some 1,300 gangs, he thus summarizes the conditions that make for group delinquency in our modern confused life:

First, those conditions in family life that make for neglect of children, such as poverty, immigrant maladjustments, disintegration, and ignorant, unsympathetic, immoral, or greedy adults.

Second, the failure of present-day religion to penetrate in any real way the experience of the gang boy.

Third, inadequate schooling that fails to interest the boy of this age.

Fourth, the lack of proper guidance for the spare-time activities that are of vital importance in developing "wholesome citizens." In short, the gang boy is handicapped throughout his entire career since those institutions and agencies that generally have meaning to the growing child have broken down in his case or, at best, are woefully weak.

On the problem of young adult crime—"juvenile delinquency" is almost a misnomer—little constructive thought has emerged, aside from the conclusions that we have thus far recorded. Youthful gang members in the metropolitan areas display a callous and indifferent attitude toward society or their own predicament. Arrest, prison terms or strong warnings seem not to deter them. The wanton slaying of a young polio victim in New York by a gang of hoodlums identified in their own vernacular as the "Egyptian Kings and Dragons" during the summer of 1957 focused public attention on the frightening problem of contemporary "mixed-up" youth. The long ensuing trial of several of the gang members was given national coverage and the episode prompted the Police Commissioner to detail a thousand extra officers to patrol the city's boroughs in an attempt to discourage city gangs from preying on society and on one another.[40]

Serious and even distressing though this problem is, it must be placed in perspective. A bare fraction of the total number of adolescents and young adults in town, city, and nation is troubled or overtly delinquent. Whether youth as a whole is frustrated, disorganized, or rudderless in a complex world without "meaningful life objectives" merits national television symposia as well as many regional conferences. Again, whether "primary group controls" have collapsed or are gradually being whittled away by disquieting international problems is another aspect of the same problem and calls for national study. The community must accept the responsibility, and short-cut methods, "remedial" measures, are certainly not the prescription that is needed.[41]

Current Confusion Regarding Crimes and Criminals

The term "crime" technically means a form of antisocial behavior that has violated public sentiment to such an extent as to be forbidden by statute. Legislators define the act and apply a penalty thereto. In other words, a crime is an act which the group regards as sufficiently menacing to warrant a decisive reaction to condemn and to restrain the offender of such an act. This is as true in our day as in primitive times. Crime, thus conceived, represents only a specialized portion of the totality of antisocial behavior.[42]

There is a large field of immorality that is conventionally regarded as

---

[40] For materials describing attempts to work with delinquent gangs, see: James R. Dumpson, "An Approach to Antisocial Street Gangs," *Federal Probation,* Vol. 13, No. 4 (December, 1949), pp. 22-29; Ernst G. Beier, "Experimental Therapy with a Gang," *Focus.* (beginning July, 1955, *"N.P.P.A. Journal"*) Vol. 30, No. 4 (July, 1951), pp. 97-102; Sam Glane, "Juvenile Gangs in East Side Los Angeles," *Focus,* Vol. 29, No. 5 (September, 1950), pp. 136-141; and William Wattenberg and James J. Balistrieri, "Gang Membership and Juvenile Misconduct," *Amer. Soc. Rev.,* Vol. 15, No. 6 (December, 1950), pp. 744-752. These materials were selected as pertinent by Clyde Vedder, *The Juvenile Offender* (New York: Doubleday, 1954).

[41] For recent books dealing with the problem of delinquency, see Frank J. Cohen, *Youth and Crime* (New York: International Universities Press, 1957); Sol Rubin, *Crime and Juvenile Delinquency* (New York: National Probation and Parole Association, 1958); F. Ivan Nye, *Family Relationships and Delinquent Behavior* (New York: John Wiley, Inc., 1958) and Sheldon Glueck, *The Problem of Delinquency* (Boston: Houghton, Mifflin, 1959).

[42] See: Howard Jones, *Crime and the Penal System* (London: University Tutorial Press, 1956), Chapter I, "The Science of Criminology"; also Clarence R. Jeffery, "The Structure of American Criminological Thinking," *J. Crim. Law,* Vol. 46, No. 5 (January-February, 1956), pp. 658-672, and "Crime, Law and Social Structure," *ibid.,* Vol. 47, No. 4 (November-December, 1956), pp. 423-435.

antisocial but is not punishable by law. Such behavior is left to the control of public opinion. The narrow borderline between crime and unethical or immoral conduct is to be seen from the fact that acts which in some countries are regarded as crimes are viewed as immoral behavior in others. Not until such practices are considered reprehensible and menacing enough to call for legislative action do they qualify as crimes in that society.

It is usually believed that the seriousness of crime, with respect to its damage to society, is the test that sets off criminality from other forms of antisocial conduct. This should be the case, but we allow the criminal code to determine what we regard as crimes. Many crimes are listed on the statute books solely because they were acts socially disapproved a century or more ago. They may not be at all serious now. Other crimes are included because of the distorted fanaticism of ignorant or bigoted groups. Many of the more disastrous forms of antisocial behavior are not even regarded as immoral, to say nothing of being criminal. This is so because the dominant groups in society still approve such behavior. Many white-collar operators fall into this category.

The majority of serious crimes are those against property. The personal damage resulting is usually incidental and unwished for by criminal and victim alike. No criminal, unless he is psychotic or mentally retarded, wishes to add liability for murder to the penalty of theft or burglary. Most physical violence is the result of haste, fear, excitement, temporary anger, or the compulsions of a deviant personality.

One exception may be noted in kidnaping. This crime was made a capital offense by the federal government with the passage of the Lindbergh Law in 1932. In that year Charles A. Lindbergh's infant child was kidnaped from his home in Hopewell, New Jersey, an act that set off one of the most thorough and dramatic manhunts in the history of American crime. The Lindbergh Law permits the F.B.I. to enter a case after twenty-four hours on the assumption that the victim has been transported over state lines. While this drastic legislation resulted in a substantial decrease in kidnapings, especially by professionals, the provision calling for the death penalty is considered unwise by some. A calculating criminal might be prompted to dispose of his victim so that he could not testify against him or identify him. In other words, the kidnaper would be less exposed to conviction if he added murder to his crime of kidnaping than if he released his victim after carrying him across state lines and perhaps slightly injuring him.

Few people realize the extent of kidnaping in this country. During the past twenty-five years over 500 cases have been investigated by the F.B.I. Almost one thousand persons have been convicted although only a few have been executed. Less than fifty have been sentenced to life imprisonment.

Aside from calculating professional criminals who, prior to 1932, included kidnaping as one of their most lucrative operations, most abductions are perpetrated by inept, bungling amateurs or by frustrated females who yearn for a baby of their own. An example of the former was the debt-ridden cab driver of New York who, in 1956, abducted and killed one-month-old Peter Weinberger. There are many cases of unhappy women who kidnap babies.

The annual homicide rate of 10,000 to 12,000 for this country is far higher than in any other civilized country, but as we pointed out earlier, it may not be higher than what we should reasonably expect. Nonetheless, steps could be taken to reduce it. But we rarely consider suicide an important social problem, though there are over 20,000 annually in this country. Homicides pale into insignificance compared to accidental deaths, which now average over 100,000 annually. Approximately three to four people are killed every year by automobiles for every person who meets death as a result of a crime or criminal negligence. Homicides are numerically unimportant when compared to the hundreds of thousands of deaths from preventable diseases and from inadequate medical care.

While there have been many books written on homicides, we have few academic studies that cover pertinent social, economic, racial, and sexual attributes of either perpetrators or victims. Such a unique study is Marvin Wolfgang's *Patterns in Criminal Homicide*.[43] This study carefully analyzes 508 cases of criminal homicide involving 621 offenders and covering a period of five years in Philadelphia. Wolfgang concerns himself with the time and place of the offenses, the weapons used, the characteristics of both the offenders and their victims as well as family relationships, the circumstances of the crimes, motives involved, and the disposition of the cases. Briefly, the author found, surprisingly enough, that one out of four of the offenses were "victim-precipitated," that two out of three offenders and almost one-half of the victims had previous arrest records, that alcohol played a prominent part in the commission of the crimes, and that only 20 per cent of the convictions for murder were in the first degree. We shall deal with the racial angle of this study on pages 170–171.

The Multiplicity of Laws

There is no real solution to crime. No doubt much of it can be eliminated —especially that portion due to economic causes and arising from the gross inequalities inherent in our social system. But the extent of petty crime depends to a large degree on the number and tyranny of our laws. Each year finds thousands of new laws and ordinances added to our statute books, which are already cluttered with outgrown or unnecessary laws. Much of

---

[43] Marvin E. Wolfgang, *Patterns in Criminal Homicide* (Philadelphia: University of Pennsylvania Press, 1958). See also, Andrew F. Henry and James F. Short, *Suicide and Homicide* (Glencoe, Ill.: The Free Press, 1954).

the time of police officers is absorbed by snooping into matters that should be left to the control of public opinion and common decency. Jails and prisons are clogged with men who are in no serious way a menace to society.[44]

Railroading of inconspicuous and friendless persons is especially glaring in our large cities in connection with members of minority nationalities or races. Negroes and Mexicans are discriminated against in some sections of the country and the foreign-born in others, so that certain statistical studies often give the impression that these groups are more criminal than the native-born white American. We must be cautious in interpreting such statistics and statements because they actually give scant support to the thesis that racial or national groups vary in the extent of innate criminality.

Many of the laws still on the statute books of some of our states reflect ideas and conditions of long ago. Some years ago the state of Delaware invoked the notorious "blue laws" of colonial days. Hundreds of citizens were prevented from engaging in business on Sunday, others were denied the twentieth-century privilege of going for a Sunday afternoon automobile ride. The Attorney General, in an attempt to have these absurd laws repealed, maintained he would be obliged to enforce them to the letter of the law, so that Delaware citizens would understand what an obsolete set of laws really meant. Accordingly, some five hundred citizens were arrested. Persons engaged in "worldly employment" on Sunday, including filling station proprietors, druggists (to fill a prescription, perhaps badly needed for a sick person), milkmen, and newsdealers were arrested, and many were fined.

The "blue laws" are attention getting, but they are probably the least harmful part of the archaic legal picture confronting modern citizens of America. Laws against blasphemy and witchcraft abound. In one state it is against the law to ride a jackass more than six miles an hour. In another state it is robbery to loot a building, but not robbery to loot a railway car, since the car is not a building. The height of absurdity is a state law in which a child may not pass from the seventh grade to the eighth unless he can recite the words of *The Star-Spangled Banner*. A few years ago it was disclosed that three Indians were sentenced to 18 years in the Idaho state prison for stealing sheep. They had broken a law enacted to protect settlers from having their stock stolen—in 1864, when this was a serious loss. Conviction of these men made it mandatory for the judge to invoke this extreme penalty. To make matters even more serious for these lawbreakers, they were not given the benefit of counsel and were urged to plead guilty.

Another offense that causes much federal government concern is the illicit manufacture and sale of liquor—moonshining—primarily in the

---

[44] *American Magazine,* now discontinued, carried for years a lively cartoon feature, "It's the Law," based on the many strange ordinances still "on the books." Collectively it represents a valuable documentation of obsolete laws.

southern mountains. For a century or more the people living in these iso-lated areas have made "moonshine" for sale, without paying the excise tax. The Treasury Department points out that this illegal business is again on the increase, approaching a "multi-million dollar industry." Little construc-tive work has been done to offset the prevailing mores of these mountain folk; thus they will no doubt persist in their occupation and we shall con-tinue to use punitive methods for behavior, illegal though it is, that is un-derstandable to social scientists.

At the same time, we are unduly hasty in creating new crimes by statute. *In 1931, 76 per cent of all the inmates of federal and state prisons had been incarcerated for committing acts that had not been crimes 15 years earlier.* Since 1900 some 500,000 new state laws have been enacted. Our state penal codes are antiquated, often absurd, and in disagreement one with another. A felony in one state is a misdemeanor in another. A crime punishable by life imprisonment in one state calls for only a few years' sentence in another. Grand larceny is interpreted to mean theft of as little as $15 in one state; in another it involves amounts of over $200.

Out of this welter of antiquarianism and confusion, a rational, modern-ized, and integrated system of criminal jurisprudence must emerge if we are to make any progress in dealing with criminals. A real effort at legal uni-formity throughout the country should be made, so that serious forms of antisocial conduct are listed as crimes, and all acts that are not a challenge to social well-being are expunged. Especially should we remove from the statute books acts that are purely a matter of taste or private morals.

The inept offender is not familiar with this maze of law. The more ex-pert criminal is well acquainted with the various penal codes and con-forms his professional acts to their provisions. Judges sentence the offender according to the provisions of the penal code, so we find that those con-victed are often the "small fry," who either are unfamiliar with the law or do not have the proper legal connections. There is a general practice by which an offender may plead guilty to a lesser offense and thus save time and expense to all concerned. The well-informed professional criminal fre-quently saves himself a long prison sentence by taking advantage of this practice.

It would not be an exaggeration to state that 50 to 75 per cent of those actually convicted and sent to prison are either ignorant violators of the law or those who do not have the financial resources to fight their cases. The gangster is protected by the "war-chest" accumulated by his superiors over a period of years; he or his bosses employ the best available lawyers, who usually see to it that he escapes a prison sentence completely or is sentenced to a nominal term with a more or less speedy discharge.

No significant change will be possible in the philosophy of correctional treatment until our penal codes are drastically amended in favor of a rational system of criminal procedures.[45] This suggestion is not made by dreamers or sentimentalists. Insistence upon a repressive philosophy is outmoded. Correctional history has demonstrated that where punishments are severe and prison sentences long, crime increases. The more progressive jurists are unanimous in their conviction that a new order must be ushered in if we are to cope with the serious problem that confronts this country. Instead of dealing with crime as we have in the past, there must be brought to each trial a knowledge of the criminal, including his background, his potentialities, his habits, and his modes of social thinking. The disposition of his case must be determined in the light of these data, not his crime.

The Status of Crime Statistics

In 1930 the Federal Bureau of Investigation assumed responsibility for gathering national crime statistics and thus began the publication of its *Uniform Crime Reports.*[46] In these *Reports* may be found a record of crimes known to the police, offenses cleared by arrest, persons held for prosecution, and persons released or found guilty of offenses.

Although in the past some police officers were reluctant to cooperate with the F.B.I., due largely to jealousy and local pride, this antipathy has declined in recent years. To offset human frailty, however, it has been suggested that all crime data might well be turned over to some other federal agency that has no active interest in the efficiency of the local police and that can publish statistics accurately and dispassionately.

Cooperation with some uniform statistical bureau, preferably governmental, is extremely important in collecting information regarding crime, as is the establishment of uniform offense categories. Robbery, for instance, may be construed differently in different jurisdictions.

Existing data released by the F.B.I. and other agencies, besides being inadequate, is often erroneously interpreted, intentionally or otherwise. The director of the Chicago Crime Commission writes:

> Unfortunately, the misuse of crime statistics is far too commonplace. On numerous occasions following the release of national crime figures,

---

[45] For the past few years the American Law Institute had been engaged on a project aiming toward the development of a model penal code which could be adopted by all forty-eight states. See "Your Chances for a Square Deal in the Courts," *Life,* Vol. 43, No. 13 (September 23, 1957), pp. 62 ff.

[46] Prior to 1930, national crime statistics were sporadic and inadequate. In 1927 the International Association of Chiefs of Police and the Detroit Bureau of Governmental Research published in 1929 an elaborate guide entitled *Uniform Crime Reporting: A Complete Manual for Police.* This set forth standardized categories of offenses. The late Bruce Smith, prominent police consultant, and Dr. Lent D. Upson of Detroit, were responsible for this first step in the development of national criminal statistics.

some political leader informs his constituents that the FBI has just completed an independent survey of crime conditions throughout the country and had found his city has less crime than others of comparable size. Actually, the F.B.I. does nothing of the kind. . . . In the Uniform Crime Reports . . . the following statement is underlined: "In publishing the data sent in by chiefs of police in different cities, the FBI cannot vouch for their accuracy. They are given out as current information which may throw some light on problems of crime and criminal-law enforcement." . . . And, says the F.B.I., "Caution should be exercised in comparing crime data for individual cities because the differences . . . may be due to a variety of factors." . . . These statements are clear and specific. . . . Yet, in the face of these warnings, some officials persist in misusing the crime figures.[47]

In addition to the F.B.I., the Bureau of the Census for many years furnished valuable data regarding commitments of convicted persons to federal and state penal institutions through its annual reports, *Prisoners in State and Federal Prisons and Reformatories.* In 1946 this service ceased for reasons of economy. As of January 1, 1950 the service was transferred to the Federal Bureau of Prisons and is now known as the *National Prisoner Statistics.*

In summing up these commendable efforts, Professor George Vold states that "actually we have now no more (possibly even less) systematically compiled and analyzed information about the people who are in our penal institutions than we had 15 or 20 years ago."[48]

The Federal Bureau of Prisons also publishes annually a sizeable pamphlet entitled *Federal Prisons,* which includes the annual report of the Director. This report is valuable for its summary of the movements and disposition of federal offenders.

So far as they go, all of the above are on a high plane of efficiency and accuracy and serve as a standard for cities and states. A few states also have excellent statewide data on crime. New York, California, Massachusetts, Minnesota, and Pennsylvania have developed statistical policies that compare favorably with those of the federal government. In addition, a number of police departments and courts furnish uniform crime reports. Aside from the great financial cost, it takes considerable integration and vision to develop an adequate over-all system of criminal data.[49]

---

[47] *A Report on Chicago Crime for 1955,* pp. 5-6.

[48] In his review of *Federal Prisons, 1954,* a publication of the Federal Bureau of Prisons, in *Federal Probation,* Vol. 20, No. 1 (March, 1956), p. 64.

[49] See Emil Frankel, "An Integrated System of Crime Statistics," *Federal Probation,* Vol. 9, No. 2 (April-June, 1945), pp. 28-32. For more thorough analyses of crime statistics, see: Emil Frankel, "Statistics of Crime," *Encyclopedia of Criminology* (New York: Philosophical Library, 1949), pp. 478-489; Ronald Beattie, "Problems of Criminal Statistics in the United States," *J. Crim. Law,* Vol. 46, No. 2 (July-August, 1955), pp. 178-186; and the entire issue of "*N.P.P.A. Journal,*" Vol. 3, No. 3 (July, 1957), pp. 230-298.

A Picture of Crime from the F.B.I. Reports

Inadequate though the annual statistics are, they do reflect the enormity of the problem. Besides what appears below—a description of crime for 1957—the reader should refer to the latest *Uniform Crime Reports,* compiled by the Federal Bureau of Investigation, for current trends. It seems to the writers that the best way to secure a picture of crime in the United States is to present a digest from these:

Estimated Major Crimes: Estimates of the number of offenses known to the police in 1957 total 2,796,400. This estimate includes the Part I or "most serious" crimes, from minor larceny to murder. The 1957 figure is higher than ever before; 9.1 per cent above 1956 and 23.9 per cent over an average of the previous five years. Changes from 1956 to 1957 among the eight "most serious" categories ranged from a decrease of less than one per cent for murder to an increase of 12.2 per cent for burglary.

Murders totaling 6,920 were down 0.7 per cent from 1956; compared with the average for the previous five years, 1957 showed a 1.1 per cent decline. Negligent manslaughters were up 1.6 per cent in 1957, and the total of 5,740 was 2.5 per cent above the average for the previous five years. Rape, including statutory offenses, increased 3.8 per cent in 1957 to number 21,080 or 13.9 per cent above the previous five-year average. The estimated 100,110 aggravated assaults in 1957 were 3.8 per cent above 1956 and 8.1 per cent higher than the average for 1952-56.

Robberies, armed and unarmed, numbered 61,410 in 1957, an increase of 8.2 per cent for the year and an increase of 1.4 per cent above the average for the previous five years. Other property crimes (burglary, auto theft, and other larcenies) increased 12.2 per cent, 9.9 per cent, and 8.4 per cent respectively in 1957, and when compared with averages for 1952-56 these classes were up 20 per cent, 26.2 per cent, and 27.3 per cent, in that order.

About 479 million dollars were lost to robbers, burglars, and thieves, but police recoveries reduced the loss to about 212 million dollars.

In terms of volume alone, crime at 56.2 per cent above the 1950 level is rising four times as fast as the total population—up 13 per cent since 1950.

Persons Arrested: Arrests for all crime classifications by police in 1,220 cities increased 4.3 per cent in 1957 as compared with 1956. In these same cities, arrests of persons under 18 years of age increased 9.8 per cent. Percentage changes from year to year indicate that arrests of persons under 18 have increased 55 per cent since 1952, the first year comparable figures were available. During the same period, 1952-57, the United States population in the 10-17 age group has increased 22 per cent. Assuming that the reporting cities have experienced a similar population growth, it appears that the percentage increase in arrests of young people is two and one-half times the percentage growth of their population group.

Persons under 18 years of age were arrested in 10.3 per cent of all arrests for the crimes against the person of murder, negligent manslaughter, rape and aggravated assault. For the crimes against property (robbery, burglary, auto theft, and other larcenies), young people under 18 represented 53.1 per cent of all arrests. Their greatest participation in Part I offenses, as evidenced by arrests, was in thefts of autos. In that category they represented 67.6 per cent of all arrests.

With the changes made by the F.B.I. through the years in recording arrests by various categories, we find a higher percentage of the recorded offenses falling into Part I, or "most serious," categories. For instance in the first edition of this book, published in 1943 and using the statistics for the year 1942, we find that of 5,256,755 arrests, only 131,161—2.51 per cent—were Part I. In that year the largest item of arrests was "traffic and motor vehicle laws": 3,766,346. In the 1957 report, this category does not appear, thus bringing the total of arrests recorded to 2,068,677 of which the "most serious" offenses—226,106—make up about 11 per cent. If the category "drunkenness" (a large item amounting in 1957 to 832,268) was dropped from the roster, the percentage of most serious crimes would again increase drastically. The figures would not change but the percentage would; it would change still more if "vices and social pathologies" and "offenses against peace and order" were no longer recorded.

In the table on page 80 we have adapted the F.B.I. figures into four groups rather than the conventional groupings known as Part I and Part II. We have added "Vices and Social Pathologies" which, as we shall point out in Chapter 5, cannot be classed with actual crimes by the progressive criminologist even though there is legislation against them. Our fourth group we refer to as "offenses against peace and order." It is from this group that thousands who are arrested are eventually released. The category referred to in the F.B.I. reports as "All other offenses" includes "all violations of state and local laws for which no provision is made" in the other categories. The category "suspicion" is defined in the *Uniform Crime Reports* as "all persons arrested as suspicious characters, but not in connection with any specific offense, who are released without formal charge being placed against them."

The percentage of serious offenses may, therefore, be raised or lowered, all depending on what categories are included or excluded. If "vices and social pathologies" and "offenses against peace and order" were excluded, as not being considered important enough to record in our annual crime rate, we find for 1957 that "most serious" offenses would actually represent 32.5 per cent of all real crime rather than the 11 per cent as computed with all categories included. And, recalling what we referred to above as the "most serious" offenses in 1942, we see that the percentage was only

2.51 per cent of all recorded offenses. It is important, therefore, to ascertain just what lies behind the annual crime rate.

We have preferred to include the important crime data in one over-all table, which includes the number of offenses by categories, by sex, by race, and by age groups. On pages 172–174 in Chapter 11 we shall use the data on Negroes and whites to indicate the ratios between the two major racial groups so far as various crime categories are concerned. In appropriate places we shall refer to various phases of the crime picture, such as female crime and offenses committed by minor racial groups and by youthful offenders.

Sources of Information Regarding the Crime Problem

Aside from statistics, students of crime and penal procedures rely heavily on many other sources of information: the published results of studies made by various governmental agencies, special commissions set up by executive order, and those private foundations or agencies concerned with the prevention of crime.[50]

Perhaps the earliest inclusive survey of crime as a national problem is that of the famous National Commission on Law Observance and Enforcement. It was appointed by President Hoover in 1929 and was charged with the task of studying the "entire question of law enforcement and organization of justice." It is best known as the Wickersham Commission after its chairman, George W. Wickersham, one time Attorney General. Although this commission owed its inception primarily to the controversy over the prohibition amendment, its reports, of which twelve were published, cover a most comprehensive field.

Another vast source of information concerning the crime problem and its many implications in the Attorney General's *Survey of Release Procedures,* consisting of five volumes published in 1939-40. Begun in 1935 at the request of Attorney General (1933-1939) Homer Cummings, it was carried on by field workers and financed by the Works Progress Administration.

The *Survey* consists of: Vol. I, *A Digest of Federal and State Laws on Release Procedures* (covering parole, probation, executive clemency, "good time" deductions, and expiration of sentence); Vol. II, *Probation* (a comprehensive picture of the legal and administrative procedures throughout the country); Vol. III, *Pardon* (a study of the law and administration of

---

[50] Bibliographies on crime are: Dorothy Campbell Culver, *Bibliography of Crime and Criminal Justice, 1932-37* (Baltimore: Wilson, 1939); A. F. Kuhlman, *A Guide to Material on Crime and Criminal Justice* (Baltimore: Wilson, 1929); Dorothy Campbell Tompkins, *Sources for the Study of the Administration of Criminal Justice, and Supplement* (Sacramento: State of California, 1949; supp. 1957). See also Kurt Schwerin, "Bibliographical Services in Criminology," *J. Crim. Law,* Vol. 41, No. 3 (September-October, 1950), pp. 254-267.

DISTRIBUTION OF ARRESTS BY SEX, RACE, AND AGE GROUPS, 1957*

OFFENSES CHARGED:	NUMBER BY SEX		NUMBER BY RACE						NUMBER BY AGE GROUPS									
	Male	Female	White	Negro	Indian	Chinese	Japanese	All Others	under 15	15-18	19-24	25-29	30-34	35-39	40-44	45-49	50 & over	Unknown
I. MOST-SERIOUS OFFENSES: approximately 11% of all offenses recorded																		
Criminal homicide:																		
2,007......a. Murder and non-negligent manslaughter	1,634	373	761	1,225	4	1	1	15	31	145	323	309	312	271	195	148	273	—
1,238......b. Manslaughter by negligence	1,134	104	973	254	3	—	1	8	2	106	273	178	178	162	99	81	159	—
11,820....Robbery	11,354	466	5,517	6,158	66	1	1	77	939	2,959	3,705	1,753	1,134	612	363	176	179	—
23,266....Aggravated assault	19,201	4,065	8,041	15,045	78	2	2	95	604	2,073	4,353	3,897	3,612	3,033	2,160	1,471	2,062	1
51,398....Burglary-breaking and entering	50,195	1,203	36,058	14,989	189	4	5	153	14,217	16,843	9,157	4,112	2,732	1,804	1,134	690	700	9
102,476..Larceny-theft	88,898	13,578	70,701	30,826	561	24	19	345	26,797	30,737	14,642	7,424	6,256	4,843	3,691	2,847	5,184	55
29,121....Auto theft	28,328	793	23,095	5,716	221	2	2	85	5,875	15,557	4,216	1,333	908	594	317	147	166	8
4,780......Rape	4,780	—	2,623	2,087	20	1	2	47	163	1,190	1,754	672	399	251	138	90	123	—
226,106....TOTAL MOST-SERIOUS OFFENSES	205,524	20,582	147,769	76,300	1,142	35	35	825	48,628	69,610	38,423	19,678	15,531	11,570	8,097	5,650	8,846	73
II. LESS-SERIOUS OFFENSES: approximately 23% of all offenses recorded																		
81,749....Other assaults	73,999	7,750	44,746	36,316	342	6	4	335	1,899	5,986	15,902	14,735	13,880	10,812	7,439	4,994	6,088	14
8,268......Forgery and counterfeiting	7,063	1,225	6,925	1,307	35	1	1	20	133	778	1,833	1,552	1,406	1,067	642	389	486	2
16,168....Embezzlement and fraud	13,834	2,334	13,030	3,037	43	—	7	48	140	489	2,466	2,988	3,185	2,579	1,761	1,175	1,383	2
3,869......Stolen property; receiving, etc.	3,560	309	2,572	1,263	19	2	1	13	527	924	749	391	364	327	224	141	222	—
16,864....Weapons; carrying, possessing	15,992	872	7,814	8,863	89	4	3	91	843	2,961	3,880	2,449	2,060	1,650	1,076	805	1,139	1
241,167..Disorderly conduct	201,562	39,605	141,057	97,628	1,673	29	20	760	8,196	27,025	50,580	34,971	32,114	27,934	21,053	15,697	23,587	10
101,099..Driving while intoxicated	96,099	5,000	84,074	15,776	864	4	16	365	15	1,529	13,900	14,260	15,878	15,254	13,494	11,082	15,683	4
469,204....TOTAL LESS-SERIOUS OFFENSES	412,109	57,095	300,218	164,190	3,065	49	50	1,632	11,753	39,692	89,310	71,346	68,887	59,623	45,689	34,283	48,588	33
III. VICES AND SOCIAL PATHOLOGIES: approximately 51% of all offenses recorded																		
12,654....Prostitution and commercialized vices	3,906	8,788	7,520	5,054	82	1	1	35	11	38	3,395	2,992	2,071	1,579	941	589	775	3
20,968....Other sex offenses (including offenses against chastity, decency, and morals)	16,064	4,904	14,492	6,226	115	10	8	117	1,331	2,674	4,132	3,280	2,727	2,229	1,564	1,177	1,854	—
22,444....Offenses against the family (non-support, neglect, and desertion)	20,433	2,011	14,624	7,639	66	1	2	112	59	627	4,232	4,645	4,363	3,360	2,306	1,376	1,475	1
7,277......Narcotics (unlawful possession, sale, and use)	6,143	1,134	3,092	4,108	18	13	—	46	17	395	2,096	2,033	1,251	631	316	253	285	—
43,347....Liquor laws (excludes Federal laws)	35,910	7,437	26,859	16,000	359	8	6	115	570	10,133	8,724	3,644	4,007	4,096	3,690	3,095	5,384	4
832,268..Drunkenness	768,849	63,419	610,051	188,323	30,026	42	103	3,723	589	12,263	68,025	77,158	102,258	117,973	118,848	111,763	223,059	332
50,462....Gambling	45,364	5,098	12,953	37,102	11	46	27	323	107	884	4,805	6,728	7,595	7,442	6,667	5,496	10,731	7
69,520....Vagrancy	62,783	6,737	51,679	16,298	1,230	9	13	291	811	6,140	10,066	6,720	7,026	7,192	7,313	7,234	17,014	4
1,058,980....TOTAL IN VICES AND SOCIAL PATHOLOGIES	959,452	99,528	741,270	280,750	31,907	130	161	4,762	3,495	33,454	105,475	107,200	131,298	144,502	141,645	130,983	260,577	351
IV. OFFENSES AGAINST PEACE AND ORDER: approximately 15% of all offenses recorded																		
84,645....Suspicion	76,612	8,033	53,789	30,277	366	6	4	203	3,739	17,756	22,429	11,929	9,194	6,616	4,605	3,379	4,993	5
229,742....All other offenses (a blanket category not including any of the above)	195,028	34,714	162,921	64,511	1,235	47	23	1,005	31,312	46,310	38,165	24,496	22,824	19,577	15,362	11,587	19,952	157
314,387....TOTAL OFFENSES AGAINST PEACE AND ORDER	271,640	42,747	216,710	94,788	1,601	53	27	1,208	35,051	64,066	60,594	36,425	32,018	26,193	19,967	14,966	24,945	162
2,068,677....TOTAL NUMBER	1,848,725	219,952	1,405,967	616,028	37,715	267	273	8,427	98,927	206,822	293,802	234,649	247,734	241,888	215,398	185,882	342,956	619

* In 1,473 cities over 2,500 population representing a total population of 40,176,379 based on the 1950 census. Figures adapted from *Uniform Crime Reports*, Vol. 28, No. 2, 1957: Table 41, p. 114 (age groupings); Table 43, p. 117 (sex); and Table 44, p. 118 (race).

pardon); Vol. IV, *Parole* (containing statistical information on approximately 125,000 parolees, and a study of parole administration in the several states); and Vol. V, *Prisons* (an attempt to give a coordinated picture of the many types of services involved in the prison treatment of that time).

The published proceeding of various national conferences held during the past three decades are also valuable. The first conference on crime was called by Attorney General Cummings and held December 10-13, 1934. There have been two national conferences on parole, in 1938 and in 1956. One should also add the reports that came from the Kefauver Committee, officially known as the Special Committee to Investigate Organized Crime, which conducted hearings in Washington and other cities during 1950-51.[51]

On the state and local levels we find several excellent surveys of crime. Perhaps the best known of these were made in Missouri in 1925, in Illinois in 1929, and in Chicago in 1945. We have already mentioned a number of reports from the state of California. By legislative act of 1947 commissions were appointed in that state to explore the following phases of crime and correction: criminal law and procedure, adult corrections and release procedures, juvenile justice, social and economic causes of crime, and delinquency and organized crime. On the international level, the proceedings of the International Penal and Penitentiary Congresses 1872-1950 present the questions and answers of these twelve Congresses of criminologists, criminal lawyers, and child welfare specialists.[52] Since 1950 the United Nations has absorbed this international body, and in 1955 held its first congress at Geneva, publishing several reports on these sessions.

---

[51] A National Conference for the Prevention and Control of Juvenile Delinquency was called by the Attorney General in November, 1946. Eighteen reports were published as a result of the conference.

[52] Negley K. Teeters, *Deliberations of the International Penal and Penitentiary Congresses, 1872-1950* (Philadelphia: Temple University Book Store, 1949).

5

Social Pathology and Crime

The familiar story, that, on seeing evil-doers taken to the place of execution, he [John Bradford, 1510-1555] was wont to exclaim: "But for the grace of God there goes John Bradford," is a universal tradition, which has overcome the lapse of time. (From the "Biographical Notice" in the Parker Society edition of The Writings of John Bradford *[1853].) We may well repeat it while reading the present chapter.*

There is a vast army of men and women in this country suffering from social inadequacies or deep-seated emotional disturbances who have traditionally been regarded as criminals. In making their acts illegal the law has been especially severe, and the attitudes of police officers, jurists and correctional officials, although scientifically outmoded, reflect the conventional attitudes of society. The jails and houses of correction and to some degree our state prisons demonstrate a lack of understanding of the problems these unfortunate groups face. In this chapter we attempt to deal with several types of individuals who are actually pathological rather than criminal, even though they violate state and federal statutes.

The Vagrant

Our laws dealing with vagrancy are a survival of the Elizabethan Poor Laws, codified in England in 1609 and brought to America by the colonists. These laws were passed to rid the streets and byways of "sturdy rogues," beggars and ne'er-do-wells who either refused to work or were marginal in their capacity to achieve self-support. Workhouses were established for their reception as early as the sixteenth century, largely for economic reasons, as we shall point out later.[1] In modern times the economic motive has disappeared, but the laws dealing with this diverse class of people remain substantially the same as those of three hundred years ago.

---

[1] See pages 330-331 for discussion of the workhouse.

Most vagrants are socially inadequate, whether the offense for which they are arrested is loitering, disorderly conduct, or drunkenness. Loitering is tantamount to crime in many cities and small towns. Sitting in a railroad station without a ticket, perhaps only in an attempt to keep warm, makes a person liable to arrest and a night in the town lock-up. Studies made during the past decade demonstrate that police and magistrates exercise little patience with the "floater," the man who is confused in answering questions, or who is found sleeping in a public park.

The following is typical in American cities which demonstrates the cursory and often heavy-handed manner in which magistrates dispose of vagrants:

> On January 3, 1954, the Philadelphia press reported that police ". . . had opened a drive against vagrants and habitual drunkards in the central city area." By February 2, the drive was at its height, and that morning 56 cases were awaiting disposition when the magistrate opened the daily divisional police court for the district which included the "skid row" and the central city area. These cases were the last items on the morning's docket, and the magistrate did not reach them until 11:04 a.m. In one of the cases there was a private prosecutor, and the hearing of the evidence consumed five minutes. As court adjourned at 11:24, that left 15 minutes in which to hear the remaining 55 cases. During that time the magistrate discharged 40 defendants and found 15 guilty and sentenced them to three month terms in the House of Correction.
>
> Four of these committed defendants were tried, found guilty and sentenced in the elapsed time of seventeen seconds from the time that the first man's name was called by the magistrate through the pronouncing of sentence upon the fourth defendant. In each of these cases the magistrate merely read off the name of the defendant, took one look at him and said, "Three months in the House of Correction." As the third man was being led out he objected, stating, "But I'm working . . ." to which the magistrate replied, "Aw, go on."
>
> The magistrate then called the name of one defendant several times and got no answer. Finally he said, "Where are you, Martin?" The defendant raised his hand and answered, "Right here." "You aren't going to be 'right here' for long," the magistrate said. "Three months in Correction."[2]

Confirmed vagrants and common drunks will have been in and out of county jails and houses of correction many times. Professor John B. Waite tells of one who was haled into court in Detroit and convicted for his one hundredth time; of another who had a petty crime record of 76. Whether these persons were vagrants or common drunks is not stated.[3] The magazine *Federal Probation* relates the case of No. 46 who, since 1925, has been booked in the District of Columbia jail 285 times; who has spent some 25

---

2 Caleb Foote, "Vagrancy-Type Law and Its Administration," *University of Pennsylvania Law Review,* Vol. 104, No. 5 (March, 1956), p. 607.

3 "Revenge Costs Too Much," *Harper's,* Vol. 292, No. 1152 (May, 1946), pp. 466-472.

years in that institution though the bulk of his sentences have been no more than 5 to 30 days at a time.[4]

Vagrancy and its attendant pathology is a serious problem to every town and city; but it is not actually a correctional problem. Rather it represents a major social problem and should be dealt with as such. Thousands of community derelicts could be kept out of town lock-ups, county jails, and houses of correction if facilities were available for scientific and rational treatment. Society has had little understanding of or patience with the marginal man. Today many social derelicts are cared for by church missions. These organizations attempt to minister to their physical needs on a relatively low level—cheap food and a "flop," sometimes in return for sawing cordwood, caning chairs, or making and peddling brooms, but they make no attempt to deal with their emotional problems.[5]

This approach is woefully inadequate despite good intentions. Vagrants, panhandlers, hoboes, and "bums," should be taken off the streets and sent to a county or regional farm and, if at all possible, be emotionally rehabilitated. If the prognosis is poor, they should be cared for in decent surroundings, free from physical restraint. We have this social debtor class with us always and it should not be permitted to create "jungles" in our cities. In fact, a forthright plan, universally adopted by our metropolitan centers, could go far in controlling the problem of vagrants and panhandlers. A study conducted by trained social workers of the Pennsylvania Prison Society indicates that few of these social bankrupts respond to social treatment. The study recommended a municipal lodging center for the majority where counseling and job placement facilities could be available for those desiring such services.[6] Clinics staffed by social workers and guidance counselors, with psychiatric services available, could supplement the municipal housing unit.

While this is a community problem, it is obvious that legislation is needed in order to implement modern concepts of social maladjustment. As Caleb Foote concludes in his study of vagrancy laws:

> The economic purposes which once gave vagrancy a function no longer exist, and the philosophy and practices of welfare agencies have so changed relief methods that a criminal sanction to enforce an Elizabethan poor law concept is outdated. To try to utilize a feudal statute as a weapon against modern crime and as a means of liberalizing the restrictions of criminal law and procedure is both inefficient and an invitation to the kind of abuses which this study has shown to be widespread.[7]

---

[4] *Federal Probation,* Vol. 21, No. 2 (June, 1957), p. 74.

[5] For a revealing analysis of the problem of the inadequate metropolitan floater, see Sara Harris, *Skid Row, U.S.A.* (New York: Doubleday, 1956).

[6] "The Petty Offender: A Philadelphia Study of the Homeless Man," *Prison Journal,* Vol. 36, No. 1 (April, 1956), pp. 3-26.

[7] "Vagrancy-Type Law and Its Administration," p. 650.

Drug Addicts and Peddlers

The use of narcotics, aside from the sheer relief of pain, is an age-old social problem. It first became conspicuous in the United States during the War between the States. It was so common after the war that drug addiction was commonly referred to as the "army disease." Addiction grew steadily, if slowly, from that period to World War I.

We are all more or less familiar with the ravages of the narcotic habit. Thomas De Quincey (1785-1859) vividly described them in his *Confessions of an English Opium Eater.* Only recently several motion pictures have depicted its tragedy.

The twofold problem of drug addiction and the illegal sale of narcotics today represents a serious one for the nation. Addiction is, in reality, a medical problem; trafficking in narcotics is and should be a police problem. Many addicts sell narcotics in order to procure money to buy them, further compounding the problem.

During Kefauver Committee hearings in June 1951, several addicts told their stories. The result was a demand for more drastic penalties instead of a complete study of existing legislation.

In 1956 Congress passed the Narcotics Control Act, which distinguishes very little between possession for use and for resale, although it is the peddler who does the harm to others.

This act calls for the following penalties for illegal possession of drugs: (1) first offense, 2 to 4 years' imprisonment and a fine not to exceed $20,000; (2) second offense, 5 to 20 years' imprisonment and a $20,000 fine; (3) third or subsequent offense, 10 to 40 years' imprisonment plus a maximum fine of $20,000. For *selling* narcotics, the first and second convictions are the same as for possession, and for selling of heroin by a person over eighteen years of age to a person under that age, a possible sentence of death. This drastic penalty is at the discretion of the jury. Here is vengeful legislation that fails to understand the medical problem of the addict, and adds nothing important to the fight against the peddler of narcotics.[8]

Much distortion and misinformation regarding the drug problem is current in this country. Serious though the use and sale of narcotics may be, it is less widespread than in the past. While the number of actual addicts can only be estimated, the Federal Bureau of Narcotics places the total at around 60,000.[9] During World War I, one out of every 1,500 draftees was

---

[8] For an analysis of this piece of legislation, see A. R. Lindesmith, "DOPE: Congress Encourages the Traffic," *The Nation,* Vol. 184, No. 11 (March 16, 1957), pp. 228 ff.

[9] John Gerrity, "The Truth about the Drug Menace," *Harper's,* Vol. 204, No. 1221 (February, 1952), p. 28. See also Harris Isbell, "What to Know about Drug Addiction," *Public Health Service Publication No. 94* (1951).

found to be an addict; this was reduced to only one out of every 10,000 in the World War II draft.[10]

Whatever the number, the United States is the only country closely associated with the peoples of western Europe or with English-speaking countries that has a narcotics problem. Very few of these countries have more than a handful of addicts. England, with a population of some 50 million, reports only slightly more than 300; New Zealand, in 1949, reported the existence of 45 addicts; northern European countries number their addicts in the tens or hundreds.[11]

There are many ways in which drug addiction may be acquired. Some get the habit when opiates are medically prescribed for the relief of their physical pain or insomnia. In the days before narcotics became more difficult to obtain, addiction frequently resulted from self-medication and from the use of patent medicines containing opiates. The drug laws have made it more difficult to market such medicines, but abuses have not been entirely eliminated.

Yet it is unlikely that normal persons will become true addicts even though they intermittently use narcotics. True addiction is unusual unless the person has a deep-seated emotional disturbance or is living under adverse conditions from which he desires to escape by means of narcotic euphoria. Many deviates, prostitutes, and criminals use drugs and alcohol to buoy them up and to make them more oblivious of the destructive conditions under which they are obliged to live. Dejected criminals are taught the drug habit by those already afflicted.

Alarmist literature and the propaganda of crusaders against addiction, most of them advocating "police control," have created a grotesquely exaggerated impression of the danger to society from the drug addict. He is commonly presented as a bold and dangerous criminal, likely to plan and execute all sorts of violent crimes. Addicts are almost never found among the organized criminals, racketeers, and gangsters:

> Drug addicts are often regarded as the most dangerous and heinous criminals and are linked up with killing and rape. This delusion has been smashed so many times that it is useless to devote serious attention to it. Suffice it to say that students of drug addiction have always been in unani-

---

[10] Gerrity, loc. cit. Dr. Kenneth W. Chapman ventures the thesis that addiction experiences periodic cycles, so far as numbers are concerned. While there seems to have been an upsurge during the past decade, he states that in 1920, with a population of 106 million, it was estimated there were 100,000 addicts in the country; today with some 170 million, the estimate by the Federal Bureau of Narcotics is only 60,000. See his "Drug Addiction: The General Problem," Federal Probation, Vol. 20, No. 3 (September, 1956), p. 42.

[11] According to a report submitted by the British government to the United Nations, there were, in 1956, 333 addicts; thirty per cent of this number consisted of medical personnel: 77 doctors, 2 dentists, and 20 nurses. Journal of the American Medical Association (hereafter, "Journal A.M.A."), Vol. 165, No. 2 (September 14, 1957), p. 181.

mous agreement that the crimes of rape and murder are rarely committed by drug users.[12]

Professional criminals rarely have anything to do with a drug addict, partly because he is at the mercy of the police when caught, and can easily be pressed into the role of informer. Besides, the addict does not have the reliability or stamina required to execute major crimes. Therefore it is evident that, if drug addiction is an evil, it is an evil in itself and not because it incites addicts to serious crimes against either person or property. When they get their drugs, "addicts as a group are anything but criminally inclined."[13] In fact, unlike the alcoholic when he has no alcohol, the addict is dangerous not when he has his drug, but when he is without it.

Many well-informed students of the narcotic problem criticize drastic legislation, such as the Harrison Anti-Narcotic Act of 1914, the Jones-Miller Act of 1922, and the Federal Narcotics Control Act of 1956, which makes a shambles of the narcotic problem rather than a solution. These students suggest, as the only sane method of procedure, adequate health education and concentrated effort to eliminate the supply of narcotics. They recognize the necessity of curbing the underworld trade in drugs, but they are opposed to making criminals out of a group of unfortunate invalids, the victims either of their own internal conflicts or of irresponsible medical quacks and patent medicine vendors.

Many eminent experts in this field hold such opinions. Dr. Ben Karpman says that until the Harrison Act was passed "things seemed to run along as smoothly as was possible under the circumstances." There was no greater increase in addiction than in the other neuroses. Few addicts were criminals. But when the law was passed, users of drugs were compelled to lie and steal to get them. Our prisons began to fill up with drug addicts. He believes that the Harrison Act produced more petty criminals than the Volstead Act, and that it created the dope racket, which is almost as serious as the liquor racket was before repeal.[14] Lindesmith believes all that is needed is a new interpretation of the narcotic acts, so that it will no longer be criminal for a physician who registers under the act to prescribe drugs for those who need them. The punitive features of the act could be applied only to the persons who handle the drug for profit and purposes of exploitation.[15]

We are emphatic in stating that the use of narcotics should not be a crime and that drug addicts should not be treated as criminals. The problem confronting the community is the care of the general run of addicts who can ill

---

12 A. R. Lindesmith, "Dope Fiend Mythology," *J. Crim. Law*, Vol. 40, No. 2 (July-August, 1949), p. 199. See also his *Opiate Addiction* (Bloomington, Indiana: Principia Press, 1947).

13 Ben Karpman, "Laws that Cause Crime," *American Mercury*, Vol. 23, No. 89 (May, 1931), p. 74.

14 *Ibid.*, p. 77.

15 Lindesmith, *op. cit.*, p. 205. See also his "Traffic in Dope," *The Nation*, Vol. 182, No. 16 (April 21, 1956), pp. 337 f.

afford to buy narcotics and who become a bane to the police and to themselves. Society must accept addicts as sick people in need of hospitalization and careful treatment, free from stigma of punishment or reproach.

There is no scientific cure for the confirmed user of narcotics except prolonged psychiatric treatment, preferably skilled psychoanalysis that will remove the motivating causes of the addiction. Merely to "get a person off the drug" is not a cure. A cure involves economic and social rehabilitation through rebuilding moral stamina and self-reliance. Unfortunately, such treatment is extremely rare. Men and women with sufficient means to pay for such treatment usually have enough money to secure all the narcotics they need. Occasionally a well-to-do addict with sufficient insight and balance to recognize his plight will turn to a psychiatrist for aid. In such cases gratifying cures have been recorded.

Much favorable publicity has been accorded the two narcotic farm-hospitals operated by the federal government at Forth Worth, Texas and Lexington, Kentucky. Addicts are received to these institutions from the courts for treatment but a large proportion voluntarily commit themselves at a modest daily rate. This latter group may leave at any time. Those committed by the courts, aside from probationers who have a determinate sentence, are kept until discharged or until their court sentence expires.

It is difficult to make studies dealing with cure or relapse since many who are discharged fail to respond to inquiry. The studies we have available show that the percentage of addicts who remain off drugs after treatment at the federal hospitals is discouragingly low.

We have discussed the topic of drug addiction in order to bring to the reader some of the conflicting attitudes toward the problem. It is the belief of some students that we need a reorientation toward the control of the use of drugs. For example, in England, where the problem is negligible, a person addicted may legitimately procure drugs from a physician if (a) he is under gradual withdrawal treatment, (b) it is demonstrated that it will not be medically safe for him to undergo withdrawal, or (c) it is demonstrated that he is capable of leading a relatively normal life under a minimum dose but not if the drug is entirely discontinued. The British Home Office recognizes that to supply an addict with minimum maintenance doses does, in some cases, constitute a medical need.

The Council on Mental Health of the American Medical Association, after a two year study of the drug problem, stated that "narcotic addiction should be viewed . . . as an illness and that there should be a progressive movement in the direction of treating addiction medically rather than punitively." Among the recommendations of the Council were:

1. Development of institutional care in cities and states with significant problems.

2. Development of programs for intensive post-institutional treatment for addicts.

3. Development of methods for commitment of addicts to institutions by civil action rather than through the criminal courts. [And, if this method is used, that] criminal sentences for addicts who are guilty only of illegally possessing and obtaining opiates be abolished.

4. Mandatory minimum sentences for addict violators [which] would interfere with the possible treatment and rehabilitation of addicts, should be abolished.

5. That the policy of voluntary admissions for the treatment of addiction should be continued, extended, and encouraged.

6. Continue support and expansion of mental health programs.[16]

It may be hard for the public to accept establishment of clinics where addicts may procure, at a nominal fee, what they "need" as prescribed by a physician, but so long as drugs are a "bootleg" commodity, controlled and sold by cunning and sinister criminals, addicts and public alike will be bedeviled with their mutual problem. The public has yet to grasp the fact that addicts are dangerous only when they cannot procure the vital drugs they need; that if they were provided their daily need and submitted to treatment through public clinics, the criminal element would no longer have a strangle hold on narcotics.

The Problem of the Alcoholic

Unlike the drug addict the alcoholic is usually accepted by the public as a sick person. In some states, if thrown out of work, he is even entitled to workman's compensation. But it has taken years to educate the public that the alcoholic suffers from a medical illness since, in the not too distant past, he was regarded as a "common drunk" and eligible only for abuse and the common jail.

It is further accepted by the public that the alcoholic needs hospitalization and psychiatric treatment. Yet there are still many "drunks" afforded nothing more than a night in the "clink." It is not unusual for persons to collapse in the street, perhaps suffering from organic or constitutional ailments, to be taken to jail or station-house by police officers rather than to a hospital. Several years ago a respectable woman, a waitress in a large club, collapsed on her way home from work. She was taken to a lockup by an ignorant officer and died in her cell from apoplexy.

Austin MacCormick, some years ago, pointed out the complete lack of intelligent insight in police and court treatment of chronic alcoholics and common drunks.[17] He enumerates about a dozen things we do and do not know about this subject.

---

[16] Abstracted from the report; *Journal A.M.A.,* Vol. 165, No. 15 (December 14, 1957), pp. 1972-1973. See also Nos. 13 (November 30, 1957), and 14 (December 7, 1957).

[17] "Penal and Correctional Aspects of the Alcohol Problem," *Quarterly Journal of Studies on Alcohol,* Vol. 2, No. 2 (September, 1941), pp. 241-259. See also Ellis G. Savides, "The Understanding and Treatment of the Alcoholic Offender," *Amer. J. Corr.* Vol. 20, No. 3 (May-June, 1958) pp. 14 f.

We do not know how many persons are arrested for drunkenness in the course of a year; we merely have sample statistics, most of which come from the reports of the Federal Bureau of Investigation, and these are not comprehensive. We do not know how many of those arrested for intoxication are chronic alcoholics. We do not know how many of those arrested are convicted or what disposition is made of their cases: suspended sentence, probation, fine, jail sentence, state farm, and so on. We do not know, except in general terms, to what institutions they are sent, what these places are like, what care and treatment they receive there, and what medical services are available. We do not know what those arrested, or even those sentenced, are like; their age, race, education, employment history, familial status, mental status, and their social and economic backgrounds. We do not know where they come from or where they go upon release, or how many, if any, show signs of having been benefited. We do not know how much the penal and correctional process of dealing with alcoholics costs.

We do know: that a veritable army of human beings charged with intoxication passes through our police stations, courts, and jails every year; that many of them are alcoholics, and that nothing but the most scientific and prolonged treatment can cure them; that they (and the common drunks) benefit little from the penal and correctional treatment they receive and are more likely to be harmed by it; that most of those sent to institutions go to county jails for very short sentences, often as low as five days and sometimes only two days, and that they return time and again; that our jails are notoriously bad, with medical services generally inadequate to care for the physical ailments and psychotherapy of any sort practically unheard of; that a few of the state farms and larger city institutions have made an attempt to treat alcoholism in a scientific way but are defeated in this program by the lack of psychiatric personnel and the shortness of the sentence.

When we survey the facilities of the country that are available for the scientific treatment of the alcoholic from the moment he is arrested until he is finally released, we find practically nothing. When the alcoholic is sentenced to jail for a few days he rarely receives any treatment and is demoralized by the conditions surrounding him. He is looked upon as a common drunk, but he may be suffering from a serious disease that only a psychiatrist can cure. MacCormick cites a situation that can occur in many cities. It may be three o'clock in the afternoon when a judge sentences a man to a day in jail, but the jail day terminates at four o'clock, so the offender spends one hour in jail; or, if he is sentenced to two days, he has already served one of them within the hour. This is sheer nonsense, especially when the case is one that calls for sober thought and scientific therapy.

Even so-called misdemeanant farms attached to city or county jail systems do little therapeutic work. It is too often assumed that merely a month or two at hard work in the open air will cure the inebriate of his habit.

The fantastic claims, so persistently made in the past, that large numbers of criminals are chronic drinkers have been exploded in recent years. In a study made some years ago by Dr. Ralph Banay of Sing Sing prisoners it was found that no more than 20 per cent had been long-term alcoholics when they committed their crimes. Chronic alcoholism, then, is much more of a social problem than it is a correctional problem.[18]

It has been estimated that of some 60,000,000 consumers of alcoholic beverages in the nation, 3,000,000 are excessive drinkers. According to Dr. Selden Bacon of Yale, about one-fourth of this number are compulsive drinkers with the remainder being labeled "chronic excessive drinkers."[19] Anyone who drinks excessively is (1) a troubled individual and in need of some sort of personality treatment, or (2) is hereditarily prone to alcoholism. About 85 per cent are males between the ages of 20 and 65. This represents about 6¼ per cent of all males in that age group.

There are two movements or organizations in the country that are accomplishing much good in this field. These are: the Research Council on Problems of Alcohol and the Yale Plan of Alcoholic Studies. The former spearheads physical, medical, and psychological research in various parts of the country; the latter explores clinical treatment areas, offers instruction of physicians, social workers, community leaders, and educators through its publications.

One should also mention the work of Alcoholics Anonymous, an organization of alcoholics that operates nationally on the basis of group therapy. Although its technique is more inspirational than medical, it has effected remarkable recoveries throughout the entire country. Organized in 1935, it is reported to have well over 200,000 members today.[20] The members, filled with missionary zeal, devote much of their time to seeking out victims of drink and encouraging them to lift themselves by their own bootstraps. They visit prisons, jails, hospitals, and any other place where chronic drunks may be found.

It is not our purpose to explore the causes of alcoholism nor to review the various excellent studies being undertaken in the field of therapy. In earlier times it was believed that all human ills were caused by drinking. Later, it was believed that persons drank in order to drift away from their social and economic problems, that if the "problem" was cleared up the alcoholic would be cured. Some students, notably the biochemists, believe that the true alcoholic may differ physically from others; that he possesses a

---

18 "Mental Health in Correctional Institutions," *Proceedings* of the American Correctional Association, 1941, pp. 377-387.

19 Selden Bacon, "Alcoholism: Its Extent, Therapy, and Prevention," *Federal Probation,* Vol. 11, No. 1 (April-June, 1947), pp. 24-32. See also the January, 1958, issue of *The Annals,* Vol. 315, "Understanding the Alcoholic," edited by Selden Bacon.

20 *Alcoholics Anonymous Comes of Age* (New York: Harper, 1957); for a gripping novel based on their work, see Thomas Randall, *The Twelfth Step* (New York: Scribner's, 1957). For a clinical study of alcoholics, see Ben Karpman, *The Hangover* (Springfield, Ill.: Charles C. Thomas, 1957).

unique metabolism that affects his response to alcohol and his ability to handle it. It should at least be obvious to the reader that the problem of chronic drunkenness is a knotty one, to which science offers no clearcut solution. Yet, it is equally obvious that punitive methods have distinctly failed. It is time to rely entirely on the best science has to offer.

The solution offered by the new correctional program includes the development of a number of colonies of the farm, forestry, or work-camp type on the indeterminate sentence basis. Scientific treatment primarily by medical men should be carefully worked out. The victims should have adequate housing, plenty of wholesome food, recreation and relaxation, a work program suited to their needs and potentialities, and medical and psychiatric analysis and treatment. Custody should be provided, but this is not a serious problem with such cases. Inmates should not be released until they are materially helped and then only under careful parole supervision.

Drunken driving is a problem distinctly different from that of chronic alcoholism. It is a serious offense, and measures should be put into effect that will discourage it. But a jail sentence or a fine will not solve the problem. A stiff fine has little meaning to a spendthrift rake or to a man of means; and it may prove financially burdensome to a poor man. A jail sentence may work even more havoc on the family of the offender since his earning capacity is affected by a sojourn in jail. Suitable cures take considerable experimentation. Impounding the driver's car or revoking his license has been attempted in some places, but this works a hardship if the man's car is used in his work. Nevertheless, we are convinced that hundreds of persons can be kept out of jails and prisons for drunken driving if the problem is attacked in a more workmanlike manner.

The Prostitute

Prostitution, like poverty, is one of the oldest social problems. It antedates the dawn of history, and in its present manifestations is more difficult to control than ever before.

As an institution it has been prevalent in this country ever since colonial days. Many prostitutes migrated from Europe along with other colonial settlers. Some were transported along with other criminals. The sermons of ministers, the criminal laws, the writings of travelers, and reports of jail conditions all attest to the existence and extent of prostitution in colonial America. How significant prostitution has been and is as a social institution is debatable. Certainly there is a vast literature on the subject. The Kinsey report seems to be skeptical of the role prostitution plays in the gratification of the sex appetite when it states: "For an activity which contributes no more than this does to the sexual outlet of the male population, it is amazing that it should have been given such widespread consideration."[21]

---

[21] Alfred C. Kinsey et al., Sexual Behavior in the Human Male (Philadelphia: W. B. Saunders Company, 1948), p. 605.

Prostitution flourished during the nineteenth century with the growth of the eastern industrial cities and the rise of western mining towns, in both of which there was a large contingent of unmarried males. The European system of frank and candid regulation of prostitution was rarely adopted in this country, although St. Louis experimented with public regulation from 1870 to 1874. The more general tendency was to herd the prostitutes into segregated areas, known as red-light districts. These were frequently very crude in western frontier towns, where the prostitutes lived in small informal dwellings known as "cribs." There was little provision for medical inspection in these segregated areas, and other vices could flourish unchecked.

About 1912 a wave of puritanical zeal, set off by the investigation of prostitution in Chicago, swept the country. Volunteer committees sprang up in urban cities, gathered information about prostitution, and campaigned against segregated districts. Chicago abolished its red-light district in 1912, and more than 30 cities soon followed, not to mention hundreds of smaller communities. Wiping out the segregated areas simply scattered prostitution, driving it into respectable neighborhoods and increasing streetwalking. The notable increase in the congestion of cities made prostitution easier to hide. The growing popularity of the automobile and the easy accessibility of the public taxi put much sexual vice on wheels.

After stamping out the red-light areas, authorities could not effectively control prostitution. It remained informal, disorganized, and more dangerous to public health than ever. The attempt to put down interstate traffic in prostitutes, through the Mann Act of 1910, accomplished little that was intended by its sponsors and turned out to be the largest single asset in today's blackmail racket.

Organized pandering is one of the most lucrative forms of organized underworld activity. The strictly businesslike organization that dominates the world of professional crime has taken control of prostitution. It is managed on a national or regional basis. The vice ring controls prostitution in most of our leading cities. The approved proprietors of houses, and the prostitutes themselves, are carefully listed and rigorously supervised by the syndicate. The girls are frequently shifted from one city to another, and even more often from one locality in a municipality to another—partly to evade discovery and prosecution, and partly to increase the novelty and appeal of the girls. They are kept under thorough control and discipline and are intimidated to prevent desertion and informing. For the first time in the history of American prostitution, many prostitutes have become prisoners in the profession—literal "white slaves."

Perhaps the only gain that has come from the gradual taking over of prostitution by the vice rings has been the increased introduction of prophylaxis. The syndicate has the sagacity to recognize that a diseased prostitute or an infected customer is poor business. So thrift and greed have been able to accomplish more in the way of social protection and the prevention of disease than the moral indignation of reformers.

Most prostitution today is either controlled by the vice ring or practiced by independent "call girls."[22] The more attractive of streetwalking girls are being picked up in ever larger numbers by the syndicate, subjected to medical treatment, and put to work. In our large cities the old-fashioned "house," dominated by the "madame," has all but disappeared. Prostitution flourishes nowadays in conjunction with taprooms, pool halls, small corner candy stores, roadhouses, and other places where men congregate. Only a few girls are available at a time but these are shifted about by those who control the vice. Tribute is levied by police officials, much of which is in the form of petty graft called a "shakedown" rather than through a businesslike contract between the syndicate and the higher-ups in the city government. This more formal type of levy, however, is prevalent in many places throughout the country.

There are two sharply divided schools of thought on handling the problem of "the world's oldest profession." The moralists and purists attack prostitution head on, without considering the underlying causes of the trouble or the complex aftermath of their campaigns. They close up segregated districts and houses of assignation and throw the streetwalkers in jail. They do nothing to remove the causes of prostitution; they simply spread it where it cannot be effectively controlled or inspected. But they cannot really suppress the practice, so long as the demand and the supply exist. Various forms of subterranean contacts with prospective customers are provided through taxi drivers, bellhops, and others, quite thoroughly removing the business from any salutary public scrutiny. These shortsighted purist campaigns have had the particularly disastrous effect of making prostitution an ideal field for syndicates.

The more practical school argues that we cannot do away with the evil without removing the causes and that such causes cannot be speedily removed. Their program involves segregated districts, public licensing, thorough medical inspection, the protection of prostitutes from exploitation by irresponsible third parties, and appropriate public health education and measures. This program will not necessarily bring a utopia, but it would surely create far more wholesome conditions than now exist. Typical of this more dynamic attitude is that of the United States army which, during World War II at least, regulated and controlled all prostitutes in the vicinity of army camps and took strict prophylactic measures in regard to soldiers who had been exposed.

The reasons why women and girls enter the profession are numerous but the economic and psychological motives lead. Often the economic pressure

---

[22] Harold Greenwald, *The Call Girl: A Social and Psychoanalytic Study* (New York: Ballantine, 1958).

does not take the form of sheer necessity. Many prostitutes, not of the "kept woman" variety, come from the domestic servant class where the ordinary material needs of life are relatively well provided, but life is drab and their vanity makes them desire better clothes and jewelry. Girls employed for long hours at low wages in the poorer classes of stores calculate that they can earn more money with less effort as prostitutes. There are many "call girls" who cater to hotel guests but who, for regular employment, work in many of the swankier women's shops of our large cities. These girls are sophisticated and alluring and sell their services high. In fact, one-hundred-dollar-a-night call girls are not averse to having their photographs in magazines and papers.

There are, of course, many instances in which women, in sheer destitution, turn to prostitution to obtain food, clothing and shelter. The number of girls who go into prostitution because they have been betrayed in their innocence by irresponsible rakes is very small, despite the popular and romantic notion that most prostitutes enter the profession for this reason. Along this line, it is of interest to note that as early as 1882, Charles Loring Brace, the great humanitarian and child-saving worker, scotched this contention by stating: "Public women of this kind are not generally, as is supposed, the victims of deception and wronged by men."[23]

Dr. Ben Reitman, pioneer student of prostitution, lists the following types: juvenile, potential, amateur, young professional, old professional, field workers or streetwalkers, bats or superannuated, gold-diggers or boulevard women, kept women, loose married women, and call girls.[24]

Prostitutes, then, are an extremely diverse lot—like other women. Some are very intelligent, others are feeble-minded. Most of them are relatively attractive, considering the class from which they are drawn; otherwise they would gain few customers in their profession. Certain elementary loyalties and virtues are particularly noticeable among them. Some have liaisons or personal paramours, to whom they are pathetically devoted. They resent any special exploitation of their associates. They are notoriously generous with the money they earn. They are usually loyal and devoted to dependent relatives. The conventional picture of the prostitute, thoroughly mercenary with a heart of stone, is true only of some who have been long in the service and have become thoroughly disillusioned and cynical.

Some of the reasons men patronize prostitutes are cogently reviewed in the Kinsey report: a lack of sexual outlets in other directions; it is substantially cheaper than marrying or supporting a girl; they can forget other responsibilities or worries such as fear of pregnancy.[25]

---

[23] Charles Loring Brace, *Gesta Christi.* (New York: Armstrong, 1882), p. 317.

[24] See Ben Reitman, *The Second Oldest Profession* (New York: Vanguard Press, 1931).

[25] Kinsey *et al., Sexual Behavior in the Human Male,* pp. 606-608.

What is to be done with prostitutes who are sent to jails and to reformatories? Prostitution in many states is a crime. But it is absurd to round up streetwalkers, arrest and send them away for treatment, unless there is some object other than to penalize them for plying a trade that is condoned by enough people to make it profitable. Arrest is not a deterrent, for most prostitutes carry on, no matter how often they are arrested.

A candid and factual work dealing with the question of arresting and penalizing prostitutes was written by Judge John M. Murtagh, chief magistrate of New York City, in collaboration with Sara Harris, the author of *Skid Row, U.S.A.,* both of whom fully understand the web and intrigue of vice and politics surrounding the pathology of prostitution. They contend that it is the public that needs to be oriented to the realities of the problem.[26]

Every effort should be made, of course, to dry up the sources of the profession by attacking the incipient causes. The problem is to discourage young girls approaching adolescence from entering the field. How best can this be done? Certainly sex education in the schools would help. Knowledge concerning the ravages of syphilis and gonorrhea as well as the physiology of pregnancy and childbirth should be honestly imparted without moralizing. Counseling services should be expanded and guidance programs for marginal girls be strengthened. Sympathetic and understanding policewomen in the larger cities should make this phase of their program one of high priority. Some integration of child-guidance clinics and the public schools could be consummated so that those young girls who show signs of infantilism of sex life might be referred to psychiatrists.

Until a program can be inaugurated that will insure the elimination of the panderer and place definite control of the problem in the hands of public welfare agencies, we can expect no adequate solution of the problem of prostitution. But prostitutes should not be sent to penal institutions, no matter how kind the superintendents are or how understanding they may be about the problem. At present, the great majority of the girls and women in our state reformitories are, or were, prostitutes. The money saved by keeping them out of these prisons could well be applied to paying the salaries of trained personnel, including case workers, psychiatrists, and medical physicians who would handle each case on its individual merits.[27]

Sex Offenders

The Homosexual

Many persons are homosexual in their behavior. The Kinsey report states that 37 per cent of American white males engage at some time of their lives

---

[26] *Cast the First Stone* (New York: McGraw-Hill, 1957; Pocket Books Cardinal Edition, 1958, 35¢).

[27] See Mazie F. Rappaport, "The Psychology of the Female Offender," *N.P.P.A. Journal,* Vol. 3, No. 1 (January, 1957), pp. 7-12.

in homosexual activity.[28] The causes of this deviation are complex.[29] Certain homosexuals, through biological factors such as inborn glandular anomalies and defects, may be irresistibly impelled to behave as they do. Others are led into this behavior through mistakes and exaggerations in family relations, faulty sex education, accidental sex experiences, the denial of normal sex experiences, and the like. There appears to be no physical foundation for their homosexual trends, but they cannot help being as they are. Homosexuals, then, have been conditioned by their physical make-up, or by peculiar types of environment and experiences.

The true homosexual is very much misunderstood by laymen and police. The public assumes that he is a degenerate rather than suffering from an affliction contracted through no fault of his own. Though many homosexuals are fine and sensitive characters, they are anathema to society in our culture, and we have no qualms about persecuting them.

Many homosexuals, if they are at all intelligent, manage to work out a satisfactory adjustment with their own kind. All of our large cities have a *demimonde* of such individuals, who have a sort of "free-masonry" of their own and, left to themselves, carry their abnormal proclivities in a way that does no harm to normal society. Those victims of this deviation whose behavior does not grow out of physiological causes, can usually be cured by psychiatric aid or the provision of normal outlets. But homosexuality resting upon a physical basis is usually less likely to be curable, though there have been some notable benefits from glandular therapy, especially if the tendency is detected early enough in the sexual life of the person.

The homosexual may either be a passive individual who assumes the role of the female, regardless of his true sex, or he (or she) may adopt the active role of the pair. In either case, they go to extremes to simulate the roles they are, by their nature or conditioning, to play. Passive male homosexuals, if under no conventional restraints, will dress as women, use rouge, paint their nails and curl their hair, even permitting it to grow long; female active homosexuals will simulate the male by wearing trousers or tailored clothes, swaggering, and otherwise attempting to pose as a male.

The true homosexuals usually found in prisons are there either for soliciting members of their own sex on the streets or for corrupting the morals of a minor. They, together with those who take advantage of their physical aberration, constitute a nuisance to all prison administrators. The average prison warden rarely possesses any scientific knowledge regarding such persons. He generally reflects the attitude of the average individual and tends to be scornful and impatient, if not brutal, in handling them. The

---

[28] Kinsey *et al.*, *Sexual Behavior in the Human Male*, p. 663.

[29] For a review of the theories of the causes of homosexuality, see George A. Silver, "The Homosexual: Challenge to Science," *The Nation*, Vol. 184, No. 21 (May 25, 1957), pp. 451 f. Dr. Silver is Chief, Division of Social Medicine at Montefiore Hospital, New York City. See also: Donald J. West, *The Other Man: A Study of the Social, Legal, and Clinical Aspects of Homosexuality* (New York: Whiteside, Inc., and William Morrow & Co., 1955).

epitome of penal sterility was found by one of the writers in a large western penitentiary where those discovered in homosexual activity were compelled to dress in women's clothes. At the time there were four undergoing this apparent humiliation. They sat by themselves in the mess room, they worked on the pick and shovel gang in their garish dress and were shamed and ignored by prisoners and guards alike. Yet it is possible the warden played directly into their hands since they probably felt rewarded in being permitted actually to assume the sex role that nature had intended for them.

In many other prisons there is a varied group into which are carelessly thrown deviates of all types; any of the wide variety of offenders who might be called "sexy." A stricter classification is more desirable if it is impossible to take homosexuals out of the prison altogether and put them elsewhere. But it is certain that they should *not* be in prison.

A word should be injected here regarding the young boy or youth who capitalizes on his charms and seeks out an adult homosexual in order to sell himself for rewards. This is a wider practice than is usually assumed. When we read of a boy being debased by a homosexual all blame is heaped upon the active participant. The writers know of several cases in which the young boy either disappeared completely from his home to live voluntarily with some homosexual—in one case for several months before he was found, in the meantime he enjoyed the movies, excursions, candy, and other gifts—or solicited older boys who had homosexual tendencies in order to bargain with them in selling favors.

The only answers to homosexuality are: (a) to encourage those who actually desire to change to take psychiatric treatment or (b) to permit them unmolested to seek out their kind to live as they wish, in the free community. Penalizing them relentlessly is a barbaric hangover from a pre-scientific age.

Alarmed by the apparent increase of this type of offense, the Quarter Sessions Courts of Philadelphia inaugurated a streamlined and progressive system of dealing with homosexuals and infantile sexual types. During the summer of 1950 an experiment was undertaken under the direction of Judge James C. Crumlish. The procedure was twofold: first, defendants were given a form to sign that waives a hearing and presentment to the grand jury, indicates the plea, indicates time for pre-trial neuropsychiatric examination, and date of trial: all rights of the defendant are safeguarded. Second, the consideration of the personal file of the defendant which has considerable bearing on the disposition of his case. A breakdown of 119 cases tried during the experimental period shows that of 102 guilty defendants, 64 were placed on probation conditional to taking psychiatric treatment. The advantages of this system indicate a vast saving in time and money, a more intelligent disposition of the cases, and the wider use of the psychiatrist in court cases.

While this system does not answer the many objections of the penologist, it is definitely a step in breaking down the traditional vindictiveness of the criminal court procedure.

Other Types of Sex Offenders

Despite the great amount of publicity given to the subject of sex offenses, those committing such acts constitute a relatively small percentage of criminals. Few are found in prison populations. True, some are executed if their crimes end in murder. But regardless of their number, they do not represent as serious a menace to society as one would gather from the public cry against them, generated by sensational newspaper and magazine articles.

There are, of course, many types of crimes associated with sex. The most serious are associated with rape, particularly forcible rape, or with assaults on young girls or elderly women. But, contrary to public opinion, there are few outright cases of this type. Most of the rape cases deal with statutory rape—sexual intercourse with a female below the "age of consent," usually placed by statute at 16 or 18 years of age. In many such cases the girl has lied about her age, and in a large percentage of the cases the girl has drawn the man into the relationship. For example, a New Jersey man was sentenced to 30 years in prison for "kidnapping" a young girl and living with her in California for over a year. The girl accompanied the older man of her own volition and resented his arrest. But she was below the statutory age and thus he was looked upon as a wanton seducer.

So far as forcible rape is concerned, it has been much overrated. In many cases the female has offered little resistance and, in others, she has "framed" the male. In other cases, the female has reported the man only after he has jilted or abandoned her. Thus what we see is a heading of forcible rape applied to all cases of rape, with or without consent, regardless of age.

Professor E. H. Sutherland has pointed out the discrepancies between the large number of arrests for forcible rape reported by the police to the Federal Bureau of Investigation and the small number of convictions in New York State. For instance, only 18 per cent of rape convictions during the decade 1930-1939 in New York State were for forcible rape. He further shows that of 324 females reported to have been murdered during three random years (1930, 1935, 1940) throughout the entire country, only 17 were reported to involve rape or suspicion of rape. This is an average of 5.7 cases of rape-murder per year in the United States. Of the 324 cases, 102 were victims of their husbands, 37 of fathers or close relatives, and 49 of lovers or suitors. In other words, 60 per cent of the victims were murdered by relatives or by intimate associates.[30]

---

[30] Edwin H. Sutherland, "The Sexual Psychopath Laws," *J. Crim. Law,* Vol. 40, No. 5 (January-February, 1950), pp. 543-554.

A definitive study of some 3,000 sex offenders of all types was made by the Department of Criminal Science of Cambridge University in England. The findings are surprisingly in line with similar studies conducted in this country. Of the 3,000 cases, a large percentage consisted of minor or unindictable cases; 10 to 15 per cent were acquitted; a large number had sex relations with willing or "semi-willing" partners; very few used actual force; many were disturbed individuals who seduced minors or mentally deficient victims because "of their inadequacy or inability to have more adult relationships"; more than a third of the heterosexual offenders were married; a great majority had little education; forty per cent were under thirty years of age; and only seventeen per cent were sexual recidivists.[31]

We are concerned in this section with the sex offender who shows signs of suffering from some emotional difficulty or who is abnormal in his lust for sexual gratification. Such persons go to shocking extremes, and may assault little girls or elderly women. These individuals have passed the borderline of normality and entered the realm of the pathological. Some are obviously infantile in their sex lives; all gain sex satisfaction in an allegedly abnormal manner. All sexual deviates are labeled *degenerate* by the public, although this term has no legal or psychiatric meaning.

Most of the terms used in our penal codes to describe "unnatural" sex offenses have been accumulated from a variety of sources within our own culture. There is considerable confusion in the use of these terms. Homosexuality is not a legal term; but such words as "sodomy," "pederasty," and "buggery" are in general use in the courtroom. Robert von Krafft-Ebing, the distinguished nineteenth-century sexologist, uses the term sodomy to cover: (1) bestiality, the "crime against nature," the "unnatural" coitus with an animal; and (2) pederasty, or buggery, which signifies anal coitus between men.[32]

Aside from the practices mentioned above, there are other deviations for which a person may be arrested and sentenced to prison. They include varieties of sexual sadism and masochism, indecent exposure or exhibitionism, incest, and the various forms of sexual intercourse that fall into categories generally considered abnormal, if not pathological.[33] Yet, as Morris L.

---

[31] Leon Radzinowicz, *Sexual Offenders: A Report of the Cambridge Department of Criminal Science,* English Studies in Criminal Science, Vol. 9, (London: Macmillan, 1957).

[32] Krafft-Ebing states: "I follow the usual terminology in describing bestiality and pederasty under the general term of sodomy. In Genesis xix, whence this word comes, it signifies exclusively the vice of pederasty. Later sodomy was often synonymously used with bestiality." *Psychopathia Sexualis* (Philadelphia: Davis, 1893), p. 404n. For a classification of sexual anomalies, see V. C. Branham, "Perversions, Sexual," *Encyclopedia of Criminology* (New York: Philosophical Library, 1949), pp. 306-314.

[33] See Nathan K. Rickles, *Exhibitionism* (Philadelphia: Lippincott, 1950); S. Kirson Weinberg, *Incest Behavior* (New York: Citadel, 1955); Ben Karpman, *The Sexual Offender and His Offenses* (New York: Julian, 1954); George W. Henry, *All the Sexes* (New York: Rinehart, 1955); James M. Reinhardt, *Sex Perversions and Sex Crimes* (Springfield, Ill.: Thomas, 1957).

Ernst and David Loth have pointed out in their analysis of the Kinsey findings, it would be extremely difficult for a large proportion of married or unmarried couples to avoid breaking the sex laws of their states, since various forms of sexual relationships are practiced widely, especially among the educated classes.[34]

Fornication, for example, is a crime in most states, yet sexual intercourse outside marriage is practiced widely. In a compilation of laws dealing with sex, Robert Veit Sherman demonstrates how many of these laws defy our mores and represent a sharply hypocritical contradiction of the commercialization of sex in this country. Drily he states that he was unable to find a prohibition of solitary, private masturbation.[35]

Distressing as violent sex offenses are, those who commit them are more in need of psychiatric treatment than imprisonment or capital punishment. Those who are sent to prison are given no treatment. But some other type of custodial treatment should be developed for a study of their cases. When they are sent to prison, they are eligible for parole and too frequently that disposition of the case is made. It is known that many of these pathological persons show signs of their affliction at an early age or after they have committed their first offense. Their later behavior is reasonably predictable, yet there is no machinery available by which they can be examined and studied with the view of segregating them before they can commit another and perhaps more horrible sex crime.

A composite picture of the sex offender has been compiled by Dr. Walter Bromberg. He states:

> He is more frequently white than Negro, native-born and a professed member of a religious denomination. He is not apt to be a recidivist, but, if he is, he has a distinctly psychopathic make-up. The sex crimes which necessitate the use of violence fall among the younger offenders, whereas the homosexual or pedophile is more apt to be a man about forty years old with serious social and psychological maladjustment. The older age group have a history of sexual deviation throughout adult life. The sex criminal is not predominantly a mental defective nor is he insane; he is of average intelligence and not a serious user of alcohol. From the standpoint of his total personality and social adjustment in other areas, he is likely to be without serious personality disturbance, aside from his abnormal sexual impulses.[36]

Some sort of psychiatric census might be of material value in controlling sexual deviates before they can commit their revolting crimes. This calls for a closer tie-up between child-guidance clinics and the public schools. After

---

[34] See Morris L. Ernst and David Loth, *American Sexual Behavior and the Kinsey Report* (New York: Greystone, 1948), Chapter 11, "Sexual Behavior and the Law."

[35] Robert Veit Sherman, *Sex and the Statutory Law* (New York: Oceana, 1949). Isabel Drummond in *The Sex Paradox, An Analytical Survey of Sex and the Laws in the U.S. Today* (New York: Putnam's, 1953), p. 121, states that Wyoming has a statute against masturbation.

[36] Dr. Walter Bromberg, *Crime and the Mind* (Philadelphia: Lippincott, Copyright 1948), p. 85; Morris Ploscowe, *Sex and the Law* (Englewood Cliffs, N.J.: Prentice-Hall, 1952).

such cases are detected, some form of police power might be invoked so as to insure expert treatment and custody until the individual is either cured or restrained from committing such acts.

Many deviant individuals are obsessed with *fetishism,* which was so aptly described years ago by Krafft-Ebing. Persons thus afflicted have an insatiable urge to satisfy their erotic sexual appetite by having contact with some part of a body or possession of a member of the opposite sex. It may be hair, eyes, shoulders, feet, or any other portion of the body; or the underthings, stockings, gloves, or hat of some person of the opposite sex. In seeking this abnormal satisfaction, if they are frustrated or detected, their escapade may end in murder. The notorious case of William Heirens, the Chicago youth, back in 1946, is a case in point. In his early years Heirens had been taught to look upon sex as wicked, but he developed a lust for female underclothing as a part of his infantile sexual pattern. His crimes included assault, robbery, and murder, among the latter being the bestial slaying of a small girl.[37] This young deviate was sent to the Illinois penitentiary. As an example of legal confusion and social frustration we submit the sentence of Heirens:

> He was arraigned in the Criminal Court of Cook County September 4, 1946. He pleaded guilty to 30 charges and received sentences on 24 burglaries of one year to life, to run concurrently; on three murders, Natural Life, to run consecutively; one robbery, 1 to 20 years to run consecutively; one burglary, 1 to life to run consecutively; one assault to commit murder, one to 14 years to run consecutively.[38]

New Jersey's Commission on the Habitual Sex Offender points out that the average citizen knows very little about the scope and nature of sex crimes, but, owing to headlines in the nation's press, is oversupplied with misinformation on the subject. Some of the popular fallacies the commission would like to correct are:

> 1. That the sex offender progresses to more serious crimes: Statistics clearly show that progression from minor to major sex crimes is exceptional, rather than the rule.
> 2. That dangerous sex criminals are usually repeaters: Actually, of all serious crimes, only homicide shows a lower record of recidivism.
> 3. That sex offenders are oversexed: Most of those treated have turned out to be physically undersexed.
> 4. That there are "tens of thousands" of homicidal sex fiends abroad in the land: Only an estimated five per cent of convicted sex offenders have committed crimes of violence.

---

[37] For an account of this bizarre case, see Foster Kennedy, Harry R. Hoffman, and William H. Haines, "Psychiatric Study of William Heirens," *J. Crim. Law,* Vol. 38, No. 4 (November-December, 1947), pp. 311-341. For a thorough case history and analysis of this case see Lucy Freeman, *Before I Kill More* (New York: Crown, 1955; also Pocket Books, Inc.).

[38] *Ibid.,* p. 341.

The commission concludes by stating that there is more danger of being murdered by a relative or other intimate associate than by some unknown sex criminal.[39]

A two-year study of 102 male sex offenders in Sing Sing prison was made by Dr. David Abrahamsen of the Governor's Commission on the Problems of the Sex Offender. The study revealed that those studied suffered from some form of mental or emotional disorder—a real or near psychosis or extreme neurosis; that every offender had experienced severe emotional deprivation in childhood from such causes as brutal, neglectful, or over-indulgent parents; and that there was little evidence of "oversexed" be-havior. There was little correlation to intelligence since the majority of the offenders compared favorably with the general population. The 102 of-fenders were classified into the following groups:

> 1. Those psychologically predisposed to violence, with fixed and overt hostility, not treatable by any known methods, likely to commit new attacks.
> 2. Those too withdrawn, emotionally solitary, old, or alcoholic to be amenable to treatment, remaining a danger to society.
> 3. Those who show a prospect of improvement, provided they receive proper psychiatric treatment, preferably in a hospital setting.
> 4. Those who could be paroled and treated in an out-patient clinic.

The report concluded that many sex offenders should be incarcerated in prison so long as they continued to be considered dangerous. It also pointed out the danger of using the term "sexual psychopath" loosely.[40]

If we wish to prevent sex crimes, we must draw up adequate legislation and improve criminal procedure to identify and segregate for treatment those who have committed revolting crimes. More important, we must give attention to the circumstances under which such individuals grow up.

Several states now have laws that attempt to cope with the sex offender. None is completely satisfactory but all are honest attempts to deal more scientifically with the problem. New Jersey passed a law in 1949 that has considerable merit. It directs that any person convicted of "rape, sodomy, incest, lewdness, indecent exposure, uttering or exhibiting obscene litera-ture or pictures, indecent communications to females of any nature what-soever, or carnal abuse" must be committed for mental examination, prior to the disposition of the case, to a special diagnostic facility located at Menlo Park. The staff of this facility is composed of psychiatrists, clinical psy-chologists and social workers. They attempt to determine whether the of-

---

[39] *Time,* Vol. 55, No. 7 (February 13, 1950), p. 85. For a more complete analysis of this commission's report, see *The Welfare Reporter,* New Jersey Department of In-stitutions and Agencies (Trenton, March, 1950), pp. 10 ff.

[40] The report is digested in *Survey,* Vol. 86, No. 4 (April, 1950), p. 200; also "Study of 102 Sex Offenders at Sing Sing," *Federal Probation,* Vol. 14, No. 3 (Sep-tember, 1950), pp. 26-32. This entire issue is devoted to the social problem of sex offenders and their disposition.

fender's conduct was characterized by repetitive-compulsive behavior or other evidence of mental, emotional, or physical aberration. If such evidence is found the clinic may recommend to the court that the person be placed on probation under psychiatric out-patient treatment or be committed to an institution, usually a mental hospital, for special treatment. Over 2,600 such offenders have been processed in this clinic since its opening.[41] The term shall in no event exceed that provided by law for the offense of which the person was convicted.[42]

Here is one of the serious problems of society in dealing with incurable or uncured liabilities. It is alleged we can no more hold the sexual psychopath in duress indefinitely than we can other offenders. Nice points of law are undoubtedly present. It is important that safeguards be set up to protect the individual, but it is vastly more important that the public be protected against the sexual psychopath who is incurable or who has not been adequately treated within an institution.

Emasculation of Sex Offenders

Castration of certain types of sex offenders has been practiced for the past fifteen years in Denmark. It is also practiced in Holland. The Denmark experience is worthy of comment at this point. All confirmed sexual offenders are sent to a special hospital-prison. The institution is rigorous but professionally staffed—all wardens or attendants have had mental hospital training. Each case is thoroughly examined as to his past record, his emotional capacity, intelligence, and social potential. Castration is "voluntary" but it is imposed only on those sex offenders who are infantile in their sex life. It cannot be performed on anyone under the age of twenty-six. Generally speaking, those who are castrated are those who have attacked young children intermittently. It is alleged that this type of person is underdeveloped in his sex life, usually unmarried, middle-aged or older, leads a life of seclusion, and is frankly troubled by his atypical sex urge. It is reported by those who have visited the establishment that no person is castrated who is likely to be treated and cured psychiatrically or who would, in this country, be described as a violent rapist.

The director of this institution is enthusiastic concerning results. Only

---

[41] See: Ralph Brancale and F. Lovell Bixby, "How to Treat Sex Offenders," *The Nation,* Vol. 184, No. 14 (April 6, 1957), pp. 293 f; "What Makes A Criminal?" *Life,* Vol. 43, No. 15 (October 7, 1957), pp. 145 ff.

[42] For an excellent article dealing with the status of sex offenders according to the various states, see: Paul W. Tappan, "The Sexual Psychopath—A Civic-Social Responsibility," *Journal of Social Hygiene* (hereafter, *"J. Soc. Hyg."*), Vol. 35, No. 8 (November, 1949), pp. 354-374. For a critique of such laws, see: Sutherland, "The Sexual Psychopath Laws," pp. 543-554. For other articles dealing with the subject of sex offenders, see: Ben Karpman, "Considerations Bearing on the Problem of Sexual Offenders," *J. Crim. Law,* Vol. 43, No. 1 (May-June, 1952), pp. 13-28; Manfred Guttmacher and Henry Weihofen, "Sex Offenders," *J. Crim. Law,* Vol. 44, No. 2 (July-August, 1953), pp. 153-175.

about two per cent of those castrated relapse to crime, and then they only engage in petty offenses; 75 per cent are happy to be released from their emotional trouble and are leading meaningful lives. A small per cent do resent their treatment. It should be pointed out that castration is not resorted to until the offender has demonstrated, through one or more periods of imprisonment, that he cannot live in a free society without getting involved in sexual trouble.[43]

The difficulty of this program is that it certainly does not prevent potential sex offenders from committing acts repugnant to society. Not until an offender has already committed such acts is he available for this type of treatment. It is extremely doubtful that the United States will ever condone emasculation, even for habitual sex offenders.[44]

The Psychopath and His Treatment

Current literature is full of descriptions of a type of behavior that cannot be explained along conventional lines. In this section we wish to describe the type known as the "constitutionally psychopathic inferior," referred to generally as "C.P.I." Reports of this type of criminal have been appearing in the newspapers and popular magazines for the past decade or so, especially when some brutal murder is committed. The professional journals have written of it for many years, and heated discussions and conflicting opinions have been aired for decades. Psychiatrists as a rule are cautious men, and many refuse to acknowledge the C.P.I., or "psychopathic personality" as a distinct clinical type.[45]

In the recent past, considerable research has been conducted on the psychopathic personality. However, it is not our purpose to review the results of this research or to enter into the conflict that still persists among psychiatrists concerning this dangerous type of person. The many articles and books on the subject may prove valuable to the reader. A perusal of

---

[43] From 1935 to 1956 some 600 offenders have been castrated in Denmark. For an analysis of the effects of this form of therapy, see Louis LeMaire, "Danish Experience Regarding Castration of Sexual Offenders," *J. Crim. Law,* Vol. 47, No. 3 (September-October, 1956), pp. 294-310. LeMaire writes: "From the experience gained in Denmark . . . there can be no doubt regarding the justification of the operation, and therefore it cannot be surprising that we who have had the opportunity of following this form of treatment and observing the results thereof must also feel it incumbent on us to raise the question for discussion in those countries where the scruples which must naturally be associated with the matter still prevail." (p. 310).

[44] For experience with castration in this country, see C. C. Hawke, "Castration and Sex Crime," *Proceedings,* Colorado Crime Conference (Boulder, Colo.: University of Colorado, June, 1950), pp. 63-65.

[45] As early as 1835, J. C. Pritchard referred to an aberration which he called "moral insanity"; in 1891, J. A. L. Koch, a German psychiatrist, labeled the type *"psychopathic personality."* See Hervey Cleckley, *The Mask of Sanity,* 3rd Edition (St. Louis: Mosby, 1955).

these works shows the state of confusion and caution existing in the psychiatric world regarding the psychopathic personality.[46]

Dr. Ben Karpman particularly cautions against glibness, not only by laymen but by psychiatrists as well, in labeling various types of emotional behavior as psychopathic. He maintains that of nearly 25,000 consecutive cases he studied at St. Elizabeth's Hospital in Washington, D.C., only 250 to 275 could be diagnosed as psychopathic personality without psychosis; and of this group, he contends, 85 to 90 per cent were of the "spurious or psychopathoid" type. Here is what Dr. Karpman says about the confused situation regarding this clinical entity: "It is perhaps more likely that in studying 100 consecutive cases diagnosed psychopathic personality, what we get is not an understanding of the patient, but a study of the mind of a psychiatrist, that is, what he means when he makes a diagnosis. . . . In his view, if he [the patient] doesn't quite play ball with others . . . if he spends money too freely, borrowing from others without repaying, indulges in behavior that runs counter to the accepted social code, then he is promptly labeled psychopathic."[47]

Although the term "psychopath" has been used as a catch-all for persons suffering from certain emotional states that at first glance baffle specific classification, certain symptomatic patterns of behavior seem to be emerging that may be regarded as signposts.

First, considerable agreement exists among authorities that many psychopaths are better than average in intelligence and, in some instances, superior. There is agreement also that psychopaths are almost completely amoral, impulsive, and have few, if any, inhibitions.

Second, the psychopathic personality may be found in every walk of life, especially in the professions: in business, in the academic world, in medicine (including psychiatry itself), and in politics.[48] These, of course, represent adult cases. They were obviously undetected as possessing dangerous potentialities in childhood, or they did not commit any acts of violence.

Third, thousands of these persons are given free rein in society either because they have not been detected or because our laws make no provision for their commitment or segregation by psychiatrists, hospitals, or courts.

---

[46] A symposium, "The Psychopathic Individual," *Mental Hygiene,* Vol. 8, No. 1 (January, 1924), pp. 174-201, presents the views of such authorities as William A. White, Loren B. T. Johnson, S. A. Silk, Lucile Dooley, and Ben Karpman. Other early works are: Eugen Kahn, *Psychopathic Personalities* (New Haven: Yale University Press, 1931); Arthur P. Noyes, *Modern Clinical Psychiatry* (Philadelphia: Saunders, 1935); and D. K. Henderson, *Psychopathic States* (New York: Norton, 1939).

[47] Dr. Ben Karpman, "Psychopathy as a Form of Social Parasitism—a Comparative Biological Study," *Journal of Clinical Psychopathology* (hereafter, *"J. Clin. Psychopath."*), Vol. 10, No. 2 (April, 1949), pp. 171-172. See also: William McCord and Joan McCord, *Psychopathy and Delinquency* (New York: Grune & Stratton, 1956) and Harold Palmer, *Psychopathic Personalities* (New York: Philosophical Library, 1957).

[48] Cleckley, *op. cit.,* Chapter 24, "The Psychopath as Psychiatrist."

Many show no signs of psychosis, however, and are thus regarded as sane.

Fourth, the psychopath is not always psychotic, nor does he ordinarily suffer from any distortion of reality—no hallucinations, for example—nor is he necessarily neurotic, although he may manifest neurotic or hysterical symptoms. As a rule he "acts out" rather than forms neurotic symptoms; such acting out usually causes unhappiness for those about him, but this unhappiness causes him little guilt feeling or remorse. His behavior usually leaves a pathetic trail of misery, unhappiness, and concern among his closest friends and relatives, but he remains blithely oblivious.

The question of what causes persons to be afflicted with this emotional pattern is debatable. Some, but not all, psychiatrists are certain that there are accompanying anomalies or stigmata in their physical makeup. Nor is there agreement concerning the parts played by heredity and environment.

Psychopaths usually manifest their asocial and egocentric tendencies in childhood or early youth, and, although they develop physically and intellectually, a specific form of emotional immaturity persists. Dr. M. J. Pescor has the following to say about causation:

> At one time [psychopathic behavior] was considered a hereditary or constitutional condition representing a type of inferiority; hence the name "constitutional psychopathic inferior" was adopted. At the present time the most popular theory is that the psychopath is the product of a poor home environment in which parental rejection plays a major role. . . . However, there is an accumulation of evidence showing that some psychopathic behavior, at least, can be explained on the basis of brain injury, infection, and other organic factors.[49]

Many classifications of the psychopathic personality have been made on the basis of symptomatic behavior. Some of these classifications cross over into other clinical entities, such as the psychoses and the psychoneuroses. Some eminent students of the type list ten or more classes; others narrow them down to three. Dr. Pescor lists three: (1) those with pathologic sexuality, a group including various sexual perverts and homosexuals; (2) those with pathologic emotionality, a motley assortment of abnormal individuals who do not fall conveniently into the psychotic or psychoneurotic classification; and (3) those with asocial or amoral trends, who are the "psychopaths proper" and may be found in our prisons and correctional institutions.[50]

Regardless of whether heredity or environment plays the more important role in the etiology of criminal psychopathy, or whether three or more types are distinguishable, we are concerned here with the fact that many of these persons commit vicious crimes while they are still children. Some young killers may be definitely psychopathic. But we must be careful not to generalize. Merely because they cannot be adjudged insane does not mean that

---

49 "Abnormal Personality Types among Offenders," *Federal Probation,* Vol. 12, No. 2 (June, 1948), pp. 3-8.

50 *Ibid.,* p. 7.

they should be executed or imprisoned. Many psychopathic personalities are in our prisons and reformatories simply because our legal machinery has thus far done nothing to identify this group. Courts can do nothing more than to send them to prisons for long terms, pending the establishment of specialized institutions for their reception and treatment.

The C.P.I. presents two serious problems to society. The first is the problem of early detection. Even if a child-guidance clinic diagnoses a child or adolescent as a psychopathic personality, no provision is made for his segregation from society. Nor is the clinic compelled to advise school authorities of the diagnosis. In our great desire to protect the child as well as his distraught parents, society is left vulnerable against a potential danger at any time during the natural life of such a person. A physician who examines a client and finds him suffering from some infectious disease must notify the authorities, and they will promptly isolate and segregate the menace. But the person who is socially or morally sick is diagnosed by the psychiatrist or clinic with no police power to demand that he be segregated.

The other problem concerns what is to be done after the psychopath has committed a violent crime—or any crime, for that matter. Again, no specialized institution or treatment is available. As we stated above, upon conviction the culprit is dealt with in the same manner as any other type of criminal. Yet he is obviously not responsible for his affliction or his acts.

It is generally agreed that the true psychopath presents a poor prognosis. Dr. Pescor states, however, that on the basis of F.B.I. reports, 63 per cent of those labeled psychopaths and released from the Federal Medical Center at Springfield, Missouri, were still out of trouble three to five years after release. Dr. Pescor attributes this to the aging process.[51]

Some experimental work with psychopaths, involving numerous therapeutic techniques, has been carried on in a few of our correctional institutions.[52] Within the past few years a new facility has been developed by the Department of Corrections in California. An establishment located at Vacaville (formerly at Terminal Island) has been developed and staffed for the treatment of various types of offenders who cannot respond to traditional correctional methods. The program is psychiatrically oriented and some 700 to 1,100 patients are actively participating. These involve sex offenders, psychotics, "psychotics in remission," geriatrics, Youth Authority cases and patients with sociopathic disturbances. Occupational therapy is employed adroitly in a similar manner as group interaction. After-treatment is provided by an outpatient clinic, located at Los Angeles, which the paroled

---

[51] Hulsey Cason and M. J. Pescor, "A Statistical Study of 500 Psychopathic Prisoners," *Public Health Reports*, Vol. 61, No. 16 (April 19, 1946), pp. 557-574.

[52] For a review of these techniques, see Robert M. Lindner, "Therapy," *Encyclopedia of Criminology* (New York: Philosophical Library, 1949), pp. 490-497. See also Edward M. Glaser and Donald D. Chiles, "An Experiment in the Treatment of Youthful Habitual Offenders at the Federal Reformatory, Chillicothe, Ohio," *J. Clin. Psychopath.*, Vol. 9, No. 3 (July, 1948), pp. 376-425.

patient is required to visit. Here further orientation is a part of the over-all program of resocialization. While this program is not exclusively for "psychopaths," it does represent a progressive development in the treatment of various types of atypical offenders, among whom are the C.P.I.'s.[53]

A specialized institution for sex offenders; first of its kind in this country, at Atascadero, California. Opened in 1954.

We have referred to the work done at the Danish hospital-prison in connection with the castration of immature sex offenders. It is located at Herstedvester, a suburb of Copenhagen. This enlightened institution is doing notable work with the psychopath. The Danish law, unfettered by the M'Naghten rules that deal with the question of insanity in the court, makes it possible to treat various types of emotionally disturbed criminals in a variety of ways. Some are sent to prison, some to mental hospitals,

---

[53] See Nathaniel Showstack, "Preliminary Report on the Psychiatric Treatment of Prisoners at the California Medical Facility, San Pedro, California," *Amer. J. Psychiatry*, Vol. 112, No. 10 (April, 1956), pp. 821-824. This institution was first opened at Terminal Island in 1950; in 1955 it was opened to new quarters at Vacaville. In addition California operates under the Department of Mental Health a large hospital at Atascadero for sex offenders and the criminal insane. This institution is the only one of its kind in this country and, like the Vacaville establishment, is psychiatrically oriented.

and the psychopaths to the hospital-prison. These individuals are sentenced indeterminately, although each prisoner has a trustee appointed to look after his welfare and, if it seems necessary, to petition for his release. In this manner no man is lost or forgotten by the court.

The psychopath institution is walled, but its interior is more like a group of hospital wards than a prison. No block is more than two stories; there are flowers everywhere. Each inmate generally has his own room, although there are small dormitories that house three or four men. Violent cases are handled as in any mental hospital with professional dignity. Every staff member has had mental hospital experience.

There is a lively recreational program with clubs, dramatics, sports, and the like. The inmates work and are paid small wages. Upon admission each patient is thoroughly examined, both physically and psychiatrically. The average inmate looks upon his incarceration as "too good to be true." But that is merely his initial reaction. After three or four months he begins to realize that he has been sentenced on an indeterminate basis and this begins to make him uneasy. As a result a violent or sullen reaction sets in. The director is convinced that this is the psychological moment to begin treatment. Staff members are trained to spot these initial moments of emotional tension and to call for the psychiatrist at once. Commonsense psychotherapy is applied. The patient is apprized of the reason for his uneasiness and of his emotional difficulty to conform to society.

Slowly the patient begins to accept the possibility of a "non-egocentric" universe. In due time he may be paroled. No deep psychotherapy is administered; the "cure" comes through individual, "straight from the shoulder" talks with the superintendent, who claims that approximately 50 per cent of the true psychopaths may be salvaged for society. Their cases are followed in society by welfare workers who have become acquainted with them while at the hospital-prison, and the superintendent places great confidence in their work. Their value emerges from the fact that they are drawn from the working classes, as are most of the inmates. The director states that his success is not based on "curing" the psychopaths but in transforming them into "nice psychopaths" who are capable of adapting themselves to ordinary life.

The work being carried on at this institution is known all over Europe and is being talked about enthusiastically. Many British, Dutch, and Belgian penologists have visited the hospital-prison and all return with an enthusiastic appraisal of the atmosphere and treatment.[54]

---

[54] The above information comes from conversations with persons who have visited it, and the following articles: Stephen Taylor, "The Psychopaths in Our Midst," *The Lancet* (British Medical Journal), Vol. 1 (January, 1949), p. 32; G. K. Stürup, *et al.*, *Danish Psychiatry* (Copenhagen: Schenberbske Forlag, 1948); Stürup, "Treatment of Criminal Psychopaths," Report on the Eighth Congress of Scandinavian Psychiatrists; H. Wulff and K. Sand, *Acta chir. scand.*, 92, 470, 1945. Also, G. K. Stürup, "Psychiatric Treatment of Offenders in Denmark," *Prison Journal,* Vol. 33, No. 2 (October, 1953), pp. 2 ff.

A similar institution has been in operation since 1954 at Utrecht, Holland, under the direction of Dr. Pieter Baan and known as the Van der Hoevan clinic. This institution, too, is psychiatrically oriented.[55]

The Defective Delinquent and his Treatment

The defective delinquent has long been regarded as an individual calling for specialized treatment. But it has been only within recent years that special institutions have been set apart for their reception and treatment. Even now only four states have such establishments to care for the troublesome mentally retarded delinquents: New York, at Napanoch and Woodbourne; Pennsylvania, at Huntingdon; Massachusetts, at the Bridgewater State Farm and Maryland, at the Patuxent facility, opened in 1955. In all other states, and even in those mentioned above, many defective delinquents and criminals may be found in reform schools and adult prisons.

No definition of a defective delinquent will satisfy everyone working with this group. For instance, a mentally retarded young offender who, by accident or peculiar force of circumstance, gets into trouble with the law may not be labeled a defective delinquent. On the other hand, there are those who may not commit a serious offense, but possess all the attributes of potentially dangerous criminals.

Since there are so few institutions for this type of offender we shall describe the work being done with this group in the Pennsylvania Institution for Defective Delinquents at Huntingdon. The basic assumption of the Huntingdon institution is that this type of offender is "capable of profiting from a rehabilitative program aimed at the development of habits of industry and obedience" provided enough time is afforded and the program is geared to the individual. The process is painfully slow but not hopeless. Each inmate is subjected to planned progression from supervised activity wherein little self-responsibility and direction are entailed, to activity with a minimum of supervision and direction wherein all objectives are capable of expression. The inmate moves along this program in accordance with demonstrated proficiency and readiness for increased responsibility. Traditional objectives are stressed since the inmates have never learned them in free society. These objectives include the formation of regular habits of work, creation of a spirit of cooperation in work and play, and training in specific situations involving self-reliance, dependability, honesty, and trustworthiness.

When the inmate is received from the court he passes through the diagnostic clinic and is given the usual tests. After a 30 day interval during which the inmate is in quarantine, he is absorbed into the population and begins the program developed through the findings and recommendations

---

[55] For details see: Derrick Sington, "Redeeming the Murderer," *The Nation*, Vol. 184, No. 6 (February 9, 1957), pp. 117 f; page 587, *infra;* Giles Playfair and Derrick Sington, *The Offenders: The Case against Legal Vengeance* (New York: Simon & Schuster, 1957), p. 312 fn 13.

of the clinic. At intervals thereafter he is reclassified by the clinic's staff, which is guided by reports on the inmate's progress.

Classification includes the following aspects: custody, guidance, housing, school and work assignments, discipline, physical and mental health, the determination of date of parole, and planning for adjustment.

The time element in an institution for defective delinquents is naturally very important. All inmates are sent by the court on an indeterminate basis. The establishment, therefore, has the grave responsibility of preparing the inmate for release to society. Since the defective learns slowly, the time spent in the institution varies, not only with his offense, but with the capacity to learn and absorb the various elements and phases of the program. Length of time spent by inmates at the Huntingdon institution ranges from 20 to 120 months, the average being about 58 months.

Aside from what trade training these inmates can absorb, the following phases of the program are vitally important in this type of institution: (1) Personal hygiene and grooming (it is apparent that most boys are badly in need of instruction in proper methods of bathing and general principles of personal cleanliness, such as care of the hair, teeth, hands, etc.); (2) Proper methods of eating (quite frequently a defective will devour food in a slovenly, crude manner); (3) Improved appearance through care of clothing and proper methods of dressing (a majority of defectives are totally unaware of the importance of personal appearance and, as a result, their slovenly appearance enhances their defectiveness. Sewing on buttons, making slight mends, proper ways of handling clothing, shining shoes, etc., are important); (4) Methods of keeping clean and tidy quarters; (5) Attempts to improve posture through military drill and exercise; (6) General rules of deportment, good manners, and interpretation of institution rules.

Release from an institution for defective delinquents sounds incongruous, but society has not yet been ready to adopt a policy of segregating for life those persons who, by reason of defective mentality or other incapacity, cannot live in a free society without getting into trouble. Until the time comes when an indeterminate sentence actually means indeterminate, institutional administrators must release those who demonstrate good work habits, who progress in their training program, and who can submit a post-institutional plan that has meaning. Let us set down here the procedure that is followed at the Pennsylvania institution incidental to release:

> The training program aims to return those individuals to their homes who can and do learn enough to behave properly and work regularly under continuing supervision in free society.
>
> Readiness for parole is determined by careful study and observation of the progress shown by each individual member of the population. The individual has this review of his institutional record at stated intervals. The classification clinic, acting for the Board of Trustees, conducts these personal interviews and evaluates the progress made.

New Institution for Defective Delinquents, Dallas, Pennsylvania. (Courtesy Pennsylvania Department of Correction.)

When the parole factors . . . so warrant, the Board initiates parole planning (job, sponsor, and home). Such planning is carried on in co-operation with the Court probation officers. When a satisfactory plan is evolved, the Board originates specific recommendations for release, forwarding such recommendations (via the Department of Welfare) to the Court of original commitment for final disposition. The final decision to release for another community test is always the exclusive authority of the Court of proper jurisdiction.[56]

No institution of this type can ever hope to improve the mentality of the defective delinquent. But much improvement can be made with most of those sent to this type of school. One advantage of a specialized institution is that there are few of the disrupting influences that constantly occur in a prison or reform school that oblige normally intelligent boys to mix with the defective types. Another advantage is that the program can be geared to low intelligence. Patience, understanding, and time seem to be the salient features of a program for defective delinquents.

Social Pathology and Crime: A Summary View

Whatever we may think of the social value of the present system of punitive criminal law that largely prevails in this country, it is clearly of least value in dealing with the offenses and offenders presently classed as criminal in the eyes of the law, but which must in any realistic appraisal be considered and handled as illnesses—social pathologies. Presumably a philosophy of punishing antisocial behavior is based on the premise that a rational person will be deterred from repeating or initiating such behavior by the knowledge that he will suffer for it. Predicated as it thus is upon the rationality of the citizen, the correctional system breaks down completely when confronted by individuals who, according to the overwhelming weight of clinical evidence, cannot avoid behavior that offends society, no matter how clearly they may see the consequences to themselves.

Diverse though the catalogue of pathological types may be, the story in each case is one of legislation and administration woefully outdated by scientific knowledge. Bookings and jailings for vagrancy keep thousands on a regular shuttle between the streets and the lockups, overtaxing to no avail the inadequate local facilities. Narcotics laws are doing for addicts what Prohibition did for alcoholics—worse than nothing. Alcoholism today is somewhat better understood by the public, but not much, and law enforcement is still prescientific on the whole. Prostitution, with the active or tacit support of large segments of the "respectable" public, flourishes in private residences through the activities of call-girls or highly organized

---

[56] From material supplied the writers by the management of the institution at Huntingdon, Pennsylvania, also from its mimeographed publication, "Coordinated Unanimous Staff Opinion on the Objectives and Goals of a Training Program for Defective Delinquents," March 14, 1956.

syndicates, having been rooted out of the old fashioned red-light districts in a series of drives more noted for high-minded energy than insight. Sex offenders, a highly diverse group, are lumped together and largely ignored by legislation that seems to reflect the opinion that willpower reinforced by heavy punishment is the best curb on such behavior. Isolated programs, here and abroad, are showing marked success in dealing with difficult types such as the pathologically amoral person, and the mentally defective juvenile. These programs enjoy the nearly universal enthusiasm of dedicated criminologists and other social scientists, but have not yet received the attention they deserve from jurists, lawmakers, and administrators.

A wider recognition that there are persons legally at liberty whom present laws cannot deter, control, or rehabilitate, may constitute the "breakthrough" so urgently needed in the field of criminology, if it is to be a usefully applied social science.

THEORIES AND FACTORS IN CAUSATION OF CRIMINAL BEHAVIOR

6 ————————————————

The Eternal Quest
for the Causes of Crime

No physical abnormality, no degree or type of insanity, no extent of mental retardation, no extremity of poor health, no degree of physical deprivation, no extreme of poverty, no filth of slum life, no lack of recreation, no stimulation of press, movie, radio or television, no hysteria or crime wave, no family discord or broken home, will surely and without exception produce crime. A crime is committed only when a peculiar combination of personal and social factors comes into juxtaposition with an utterly unique physical structure of a human being, to create a specified crime situation. *And, viewed in a merely external fashion, the same* apparent *concatenation of factors might not produce a crime the next time they merge simply because that precise* sameness *can never absolutely recur. In the particular situation where the crime is committed, it is the inevitable outcome of* all *the elements in the situation —as inevitable an outcome as any physical occurrence could possibly be. But external and apparent similarities and repetitions in social situations may not be actual and complete repetitions. (From* New Horizons in Criminology, First Edition, *Englewood Cliffs, N.J.: Prentice-Hall, Inc., 1943.)*

THE AUTHORS

The Dilemma of Causation

The two phases of the crime problem that elicit more discussion, initiate more investigation and research, and call for more community action are the causes of crime and the prevention of crime. Every individual has his own explanation for crime and his own ideas of prevention. Millions of words, making up hundreds, even thousands, of books, monographs, articles, and reports by hundreds of scholars, moralists, reformers, journalists, legislators, and jurists have been published during the past two centuries on the subjects of causation and prevention. Community forums, luncheon clubs, radio and television broadcasts, college and university discussion groups, as well as regional and national conferences, have debated these subjects on innumerable occasions.

What causes a specific individual to break a taboo, a social sanction, or a law has ever been an enigma to society. Crime, as well as immorality, is the backwash of a culture. As Professor Frank Tannenbaum has eloquently expressed it:

> Crime is eternal—as eternal as society. So far as we know, human fallibility has manifested itself in all types and forms of human organization. Everywhere some human beings have fallen outside the pattern of permitted conduct. It is best to face the fact that crime cannot be abolished except in a non-existent utopia. Weakness, anger, greed, jealousy—some form of human aberration—has come to the surface everywhere, and human sanctions have vainly beaten against the irrational, the misguided, impulsive, and ill-conditioned. For reasons too subtle and too complex to understand, the ordinary pressures and expectancies that pattern the individual's conduct into conformity break down in given instances. They have always done so: they always will. No way of drawing the scheme of the good life has yet been discovered which will fulfill the needs of all human beings at all times.
>
> Crime is therefore an ever-present condition, even as sickness, disease, and death. It is as perennial as spring and as recurrent as winter. The more complex society becomes, the more difficult it is for the individual and the more frequent the human failures. Multiplication of laws and of sanctions for their observance merely increases the evil. Habituation becomes more difficult in a complex society, and the inner strains grow more obvious.[1]

Beyond Professor Tannenbaum's contention that crime is inevitable in society, the Italian criminologist, Giorgio Florita, is even convinced that, like sin, crime is normal in society and it is our sanctions and laws, made by man, that are abnormal.[2] This fascinating, but not necessarily novel, thesis (Durkheim, the French sociologist, held a similar view) cannot be developed here since it is unrealistic in our present stage of social control.

It has been popular to attribute criminal behavior to one cause or factor, or to one set of factors. It is equally popular, especially among the "experts," to construct a "system" or "frame of reference" that will explain delinquency and crime, suicide, desertion and divorce, or any other form of human pathology. Students in the behavioral sciences are constantly on the alert, hoping to discover the "open sesame" to the riddle of human misconduct. Yet this riddle of crime causation remains to bedevil society. No

---

[1] "Foreword," to *New Horizons in Criminology*, 1st. Edition (1943), p. v.

[2] Giorgio Florita, "Enquiry into the Causes of Crime," *J. Crim. Law*, Vol. 44, No. 1 (May-June, 1953), pp. 1-16. See Walter A. Lunden, *Pioneers in Criminology— Emile Durkheim (1858-1917) J. Crim. Law*, Vol. 49, No. 1, (May-June, 1958), pp. 2-9.

unilateral theory, however profound, whether it be nurtured and expounded by biologists, psychologists, psychiatrists, or sociologists, can ever hope to answer the question of the *totality* of criminal behavior.

The traditional approach to causation in this country has been the socio-economic. Yet many countries of the world, notably those on the European continent and in Latin America, have insisted that biological or constitutional factors are more influential in eliciting antisocial behavior. The geographical determinists have espoused weather, the seasons, topography, humidity, and the like to explain why people behave as they do. The psychiatrists, the sociologists, the endocrinologists, the psychologists and the religionists, as well as the racists and the eugenists, have all attempted to give us the answer. The problem is indeed complex.

It is incumbent on the writers of a textbook on crime to review some of the conventional theories of criminal behavior. In this section we are only following the familiar pattern of setting down the most discussed theories or hypotheses that attempt to explain aberrant behavior that is criminal in degree.

The result may disillusion those who would like to have a "packaged theory." The writers of this book have no such theory to advance, but will make whatever attempt to synthesize various theories as these may justify.

In treating the causes of crime, it is necessary to bear in mind just what we mean in this connection by "causes." As set forth in the epigraph to this chapter, we can never be sure that a given set of factors will always produce a crime or a delinquent act. As a result, some writers on criminal behavior state that it is futile to discuss causes at all. This is as untenable as the position of others who write assuredly about causes, as if they were discussing the metered energies in a physical experiment.

As a first step to a better understanding of crime—and its causes—it is well to cease discussing crimes and criminals in any general sense and concentrate solely on individual crime situations and individual criminals. We must build up our knowledge of crime and criminals by studying particular crimes and individual criminals. But such study will have little permanent value unless we discover from it the personal and social situations *most likely* to produce crime or most favorable to crime. Unless we find what these conditions are, we can do little to reduce and prevent crime. Investigation of an individual prison inmate may help to rehabilitate him, but it will be of no general social importance unless it helps to tell us what personal and social factors encourage antisocial behavior.

Without contending that they will inevitably cause crime, it is apparent that certain conditions are more favorable to crime than others. Bad heredity, physical defect, mental imbalance, mental defect, emotional insecurity, a slum environment, poor education, criminal associates, extreme poverty, an environmental stimulation to crime are obviously more favorable to crime than their opposites. It is true that any or all of these unfavorable conditions will not inevitably drive a given person to commit a crime under all circumstances. Conversely, any one or all of the favorable conditions listed above may not be gilt-edge insurance against a person committing a crime. Hidden factors that may tip the scale either way can never be eliminated from specific situations by all the theory in the world.

Prescientific Theories

The oldest theory advanced to explain criminal conduct was "diabolical possession and instigation." Crime was the result of a person succumbing to the blandishments of the evil spirits. This view flourished in primitive, oriental, and medieval societies. Since evil spirits infested the person and had to be driven out, the conventional notion of primitive punishment was either to exorcise the evil spirits or to get rid of the one possessed by death or exile—"social death." In part, this doctrine was based on the concept of protecting the community or family group against further outrages by the offending individual, but far more important was the belief in the necessity and the desirability of placating the gods.

An excellent example of demonism as a theory was the prosecution, in 1692, of witches in Massachusetts. These harmless women were thought to be possessed by the Devil, so they had to be eliminated. Marion L. Starkey, in a carefully documented book, *The Devil In Massachusetts*,[3] describes how this tragic episode in American colonial history was precipitated. Betty Parris, nine-year-old daughter of the Rev. Samuel Parris, and Abigail Williams, her twelve-year-old cousin, became "spiritually sick" from Puritan prudery and from "the inhuman strain of coping with an adult world which had been arranged without understanding of the needs and capacities of children." Learning the art of voodoo from an old Negro slave servant, they maliciously involved an increasing circle of harmless persons of the community until the entire colony was enmeshed in an orgy of religious frenzy that remains today a blot on the Bay State's history.

The belief in demonism still persists in many cultures and, to a great degree, in our own traditions. Many still believe, for example, that willful persons are "thinking on the Devil's side" and must be chastened. The actual belief in witches and demons is practically extinct in most sections of this country, but there are still small culture pockets where it is prevalent. Hexing may be found, even today, among the "Pennsylvania Dutch." The belief in charms to ward off evil spirits is still current among these people and among groups in the more isolated parts of the South.

Many of those who believe in these relics of a bygone era of superstition are definitely pathologic. We must not forget, however, that the belief in witches, evil spirits, and the efficacy of charms is deeply rooted in the

---

[3] New York: Knopf, 1949.

culture of many sects; within such a framework, the individual may function in a perfectly "normal" way.

The belief in demonology as a cause of crime is waning so rapidly that we need not discuss it further. It is important, however, to keep it in mind, since it forms the basis of many subsequent human aberrations.

Freedom of the Will

The concept of "freedom of the will" has had far-reaching effects in Western theology and education. This thesis states that since man is free to do as he pleases or to make his own choices, it is of prime importance to prevail upon him, through teaching or by threats and rewards, to conform to the sanctions of society. By precept and example, he must be taught to emulate the good and true and to shun those who would tempt him to violate the prevailing morality and legal code. The scheme is not consistent throughout, for various rationalizations are manifested by courts, juries, and moral groups when more privileged persons commit crimes against the social good.

It is obvious that from the maintenance of this doctrine of free will many fictions are created and impossible virtues and questionable hero-worship are encouraged. The individual is reminded constantly of his responsibility to others, so that he will develop a conscience to help him decide in the socially approved manner. Discipline of the will is the natural purpose of all these injunctions. One who violates the law or commits a sin has apparently not "disciplined" himself to behave in the approved manner; so society must "teach him a lesson."

This, as we shall see later, is the philosophy behind our traditional criminal code and it is deeply rooted in our social fabric. Certain concessions have been made by law from time to time—to children in the early "age of indiscretion," and to others who are irresponsible and come within the doctrine of limitation, particularly the insane, the defective and, more recently, the chronic alcoholic.

With the development of the biological sciences and the growth of new philosophies regarding life and social responsibility, other concepts of crimes and criminals developed, which somewhat mitigated the doctrine of free will. Notably, there was a shift from theories of revenge, retribution, expiation, and deterrence, to a theory of reformation of the offender with its ultimate aim the *protection of society*.[4] The emphasis was thus no longer focused on the crime but rather on the offender. Modern correction, as we shall see later, is concerned with the individual offender and his "treatment."

---

[4] Note, however, Nathan Leopold's tacit acceptance of the idea of expiation in the statement quoted above (p. 60), in which he considers whether his punishment was "sufficient" to his crime.

The conflict between the advocates of free will and the determinists will probably never be resolved. It represents a challenging rhetorical exercise, but few persons are swayed from their original convictions. The reader is referred to an excellent summary of the two points of view by Professor Robert G. Caldwell. In his concluding statement he writes:

> Now the fact is that science can neither prove nor disprove that there is freedom of will. All that science thus far has been able to show is that the area in which determinism operates is greater than it was formerly believed to be. Any statement beyond this carries one out of the realm of science into the field of speculation. Nevertheless, it seems safe to conclude that the area of the unknown in human behavior is still so great that, as far as we now know, there may well be an element of free will in every human act. . . . Even the determinist . . . agrees that the criminal is responsible at least in the sense that he has in him the causes that produce criminality.[5]

Popular Explanations

We still find a number of "crackpot" theories to explain crime. Each has its adherents among superficially informed but well-meaning people. Reformers, moralists, pulpiteers, and civic leaders all advance their favorite themes to explain criminal behavior. News articles reporting speeches by thoughtful people give these causes of crime: (1) we have forgotten God; (2) the family is breaking down, and children no longer respect their parents; (3) the movies, dance-halls, radio and television programs, and the comics are responsible; (4) we lack moral discipline or we have strayed from the ideals of the Founding Fathers. Somewhat facetiously, Professor Jeremiah P. Shalloo has listed a number of pet theories, all of which have their adherents. It is little wonder that we are confused:

> Crime and delinquency have been and are currently being explained by: the exploitation of the workers, lack of education, inadequate recreational facilities, defective glandular functioning, biological inferiority, police corruption, neglect in religious training, psychometric deficiency, emotional instability, frustration of the fundamental satisfaction drives, adult insufficiency, broken homes, lack of love, poverty, alcohol, narcotics, lack of intelligent parental control, the persistence of a frontier psychology, the doctrine of easy money, an unequal distribution of wealth and income, defective moral and social conditioning, exhausted nervous systems, focal

---

[5] Robert G. Caldwell, *Criminology* (New York: Ronald, 1956), p. 401. Copyright 1956 The Ronald Press Company. See also: a provocative article dealing with "indeterminism vs. determinism," by a professor of criminal law in Norway, Jos Andenaes, "Determinism in Criminal Law," *J. Crim. Law,* Vol. 47, No. 4 (November-December, 1956), pp. 406-413; Donald Taft's cogent discussion of the subject in his *Criminology,* 3rd Edition (New York: Macmillan, 1956), p. 343-348; Arthur Koestler, *Reflections on Hanging* (New York: Macmillan, 1957), Chapter 6, "Free Will and Determinism." Koestler accepts determinism. For an analysis of the question by a strong advocate of free will, see John Howard Coogan, "Free Will and the Academic Criminologist," *Federal Probation,* Vol. 20, No. 2 (June, 1956) pp. 48-54.

infections, temporary insanity, social inadequacy, just plain stubbornness, incorrigibility and perverseness, and lastly, the modern doctrine of individual liberty . . .

He then proceeds to enumerate the many solutions that have been and are being offered to cope with the problem:

The solution . . . is offered in terms of more drastic punishment, larger penitentiaries, military training in reformatories, the development of the family system in juvenile institutions, improvement of the placing-out in foster homes, vocational guidance, child-guidance clinics, visiting teachers, co-ordinating councils, penal colonies, immigration restriction, slum clearance, prohibition, abolition of gangster pictures, radio censorship, appointment of judges, public defenders, classification, boys' clubs, sex instruction, mothers' assistance, abolition of parole, and a return to the tested truths of the founding fathers.[6]

The list of causes and the proposed remedies presented above does not "exhaust the array that has been suggested or is now being described, analyzed, delineated, suggested, supported, or vigorously maintained by anthropologists, psychologists, the clergy, social workers, statisticians, economists, sociologists, legislators, and editorial writers."[7]

Personal opinions are glibly set forth as scientific facts by men and women who have worked in the crime field for many years and what they say is usually accepted without criticism. As Professor Shalloo says, ". . . depending on the specific orientation of the writer (or speaker), each is fully convinced that *his* particular brand of etiological explanation is correct and *his* particular solution would contribute immeasurably toward ending this highly important problem."[8]

In succeeding chapters we shall assay the thinking, the claims, and conclusions of many who have spent their lives in attempting to solve the riddle of crime and delinquency. It is important that this be done; no one working in the field should be ignorant of the findings of science in its attempts to understand human behavior.

---

[6] Jeremiah P. Shalloo, "Probation versus Parole," *Federal Probation,* Vol. 4, No. 2 (May, 1940), p. 28.

[7] *Ibid.*

[8] *Ibid.*

The Constitutional

School of Criminology

We have exhaustively compared, with regard to many physical characters, different kinds of criminals with each other, and criminals, as a class, with the law-abiding public. . . . Our results nowhere confirm the evidence of a physical criminal type, nor justify the allegation of criminal anthropologists. They challenge their evidence at almost every point. In fact, both with regard to measurements and the presence of physical anomalies in criminals, our statistics present a startling conformity with similar statistics of the law-abiding class. Our inevitable conclusion must be that there is no such thing as a physical criminal type. (From "The English Convict," Federal Probation, *Vol. 19, No. 4, December, 1955.)*

CHARLES GORING

Its Emergence through Phrenology

Biology was the first of the sciences to deal with *Homo*

sap   AUTHOR
ple   it would answer their per-
cau   Barnes, Harry Elmer & Teeters, Negley K.
tio   TITLE

dis   FEB 2 '62   New Horizons in Criminology
st   CATALOG   3rd edition
G   O + R
Q   BIP   REF.C   (If for RESERVE state Quarter:)
o
       CBI   PUBLISHER & PLACE
1   PUB. WKLY   Prentice-Hall, Inc.
                        EDITION & DATE   VOLUMES   PRICE
       LC   1959 $7.50
                        BIBLIOGRAPHIC SOURCE   FUND
       REQUESTED BY
                    Florence Jones   APPROVED BY   DATE REQUESTED
       CHECK BOX IF YOU WISH BOOK HELD FOR YOU ☐

While phrenology today is not too far removed from pure quackery, it was regarded as highly respectable a century or more ago. The great sociologist, Auguste Comte, was influenced by it, as was also the psychologist, William James. Wrote Professor James in his *Principles of Psychology:* "A hook nose and a firm jaw are signs of practical energy . . . a prominent eye is a sign of power over language and a bull-neck is a sign of sensuality."[2]

As a young medical student, Gall noticed that some of his fellows with pronounced characteristics had certain head configurations. He asked himself why people had ". . . such different faces and such different natures; why one was deceitful, another frank, a third virtuous?" In attempting to answer these questions, he made it a point of his life to examine every head he could find. He haunted medical laboratories, he visited prisons and lunatic asylums; his fingers fairly "itched" to measure the bumps and inequalities of the skulls he found. He thought he discerned a relationship between head "knobs" and certain propensities and character traits, to which he gave fancy names. In this manner phrenology launched itself upon a world eagerly waiting to receive it.[3]

A number of phrenological studies were made by prison administrators during the heyday of the fad. M. B. Sansom's *Rationale of Crime* (1846), a book avowedly based on phrenology, was prefaced by Eliza Farnham, Superintendent at the time of the women's division of Sing Sing prison in New York.

The inability of phrenology to maintain itself against its critics is due to the fact that it did not recognize that no single portion of the brain can create that complexity which we call human behavior. As soon as psychology came into wide use—even though its early phase, known as "faculty" psychology, was an attempt to associate traits with physical structure—phrenology had to go, slow though its total eclipse was. No explanation of human behavior can possibly be as simple as the tenets of this fantastic pseudo-science. It finally succumbed and is used today only for exploiting the gullible who crave oversimplified explanations of character.

Cesare Lombroso and the Positive School of Criminology

The putative father of the constitutional approach to criminal causation was Cesare Lombroso (1836-1909), the Italian physician. While followers of this approach working today in the biological field may not wish to be identified with the great Lombroso, his dogged determination to appraise criminality in scientific terms foreshadowed the later work of the anthropologist, Earnest Hooton, and his contemporary William H. Sheldon, who espouses body-build as a correlative of behavior.

---

[2] William James, Principles of Psychology, Vol. I (New York: Holt, 1890), p. 28.
[3] For a good account of the popularity of phrenology in this country, see John D. Davies, *Phrenology, Fad and Science* (New Haven: Yale, 1956), especially Chapter 2.

Cesare Lombroso (1836–1909). (Courtesy Robert H. Gault.)

Lombroso may have been wrong in insisting on his "born criminal" type but his contribution to the science of criminology can never be forgotten. From his studies there arose a science of criminal anthropology which had a very popular reception during the latter part of the nineteenth century.

Lombroso's thesis was that the typical criminal can be identified by certain definite physical characteristics or stigmata, such as a slanting forehead, long ear lobes or none at all, a large jaw with no chin, heavy supraorbital ridges, either excessive hairiness or an abnormal absence of hair, and an extreme sensitivity or nonsensitivity to pain.

Lombroso started his education in medicine. While serving in the army as a physician he noted that troublesome soldiers tended to be atypical in their physical characteristics, whereas those who were orderly or predictable in their behavior usually did not possess any outstanding physical characteristics. He noticed also that many soldiers practiced tattooing on the arms, chest, and other parts of the body; that the tattooing on the hardened or troublesome soldiers was often coarse and indecent, and that the more harmless and simple pictures were on the bodies of the sober and dependable soldiers. From this observation he concluded that the pictures one makes on his body might be an indication of his nature. Though the notion in itself was not significant at the time, today we see that he anticipated by almost a century the latest psychological research. Several of the modern projection methods derive intelligence levels and personality characteristics from such subjective drawings, notably Rorschach techniques. Lombroso wished to substantiate the idea that psychological attitudes had a high correlation with physical stigmata. This tentative conclusion stimulated him to make a more thorough study of criminals in Italian prisons.

In his prison work he found similar phenomena. His contact with a notorious brigand, Vilella, convinced him even further that he was on the right track in the explanation of criminal behavior. He found Vilella a man of great strength and agility, possessing the traditional boastfulness of the inured criminal. Upon his death, Lombroso was called to perform a postmortem. He found in this criminal's brain certain evidence of *atavism*— "an enormous occipital fossa and an hypertrophied vermis like that in the lower vertibrates."[4] Here, then, was the answer: the criminal was a *throwback* or biological reversion. The somatological characteristics of the criminal, he thought, were also the physical traits of primitive man.[5]

Lombroso was not the first to suggest the importance of these atypical physical characteristics.[6] Neither can it be stated that he was the first to spin a theory regarding the causes of crime. But it was he and his novel theories that did startle the world of scientific criminology into focusing attention upon the individual for further study. From his studies there arose a science of criminal anthropology which had broad support during the latter part of the nineteenth century.

Lombroso did not stop with the discovery of a criminal type. He called his "born criminal" morally insane, *"fou moral."* It is debatable what he meant by this term, but he probably thought of moral imbecility rather than insanity. He observed that many of his "born" criminals were potentially epileptic. Lombroso developed a classification of criminals, a practice quite popular in his day. The types were: (1) born criminal; (2) the insane criminal; (3) the criminal by passion, including the political "crank"; (4) the occasional criminal, with three subtypes: (a) pseudo-criminals,

---

[4] Robert H. Gault, *Criminology* (New York: Heath, 1932), p. 82.

[5] Cesare Lombroso, "Introduction," in Gina L. Ferrera, *Lombroso's Criminal Law* (New York: Putnam's, 1911), pp. xiv-xvi.

[6] See Alfred R. Lindesmith and Yale Levin, "The Lombrosian Myth in Criminology," *Amer. J. Soc.*, Vol. 42, No. 5 (March, 1937), pp. 653-671.

who are not dangerous, and whose acts might be in defense of honor, for mere subsistence, or committed under unusual circumstances; (b) habitual criminals, conditioned to crime by unfavorable environmental circumstances, though free from inherent criminal taint; and (c) the criminaloid, who hovers between the born criminal and the honest man, and who shows, upon examination, a touch of degeneracy.

The wide acceptance of Lombroso's theory of criminal causation and the reputation he achieved, due to his break with the dogmatic moralistic views of the past, gave rise to the Italian or Positive school of criminology. The most prominent in this group, aside from Lombroso, were Enrico Ferri (1856-1928), professor of criminal law and sociologist, and Raffaele Garofalo (1852-1934), sociologist and jurist. The three differed in many particulars regarding the causes of crime, but all were in agreement that the problem involved scientific understanding of the offender. Their real contribution was the transference of emphasis from the crime to the criminal. Prior to their work and writings, free will and moral responsibility were universally accepted in accounting for crime. It remained for the Italian school to shatter this belief among most students of crime causation.

Lombroso's theories received much less attention in this country than abroad, where they attracted many scholars and penal lawyers. To penologists here the development of custodial treatment seemed more important than crime causation. Lombroso suggested that the anthropologists make a test of the validity of his theories, but this challenge was not accepted for many years.

Around the turn of the century, however, Dr. Charles Goring, physician of His Majesty's Prisons, enlisted the services of Dr. Karl Pearson, the eminent statistician, and others, and made an exhaustive study of the physical types of convicts. In 1913 he published his findings in his work *The English Convict*. The results were distinctly contrary to the Lombrosian contentions:

> On the side of physical anthropology the study is an effective answer to the claims of Lombroso and the adherents of his doctrine. . . . Goring and his associates had spent twelve years making greatly detailed studies of 3,000 prisoners. All the prisoners were recidivists. . . . They, if any, therefore, according to Lombroso's views, should reveal the criminal type. These studies included measurements in almost infinite detail of certain physical features of the prisoners, and he worked out correlations between figures obtained from one group of prisoners and corresponding figures obtained from each of several other groups. . . . There were no striking differences to be found between those of one group and those of another. Such as were discovered could be attributed to the manner of life of the criminals . . . and to a natural selection among those who follow different occupations in normal life. . . . The evidence for a criminal "physical type," based on anthropometric data relating to the skull and face, and based, too, upon certain descriptive data concerning

facial and other features, is nothing. No such physical characteristics can be accepted as signs of the criminal or any other sub-group of criminals.[7]

After Goring's study, the Lombrosian doctrine was seemingly repudiated. In this country, for example, very few followers made themselves known by writing. One early exception was a criminal anthropologist, Arthur Mac-Donald (1856-1936), but his works are all but forgotten.

Yet the successor to the Lombrosian thesis, known abroad as the "constitutional" school, has been widely accepted not only in Continental Europe but throughout Latin America as well. This school contends it is the morphological and psychological propensities of the individual that must be examined in order to understand his environment, as well as his behavior. It is, in essence, the heredity of the individual that motivates him rather than the disturbing factors in the social situation.

As stated above, this thesis is dominant in Latin America. In many of the prisons in that part of the world anthropologists may be found on the clinical staffs. In no prison in the United States is there an anthropologist. The constitutional approach, sometimes referred to as the "Neo-Lombrosian," is beginning to be more carefully scrutinized in our own country, although most students of the problem of causation are still skeptical of its value.

The American School of Constitutional Criminology

Endocrinology

The names of Hooton and Sheldon are closely allied with the constitutional approach in this country, the former with his monumental *Crime and the Man*,[8] and the latter with a series of works dealing with body-build.[9] We must certainly also identify the endocrinologists with this approach, however. Endocrinology, the science of the endocrine or ductless glands, is taken by a number of people in this country to be the clue to personality and criminal causation. According to this theory, some enthusiasts, at least, contend that persons become criminals because of the malfunctioning of one or more of the ductless glands.

The impressive literature in this field makes the layman feel that much has been accomplished in resolving behavior problems to malfunctioning

---

[7] Robert H. Gault, *Criminology* (Boston: Heath, 1932), pp. 86-87. For a critique of Goring's study, in which Lombroso's contributions and errors are appraised, see Maurice Parmelee, *Criminology* (New York: Macmillan, 1922), pp. 495-522.

[8] Earnest A. Hooton, *Crime and the Man* (Cambridge: Harvard University Press, 1939).

[9] William H. Sheldon, S. S. Stevens, and W. B. Tucker, *Varieties of Human Physique* (New York: Harper, 1940); William H. Sheldon, *Varieties of Temperament* (New York: Harper, 1942); and W. H. Sheldon, Emil M. Hartl, and Eugene McDermott, *Varieties of Delinquent Youth* (New York: Harper, 1949).

glands. But we find many criminals whose glands function normally and many law-abiding citizens who suffer from certain glandular aberrations. Some years ago, M. G. Schlapp, writing in the *Journal of Heredity,* contended that probably one-third of all prisoners were sufferers of emotional instability, due to glandular or toxic disturbances.[10] Schlapp and E. H. Smith, in *The New Criminology,*[11] describe the thief and the murderer solely in glandular terms. While these authors represent the most enthusiastic exponents of the role of endocrinology as it is applied to criminal behavior, there are others whose claims are more subdued as to the application of endocrine determination to the etiology of crime.[12]

Professor R. G. Hoskins, an outstanding authority in the field of endocrinology, recognizes its limitations in these words:

> Before psychology, sociology, and criminology can be convincingly rewritten as merely special aspects of endocrinology, many more facts than are now available will have to be collected and integrated.[13]

Professor M. F. Ashley-Montagu, a noted anatomist, makes the following statements:

> I should venture the opinion that not one of the reports on the alleged relationship between glandular dysfunctions and criminality has been carried out in a scientific manner, and that all such reports are glaring examples of the fallacy of *false cause.* . . . The fact is that as far as the endocrine system and its relation to personality behavior are concerned, we are still almost completely in a world of the unknown, and to resort to that system for an explanation of criminality is merely to attempt to explain the known by the unknown.[14]

It is not our purpose here to discount the excellent and even startling work that has been done in relieving certain tensions and discomforts of persons suffering from a glandular imbalance. The alert school teacher or parent may be suspicious of such imbalance in their children and may call on their school or private physician for advice. Glandular treatment often clears up such atypical behavior problems. Our point here is that the delinquent suffers no more from such dysfunctioning than do members of the general population, and that such anomalous physical tendencies need not necessarily lead to overt or incipient delinquency.

The general impression among laymen, as well as among many with leanings toward the biological sciences, is that heredity plays a leading role

---

[10] M. G. Schlapp, "Behavior and Gland Disease," *Journal of Heredity,* Vol. 15 (1924), p. 11.

[11] (New York: Liveright, 1928).

[12] For example, see Louis Berman, *The Glands Regulating Personality* (New York: Macmillan, 1921).

[13] R. G. Hoskins, *Endocrinology* (New York: Norton, 1941), p. 348.

[14] M. F. Ashley-Montagu, "The Biologist Looks at Crime," *The Annals,* Vol. 217 (September, 1941), p. 55.

in criminal behavior. Ashley-Montagu, however, insists to the contrary: "There is not the slightest evidence to believe that anyone ever inherits a tendency to commit criminal acts. Crime is a social condition, not a biological condition."[15]

Identical Twins and Their Behavior

Criminality in twins, especially in identical or one-egg twins, has long been paraded as obvious proof that heredity is of the utmost importance in explaining antisocial behavior. While the evidence accumulated both here and abroad shows conclusively that in a vast majority of cases both identical twins are either delinquent or nondelinquent, it still does not rule out the influences of the social environment.

There are two kinds of twins—those genetically alike, one-egg or identical twins; and those genetically unlike, or ordinary twins from separate eggs. When both members of a twin pair are found to be similar with respect to the commission of one or more crimes, they are labeled "concordant." When they are dissimilar in their behavior, when one is criminal and the other is not, they are labeled "discordant." Ashley-Montagu sets down the findings of several investigators on the criminal behavior of twins, both identical and ordinary.

Criminal Behavior of Twins[16]

Author	One-Egg Twins		Two-Egg Twins	
	Concordant	Discordant	Concordant	Discordant
Lange (1929)	10	3	2	15
Legras (1932)	4	0	0	5
Kranz (1936)	20	12	23	20
Stumpfl (1936)	11	7	7	12
Rosanoff (1934)	25	12	5	23
Total	70	34	37	75
Per cent	67.3	32.7	33.0	67.0

We see from the table that of 104 pairs of one-egg twins examined in these five studies, 70 were concordant (criminal together), and 34 were discordant (with one member law-abiding). But the two-egg twins were in reverse, 67 per cent being discordant and 33 per cent concordant. According to H. H. Newman, an authority on twinning:

> The only serious criticism I have found to be aimed at the twin method of studying the factors of crime is that one-egg twins far more than two-

---

[15] *Ibid.*

[16] *Ibid.,* from J. Lange, *Verbrechen und Schicksal: Studien an kriminellen Zwillingen* (Leipzig: Georg Thieme, 1929). Trans. Charlotte Haldane as *Crime and Destiny* (New York: Boni, 1930); A. M. Legras, *Psychose en Criminaliteit bij Tweelingen* (Utrecht: 1932; in German, in "Psychose und Kriminalität bei Zwillingen," *Zeitschrift für gesamte Neurologie und Psychiatrie,* Bd. 144, 1933); H. Kranz, *Lebenschicksal krimineller Zwillinge* (Berlin, 1936); F. Stumpfl, *Die Ursprunge des Verbrechens* (Leipzig, 1936); and A. J. Rosanoff *et al.,* "Criminality and Delinquency in Twins," *J. Crim. Law,* Vol. 24, No. 5 (January-February, 1934), pp. 923-934.

egg twins are close companions in their social activities and are therefore more likely to encounter together such social influences as might lead to criminal behavior. This is one more instance of lack of control features in nature's scientific experiments, for it can hardly be maintained that the social environment of two-egg pairs is as closely similar as that of one-egg pairs. Therefore, environmental similarities may to some extent account for the close concordance in crime of one-egg twins, while lack of any such similarity in environment may to an equal extent account for lack of concordance in crime of two-egg twins. Undoubtedly the study of crime by means of the twin method is less than it seemed at the outset.[17]

Ashley-Montagu offers the argument that "the factor of environment has been virtually completely omitted from [consideration in] these studies of criminal behavior in twins."[18]

The Work of Hooton

In 1939 Earnest A. Hooton, anthropologist of Harvard University, published his *Crime and the Man*. This book was the result of an intensive twelve-year study that in many respects seemed to vindicate Lombroso.

Hooton disposed of Goring's study (p. 127, *supra*) as unscientific, saying that he distorted "the results of his investigation to conformity with his bias."[19] Hooton attempted to show that crime and other forms of antisocial behavior are due almost exclusively to physical and racial factors. He stated that the criminals he and his associates studied, all of them prisoners (which is an important limitation), differed from noncriminals in various physical particulars that composed a definite pattern of physical inferiority. He considered it necessary that the "criminal stock" be eliminated. Only by sterilizing these defective types and breeding a better race did he think it possible to check the growth of criminality.

Hooton's study was based on measurements of 13,873 male prisoners in ten states, and of 3,203 persons, including patrons of a bathing beach in Massachusetts, firemen in Nashville, members of a militia company, and a few out-patients of a hospital. The measurements were admirably taken; some 107 different anthropometric characteristics are represented. But despite this careful work, many assumptions made in the course of the study are not borne out if one views the problem of crime as a whole.

The work was not well received by sociologists, criminologists, or anthropologists; most of the reviewers were distinctly unfavorable in their evalua-

---

[17] Quoted by Ashley-Montagu, *op. cit.,* from Horatio Hackett Newman, *Multiple Human Births: Twins, Triplets, Quadruplets, and Quintuplets* (New York: Copyright 1940 by Horatio Hackett Newman. Reprinted by permission of Doubleday & Co., Inc.), p. 160.

[18] *Op. cit.,* pp. 54-55. For more recent criticism regarding "twinning," see George B. Vold, *Theoretical Criminology* (New York: Oxford University Press, 1958), pp. 96-98.

[19] Hooton, *Crime and the Man* (Cambridge: Harvard University Press, 1939), p. 11.

tion of it as a scientific contribution.[20] Some, however, praised it for the interesting and witty style that endeared it to the public—a public that was gratified to find in the book an apparent justification for its own deep-seated prejudice against "the criminal."

Hooton's work included a minute anthropological study of 5,689 prisoners. He divided them into nine "racial" types and proceeded to ascertain what offenses the criminals from these groups tended to commit.[21] He makes the following conclusion:

> . . . these data seem to me to prove that crime is not due to race, but merely varies in conformity with racial predilection. Every race is criminal-istic, and within every race it is the biologically inferior—the organically unadaptable, the mentally and physically stunted and warped, and the sociologically warped—who are responsible for the majority of the crimes committed. Each race has its special abilities and its quota of weaknesses. Each produces its pitifully few men of genius, its hordes of the mediocre, its masses of morons, and from the very dregs of its germ plasm, its regiments of criminals.[22]

There can be no serious quarrel with the conclusion expressed by Professor Hooton, since he allows plenty of room for differences in native ability within his nine racial types. What he is saying does not actually go beyond the fact that the biologically weak find it difficult to adjust to a competitive society. He does not exclude the "sociologically warped" as one source of criminal behavior, although he perhaps too easily includes these cases among the "biologically inferior." His analysis of the law-abiding groups from the nine racial types indicates that they seem able to keep out of difficulty mainly through their biological superiority. These persons are superior specimens compared to the prison inmate population that he studied and evaluated criminologically. Hooton's general thesis turns out to be the notion of a special proclivity to particular types of crime by given races and sub-races rather than any theory that a given race is specially prone to criminality.

Another serious criticism of Hooton's work is that he completely over-looked white-collar criminals who are admirable mental specimens in many cases, and biologically superior—hardly the "dregs of society." However, at the time Hooton made his study, little attention had been given to the role of the white-collar criminal.

The Body-Build Thesis as a Clue to Behavior

Shortly after the publication of Hooton's book, another phase of constitutional research began to attract attention. This was the work of Pro-

---

[20] For cogent criticisms see: E. H. Sutherland, *J. Crim. Law,* Vol. 29, No. 6 (March-April, 1939), pp. 911-914; Robert K. Merton and M. F. Ashley-Montagu, "Crime and the Anthropologist," in *American Anthropologist,* Vol. 42 (July-September, 1940), pp. 384-408.

[21] *Op. cit.,* Chapter 7, pp. 204-252.

[22] *Ibid.,* p. 252.

fessor William H. Sheldon and his associates at Harvard University. Following and perhaps paralleling the earlier work of the German, Ernest Kretschmer, Sheldon and his associates contend in their first book, *Varieties of Human Physique,* that all persons may roughly be divided into three categories by which their personalities, as well as potentialities, can be predicted. These three groups are: (1) endomorphic, involving a soft roundness throughout the various regions of the body, short tapering limbs, small bones, soft, smooth velvety skin; (2) mesomorphic, with relative predominance of muscle, bone, and connective tissue, trunk large and well muscled, heavy chest, wrists and hands; (3) ectomorphic, characterized by linearity, fragility and delicacy of body, small, delicate bones, droopy shoulders and prominent ribs, small face, sharp nose, and fine hair.

These three physical types have their accompanying component temperaments: (1) viscerotonic, characterized by a "general relaxation of the body . . . a comfortable person. He loves comfort; soft furniture, luxurious surroundings"; is the eternal extrovert; (2) somatotonic, an active dynamic person who walks "assertively, talks noisily, behaves aggressively"; (3) cerebrotonic, is an "introvert" whose history "usually reveals a series of functional complaints: allergies, skin trouble, chronic fatigue, insomnia . . . sensitive to noise and distractions . . . not at home in social gatherings and shrinks from crowds . . . meets his troubles by seeking solitude."[23]

This initial work by Sheldon and his associates was followed shortly by their *Varieties of Temperament.*

But it is in Sheldon's latest work, *Varieties of Delinquent Behavior,* that we find the true thesis of the constitutionalists. This book describes, in a lively and witty style, the lives and adventures of 200 young adults who had been referred to the Hayden Goodwill Inn, a South Boston social center, by various social agencies. The period covers the years 1939 to 1946. The authors' thesis is that behavior is a function of body structure and, with careful measurement and interpretation, fairly accurate predictions concerning the individual's behavior can be made. The hypothesis advanced by the constitutional school is the direct antithesis of the psychiatric approach. In fact, the authors delight in needling the psychiatric jargon as well as the findings of those who use this technique in delinquent behavior treatment.

Sheldon and his associates have found a bit of support from Sheldon Glueck and Eleanor Glueck, contemporary masters in delinquent research. In their earlier *Unraveling Juvenile Delinquency,*[24] and in their later *Physique and Delinquency,*[25] they have intimated in their conclusions that body-build plays more of a role in precipitating delinquent behavior than has heretofore been acknowledged. In the latter work they state that: (a)

[23] Sheldon *et al., Varieties of Human Physique* (New York: Harper & Bros., 1940), p. 236.
[24] (New York: Commonwealth Fund, 1950), Chapter 15, "Bodily Constitution."
[25] (New York: Harper, 1956).

differences in physique type are accompanied by differences in the incidence of traits associated with delinquency; (b) differences in body type produce differential responses to environmental pressures; and (c) differential incidence of traits and reactions to environment among physique types is related to differences among them in the etiology of their delinquencies.[26]

Although studies such as these have a real classificatory value in establishing types of trait-categories of either a biological or behavioral nature, limitations do exist in the assumption of a causal relationship between the different levels of classification involved—namely, the biological constitution and the personality or behavioral constitution. The existence of such a relationship depends in great part on statistical correlations between the constructed biological types and the constructed personality types. Personality is a social phenomenon derived from and construed through the interaction of the individual with others in his social world; thus attitudes and modes of behavior grow out of the cultural medium, in time or place.

Admittedly, the individual interacts socially as a biological organism and it may well be that his unique organic structure bears a direct causal relation to the manner in which he plays his role within the social group. Yet no social group revises its complex of attitudes completely to fit each generational change or each new set of members; in fact, the reverse seems true. It is therefore possible that, rather than the constitution of any single member of the group determining his attitudes toward that group, the attitudes of the group, already in existence and solidified, predisposes certain types of behavior toward, and consequently from, individuals of that biological constitution.

If correlations could be shown to exist cross-culturally, irrespective of the attitude-system of any particular society, it then would be quite apparent—even to the sociologists—that a causal relationship does exist between biological constitution and personality. With the evidence as it exists, however, the constitutional approach to social behavior—particularly as it applies to criminality and delinquency—is a rather sterile one to a dynamic problem.

As Professor George Vold states the case against theories of the Sheldon type:

> Physical type theories turn out to be more or less sophisticated forms of shadow-boxing with a much more subtle and difficult to get at problem, namely, that of the constitutional factor in human behavior. There is no present evidence at all of physical type, as such, having any consistent relation to legal and sociologically defined crime.[27]

---

[26] Digested from a critique of the Gluecks' study by Albert Morris of Boston University in *Harvard Law Review,* Vol. 70 No. 4 (February, 1957), pp. 753-758.

[27] George Vold, *Theoretical Criminology* (New York: Oxford University Press, 1958), p. 74.

Mental Retardation and Crime

The constitutional approach to crime causation also embraces the area of mental retardation. While the word "feebleminded" is widely used, it is not highly regarded in professional or medical circles.

A clear-cut distinction between mental defect and mental disease was first made in 1838. In that year the famous French alienist, Jean Esquirol, published his work, *Mental Diseases*. On this side of the Atlantic, Isaac Ray, a contemporary of Esquirol, is generally given credit for making the same distinction, although in his early works he did not lift the condition of idiocy from the category of insanity. It was Isaac Ray, however, who first posed the question whether we could impute crime to one who knew nothing regarding intent or consequence of injury. The question of the moral responsibility of the mentally retarded may still be unsettled in legal circles, but it is closed in medical science, psychology, sociology, and criminology.

The ensuing 50 years witnessed many interesting developments in the understanding and treatment of the mentally retarded. The attention of the public was first focused on a wave of studies dealing with alleged mentally retarded families such as *The Jukes* by Richard Dugdale, published in 1877, and *The Kallikaks* by Henry H. Goddard, published in 1912. It was assumed, not so much by the authors of these works as by those who read into them what they wanted to believe, that there was a close relationship between mental retardation and crime. Yet there is little evidence that these families committed actual crime. For example, Goddard found only three actual cases of crime in the long run of 480 members of his Kallikak family. These families were more socially and culturally bankrupt than they were criminally inclined.[28]

The widespread notion that criminals are mentally retarded or at least below the average in intelligence, derives in our culture from religious dogma, literary and dramatic license, and speculative hypotheses, according to Carl Murchison. He finds no scientific justification whatsoever for the notion.[29]

Out of these early folk studies also appeared the idea that most retardation is hereditary. Dr. Goddard postulated in 1914 that mental defectiveness is probably transmitted in the Mendelian manner. Today this is considered doubtful, and it is generally accepted that much of what passes for congenital mental retardation is caused *in utero* or by early illness, bodily injury, malnutrition, or some other environmental cause.

---

[28] Abraham Myerson, *The Inheritance of Mental Diseases* (Baltimore: Williams & Wilkins, 1925), pp. 78-79. The evidence for inherent mental retardation is probably not as clear-cut as it seemed to the original investigators.

[29] "American White Criminal Intelligence," *J. Crim. Law,* Vol. 15, No. 2 (August, 1924), pp. 254-257.

Goddard also stated that the mentally retarded make the best material out of which delinquents and criminals are made. We know that those who are not blessed with good intelligence are highly suggestible and are easily led, either into delinquent behavior, or into law-abiding pursuits. Mental retardation should be detected early in the life of the person so that he may have the benefit of a useful educational environment.

Studies by Murchison, Doll, Root, and Tulchin show that criminals as a class compare favorably with the general population in respect to intelligence, whether we assay them after arrest, or within correctional institutions.[30] It is true that most children sent to state reform schools are deficient in intellectual capacity, but this is accounted for by the fact that the brighter delinquent children are disposed of in some other manner— usually placed on probation or sent home for parental supervision. The statement made by Dr. William Healy, psychiatrist, that although the retarded appear among serious delinquents from five to ten times more frequently than in the general population, the most serious delinquents are not below the average in intelligence. Healy made a study of 4,000 cases of habitual delinquents in Chicago and Boston and discovered that 72.5 per cent were mentally normal, and only 13.5 per cent were retarded. In an intensive study of 400 delinquent cases, he found only two cases that could be labeled morons.[31]

A study made by Weiss and Sampliner of 189 adolescents, aged 16 to 21, who were first offenders, reveals that the distribution of intelligence follows the distribution in the general population.[32] As L. G. Lowrey puts it: "Feeblemindedness is not the universal factor in delinquency it was once thought to be. Even inferior intelligence, broadly conceived, cannot be considered an outstanding feature of the personality of delinquents."[33]

Mentally retarded children need special school curricula that have some meaning for them. We should not insist that they absorb the conventional skills to which average children are exposed. Yet we cannot expect the schools to do all the work. So far as defective delinquents are concerned, especially those who are older, they should be segregated in specialized institutions for treatment. Some should be segregated for life, but it has been

---

[30] For a résumé of the studies in this field up to 1933 (when these studies were popular), see L. D. Zelany, "Feeblemindedness and Criminal Conduct," *Amer. J. Soc.,* Vol. 38, No. 4 (January, 1933), pp. 564-576. For a recent statement regarding this subject see Edward J. Ferentz, "Mental Deficiency and Crime," *J. Crim. Law,* Vol. 45, No. 3 (September-October, 1954), pp. 299-307.

[31] William Healy and Augusta Bronner, *Treatment and What Happened Afterward* (Boston: Judge Baker Guidance Center, 1939), p. 22.

[32] H. R. Weiss and R. Sampliner, "A Study of Adolescent Felony Violators," *J. Crim. Law,* Vol. 34, No. 6 (March-April, 1944), pp. 377-391.

[33] L. G. Lowrey, "Delinquent Criminal Personalities," J. McV. Hunt (ed.), *Personality and the Behavior Disorders,* Vol. 2 (New York: Ronald, 1944), p. 808. Copyright, 1944, The Ronald Press Company.

demonstrated by superintendents of such institutions that many can be paroled safely to the community.

As a result of the early alarm regarding the alleged role of the mentally defective in committing delinquencies and crimes, a wave of near-hysteria called for sterilization and emasculation as the only way to cut down the increasing number of socially and mentally bankrupt persons.

Sterilization, known as vasectomy in males and salpingectomy in females, was first performed in this country by Dr. Harry C. Sharp as early as 1899 among inmates of the old Jeffersonville, Indiana Reformatory. This was many years before it was considered legal. The first eugenic sterilization law was introduced in the state of Michigan in 1897 but it failed to pass. Pennsylvania passed a law but it was vetoed by the governor. The first actual law to get on the statute books was that passed by Indiana in 1907. Altogether some 27 states have passed such laws. They run the gantlet of the courts, which have declared invalid some acts introduced and passed. The issue was carried to the Supreme Court in 1927 in order to test the constitutionality of the Virginia law. The court upheld the statute, with Justice Oliver Wendell Holmes making his famous statement, "Three generations of imbeciles are enough."[34] Despite this ruling, states have been very conservative in enforcing their sterilization laws or in passing new ones. This is probably due to widespread repugnance of practices that seem to violate human rights. Through 1956, about 50,000 sterilizations had been reported in all; these included close to 20,000 in California, 6,956 in Virginia, and 4,256 in North Carolina. What is needed most is a sensible and scientific program of education regarding both sterilization and castration of defective individuals.

It will not be possible, however, to eliminate the retarded by sterilization regardless of how much we may wish to do so. Professor H. S. Jennings states that, since only half of the defective are thus by heredity, preventing all propagation among the hereditarily retarded in a given generation would reduce the total of mentally defectives by only about 11 per cent, under normal conditions. Professor Samuel J. Holmes calculates that to get rid of the larger part of hereditary defectives it would be necessary to sterilize at least the lowest 10 per cent of the total population. Furthermore, sterilization as applied to the criminal population has no practical significance as a punitive device or as a means of reducing the number of delinquents.[35]

So far as voluntary sterilization of prisoners is concerned it is of interest to note that it was practiced in California for many years. Between 1934 and 1941, 628 men requested sterilization in San Quentin prison. About

---

[34] Buck v. Bell, 274 U.S. 200, 207 (1927).

[35] See Amram Scheinfeld, *The New You and Heredity* (Philadelphia: Lippincott, 1950), especially Chapter 46, "Eugenics and the 'Unfit.'"

47 per cent of the group gave as their reason "Do not want children or more children." About 42 per cent asked for the operation to improve their health. About half of the 628 who volunteered were in prison for robbery and burglary.[36] It must be emphasized that in no way was this practice introduced as a punitive device. A notice was placed on the bulletin board of the prison stating that if any prisoner wished the operation he could apply to the medical department.

Physical Handicaps as a Contributing Factor to Crime

Many personality problems of children and adults are developed by physical handicaps or other such anomalies. Unusually short stature, skin blemishes, crippled arms and legs, oversized ears, poor eyesight, abnormal obesity, encephalitis, to mention only a few, are likely to cause serious personality or emotional difficulty among those thus afflicted. Compensatory behavior often develops in such a manner that forms of delinquency result. But we must not overlook the fact that the vast majority of persons thus handicapped make an adequate adjustment to their world. This is particularly true of those who have grown through childhood with a set of understanding parents who have not rejected them.

Physical disfigurement is cited by Ralph Banay as a factor in delinquency and crime. He describes the present-day criminal whose career is determined by his appearance as "a shy youngster with a slight physical defect; a nose that's too big, eyes of two different colors, crossed eyes, acne, a disfiguring birthmark, or a club foot." Such disfigurement often produces a complex in the individual that the world will not accept him. "Billie" Cook is a case in point. This young killer, who murdered the entire Carl Mosser family in early 1951, was born with a deformed right eyelid which, after an operation, forced him to sleep all his life with one eye open. This deformity alone was not responsible for his criminality, which ended at 18 years of age when he was captured in Mexico, but it doubtless was a contributory factor. Others, who are equally disfigured, may devote their energies into socially constructive channels, but still others may try to escape the problem by retreating into a psychosis or may develop an antisocial attitude and commit crime.[37]

Aside from the application of therapy by psychiatrists or social case workers, plastic surgery and physical treatment of specific handicaps have

---

[36] Leo L. Stanley, "Voluntary Sterilization in Prison," *Proceedings,* American Correctional Association (New York: The Association, 135 East 15th St., 1941), pp. 345-352.

[37] Ralph S. Banay, "Physical Disfigurement as a Factor in Delinquency and Crime," *Federal Probation,* Vol. 7, No. 1 (January-March, 1943), pp. 20-24; see also: Harry J. Baker and Virginia Traphagen, *The Diagnosis and Treatment of Behavior-Problem Children* (New York: Macmillan, 1935); E. W. Wallace, "Physical Defects and Juvenile Delinquency," *New York State Journal of Medicine,* Vol. 40, No. 21 (November 1, 1940), pp. 1586-1590.

achieved interesting results in adjusting such persons to their workaday world. Plastic surgery, for example, has been applied to inmates in the Illinois State prison for many years.

Concluding Remarks

Our knowledge of human behavior has many gaps, but we certainly cannot know what makes a person behave as he does until we study him through all available discipline. Only in the circles of enthusiastic environmentalists do we find a complete repudiation of the constitutional approach. The resources of psychology, biology, sociology, and their allied fields, social psychology and psychiatry, as well as medical science, with its ally endocrinology, have all been devoted to the eternal riddle of crime causation. The facts show that the role the individual plays in life is determined by his structure and his conditioning. Certainly this rules out the old bugaboo of freedom of will or of "perverse will." Psychiatrists do not "blame" people for being mentally ill, nor do physicians blame their patients for "catching" a cold or being afflicted with a kidney ailment. Yet in the field of delinquency and crime, many authorities still cling to concepts associated with "blame." Not until we recognize the fundamental truth that the individual is *made* a delinquent by forces beyond his control, operating upon his inherited structure, can we make much progress in understanding and correcting such behavior. The constitutionalists have made and are still making contributions to human behavior. They are needed to the same degree as the environmentalists.

8

Geography and
Criminal Causation

By the statistics of previous years one can foretell with astonishing exactness the number of crimes to be committed during the following year in every country. . . . Through a very simple mathematical operation we can find the formula that enables us to foretell the number of crimes merely by consulting the thermometer and the hygrometer. (From Les Prisons, *Paris: 1890—quoted by Bernaldo De Quiros,* Modern Theories of Criminality, *Boston: Little, Brown, 1911.)*

PRINCE PETER KROPOTKIN

Introductory Statement

Even the layman is likely to ask for a more adequate explanation of criminal behavior than merely "heredity" or "environment." These two words are grossly overworked; yet all behavior is caused by a combination of the conditions represented by the words, construed broadly enough. We have dealt with hereditary factors in the preceding chapter. In the following several chapters we shall attempt to appraise the environmental factors, which include not only the most intimate interrelationships but also the apparently remote influences on the personality through topography, the weather, and humidity. A word about the whole subject of environment is of some importance first.

We usually think of the environment as merely the more apparent factors in the life of an individual—his neighborhood, the house he lives in, his family life, church and school, and the interactions he has with his friends. These are, of course, part of the environment. But as a scientific concept, environment must include every stimulus that impinges on the individual's structure from the moment of conception—the moment the new life begins. The environment is represented by every possible interaction between the individual and every other individual with whom he comes in contact, not only physically but through every cultural medium. What he absorbs from his reading, from the motion pictures, the radio, television— all are a potent part of his environment. Beyond persons, beyond man-made influences, physical or mental, it also includes those of the weather and other geographical phenomena.

Innuendo, nuances, whisperings, off-the-record remarks, adult repartee, gossip—in short, every interaction, regardless of intensity, has its effect in forming character and affects human behavior. Walt Whitman cogently expresses this situation:

> There was a child went forth every day,
> And the first object he look'd upon, that object he became,
> And that object became part of him for the day or a certain part of the day,
> Or for many years or stretching cycles of years.
>
> The early lilacs became part of this child,
>
> . . . . .
>
> And the old drunkard staggering home from outhouse of the tavern whence he had lately risen,
>
> . . . . .
>
> And the friendly boys that pass'd, and the quarrelsome boys,
> And all the changes of city and country wherever he went.
>
> . . . . .
>
> The mother at home quietly placing the dishes on the supper-table,
>
> . . . . .
>
> The father, strong, self-sufficient, manly, mean, anger'd, unjust,
>
> . . . . .
>
> The blow, the quick loud word, the tight bargain, the crafty lure,
> The family usages, the language, the company, the furniture, the yearning and swelling heart,
>
> . . . . .
>
> The doubts of day-time and the doubts of night-time, the curious whether and how,
>
> . . . . .
>
> Men and women crowding fast in the streets, . . .
>
> . . . . .
>
> The streets themselves and the façades of houses, and goods in the windows,
> These became part of that child who went forth every day, and who now goes, and will always go forth every day.[1]

The early environmentalist school of criminologists vaguely anticipated the modern sociological, or more precisely, ecological, approach to crime causation. However they did not define "environment" very sharply and did not use the cartographic or statistical methods that are so distinctive

---

[1] "There Was a Child Went Forth," from F. O. Matthiessen (ed.), *The Oxford Book of American Verse* (New York: Oxford, 1950), pp. 276-278 *passim*.

of the ecological approach. Henry T. F. Rhodes, an American living in France, in *The Criminal in Society,* outlines the criminological schools as statistical, positivist, environmental, bio-sociological, and indeterminate, or moralistic.[2] Among the environmentalists he places: Tarde, who insisted that concentration of population increased opportunities for imitation, so that crime (like other occupations) became standardized in large cities with heightened skills; Corre, who thought crime was caused either by great wealth or great poverty; Manouvrier, the anthropologist, who reckoned environment as a restraining influence that holds in check some of the universal criminal tendencies; and the three Italians, who saw the social environment as the sole cause of crime: Turati, who was sure crime could be abolished if society and economic conditions were remodeled; Battaglia, who considered that the complexity of an ill-adjusted economic system was the cause of crime; and Colajanni, who claimed as early as 1887 that crime *does* pay and that the honest man earns less and is more likely to be injured or killed than the thief; and Bonger, who saw almost all crimes as normal acts, and the capitalist system as the force that makes man more capable of crime. It is obvious that this group of environmentalists had an economic rather than an ecological approach, and that the economic angle was Marxian.

De Quiros, the Spanish criminologist, listed a few sociological theories as early as 1910. He includes Lacassagne who, borrowing a simile from biology, said: "Social environment is the heat in which criminality breeds; the criminal is the microbe, an element of no importance until it meets the liquid that makes it ferment; communities possess the criminals they deserve"; Vaccaro, who maintained that the criminal was a rebel against the complicated system of domestication by which the winners try to develop only the aptitudes of the domesticated which they can better utilize for their ends; and Aubert, who developed his idea of the "Social Center," by which the criminal exaggerates "the characteristics of the human center," that is, the frustrations, phobias, and psychoses arising from the conflict between the individual and the environment.[3]

While few students today read much of the works of the earlier criminologists, it is important to suggest that the problem of crime has attracted the attention of diverse scholars in all lands for many decades.

The Geographical Determinists

Aside from a few Americans who have studied geographical factors as possible causes of crime, the bulk of the research in this field has been

---

[2] Henry T. F. Rhodes, *The Criminal in Society* (London: Lindsey Drummond, 1939).

[3] C. Bernaldo de Quiros, *Modern Theories of Criminality* (Boston: Little, Brown, 1911), pp. 61-64.

pioneered and developed by scholars abroad. It has been the feeling in this country that such research has little to offer in explaining behavior. Yet, as one surveys the field, he is impressed by the variety of studies that have been undertaken.

Geographers have long contended that climate and topography exercise considerable influence on behavior. Montesquieu, in his *Spirit of Laws,* contended that criminality increases in proportion as one approaches the equator, and drunkenness is more prevalent as one approaches the poles.

Adolph Quetelet, the "father of statistics," claimed that crimes against the person were more prevalent in warm climates and crimes against property more numerous in cold areas. He called this the "thermic law" of crime. Quetelet's thesis was substantiated by Mayo-Smith, in his work *Statistics and Sociology.*[4] A study made in France by M. DeGuerry de Champneuf showed that from 1825 to 1830, for every 100 crimes against the person in the northern part of the country there were 181.5 against property, whereas in southern France, for every 100 crimes against the person, there were only 48.8 crimes against property. Other scholars, notably the Italians Lombroso and Ferri, and the German Aschaffenburg, made similar observations in their respective countries—north against south, cold weather against hot. Lacassagne, in his analysis of crimes in France between 1825 and 1880, ranked December as high month for property crimes, with January, November, and February next, followed by October and March.

Prince Peter Kropotkin (1842-1921), the great Russian anarchist, went so far in this climatic approach as to set up a formula by which he contended that the number of homicides could be predicted. Here it is: "Take the average temperature of the month and multiply it by 7; then add the average humidity, multiply again by 2 and you will obtain the number of homicides that are committed during the month."[5] This is absurd as there is no factor in the formula to represent the population of the country; in other words, for a typical January the formula would predict 594 homicides for New York City—quite a carnage in itself, but nothing compared to the effect on Reno, Nevada, with nearly the same climate, and 580 deaths!

The extreme climatologists have tried to discover a spring urge, or biological sex-periodicity, to explain the maximum of sex crimes in May and June; but modern science denies the survival today of a natural season for human mating.

More scientific and less concerned with mathematical formulas, the American, Edwin Grant Dexter, contributed more outstanding and thoroughgoing studies of the effects of climate and meteorological conditions on human behavior. In a statistical study of certain criminal cases in

---

[4] (Cambridge: Cambridge University Press, 1907).

[5] *Les prisons* (Paris: 1880); quoted by Bernaldo de Quiros, *Modern Theories of Criminality* (Boston: Little, Brown, 1911), p. 34.

New York City and Denver, he correlated them with the records of the weather. He found that cases of assault and battery (from 1891 to 1897) in New York City were most numerous in the warmest months of the year, during periods of low barometer and low humidity, during winds, on cloudy days, and during periods of some precipitation. The divergence in his findings is explained by the differences in the number of cases studied and in the differences in altitude of the two cities. Dexter's findings possibly got more attention than they intrinsically deserved because he made no unwarranted claims and was careful not to exaggerate his interpretations—in short, he avoided some of the more common indications of being unscientific.

In studies of the influence of climate upon criminality there is the same danger as in any particularistic theory of crime causation—that of oversimplifying the problem. It must always be kept in mind that while all persons in a similar area are exposed to the same geographical conditions, they show different behavioral patterns. Nor should it be overlooked that geography is but one of the almost infinite number of environmental influences impinging on the individual and conditioning his behavior.

Rural and Urban Crime Rates

It has long been accepted that the city outdistances the country in delinquency and crime rates. In general we can only rely on the number of arrests by police and other peace officers for statistical material that will yield crime indexes. Hence the efficiency of the apprehending authorities is a serious variable in comparing crime rates in urban and rural areas.

Among earlier studies Professor George Vold found that in the case of murder, manslaughter, and the serious sex crimes there was little difference between rural and urban areas in Minnesota; but for other offenses the urban crime rates were much higher. However, since 1945 the figures show that rural crime has increased and that offenses against property have been steadily mounting.

The figures for 1957 show that rural crimes known to the police increased 11.1 per cent over 1956, due largely to burglary and theft. There was a decrease of 4.5 per cent in murders and the highest percentage of increase was in burglaries, 12.3 per cent.

The crime problem in areas outside cities is about half that inside cities per unit of population. The table summarizes known data from areas outside cities. These "rural areas," including the urbanized fringe areas, report more crimes per capita in certain categories than some cities. The rural areas represented in the table reported more murders per 100,000 inhabitants than city groups with less than 100,000 inhabitants. More robberies and aggravated assaults occurred per unit of population in rural areas than in the smallest cities, those with less than 10,000 population.

Rural Crime Rates, 1957[6]

[Offenses known and rate per 100,000 inhabitants, as reported by 1,638 sheriffs, 167 rural village officers, and 13 State police; total rural population 42,600,567 based on 1950 decennial census]

| | Offenses known | |
Offense	Number	Rate
Murder and nonnegligent manslaughter	1,951	4.6
Manslaughter by negligence	2,508	5.9
Rape	5,844	13.7
Robbery	8,069	18.9
Aggravated assault	17,355	40.7
Burglary—breaking or entering	120,412	282.7
Larceny—theft	186,255	437.2
Auto theft	32,918	77.3

No explanation for the upward swing in rural crime rates has thus far been totally satisfactory. It is possible that the gradual urbanization of rural and less populated areas may be the answer. Transportation, means of communication, and the prevalence or scarcity of jobs vary little nowadays between small towns and urban centers so we must look elsewhere for an explanation. Increased efficiency of peace officers together with a more cooperative attitude in reporting offenses may be part of the answer.

Several regional studies have been made within the past decade that embrace rural communities, but no categorical statements can be made relative to the rarity of crime and delinquency in sparsely settled areas. Professor M. G. Caldwell shows that in Wisconsin, counties in which there are cities tend to have a high rate of delinquency. But isolated logging counties of the northern tier of the state, in which there were no cities of any size, also tend to have a high rate of delinquency.[7] Paul Wiers made a study of delinquency in rural areas in Michigan. He found that the rate of delinquency for sparsely settled northern counties was higher than in the southern agricultural counties. This rate was less, however, than urban Wayne county, which embraces the city of Detroit.[8] In another study, Wiers points out that low delinquency rates are highly correlated with "favorable conditions of population density and urbanization" rather than mere rural or urban conditions alone.[9]

In the past, certain obvious reasons were advanced for the greater

---

[6] *Uniform Crime Reports*, Vol. 28, No. 2, April 23, 1958, Table 34, p. 96.

[7] M. G. Caldwell, "The Extent of Juvenile Delinquency in Wisconsin," *J. Crim. Law*, Vol. 32, No. 2 (July-August, 1941), pp. 148-157.

[8] Paul Wiers, "Juvenile Delinquency in Rural Michigan," *J. Crim. Law*, Vol. 30, No. 2 (July-August, 1939), pp. 211-222.

[9] Paul Wiers, *Economic Factors in Michigan Delinquency* (New York: Columbia University Press, 1944), p. 47. See also: Herbert A. Bloch, "Economic Depression as a Factor in Rural Crime," *J. Crim. Law*, Vol. 30, No. 4 (November-December, 1949), pp. 458-470; "Rural Criminal Offenders," *Amer. J. Soc.*, Vol. 50, No. 1 (July, 1944), pp. 38-45.

amount of general criminality in the larger cities: better opportunities for gainful crime and for secrecy and safety, the prevalence of children of the foreign-born, the breakdown of family life, and the increase of juvenile gangs. Especially important with respect to juvenile crime in the cities is the inadequate provision for normal physical expression on the part of youth.

Organized vice and gambling are more general in metropolitan centers than in small towns. The reasons are clear enough. There is a larger clientele interested in such matters, among other reasons, because they do not have many normal and healthful outlets. There is also a greater proportion of unmarried persons in cities. The tense nervous life of cities favors indulgence in more artificial and extreme forms of diversion and excitement. There is less normal family life in the city than in the country. Those who manage vice and gambling have a far greater prospect of remunerative returns in city communities. Liquor consumption and drinking habits are relatively far more common in the cities. Also, the places in which drinks may be obtained are more attractive and drinking is more respectable. Drug addiction is more prevalent in urban areas. Many resort to drugs in order to dull the strains and distractions of city life. Drugs are more easily obtained in the city than in the country.

Professor George B. Vold explains the scarcity of rural crime in another way. He considers that the traditional American rural culture serves as a sort of insulation against crime:

> Rural culture also has its effects on the individual. The value and respectability of work, of family stability and continuity, of land as insurance against want and as an indicator of status, and a general scorn for pleasure seeking and the "soft life," are all part of the traditional rural pattern. Its effect on the individual is, among other things, to provide a pattern of conformity, an acceptance of the regulations and controls of the settled community.[10]

One other phase of this contrast between the amount of crime in urban and rural areas is its detection by the police. Many minor delinquencies and other forms of boisterous behavior are unreported in rural areas because they are not detected or because the perpetrators are known personally to the officers of the law and dealt with in sympathetic fashion. The impersonal relation of city police to small-time offenders, especially to older boys engaged in pranks, prompts them to place such offenses on the police blotter. These build up a higher crime rate for the city.

---

[10] George B. Vold, "Crime in City and Country Areas," *The Annals,* Vol. 217 (September, 1941), pp. 38-45.

9

Economic Factors

and Crime Causation

Governments do not see the future criminal or pauper in the neglected child, and therefore they sit calmly by, until roused from their stupor by the cry of hunger or the spectacle of crime. They erect almshouses, the prison, and the gibbet, to arrest or mitigate the evils which timely caution might have prevented. The courts and the ministers of justice sit by until the petty delinquencies of youth glare out in the enormities of adult crimes; and then they doom to the prison or gallows those enemies to society who, under wise and well-applied influences, might have been supports and ornaments of the social fabrics. Horace Mann (quoted in Federal Probation, *Vol. 20, No. 3, September, 1956).*

Divergent Theories

One of the oldest and most widespread theories advanced for the cause of crime is poverty. The economic determinists, the Marxists, the early social workers, and the humanitarians all stressed economic factors as an expression of their own individual convictions. The most widely accepted by the public, support for poverty as a cause can easily be garnered from innumerable studies and conclusions submitted by scholars both here and abroad.

One of the earliest students of crime who stressed the economic approach was the Italian Ettore Fornasari di Verce, who as early as 1894 pointed out that the poorer classes of Italy, amounting to 60 per cent of the total population, furnished 85 to 90 per cent of the convicted criminals.[1] William Bonger, the Dutch criminologist, argued that poverty disposes to

---

[1] *La Criminalita e le Vicende economiche d'Italia* (Turin: Bocca, 1894), pp. 3-4.

crime and furnishes the motive for it because the capitalistic structure of society is responsible for innumerable conflicts. He further commented on the excessive use of alcohol by the poor, which indirectly causes a great amount of crime.[2]

Early American writers on economic classes vividly described the misery of the masses, especially in the large cities, and constantly stressed the close relationship between their low incomes and delinquency and crime. Early literature is full of exhortation against an economic system that compelled large numbers of people to live in misery and thus become criminals. The writings of Charles Loring Brace, Jacob Riis and others point up this thesis in graphic terms.

In our discussion of poverty as a cause of crime we deal only with that group of criminals that falls within the category of traditional or "small fry"; for few of the big time criminals, whether professional or "respectable," are ever made criminal by poverty.

The statement attributed to the former mayor of New York City, the astute Fiorello LaGuardia, that there is a lot of crime prevention in a T-bone steak, may serve as a springboard to our discussion of poverty as a cause of crime.

Poverty alone is rarely a cause of crime. This is evidenced by the courage, fortitude, honesty, and moral stamina of thousands of parents who would rather starve than do wrong and who inculcate this attitude in their children. Even in blighted neighborhoods, where poverty and wretched housing conditions prevail, crime and delinquency in the majority of residents is nonexistent. True, much lawlessness is conspicuous, but most of the children in these neighborhoods do grow up to be reasonably good citizens.

Many crimes other than those of violence (and even some of these) may be traced to economic causes—*greed,* not *need*—but it is not true that most of those who commit such offenses are driven to their commission by hunger. Such crimes have an economic basis, for the perpetrators are not satisfied with their meager income from lawful pursuits. In a broad sense, then, economic causes are the most potent for the commission of *petty traditional* crimes, especially larceny.

Many of the earlier studies of blighted neighborhoods show that the congestion of poverty-stricken areas and the lack of adequate recreational facilities have fostered a quasi-delinquent attitude on the part of many youngsters. They are proud of their toughness and tend to mimic the "big shots" who frequent such neighborhoods. These congested areas attract disorganized persons as residents; hence there are more persons to arrest in these localities. The recorded arrests, therefore, are out of proportion to the inherent criminality of the neighborhood.

---

[2] W. A. Bonger, *Criminology and Economic Conditions* (Boston: Little, Brown, 1916), p. 643.

An American slum. (Courtesy Edward Gallob.)

It is annoying and depressing to be poor. To go through life forced to submit to substandard living is monotonous and painful. Yet millions of people in this country know nothing but just this monotony and misery. Even in years of so-called prosperity, over three million families and unattached individuals received less than a thousand dollars a year, and over five million families and unattached individuals received less than $2,000 per year.[3]

Neither poverty nor the delinquency encouraged by poverty is a respecter of geography. Regardless of where they live, many poverty-stricken children

---

[3] Department of Commerce, *Survey of Current Business,* as quoted in *Statistical Abstract of the United States, 1957* (Washington, D.C.: U.S. Government Printing Office, 1957), p. 309. The figures are for 1954.

hover between delinquency and constructive adjustments. This is particularly true since it is the daily experience of the poor to contrast their economic lot with those who can enjoy good food, comfortable clothes, an occasional movie, perhaps a car, and a "date" with a girl. Actually it is not the poverty but the contrast that is the disturbing element. In earlier times the poor could scorn the rich as worldly and ineligible to the joys of the good afterlife because of their pride and greed. But the low economic groups of today derive less rationalizing aid from these beliefs; many do deviate from a law-abiding life and, largely through envy and the urge to emulate the rich and comfortable, commit crimes.

Juvenile delinquency, as related to poverty, was studied by Dr. Cyril Burt, British author of *The Young Delinquent;* he contends that 19 per cent of the delinquent children he studied came from the homes of the very poor, whereas only 8 per cent of the total population of London came from such homes; 37 per cent came from the "moderately poor," though the population percentage of this class was only 22. In brief, over one-half of the total delinquency was from the very poor and "moderately poor" families.[4] Burt hastens to add, however, that most of the delinquents coming from the comfortable groups succeed in avoiding "official inquiry and action." He concludes that poverty alone does not produce crime. Succinctly, he remarks: "If the majority of the delinquents are needy, the majority of the needy do not become delinquent."[5] Once again we are reminded of the differential treatment accorded rich and poor by the police, to the detriment of the poor.

Turning to our own country we find two early definitive studies. The first, by Dr. William Healy, shows that in only 0.5 per cent of his cases was poverty a major cause and in only 7.1 per cent was it a minor cause.[6] The second study, by Breckinridge and Abbott, states, according to Dr. Burt, "that in round numbers, nine-tenths of the delinquent girls and three-fourths of the delinquent boys come from homes of the poor."[7] Commenting on the Healy study, Dr. Burt says: "English readers are astonished to find how small an emphasis is placed on poverty. . . . In a volume of over eight hundred pages the short paragraph devoted to poverty occupies no more than seventeen lines."[8]

In a later study, Healy and his wife and co-worker, Dr. Bronner, resorted to a scale of economic standards like that used by Burt: (a) *destitution,*

---

[4] Cyril Burt, *The Young Delinquent,* 1st ed. (London: University of London Press, 1938), pp. 68-69.

[5] *Ibid.,* p. 92.

[6] Dr. William Healy, *The Individual Delinquent* (Boston: Little, Brown, 1915), p. 134.

[7] Sophonisba Breckinridge and Edith Abbott, *The Delinquent Child and the Home* (New York: Russell Sage Foundation, New York, 1912), p. 74. Above quotation from Burt, *op. cit.,* p. 69.

[8] *Ibid.*

(b) *poverty,* a constant struggle to make "ends meet," (c) *normal,* (d) *comfort,* and (e) *luxury.* Of 675 juvenile delinquents whom they studied, they found 5 per cent in the *destitute* class, 22 per cent in *poverty,* 35 per cent who registered *normal,* 34 per cent in the *comfort* class, and 4 per cent in *luxury.* In other words, 27 per cent of the cases came from poverty-stricken homes. In their summaries regarding the results of this study, they state:

> Thus it is clear from the figures (73 per cent coming from normal, or better homes) that great importance cannot be attached . . . to the effects of economic status on cure of delinquent trends.[9]

These same writers, in their 1936 study, *New Light on Delinquency and Its Treatment,* formulated the situation in a statement that minimizes poverty *per se,* but they correctly emphasize the unsatisfactory human relationships that usually flow from destitute and poverty-stricken homes and neighborhoods (italics added):

> It is commonly held that neighborhood conditions, bad associates, poor recreation, etc., are accountable for the production of delinquency. *In truth, these are destructive influences,* but seeking further it appears that at some varying distance upstream in the sequence of delinquent causation there are almost always deeply felt discomforts arising from unsatisfying human relationships. Herein we have found an answer to one of our prime questions: why, living under the same environment conditions, often inimical, is one child non-delinquent and the other delinquent? The latter we almost universally found to be the one who at some stage of his development had been blocked in his needs for satisfying relationships in his family circle. On the other hand, the non-delinquent had nearly always been without any such acute frustrations. His relationships with those in his immediate social environment had been much more satisfying.[10]

Differential treatment by law-enforcement agencies to the detriment of minority groups and to the poor is almost universal, but it is most marked in the urban centers. Rates for actual arrests—those entered on the books—are highest for the lowest-income groups. This is pointed out by Professor Sutherland as follows:

> First, the administrative processes are more favorable to persons in economic comfort than to those in poverty, so that if two persons on different levels are equally guilty of the same offense, the one on the lower level is more likely to be arrested, convicted, and committed to an institution. Second, the laws are written, administered and implemented primarily with reference to the types of crimes committed by people of lower economic levels.[11]

---

[9] William Healy and Augusta F. Bronner, *Delinquents and Criminals* (New York: Macmillan, 1926), p. 121.

[10] *New Lights on Delinquency and Its Treatment* (New Haven: Yale University Press, 1936), p. 301.

[11] E. H. Sutherland, *Principles of Criminology* (Philadelphia: Lippincott, Copyright 1939), p. 179.

Inadequacy, frustration, and emotional insecurity play an important part in delinquency, and they cannot be minimized; but poverty-stricken homes are very drab places to inspire socially acceptable behavior, especially when children from such homes come in contact with poor frustrated individuals living in the same neighborhoods who resent any display of wealth or comfort on the part of more fortunate people. For example, the ostentatious display of wealth, as portrayed in the movies and on television, builds up resentment within certain individuals who feel they can never hope to obtain such comfort. It may be argued that poverty alone does not force a person to commit a specific crime, but it does produce the conditions most conducive to crime, both personally and socially.

It is envy and ambition, rather than hunger or cold, that stimulate many petty crimes, in the same way that greed urges on the big-time criminals. It is not lack of clothing, but perhaps the lack of expensive clothing, that tempts hundreds of girls, for example, to become prostitutes. We may say the same thing concerning the 16- to-21-year-old group of boys who are arrested for stealing automobiles and holding up filling stations. Obviously, economic conditions lie at the base of such crime. It may be argued that many of these juvenile crimes are due in part to sex. In most cases these young hoodlums want money and cars to make an impression on their girl friends.

The eradication of poverty is not an easy task, but it is obvious that its presence in a land of abundance is a tragic anomaly. Economic insecurity, undernourishment, inadequate clothing, and lack of necessary medical care are bound to create attitudes dangerously close to recalcitrance and incorrigibility. There is little wonder, then, that delinquency and crime are so frequently associated with poverty.

Nevertheless, we are convinced that crime and delinquency cannot be completely eliminated even in a utopia. No doubt many who might become delinquent under adverse economic conditions might then become decent law-abiding citizens, but there is evidence in the etiology of crime, as interpreted by present-day statistics, that many crimes might be committed in any utopian state. Certainly, many crimes of violence committed at present have no economic basis. Personality problems, which end in some overt criminal tragedy, are cases in point. Perhaps the most we can admit is that poverty is a concomitant of much crime.

It is generally believed that unemployment plays a large part in crime rates. There are few studies that deal with this subject. One by Mary Van Kleeck involving the work status of 300 persons sent to Sing Sing prison showed that 52 per cent were out of work when they committed their crimes.[12] During the Great Depression a study was made in a prison in

---

[12] "Work and Law Observance in the Histories of Men in Sing Sing Prison," *Report* of the National Commission on Law Observance and Enforcement, Vol. 1 (1931), pp. 193-218.

Illinois by Donald Clemmer. Employment records of some 800 men were studied and it was found that at the time of arrest for their crimes, only 11 per cent were unemployed.[13]

However, the mere fact that a man was employed or unemployed at the time he committed his crime does not tell the entire story. It is regular employment and satisfaction in that employment that are more important for the individual.

Conclusion

What we see from the material presented in this chapter is that poverty, tragic as it is in this land of plenty, plays a far smaller role in crime causation than has been previously contended. Yet it is not honest for us to ignore the relationship that does exist. Poverty leads to slums with their attendant miseries in which children, as well as adults, are led into differential association with delinquent types. Certainly, indirectly, poverty can easily develop envy and bitterness that may result in crime and violence. The same can be said of unemployment. Every effort should be made to supply a decent standard of living for all and a job for every one capable of holding one. Even when that day is reached, however, there will always be some persons who run afoul of the law because of greed, envy, and emotional compulsion.

---

[13] *The Prison Community* (Boston: Christopher Press, 1940), p. 52.

10

Modern
Sociological Theories

The sociologist . . . views the typical criminal as no more or less disorganized emotionally than the typical noncriminal. Criminal behavior is learned through group associations just as politeness, tennis, or banking are learned through other types of association, and for the same reasons—the common desire for companionship, participation, and approval. (From Criminology, *2nd edition, New York: Crowell, 1955.)*

RUTH S. CAVAN

The Ecological Approach

Most of the courses in criminology taught in our colleges and universities are found in departments of sociology and most of the texts are written by sociologists. Since approximately 1920, with the emergence of the scientific school of sociology developed under Professors Park and Burgess, the sociologists have focused much attention on the community and its impact on the individual. Starting with the ecological approach at that time, out of which emerged the "delinquency area" thesis, several sociologist-criminologists have attempted to explain crime in scientific terms.

Briefly stated, the conceptual framework of social ecology assumes that man is an organic creature and therefore behaves according to the general laws of the organic world. Human ecology deals with the relations of people to their spatial environment and to their various reactions to the various environmental stresses and strains.

Professor Robert E. Park of the University of Chicago, a pioneering sociologist, looked upon the community as his unit of investigation. Even in limiting ecology to the community, it is a broad subject. And, of course, there are limitations in carrying out the analogy from the plant world. But out of this general thesis there grew a brilliant series of studies that have made their mark on sociology and social investigation.

Frederic Thrasher's *The Gang,* one work growing out of this era, involves a study of 1,313 gangs in Chicago and indicates their concentration in a twilight zone of factories and railroads radiating from the central Loop district. According to Thrasher: "Gangland represents a geographically and socially interstitial area" between the Loop and the residential districts and in other mid-positions. He made it plain that village gangs do not usually become a social problem and that not all city gangs are delinquent, but many gangs are training schools for crimes. His research, though it was not primarily concerned with delinquency, made important contributions to the natural history of regions that are the habitat and nursery of delinquency, for he pointed out the location and characteristics of these regions. The concept of a transitional or interstitial area was important, for it suggested that crime originates on the edges of civilization and respectability and in communities imperfectly adjusted to normal conditions. Thrasher's work still has considerable meaning in attempting an appraisal of the sudden upswing of young adult delinquent gangs in our metropolitan centers.

It may be noted that Thrasher's work called attention to the appeal of the gang for adventurous recreation, to the gang's importance as a social group, to its cultivation of an intense spirit of loyalty, and to other qualities it possesses that have social values. His studies revealed some of the requisites that must be built into activities to remedy anti-social effects of gangs —requisites that are adequate substitutes for gang traits. But the outstanding and immediate ecological factor of Professor Thrasher's book was the mapping and description of gangland.

Much the same regions were the subject of the late Clifford R. Shaw's *Delinquency Areas,* which developed the fact that delinquency rates are high in the center of Chicago and progressively lower at greater distances from the center and from industrial areas. This picture of Chicago, where the "delinquency area" concept was born, was paralleled by similar findings for other cities.

Shaw and his colleagues demonstrated that conduct symptomatic of pre-delinquency, including habitual truancy, incipient delinquency, and the like, tends to be clustered or concentrated in certain well-defined areas. The highest rates are usually found in the congested and disorganized urban sections lying adjacent to the central business and warehouse areas, and the lowest rates in the outlying suburban and residential districts. As any student of urban problems is aware, there are many characteristic differences between such areas. The downtown and warehouse districts were once residential areas but are now slums. The older inhabitants moved out as business began to encroach upon their neighborhoods. These old neighborhoods take on a certain sordidness that causes a breakdown in the traditional social controls.

The people who move into these more or less abandoned neighborhoods are marginal; they hover around the bare subsistence level. Immigrants, Negroes from the South, migrant families, and others attempting to "break in" to the precarious economic life of a great city, of necessity gravitate to these neighborhoods. These elements differ markedly in their cultural backgrounds and often clash among themselves. Few, if any, stabilizing community forces remain; the families are often financially impoverished and find it extremely difficult to carry on any policy of neighborhood reconstruction. The disintegrative forces overpower the feeble efforts of community agencies to bring about the stabilization that is fundamental to guiding normal social behavior.

So, instead of thinking in terms of the significance of poor housing, overcrowding, low standards of living, low educational standards, and other such conditions, this approach looks upon these merely as symptomatic of more basic degenerative processes. That is, delinquency areas are due to the deterioration of the fundamentals and characteristics of social control.

Shaw made it quite plain that there is a definite relation between his *delinquency areas* and juvenile delinquency. In addition, he demonstrated a serious relation between delinquency and recidivism, showing that an area that has a high rate of delinquency also has a high percentage of recidivism among the delinquents. Yet, as Warner and Lunt showed in their *The Social Life of a Modern Community,* there is a differential treatment of persons in slum areas, a fact frequently ignored when accepting the delinquency area concept.[1] They show that the high arrest rate by the police in these areas is not necessarily an index of real delinquency and crime for a city. Police are not "fixed" to refrain from arresting in such neighborhoods, as is the case in many well-to-do areas of a large city. This salient fact must always be borne in mind when we appraise the high delinquency rate of blighted neighborhoods.

In their later book, *Juvenile Delinquency and Urban Areas* (1942), Shaw and his colleagues expanded their research to include several other cities where conditions exist similar to those reported and described in their earlier work. The conclusions reached by Shaw and his associates are that group delinquency, which characterizes much of our modern crime, is deeply imbedded in the roots of modern community life; that attitudes prevailing in metropolitan centers seem to sanction delinquency through the conduct, speech, and gestures of adults with whom city juveniles come in contact; that the competing values of modern life confuse the growing boy and encourage him to seek a life of excitement in which he can gain a satisfying status with his kind; and that year after year this situation grows more serious. Any solution, these writers feel, must come from community agen-

---

[1] W. Lloyd Warner and Paul S. Lunt, *The Social Life of a Modern Community* (New Haven: Yale University Press, 1941), p. 376.

cies focusing their attention on the setting or neighborhood life from which these young delinquents emerge.

In general, studies of other cities confirmed his findings for Chicago—that major ganglands and delinquency areas had bad housing conditions; no play spaces except streets and vacant lots and railroad yards and track; heavy employment of under-age children; much truancy; heterogeneous racial make-up; a population that is decreasing, mobile, and dense; industrial and commercial crowding; poverty and family dependence—in addition to locations that are centralized and interstitial or transitional. Intensive studies of specific crimogenic areas inside certain cities now further developed these characteristics.

A Critical Note on the Delinquency Area Concept

The 1920's and 30's witnessed a plethora of ecological studies of crime, delinquency, and other forms of social disorganization, all by competent students of the Park and Burgess school. Many of these were accompanied by charts and maps of cities and areas of cities. In more recent years the emphasis seems to be away from the ecological approach. There have been very few studies in the past twenty years emphasizing this once popular approach. This is due to two factors, at least. One of these, pointed out by Sophia Robison in her provocative book, *Can Delinquency Be Measured?* is that the term "delinquency" is so variously defined and considered, and so subjectively interpreted, that it cannot be used as a unit of measurement. As she states:

> Although the *delinquency* area technique of study, developed in Chicago and later extended to an examination of the locus of delinquency in other cities, has received official recognition, the suspicion persists that this method is not only essentially invalid to indicate the extent of juvenile delinquent behavior but that it does not furnish any very useful approach to the problem of understanding or preventing delinquent behavior.[2]

Her analysis of delinquency in New York City failed to validate an index of delinquency said to be established by studies in Chicago and other cities.[3] Miss Robison further cannot accept the thesis that delinquency decreases from the general business district outward.

The second factor that tends to militate against the area concept is the psychiatric approach to the problem of maladjustment and delinquent behavior. Although no one can deny that the great bulk of delinquency comes from the blighted areas of our large cities, this fact cannot obscure the existence of much delinquency in the homes of the economically favored. It is just not recorded so frequently. The psychiatrists are not much impressed

---

2 Sophia Robison, *Can Delinquency Be Measured?* (New York: Columbia University Press, 1936), p. 4.
3 *Ibid.*, p. 210.

with the ecological or statistical studies made in most cases by the sociologists. For example, Alexander and Healy in their work, *Roots of Crime,* maintain that such studies do nothing more than call attention to a phenomenon that every policeman knows, and needs no further explanation.[4]

Another interesting approach to the dilemma of delinquency in working-class society is that presented by Whyte in *Street Corner Society*[5] and, more recently, by Cohen in *Delinquent Boys.*[6] In the former work, contrasts are made between values of college boys who come from working-class homes and the "corner boys" who also belong in similar homes. Cohen, while criticizing the "delinquency area" concept in stating that "interstitial" and "slum" areas "have a vast and ramifying network of informal associations," does maintain that delinquent behavior, that is, systematic delinquent behavior, is found primarily in lower economic groups. His thesis is that there is a pattern of delinquency that persists in certain social classes which he aptly describes as "non-utilitarian, malicious and negativistic."[7]

While criticisms of the "delinquency area" concept may not be fatal, or even serious, they are at least worthy of record. Another qualification of the concept would arise if studies of any quantity were available that would show just how many children grew up in blighted neighborhoods and do not become delinquent.

Another criticism of the area concept is that, while the work done, and the approach itself, are valuable, they have drawn attention away from the individual's maladjustment. This maladjustment must always be considered in diagnosing and treating the delinquent. The area concept also permits divers people and reform organizations to ride the popular bromides of poverty, broken homes, poor housing, and the like as *the* basic causes of delinquency and crime.

An ecological approach to delinquency in urban areas can be of practical value to agencies such as: city departments in charge of public safety, public welfare, and health; private and public agencies engaged in developing low-cost housing projects; the board of education, recreation bureau, and social agencies. Neighborhood and community councils as well as other civic organizations devoted to improvement of specific areas need factual information along these lines. State and city planning commissions, preparing blueprints for improved living conditions or for redevelopment projects, will find such studies of inestimable value in determining needs.

The delinquency area concept has served a useful purpose in assisting in the fight against delinquency and crime. But there is a danger that it will

---

[4] Franz Alexander and William Healy, *Roots of Crime* (New York: Knopf, 1935), p. 274.

[5] William F. Whyte, *Street Corner Society,* rev. ed. (Chicago: University of Chicago Press, 1955).

[6] Albert K. Cohen, *Delinquent Boys* (Glencoe, Illinois: Free Press, 1955).

[7] *Ibid.,* p. 25. See later discussion of Cohen's thesis, p. 161.

continue to be used, as it has all too frequently in the past, as an over-all simplification of the problem.

Current Sociological Theories of Causation

Since the heyday of the ecological approach to causation a number of academic sociologists have advanced theories that merit review. The texts and professional journals bear evidence that, regardless of their apparent adequacy, they fall short of answering the over-all question of criminal behavior.

Perhaps the first sociologist to develop a theory of causation, aside from the ecologists, was the late Edwin H. Sutherland, a most competent criminologist. His thesis, known as the "differential-association" theory when first developed, stated that most criminal behavior is learned through contact with criminal patterns which are present, are acceptable, and are rewarded in one's physical and social setting. As this theory is stated in the latest Sutherland text we find:

> A person becomes delinquent because of an excess of definitions favorable to violation of law over definitions unfavorable to violation of law. . . . When persons become criminal, they do so because of contacts with criminal patterns and also because of isolation from anti-criminal patterns. Any person inevitably assimilates the surrounding culture unless other patterns are in conflict.[8]

In other words, persons become criminal because they apparently have more contacts in their daily lives with criminals or quasi-criminals than with non-criminals; that their milieu is heavily weighted by individuals who are crime-motivated. There can be no doubt that individuals become what they are largely because of the contacts they have, but both the constitutional or inborn hereditary structure and the *intensity* of the environmental stimuli must be appraised as well. While Sutherland did not advance this theory to explain *all* criminal behavior, it has been assumed by many that it more adequately explains dissident behavior than most others. But it fails to give any importance to the biological makeup of the individual nor does it make any provision for evaluating or weighting the influences in the environment that impinge upon the structure of that individual.[9]

Another distinguished sociologist-criminologist who has given much thought to the etiology of crime, at least so far as it appears in our western culture, is Donald R. Taft. His thesis is that in any culture that is highly

---

[8] Edwin H. Sutherland and Donald R. Cressey, *Principles of Criminology*, 5th edition (Philadelphia: Lippincott, Copyright 1955), p. 78.

[9] For more detailed analyses of the "differential-association" theory see: Robert H. Caldwell, *Criminology* (New York: Ronald, 1956), pp. 181-185; Sheldon Glueck, "Theory and Fact in Criminology," *The British Journal of Delinquency*, Vol. 7, No. 2 (October, 1956), pp. 92-109; and George B. Vold, *Theoretical Criminology* (New York: Oxford University Press, 1958), pp. 194-198.

competitive and materialistic, in which the striving for prestige and status is so strongly impelled by social forces, much crime must inevitably take place. His "tentative" thesis is: given a dynamic culture, complex, highly materialistic, with inconsistencies between precept and practice, with a high degree of differential treatment between members of dominant groups and members of the underprivileged and minority groups, with "success" measured more by "what you show" rather than "what you have," or in conspicuous consumption rather than in integrity, such a culture is bound to have much crime.[10]

While one may concur with what Professor Taft states insofar as the total crime picture is concerned—although as with all hypotheses of crime there is some truth—his thesis does not explain why the majority of the American people do not succumb to the materialistic and almost arrogant culture that he describes. Regardless of how low the moral tone of business, professional and political life may be, we realize that millions of citizens rise above these culture-deviations and not only refrain from crime and immorality, but resist venality wherever present. Taft's thesis, like Sutherland's, unquestionably explains much of our crime but it cannot be accepted in hundreds, even thousands, of instances.

Another thesis has been suggested by Professor Walter Reckless although much of his work in the field refrains from espousing any specific theory. He contends that crime is "not a generic behavioral phenomenon" but can be explained only "as different orders of behavior." He suggests what he calls "social vulnerabilities" of the individual and "categoric risks for official action." The former are the weaknesses of the individual when confronted with a situation which may precipitate delinquent or criminal behavior. He lists these as the situation, the social order, pressures and weaknesses in the milieu. Categoric risks are persons who are most likely to engage in crime, become arrested, or admitted to institutions.[11] In essence, this approach fits into what has sometimes been referred to as "actuarial sociology."[12]

Another attempted explanation that might be mentioned is what is referred to as the "closure" theory suggested by Professor Edwin M. Lemert. In a study of forgers Lemert found a certain type of offense which he called "naïve forgery." Those committing such offenses find themselves in a crisis situation and, usually without any crime experience, turn to forgery as a way out of their dilemma. This selection of the alternative is called "closure."[13] Reckless, in appraising the closure theory states that it might

---

[10] See Donald Taft, *Criminology,* 3rd edition (New York: Macmillan, 1956), Chapter 18.

[11] Walter C. Reckless, "The Etiology of Delinquency and Criminal Behavior," *Social Science Research Bulletin,* No. 50 (1943), pp. 131-137.

[12] See Marshall B. Clinard, "Sociologists and American Criminology," *J. Crim. Law,* Vol. 41, No. 5 (January-February, 1951), pp. 549-589.

[13] "An Isolation and Closing Theory of Naïve Check Forgery," *J. Crim. Law,* Vol. 44, No. 3 (September-October, 1953), pp. 296-307.

well be applicable in explaining certain offenses such as unorganized auto theft, noncompulsive shoplifting, criminal abortion, arsenic poisoning, and even gangsterism and racketeering. He adds:

> The closure theory has several advantages. It brings together in a centripetal series of concentric circles the pressures of the situation and the pressures within the person, so that it explains how the individual takes a certain route in his behavior among available alternatives. The closure theory does not require a control group, that is, a comparison of delinquents with nondelinquents. . . . The closure theory approach needs primarily the selection of the cases that according to certain criteria fall within a certain order of delinquent or criminal behavior. In other words, it merely requires the determination of a reasonably feasible homogeneous offender group or behavior problem group.[14]

Other sociologists have suggested hypotheses that attempt to explain crime but, on the whole, there is general agreement that it may be best explained within the group experiences or environmental milieu of the individual. Some maintain that the "typical criminal [is] no more or less disorganized emotionally than the typical noncriminal."[15] However, others maintain that the criminal is a deviant who has been unable to adjust himself to conventional sanctions imposed by society due to inadequate and unfortunate interactions with parents and primary group contacts. It is in this latter area that we find some agreement between the sociologists and the psychiatrists.[16]

We referred earlier to the thesis suggested by Professor Albert K. Cohen. In his analysis of the culture of the gang, he sees much delinquency flowing from a subculture that persists primarily in many urban areas, generation after generation. He maintains that most middle-class and working-class children grow up in "significantly different social worlds" and, due to the difficulty experienced by working-class boys in measuring up to the conventional materialistic standards and socially acceptable behavior norms of conduct, they tend to engage in "negativistic" delinquency. His is a fruitful hypothesis in explaining the gang spirit in metropolitan areas but, as he himself states, does not explain random, individual delinquency or crime which is based largely on personal conflict or disorganization. Nor does it explain compulsive crime.[17]

It may be too much to expect the sociologist to explain all crime by means of any unilateral theory. "All crime" is a concept as broad as that of "all social behavior"; for all constructive organized activity will have its perversions in some type of criminal activity. And it is clearly unreasonable

---

[14] See his appraisal in *The Crime Problem* (New York: Appleton, 1955), pp. 111-114.

[15] Ruth Cavan, *Criminology*, 2nd edition (New York: Crowell, 1955), p. 708.

[16] See Chapter 13 for the position of the psychiatrists.

[17] Cohen, *Delinquent Boys*, p. 25.

to expect sociology at its present stage of its development to have a "unified theory" embracing every social event. Certainly all such attempts have failed whether developed by sociologists or by those involved in other disciplines.

But the sociologist can be helpful in pointing up the difficulties of the individual in his quest for status, in measuring up to family controls, adjusting to primary group norms, and achieving recognition, affection, and prestige. The sociologist has been especially helpful in describing the conditions of the environment that tend to precipitate personal hostility, conflict or frustration. The theories briefly outlined above are at best only segmental, and emphasize clearly that the answer to criminal behavior is too complex to be settled by the knowledge derived from any one behavioral discipline.

11

Minority Tensions
as Factors in Crime

Adult attitudes of the larger community toward minority groups, particularly racial groups, add to the difficulties in dealing with juveniles. The tendency toward relatively high juvenile delinquency rates in certain minority groups may be explained by a number of conditions, including low economic status, bad housing, overcrowding, restricted employment opportunities, racial proscriptions, and intensified conflict between the older and newer generations. (From The Juvenile Offender, *Garden City, N.Y.: Doubleday, Copyright 1954.)*

CLYDE B. VEDDER

Ethnic Groups

Tensions exist in any area where a minority threatens the status or way of life of the dominant group. In relatively homogeneous countries or cultures small numbers of "out group" people present little or no threat; thus they are not intimidated or discriminated against. There seems to be a saturation point beyond which suspicion, mistrust and then fear develop. Then, in times of crisis or general tension, it is not uncommon to use the minority group as a scapegoat or "whipping boy" on which to lay the blame for any trouble.

As early as 1835 arguments against the first wave of immigrants (mostly Irish) included a warning that they would increase the crime rate. These arguments were enunciated especially in the large cities along the eastern seaboard. However, with the gradual assimilation of these early immigrants in a potentially rich country and the relatively slow rate of immigration for the following fifty years, the situation remained somewhat dormant as far as tension was concerned.

The Old American or Anglo-Saxon stock dominated this country for the first hundred years after its establishment. It practically monopolized the fields of politics, education and business. This was a land of north European background and, as such, was supposed to be destined not only to greatness but to moral and religious vigor as well.

But with the impetus given to immigration of south Europeans from the 1880's on, the so-called New Immigration, this country changed gradually its Anglo-Saxon complexion and merged into an amalgam of many ethnic or national groups and races. Many people complain that the newer elements are directly or indirectly responsible for the social ills of the country; they wish to return to the "golden age" of the Founding Fathers when, it is believed, everything was wholesome under the leadership of substantial north Europeans.

Today the United States is a country composed of descendants of immigrants from many European countries, Asia, Latin America, and Africa. No longer can we make valid distinctions between the Old American stock and the immigrant. The children of these immigrants are slowly becoming today's Old American stock. The diverse cultures of diverse peoples have become merged into a new culture that differs in many substantial ways from that of a century and a half ago.

Aside from the relatively few immigrants who have come to this country under the Quota Law or from non-Quota countries, most non-native Americans are well past middle life and their children and grandchildren are mostly native born. Many of the fears that were expressed during the period between 1880 and 1920 concerning the criminality of the immigrant and his native-born children are no longer of much consequence, even though there may have been some justification for such apprehension at that time. These latest immigrants have become well assimilated and their children have assumed most, if not all, of the conventional traits usually assigned to Americanism. Large numbers of them have become leaders in all professions and business enterprises.

Some day, in the not too distant future, we are going to find that we are, in truth, an amalgam people, that the Old American stock has vanished, that we have very few immigrants in our midst, and that even second generation immigrants are disappearing. When that time comes all the studies of the relationship between immigrants and crime will have only an historical interest. One evidence of this is that no longer do the *Uniform Crime Reports* differentiate between native-born and foreign-born criminals. We shall still have studies dealing with race and crime embracing the Negro, the Chinese, the Japanese and the Indian, but Europeans will be merged with the dominant white population in such a manner that they can be no longer identified. Some segments of the white race may not be so merged; we refer primarily to the Mexican-American of the Southwest and West, and Puerto Ricans in some of our large eastern cities, notably in New York.

At this point it is best that we define nationality and race. "Race" is a biological concept; the term is used to differentiate mankind on the basis of common hereditary traits possessed on a relatively large scale. Since all large groups possess, more or less, all the traits common to mankind, it can be stated that theoretically there are no *pure* races. All people have some mixed ancestries.

But aside from this biological concept, we are constantly confronted by a sociological or cultural confusion of the terms *race* and *nationality*. People speak of the Italian or German races; of the Jewish or the English races. All of these are or have been nationality groups, but the Jews maintain their identity primarily through their cultural and religious background. So, when we hear that Italians or Irish commit more crimes than others, this would not claim a racial trait, but rather amounts to an allegation that it flows from their culture.

During the early days when millions of immigrants were entering the country, there was a sharp difference of opinion among students about the relationship of the immigrant to the increase of crime and vice. The earlier studies indicated a high correlation between immigration and crime, but later the opposite result has been reflected in the statistical evidence collected.[1] The first studies failed to take into consideration certain variable cultural factors, such as language difficulties, culture traits peculiar to specific groups, and lack of understanding of our manners and ways of doing things. In addition, police were prone to arrest recent immigrants on slight provocation, which increased their normal rate of delinquency and criminality. As patrolmen became more accustomed to immigrant types, much of their cultural behavior, formerly considered strange and bad, came to be overlooked.

Subsequent studies showed that the most markedly criminal element in the population included a large proportion of children of the foreign-born. The problems in cultural adjustment for the second generation immigrant are always serious. The customs and folkways of the parent culture have a less potent influence over this group and are less effective in restraining them from criminal conduct. Not finding themselves entirely "at home" in either the old culture or the new, they seek an accommodation of their own, too often in gangs that may turn their energy into antisocial activity. However, some groups have more difficulty in making the necessary adjustment to American life than others; and we must not make the mistake of lumping all second generation immigrants into one large mass of maladjusted individuals.

---

[1] H. H. Laughlin, "Analysis of America's Melting-Pot," *Hearings* before the Committee on Immigration and Naturalization, 67th Congress (November 21, 1922); this study indicated a great deal of crime and vice among immigrants. For an opposite view, see Alida C. Bowler, "Recent Statistics on Crime and the Foreign-Born," National Commission on Law Observance and Enforcement, *Report on Crime and the Foreign-Born,* Report No. 10 (Washington, D.C.: U.S. Government Printing Office, 1931).

We can safely conclude, then, that this problem resolves itself into one of cultural nonassimilation and social maladjustment, affected to some degree by pure prejudice and differential treatment, rather than to any inherent criminal tendencies residing in any specific foreign group or combination of groups.

It has been taken for granted by the public that members of nationality groups other than the Old American stock make up the preponderant element among the more dangerous gangs of racketeers and stick-up men. It is often difficult to ascertain the nationality of persons committing crimes since many of them change their names. But aside from this qualification, it is interesting that the notorious "bad men" of the 1930's—Dillinger, Van Meter, Floyd, Nelson, Barker, Kelly—had traditionally Old American names. In addition to these criminals, many "white collar" criminals are American-born; some of long American genealogies.

We cannot overlook the fact, however, that many of the overlords of national gangdom—the syndicates—are either Italian born or members of second-generation stock. They are the modern leaders of the old Black-Hand or Mafia (Unione Siciliana) gangs that laid tribute on Italian immigrants and otherwise terrorized hundreds of communities during the days of south European immigration. One is reminded of the names of Capone, Luciano, Anastasia, and Vito Genovese, often referred to as the chief of the terroristic society.[2] The Kefauver Committee report, as well as the McClellan Senate subcommittee in 1958, dealt at some length with the role of the Unione Siciliana in its control of the nationwide empire of crime.

The Mexican-American

There are two areas of this country where there exist tensions between old, established, and dominant groups and minority peoples; in California and parts of the Southwest where relatively large numbers of Mexicans have settled and where many more have infiltrated; and in the large cities of the eastern seaboard, especially in New York City which has witnessed a large influx of Puerto Ricans during the past two decades. In both areas we find the stock complaints that these groups swell the crime rate out of proportion to their numbers.

The Mexicans are an ethnic group of mixed Spanish and Indian descent, the relative proportion of these stocks depending upon the particular area of their origin. Nine-tenths of the Mexican population of this country live in five states—Texas, California, Arizona, New Mexico, and Colorado. Because of their aloofness, it has been easy for the dominant stock to segregate them, especially in areas where they have settled in large numbers. It should be mentioned, however, that in many areas the Mexicans have settled for a much longer period than their dominant group neigh-

---

[2] *New York Times* (November 17, 1957).

bors. It is estimated that there are about 3,500,000 Spanish-speaking people in this country, most of whom are of old Mexican stock or more recent arrivals. Mexicans do not come under the Quota Laws and if they can pass the tests set up by the immigration authorities may come into the country legally. However, thousands of others, mostly peons who wish to better their economic condition, enter the country illegally. Known as "wetbacks," they stream across the boundary and present not only a serious economic problem, but fan the latent prejudice of some members of the dominant groups of the area.[3]

Delinquency, immorality and vice especially flourish along the Mexican border. Considerable testimony substantiated such conditions before the Senate subcommittee on juvenile delinquency. According to the committee: "These border towns beggar description to the average person in the United States. Prostitution, drunkenness, sale of narcotics and pornography, perversion, and all forms of vice flourish openly." It was further testified that "kids coming from Whittier, Los Angeles, Santa Monica, Ventura, Santa Barbara, Riverside, Orange County, El Centro, and all over southern California" frequent the dives in Tiajuana.[4]

It was further stated that "substantial numbers of juveniles between 13 and 18 years of age going unescorted into Mexican border towns such as Tiajuana, Juarez, Laredo, and Nogales," from our own states "seeking recreation, thrills, and excitement."

Earlier studies made for the Wickersham Commission (1931) showed that the Mexican was regarded by members of the dominant group as "a bad lot," as well as a "natural thief" who had no "idea of private property."[5] The studies further indicated that the nomadic life many Mexicans were forced to lead after their arrival in this country tended to demoralize them. Handman states: "There is no evidence to show that the Mexican runs afoul of the law any more than anyone else, and if the complete facts were known they would most likely show that he is less delinquent in Texas than the non-Mexican population of the same community."[6]

More recent studies, however, show that Mexican youth are more delinquent than the youth of the dominant groups in the South-West. There are more Mexican children in reform schools in proportion to their numbers in the general population. But such a situation may be due primarily to

---

[3] For a discussion of this problem, see Bea Casey, "County Jails Along the Mexican Border," *Prison World* (September-October, 1949), pp. 10 f.

[4] Subcommittee on Juvenile Delinquency, Committee on the Judiciary, U.S. Senate, *Report,* 85th Congress, 1st Session (Washington, D.C.: U.S. Government Printing Office, March, 1957), p. 59.

[5] Paul S. Taylor, "Crime and the Foreign-Born: The Problem of the Mexican," National Commission on Law Observance and Enforcement, *Report on Crime and the Foreign-Born,* pp. 199-243.

[6] Max Handman, "Preliminary Report on Nationality and Delinquency: The Mexican in Texas," National Commission, *op. cit.,* pp. 245-264.

the differential treatment they receive as well as to their confusion in adapting to local discrimination.

In southern California, especially in the larger cities, the Mexican delinquent boy is considered quite apart from the other young white malefactor. Aside from what differential treatment he may receive from the authorities, it is alleged by many that he and his gang represent a real problem. This certainly was true prior to World War II and up to the time that a serious flare-up occurred in downtown Los Angeles (1943). Psychiatrists, as well as newspapers in and around Los Angeles, made much of the eccentric characteristics of the adolescent Mexican boys' gangs. The newspapers whipped up and fanned the latent prejudice by calling attention to the peculiar garb worn at that time by the Mexican youth as well as to his *pachuco* hairdo. The psychiatrists explained such dress as symbolic of the disorganization of a discriminated group.

In the reform schools and Youth Authority camps of California the Mexican boys naturally tend to be clannish. Many of them persist in wearing their distinctive hairdos, which are a badge of solidarity against the dominant group. But with understanding of the handicaps under which they live, especially in the large urban cities, they cause no more difficulty than others.[7]

We have somewhat of a counterpart to the problem of Mexican assimilation in the West and Southwest in the tremendous influx of Puerto Ricans, most of whom are from the lowest economic groups, to the large cities of the East. New York and more recently, Philadelphia, are confronted with an enormous increase of delinquency among children from this group. Suddenly transplanted from a poverty-stricken tropical economy to a heavily congested urban area has brought about widespread demoralization among thousands of these families. Not only do they clash with the dominant white groups living among and contiguous to them, but also with the Negroes who for so many decades have been obliged to bear the brunt of the hostility and discriminaton of the dominant group.[8]

The Negro and Crime

It is lamentably true that figures of arrests, convictions, and commitments to institutions show up the Negro to a decided disadvantage. The disparity

---

[7] For cogent analyses of the Mexican-American and his problems, see: Ruth Tuck, *Not with the Fist: Mexican-Americans in a Southwest City* (New York: Harcourt, 1946); Beatrice Griffith, *America Me* (Boston: Houghton, 1948). This latter book deals with differential treatment afforded Mexican youth; also with social cleavage between middle and lower class Mexican-Americans.

[8] For discussions of the problem of Puerto Rican assimilation see C. W. Mills, Clarence Senior, and R. K. Goldsen, *Puerto Rican Journey* (New York: Harper, 1950); also, J. Milton Yinger and George E. Simpson, "The Integration of Americans of Mexican, Puerto Rican, and Oriental Descent," *The Annals*, Vol. 304 (March, 1956) pp. 124-131.

of arrests, as gathered from the *Uniform Crime Reports* of the F.B.I. is substantially increasing in most categories of offenses. This may be seen from the tables on pages 172–174 which show the number of arrests of Negroes and whites with their respective rates, and the ratio of Negro arrest rates to white arrest rates, per 100,000 for the years 1940, 1950, and 1957. Let us take a few crimes to illustrate: Homicide—Negro arrests in 1940 were 7.16 times as frequent as white; in 1950, 8.26 times as frequent; and in 1957, 8.09 times as frequent. Robbery: in 1940, 4.69 times as frequent; in 1950, 5.44 times as frequent; and in 1957, 10.55 times as frequent. Gambling: in 1940, 8.12; in 1950, 9.49; and in 1957, 27.09. Rape: in 1940, 3.02; in 1950, 4.05; in 1957, 7.53.

In *Category A* (for each of the three years represented) which we label "High Ratio Offenses," the disparity is especially noticeable. In *Category B,* which is labeled "Medium Ratio Offenses," the disparity between the rates of the two races is somewhat less and in *Category C,* labeled "Low Ratio Offenses," the differences are still less, though still significant.

It should also be noted that there is some shifting of offenses from one category to another during the three sample years 1940, 1950, and 1957; for instance, larceny which is a "High Ratio Offense" in 1940 became a "Medium Ratio Offense" in both 1950 and 1957. On the other hand, rape, which falls within our *Category B* for 1940 and 1950, becomes a "High Ratio Offense" in 1957.

Any appraisal of crime among Negroes must be prefaced by the evidence, amassed through the years in innumerable studies primarily of sociologists, but also provided by the authorities dealing with Negro-white relationships, of prejudice, discrimination, and differential treatment in law-enforcement. This is true in both the North and the South. Although there are many factors to be considered, there is no doubt that differential treatment of Negroes by arresting authorities, magistrates, juries, and judges, plays an important part in assaying the crime rate of this vast minority group.

Gunnar Myrdal, in *An American Dilemma,* states that if Negroes commit more crimes than whites it is because of: (1) discrimination; (2) poverty, ignorance of the law, and lack of influential connection; and (3) the existence of the slave tradition and the caste situation.[9]

Professor Donald Taft sums up the differential treatment accorded the Negro, so far as his crime rate is concerned:

> Negroes are more likely to be suspected of crime than are whites. They are also more likely to be arrested. If the perpetrator of a crime is known to be a Negro the police may arrest all Negroes who were near the scene—

---

[9] Gunnar Myrdal, *An American Dilemma* (New York: Harper, 1944), pp. 974-975. For an analysis of the psychological cost the Negro pays in our culture, see Abram Kardiner and Lionel Oversey, *The Mark of Oppression* (New York: Norton, 1951).

a procedure they would rarely dare to follow with whites. After arrest Negroes are less likely to secure bail, and so are more liable to be counted in jail statistics. They are more liable than whites to be indicted and less likely to have their cases nol prossed or otherwise dismissed. If tried, Negroes are more likely to be convicted. If convicted they are less likely to be given probation. For this reason they are more likely to be included in the count of prisoners. Negroes are also more liable than whites to be kept in prison for the full terms of their commitments and correspondingly less likely to be paroled.[10]

With more impartial treatment by all law-enforcement units, Negro crime might be considerably reduced. But differential treatment by the police and courts is not the whole story by any means. Into the hopper of perplexities must be thrown faulty education, discrimination by a large segment of the dominant white group, family disorganization induced by the competitive struggle to survive in a market-place where exist few equal opportunities, and despair in attempting to accommodate to urban life.

When we recognize the large number of major crises in the history of the Negro group since its arrival on American shores, we are not surprised that so many have wandered into trouble. Brought here against their will, each succeeding generation had nothing to look back upon but slavery or, at best, membership in a subject race and little to look forward to but a life of social, political, and economic handicaps. In the South, the Negro has been obliged to ingratiate himself, and through self-effacement to carry the unjust burdens of his race. In the North, with much more prejudice—subtle and adroit—than is usually admitted, the Negro has been forced to develop a more militant stand in order to survive. Many of them have been forced to live by their wits and have turned to a life of crime. In any treatment of crime among Negroes, economic factors loom large, as is the case with any aspect of the crime problem. The federal government's persistent efforts to expand civil rights as well as the programs of Fair Employment Practices, now operating in some states and large cities, have had a salutary effect on racial relations.

Two recent studies dealing with the problem of Negro crime are worthy of brief comment. The first, by Marvin Wolfgang, dealing with criminal homicides in Philadelphia during a five year period, indicates that Negroes were involved in a much higher ratio than were whites. He found that of the 588 victims of murder included in the study covering the years 1948-1952, 73 per cent were Negroes, and of the 621 offenders, 75 per cent were Negroes. He adds that in 1950, only 18 per cent of the city's population was Negro; thus they greatly exceeded their quota of homicides. His data show that Negroes had over three times their "share" of victims and four

---

[10] *Criminology,* 3rd edition (New York: Macmillan, 1956), p. 134. See also Sidney Axelrad, "Negro and White Male Institutionalized Delinquents," *Amer. J. Soc.,* Vol. 57, No. 6 (May, 1952), pp. 569-574.

times their "share" of offenders. He concludes that there is "a significant association between race and homicide." But he does not offer an explanation.[11]

The second study, *Racial Factors and Urban Law-Enforcement,* by William Kephart of the University of Pennsylvania, also has Philadelphia as its locale. It deals primarily with some of the major racial problems confronting a large metropolitan police force. He found little differential treatment on the part of the police and thus minimizes it as a significant cause for the high Negro crime rate. The author of the study is alarmed at this high Negro crime rate and maintains that it presents more of a challenge to the "Negro community" than does fair employment or educational opportunity. In fact, he contends that the responsibility for Negro crime "falls primarily on the Negro community itself" and adds that "thus far acceptance of the responsibility has lagged." He also recommends that white organizations—friends of the Negro—should turn some of their attention from "equality-of-opportunity for Negroes" to the problem of Negro crime.[12]

Apparently there is a trend developing which attempts to place responsibility for the high Negro crime rate on the "Negro community." It is alleged that Negro leaders and friends of the Negro among the white race are engaging in a "conspiracy of concealment" of the facts in order to "play down" or minimize the *real* crime rates of the Negro population.[13]

Attempts to identify the "Negro community" are futile and naïve. Just as among members of the white race, there are many status groups among Negroes in a large community. It is just as difficult for Negro religious or political leaders to focus their attention on a specific problem as it is for their counterparts in the white race. In short, to place responsibility for Negro crime solely on the Negro community oversimplifies the task.

Whether or not there is a "conspiracy of concealment" among Negro and white leaders in order to cover up arrests, discourage the gathering of Negro statistics, or of failing to report instances of discrimination, such a policy is shortsighted. It represents a repudiation of facts which are there whether "we will or no." No real understanding of a problem can ever crystallize if facts are to be ignored, repudiated, or "uncollected." On the other hand, if we are to understand and cope with the problem of crime—either white or Negro—we must continue to engage in efficient and relentless studies of the facts as they exist, and to make intelligent and courageous interpretations of those facts. Only then can white and Negro leaders work together

---

[11] Marvin E. Wolfgang, *Patterns in Criminal Homicide* (Philadelphia: University of Pennsylvania Press, 1958), pp. 31-32.

[12] William M. Kephart, *Racial Factors and Urban Law-Enforcement* (Philadelphia: University of Pennsylvania Press, 1957), pp. 174-175.

[13] See editorial, "The Negro Crime Rate: A Failure in Integration," *Time Magazine,* Vol. 71, No. 16 (April 21, 1958), p. 16.

ARREST RATIOS OF THE NEGRO AND WHITE POPULATIONS
OF THE UNITED STATES, 1940

	NEGROES		WHITES		NEGRO/WHITE RATIOS
	Arrests	Rate	Arrests	Rate	(White = 1)
Category A:					
High-Ratio offenses					
Liquor Laws	4,700	52.27	5,074	5.68	9.20:1
Carrying Weapons	2,606	28.98	2,857	3.20	9.06:1
Assault	14,978	166.58	17,331	19.39	8.59:1
Gambling	5,563	61.87	6,809	7.62	8.12:1
Homicide	2,549	28.35	3,539	3.96	7.16:1
Robbery	4,077	45.34	8,643	9.67	4.69:1
Disorderly Conduct	8,255	91.81	19,583	21.91	4.19:1
Larceny	17,763	197.56	42,430	47.47	4.16:1
Category B:					
Medium-Ratio offenses					
Receiving	977	10.87	2,519	2.82	3.85:1
Suspicion	16,806	186.92	43,485	48.65	3.84:1
Prostitution	2,283	25.39	6,374	7.13	3.56:1
Burglary	8,531	94.88	25,269	28.27	3.36:1
All Other	8,899	98.97	27,686	30.97	3.20:1
Narcotics	968	10.77	3,118	3.49	3.09:1
Rape	1,333	14.83	4,385	4.91	3.02:1
Sundry Traffic	3,282	36.50	11,441	12.80	2.85:1
Category C:					
Low-Ratio offenses					
Vagrancy	10,485	116.61	39,500	44.19	2.64:1
Not Stated	836	9.30	3,228	3.61	2.58:1
Arson	188	2.09	867	.97	2.15:1
Offense Against Family	1,247	13.87	6,431	7.20	1.93:1
Auto Theft	1,983	22.05	10,928	12.23	1.80:1
Other Sex Offenses	1,426	15.26	7,800	8.73	1.75:1
Drunkenness	14,192	157.84	92,629	103.63	1.52:1
Embezzlement	2,206	24.54	16,475	18.43	1.33:1
Forgery, Counterfeiting	646	7.18	6,332	7.08	1.01:1
Driving While Intoxicated	1,967	21.88	24,962	27.93	.78:1

Note: Arrest data from *Uniform Crime Reports,* Vol. 11, No. 4 (January, 1941), Table 104, p. 223; arrest rates are based on 100,000 adult population. Populations from 1940 census, 15 years of age or older: Negroes, 8,991,232; Whites, 89,387,375.

ARREST RATIOS OF THE NEGRO AND WHITE POPULATIONS
OF THE UNITED STATES, 1950

	NEGROES		WHITES		NEGRO/WHITE RATIOS
	Arrests	Rate	Arrests	Rate	(White = 1)
Category A:					
High-Ratio offenses					
Narcotics	4,262	41.57	3,939	3.99	10.42:1
Carrying Weapons	5,198	50.69	5,082	5.14	9.86:1
Gambling	7,462	72.77	7,584	7.67	9.49:1
Liquor Laws	5,306	51.75	5,841	5.91	8.76:1
Assault	27,619	269.35	31,277	31.65	8.51:1
Homicide	2,889	28.18	3,372	3.41	8.26:1
Prostitution	3,260	31.79	5,190	5.25	6.06:1
Robbery	7,060	68.85	12,517	12.66	5.44:1
Category B:					
Medium-Ratio offenses					
Receiving	1,050	10.24	2,209	2.24	4.57:1
Larceny	20,672	201.60	44,776	45.30	4.45:1
Disorderly Conduct	13,610	132.73	31,217	31.58	4.20:1
Rape	2,717	26.50	6,473	6.55	4.05:1
Suspicion	13,054	127.31	32,751	33.14	3.84:1
Burglary	11,534	112.49	31,776	32.15	3.50:1
Sundry Traffic	7,191	70.13	20,419	20.66	3.39:1
All Other	9,454	95.20	28,346	28.68	3.15:1
Category C:					
Low-Ratio offenses					
Arson	241	2.35	806	.82	2.87:1
Offenses Against Family	3,415	33.30	11,708	11.85	2.81:1
Vagrancy	10,657	103.93	37,157	37.59	2.76:1
Not Stated	1,555	15.17	6,179	6.25	2.43:1
Auto Theft	3,500	34.13	14,695	14.87	2.30:1
Other Sex Offenses	3,473	33.87	16,057	16.25	2.08:1
Drunkenness	30,040	292.97	143,867	145.56	2.01:1
Forgery, Counterfeiting	1,689	16.47	9,927	10.04	1.64:1
Embezzlement, Fraud	2,962	28.89	18,346	18.56	1.56:1
Driving While Intoxicated	5,706	55.65	44,911	45.44	1.22:1

Note: Arrest data from *Uniform Crime Reports,* Vol. 21, No. 2 (January, 1951), Table 44, p. 112; arrest rates are based on 100,000 adult population. Populations from 1950 census, 15 years of age or older: Negroes, 10,253,780; Whites, 98,836,950.

**ARREST RATIOS OF THE NEGRO AND WHITE POPULATIONS
OF THE UNITED STATES, 1957***

	NEGROES		WHITES		NEGRO/WHITE RATIOS
Category A:	Arrests	Rate	Arrests	Rate	(White = 1)
High-Ratio offenses					
Gambling	37,102	326.69	12,953	12.06	27.09:1
Narcotics	4,108	36.17	3,092	2.88	12.56:1
Weapons, Carrying and					
Possessing	8,863	78.04	7,814	7.27	10.73:1
Robbery	6,158	54.22	5,517	5.14	10.55:1
All Assaults	51,361	452.25	52,787	49.14	9.20:1
Homicide	1,479	13.02	1,734	1.61	8.09:1
Rape	2,087	18.38	2,623	2.44	7.53:1
Disorderly Conduct	97,628	859.64	141,057	131.31	6.55:1
Prostitution	5,054	44.50	7,520	7.00	6.36:1
Category B:					
Medium-Ratio offenses					
Liquor Violations	16,000	140.88	26,859	25.00	5.64:1
Suspicion	30,277	266.60	53,789	50.07	5.32:1
Offenses Against Family	7,639	67.26	14,624	13.61	4.94:1
Receiving Stolen Goods	1,263	11.12	2,572	2.39	4.65:1
Larceny—Theft	30,826	271.43	70,701	65.82	4.12:1
Sex Offenses Other Than					
Rape	6,226	54.82	14,492	13.49	4.06:1
Burglary—Breaking and					
Entering	14,989	131.98	36,058	33.57	3.93:1
All Other Offenses	64,511	568.04	162,921	151.66	3.75:1
Category C:					
Low-Ratio offenses					
Vagrancy	16,298	143.51	51,679	48.11	2.98:1
Drunkenness	188,323	1,658.24	610,051	567.90	2.92:1
Auto Theft	5,716	50.33	23,095	21.50	2.34:1
Embezzlement, Fraud	3,037	26.74	13,030	12.13	2.20:1
Forgery, Counterfeiting	1,307	11.51	6,925	6.45	178:1
Drunken Driving	15,776	138.91	84,074	78.26	1.77:1

* Traffic Violations and Arson no longer Recorded.

Note: Arrest data from *Uniform Crime Reports,* Vol. 28, No. 2 (April 23, 1958), Table 44, p. 118; arrest rates are based on 100,000 adult population. Populations estimated, 15 years of age or older: Negroes, 11,356,811; Whites, 107,422,720.

in attempts to improve all facets of community life, one of which is the high crime rates of both races.

The Supreme Court decision on segregation in schools and transportation facilities has precipitated another crisis in the field of race relations. The vast majority of the white people in the South accept the decision and its ramifications as inevitable but are concerned as to its implementation. Extremists have muddied the water and through their uncompromising position have militated against the complete acceptance of the court's action. While this serious dilemma is outside the province of this book, it is obvious that the age-old prejudice against the Negro has been intensified in many quarters. This, in turn, is undoubtedly reflected in the crime rate.[14]

Races other than the Negro

Sections other than the North and South also have their "racial conflict" problems. Wherever a minority group is large enough to disturb the *status quo* of white or Old American supremacy, there we shall find friction. And this friction will call forth from the dominant group all the bugaboos of inferiority of the racial minority.

The South and North are not disturbed, for example, concerning the Japanese population, but the states along the West Coast are. Thus we shall make a few comments about the crime rates of the Japanese, Chinese, and Indians although they are insignificant so far as the over-all picture of crime is concerned in this country.

Much as the Oriental has been shunned in the United States it is essentially true that neither he nor his children can be charged with a high delinquent or criminal rate. The Japanese immigrant, especially, has been discriminated against prior to and during World War II. But the war merely aggravated a deep-seated antagonism that has its roots in economic competition. The familiar "red herring" of race has been used to rationalize the fear that the whites, especially in agriculture, have felt when they have been confronted by the Japanese worker. In more recent years, however, a noticeable attitude of tolerance has developed along the West Coast for both the Japanese-American and the Mexican-American.

In 1957 the total number of Japanese arrested for all offenses was 273, according to the *Uniform Crime Reports*. Of this group only 1 was arrested for homicide. There were 103 arrested for drunkenness, 5 for burglary, 19 for larceny, and 2 for rape. These small rates are ample proof that the Japanese are a relatively law-abiding group.[15]

The Chinese have been a traditionally law-abiding people. In China native villagers were controlled primarily by the group customs, particularly

---

14 See *The Annals,* "Racial Desegregation and Integration," Vol. 304 (March, 1956) for a symposium dealing with this knotty problem.
15 Vol. 28, No. 2 (April 23, 1958), Table 44, p. 118.

the patriarchal system, which looked upon a digression from the family traditions as a disgrace, not only to the individual but to his entire family, living, dead, and unborn. Aside from the banditry indigenous to China, much of which was precipitated by political unrest, the Chinese boy grew up knowing little about delinquent behavior. The impact of Western culture, however, has wrought changes in the old familial pattern. In recent years China has codified as much law as is possible in a country experiencing so much political upheaval. In the postwar era, it will doubtless experience much of the crime and delinquency that has characterized America.

In this country, where, according to the 1950 census, there are 117,629 Chinese, they are of necessity quite clannish. Much of their community behavior is regulated by self-appointed control groups. Tong warfare was formerly characteristic of metropolitan life among American Chinese, but today few of this race carry on any outside criminal activity.

During 1957, 267 Chinese were arrested. There was 1 arrest for homicide, 4 for burglary, and 24 for larceny. Thirteen were arrested on narcotics charges.[16]

Another minority group is the American Indians, most of whom still live on reservations. The 1950 census tallied 343,110 Indians. Their 1957 arrests amounted to 37,715 the bulk of them for drunkenness—30,026. There were 7 arrested for homicide, 189 for burglary, and 20 for rape.[17]

Although there is still some discrimination against the Indian, this group is protected today by a much more sympathetic governmental policy. There is still a great deal to be done, however, to make up for the exploitation and neglect of these people throughout the past two centuries.

---

[16] *Ibid.*
[17] *Ibid.*

12

Home and
Community Influences

The home is the cradle of human personality. Each person, from the moment of birth is deeply influenced by the people around him. The baby is born not knowing what to think or how to feel about life, but ready to learn, and learn he does, willy nilly. From a warm, loving, stable family, the child learns that people are friendly, worth knowing, and can be depended upon. When a family is cold, despairing, rejecting, or neglectful, the child learns distrust, hostility, or downright hatred of people. Such families are to be found in all economic, cultural, racial, national, and educational backgrounds. (From Report on Home Responsibility, *Washington, D.C.: U.S. Government Printing Office, 1947.)*

NATIONAL CONFERENCE ON PREVENTION AND
CONTROL OF JUVENILE DELINQUENCY, 1946

Poor Housing

Through the years perhaps the most frequently heard cause of delinquency and crime is parental inadequacy. It is easy to make the home and the parents scapegoats and "whipping boys" for the high crime rate. The hue and cry to bring parents "to book" for the depredations of their children is well-nigh pathological as well as universal. Ordinances have been passed in many communities to hold parents financially responsible for the mischief of their children; compulsory courses have been inaugurated for parents whose children have run afoul of the law. Such measures indicate how desperate and frustrated the authorities are in their attempts to reduce the amount of delinquent behavior.

In the last analysis, lack of parental control and absence of parental insight provide the basis for the aberrant behavior of children, many of whom grow into adult criminals. The reason there is not more delinquency is that millions of families, living in congested and blighted neighborhoods, with large numbers of children, somehow manage to rear their children into reasonably good citizens. Crude and perhaps ignorant of the more sensitive values of life they may be, they somehow possess some knowledge of what is expected of parents in order to rear children who are to become our future citizens.

In previous chapters we discussed the emotional factors that sometime arise in the rearing process, in homes of varying scales of living. We have also dealt with the problem of low income and poverty generally. In this chapter we deal with inadequate home life, especially among low income groups, and also with the hazards of the neighborhood and community. But we cannot overlook the fact that large numbers of delinquents come from both middle class groups and, to a larger degree than is willingly admitted, from the wealthier elements of our population.[1]

It is almost too much to expect parents to rear wholesome children in areas of tumbledown housing where the minimum decencies of physical existence are lacking or, at best, are crude and defective. In hundreds of urban centers, as well as in rural areas, millions of families still are deficient in their modern sanitary needs. Outside toilets and outdoor water taps, with no hot water except what can be heated on crude oil stoves, are much more prevalent than the opponents of public-supported housing are willing to admit. High rentals per room are characteristic of slum areas; this means doubling up of families or members of families and destroying all vestiges of privacy. Landlords defy housing regulations and manage to get away with such selfish neglect because so many municipalities cannot or will not enforce the laws. Much of this is due to graft or collusion between venal politicians and greedy landlords.

A report from Birmingham, Alabama, states that a study of the city's 52 census tracts revealed that "if high population density, low economic value of houses, physical deterioration and lack of necessary equipment, and high rate of tenancy can be evaluated as indicators of poor housing, the findings of this investigation present objective confirmation of the general principle that poor metropolitan housing tends to be associated with high rates of delinquency, and *vice versa*."[2]

A study covering children referred to the Municipal Court of Philadelphia as delinquents revealed that 32 per cent lived in neighborhoods that had relatively high delinquency ratios, *i.e.*, at or over twice the average for

[1] See Alison W. Davis and Robert J. Havighurst, *Father of the Man* (Boston: Houghton, Mifflin, 1947).

[2] Quoted in *Focus*, Vol. 28, No. 2 (March, 1949), p. 59. See also Gordon H. Barker, "Juvenile Delinquency and Housing in a Small City," *J. Crim. Law*, Vol. 45, No. 4 (November–December, 1954), pp. 442–444.

the city as a whole. They live in ten per cent of the census tracts of the city that have a child population amounting to 11 per cent of the total city-wide child population. The area of highest delinquency extends about 2¾ miles north from the center of the city, 1⅛ miles to the south, 1¼ miles to the east, and from 1 to 2 miles to the west (following the Schuylkill River). Although this area covers only one-twelfth of the city's territory, it contains almost one-fourth of its population and almost one-half of all the city's delinquent children. The delinquency ratio of this area is 40 per 1,000; that is, almost twice as high as the city-wide average (21 per 1,000). More than half of all census tracts with highest population density (121 to 150 persons per acre), and exactly one-half of all census tracts falling into the second highest category of population density (91 to 120 persons per acre) are to be found in this area.

In this area are located almost two-thirds of all census tracts in the city that show less than 20 per cent owner-occupied dwelling units; in other words, 80 per cent or more of the dwelling units are rented homes, apartment houses, or rooming houses. The percentage of sub-standard housing (according to the U. S. Census report, a dwelling unit that is in need of major repair or has no private bath) in this area was more than twice the average for the whole city; namely 39.7 as compared to 17.8 per cent. Almost three-fourths of all the geographical units (or combinations of census tracts) in this area show above average figures for both sub-standard housing and juvenile delinquency.[3]

There are reports from several urban communities pertaining to the wholesome low cost housing projects that replace slum areas. As one of the results of such public housing projects (federal or municipal) the lower delinquency rate is often mentioned. For instance in New Haven, an area which from 1924 to 1940 produced a juvenile delinquency rate of 3.18 per hundred children annually reported an annual average of 1.64 per hundred children during 1940 to 1944 after its conversion into a new housing project area.[4]

Studies conducted in Chicago during 1948 pointed out some of the results of public housing, viz., the fact that school attendance of children in the housing projects, as compared to the surrounding slum areas, had been decidedly better. In this study, which covered 4,400 children in the housing projects and a total of 13,300 children in the surrounding neighborhoods, it was found that whereas 33 per cent of the children in all of the schools studied came from the projects, only 27 per cent of all unexcused absences, particularly truancy, were caused by children in the projects. In some of the projects the difference was even more marked; for example, in

---

[3] For further details, see J. O. Reinemann, "Where Do Philadelphia's Delinquent Children Live?", Thirty-Second Annual Report, Municipal Court of Philadelphia, (1945), pp. 379-388.

[4] Reported in the Journal of Housing, Vol. 3, No. 1 (January, 1946), p. 27.

one of the schools studied, 43 per cent of the children came from the housing project but only 28 per cent of the unexcused absences recorded were of children from this project.[5]

It is not always fair to compare delinquency rates in housing projects with the city average; rather a comparison should be made of these rates with the figures obtained for the surrounding slum areas. Thus in Cleveland, out of a total of 4,018 children living in six low-rent housing projects, the rate of delinquency cases in 1947 averaged 1.57 per cent. The rate for the city as a whole was 1.8. And in the slum areas adjacent to the housing projects, the average rate was 2.26 per cent.[6]

There can be no doubt that poor housing can be correlated with delinquency and crime rates, but once again we can only see an indirect relationship between the two. Thousands of parents, living in sub-standard homes, manage somehow to rear their children to be useful citizens.

Broken and Disorganized Homes

The well-integrated and socially mature home cannot be duplicated, but that utopian ideal is rare in these confused days when the stresses and strains of modern life make it extremely difficult to attain "peace of mind." Of course, not all confused homes produce delinquents but they are not especially healthy places in which to rear children. What constitutes a "normal" home? Many years ago Dr. Miriam Van Waters set down what she thought the home should furnish the child:

> The home has primary tasks to fulfill for its young: to shelter and nourish infancy in comfort, without inflicting damage of premature anxiety, enable the child to win health, virility and social esteem; to educate it to meet behavior codes of the community, to respond effectively to human situations which produce the great emotions, love, fear and anger; to furnish practice in the art of living together on a small scale where human relationships are kindly and simple; finally the home has as its supreme task the weaning of youth, this time not from the breast of the mother, but from dependence, from relying too much on that kindliness and simplicity of home, so that youth may not fail to become imbued with joy of struggle, work and service among sterner human relationships outside.[7]

How many homes can measure up to such standards? What Dr. Van Waters wrote in 1925 cannot be improved on today. Juvenile maladjustment and, to some degree, delinquency, occur not only in the inadequate home but wherever little insight into the needs of the child is found. We are

---

[5] Quoted from James S. First, "Public Housing Measured," *The American City,* Vol. 64, No. 2 (February, 1949), pp. 97-98.

[6] Quoted from *Journal of Housing,* Vol. 5, No. 10 (October, 1948), p. 265.

[7] Dr. Miriam Van Waters, *Youth In Conflict* (New York: The New Republic, 1925), p. 64.

concerned with the broken home, whether it be a *psychologically* broken home, or a *physically* broken home. The former is well described as "a tyranny ruled over by its meanest member."[8] It is the home where both parents reside physically, but where there is constant bickering, little respect for the rights of each individual, and where the child is "pushed around" or ridiculed. It is the authoritarian home, in which the father assumes the old-fashioned patriarchal role and the wife and children are relegated to a passive status. In such homes the child is too often rejected, never having the genuine experience of "belonging," and becomes, as a result, desolate, anxious, restless, or often hostile. Our child-guidance clinics are full of such cases and there is evidence that thousands of others unfortunately never get to the clinics. They are supposed to "outgrow" their peculiarities.[9]

There are, then, two types of broken homes; the type where one parent is missing, making home life atypical in carrying on its responsibilities toward the individual members; and the psychologically broken, or disorganized, home where cross purposes are more conspicuous than harmony.

Is there a high correlation between the physically broken home—one in which one or both parents are missing, dead, divorced, or deserted—and delinquency? There are many studies of this angle of the problem and, as in other phases of the subject, there is not complete agreement. We shall cite some of the traditional studies to show this disagreement.

The early study of Breckinridge and Abbott, *The Delinquent Child and the Home,* in an analysis of 13,000 cases of delinquent children, shows that 34 per cent came from broken homes.[10]

Figures compiled annually by the Municipal Court of Philadelphia, regarding home conditions of delinquent children, show that these children come from broken homes in approximately 47 per cent of all cases. The reports further show that lack of normal home life among girls is manifest in a greater percentage of cases than among boys.[11]

A four year study by the California Youth Authority showed "that 62 per cent of the states' juvenile delinquents were the result of broken homes."[12] Figures from various sections of New Jersey show that 50 per cent of the children coming before juvenile courts were from broken homes. Homes in which fathers were deceased led the list. Following in order were homes deserted by fathers, separated parents, mothers deceased, and lastly, divorced parents.[13]

---

8 *Ibid.,* p. 48.

9 Cf. David Abrahamsen, "Family Tension, Basic Cause of Criminal Behavior," *J. Crim. Law,* Vol. 40, No. 3 (September-October, 1949), pp. 330-343.

10 Sophonisba Breckinridge and Edith Abbott, *The Delinquent Child and the Home* (New York: Russell Sage Foundation, 1912), pp. 91-92.

11 *Forty-Third Annual Report,* 1956, p. 44.

12 Quoted in *Federal Probation,* Vol. 12, No. 4 (December, 1948), p. 59.

13 Report of the Governor's Committee on Youth, *The Welfare Reporter,* (September, 1950), p. 8.

On the other hand, a study cited by the Wickersham Commission of 40,503 children appearing in 93 different courts showed 64 per cent living with both parents, whereas the percentage coming from broken homes was only 36.[14] The same report includes other significant figures: "The statistics for the year 1927, from the Children's Bureau, U. S. Department of Labor, showed that 67 per cent of 16,258 delinquent boys and 48 per cent of 3,040 delinquent girls were living with both parents. Thus 33 per cent of the boys and 52 per cent of the girls came from broken homes."[15] Students of this problem generally agree that the percentage of female delinquents from broken homes is higher than that of males, but as H. Ashley Weeks contends, this difference is more apparent than real.[16] Nevertheless, Weeks finds a positive relationship between delinquency and the broken home.

Sheldon and Eleanor Glueck, in their evaluation of 1,000 cases of juvenile delinquency, found that the broken home or "poorly supervised home" plays a large role. They state: "Our delinquents come largely from homes which were for one reason or another broken or distorted. There can be no doubt that these boys had an unwholesome home life. Even the possibility of nondelinquents having so high an incidence of inadequate homes would not make it less necessary to take into account the home background in developing any treatment program for delinquency."[17] This study, of course, takes both types of broken homes into consideration. The Gluecks' study of 500 delinquent and 500 nondelinquent children points up the importance of an adequate home life in steering the child along socially approved avenues.[18]

The White House Conference study of the delinquent child states: "Estimates as to the prevalence of this condition [broken home] in the histories of juvenile delinquents range from about 20 to nearly 50 per cent.[19] Professor Lowell Carr minimizes the role of the physically broken home so far as delinquency is concerned. He points out that perhaps one-eighth of the homes in America are physically broken and their contribution to delinquency is slight indeed. He further states that "most of our actual delinquents" continue to come from homes that are not so broken.[20]

[14] Report, National Commission on Law Observance and Enforcement, "Report on the Causes of Crime," Vol. 1, p. 67.

[15] Quoted ibid., n., from Third Annual Report, Children's Bureau, Table 4, "Juvenile Court Statistics for 1929."

[16] H. Ashley Weeks, "Male and Female Broken Home Rates by Types of Delinquency," Amer. Soc. Rev., Vol. 5, No. 4 (August, 1940), pp. 601-609.

[17] Sheldon and Eleanor Glueck, One Thousand Juvenile Delinquents (Cambridge: Harvard University Press, 1934), p. 76. Reprinted by permission of the President and Fellows of Harvard College.

[18] Sheldon and Eleanor Glueck, Unraveling Juvenile Delinquency (New York: Commonwealth Fund, 1950), p. 287.

[19] The Delinquent Child, Report of the Committee on Socially Handicapped Delinquency, (New York: Century, 1932), p. 351.

[20] Lowell Carr, Delinquency Control (New York: Harper, 1940), p. 140.

Studies dealing with the family status of children incarcerated in institutions show almost universally that the bulk of the inmates come from broken homes. This is especially true of girls. This phenomenon is not unusual since juvenile court judges are traditionally obsessed with the broken-home-causes-delinquency thesis or they despair of any other disposition of such cases; thus they feel constrained to send their charges to institutions. Of course, hundreds of children from broken homes are placed in foster homes also, whether they are delinquent, neglected, or dependent.

Is there a positive relationship between the broken home and delinquency? Apparently there is no open and shut answer. It might help, however, if we knew how many homes in the nation are broken or what percentage of broken homes manage to rear children without having a history of delinquency to mar them.[21]

There are very few studies that will give an answer to such questions. Shideler, on the basis of the census of 1910, estimated that approximately 25 per cent of all children in the United States lived in homes that might be called "broken."[22] Slawson found that 45 per cent of the delinquent boys sentenced to New York State correctional schools, but only 19 per cent of a group of 3,198 boys from three public schools in New York City, came from broken homes.[23] Shaw and McKay, in a study for the Wickersham Commission, concluded that there "is no consistent relationship between rates of broken homes and rates of delinquency."[24] The White House Conference, in evaluating this study, felt that the "rate of broken homes in the general population had probably been under-estimated."[25]

One can probably leave this phase of the subject by stating that the phenomenon of the physically broken home as a cause of delinquent behavior is, in itself, not so important as was once believed. In essence, it is not the fact that the home is broken but rather, that the home, or family life, is inadequate.

Many mothers, and to some degree fathers, separated or divorced from their mates, or who are attempting to operate their homes with their mates deceased, are doing a splendid job of rearing their children without delinquency manifesting itself. It is not so much the physically broken home, as it is the home that is confused or inadequate because of ignorance, indifference, or a faulty conception of the child's place in the general scheme of things, that contributes to the high delinquency rates. Professor Charles W. Coulter lists such disorganized homes as follows:

[21] For further analysis of this problem see Harry M. Shulman, "The Family and Juvenile Delinquency," *The Annals,* Vol. 261 (January, 1949), pp. 21-31.

[22] E. H. Shideler, "Family Disintegration and the Delinquent Boy in the United States," *J. Crim. Law,* Vol. 8, No. 5 (January, 1918), pp. 709-732.

[23] John Slawson, *The Delinquent Boy* (Boston: Badger, 1926), p. 359.

[24] Report, National Commission on Law Observance and Enforcement, "Report on Causes of Crime," Vol. 2, p. 67.

[25] *The Delinquent Child,* Report of the Committee on Socially Handicapped Delinquency, (New York: Century, 1932), p. 351.

1. Homes with criminal patterns;

2. Homes in which there are unsatisfactory relations because of domination, favoritism, nonsolicitude, overseverity, neglect, jealousy, a stepparent, or other interfering relative;

3. Homes in which one parent has a physical or mental disability—invalidism, feeblemindedness, blindness, deafness, psychoneurosis;

4. Homes socially or morally maladjusted because of differences of race, religion, conventions, and standards, or an immoral situation;

5. Homes under economic pressures—unemployment, low income, homes in which mothers work out.[26]

This is, of course, the thesis of the Gluecks' definitive work, *Unraveling Juvenile Delinquency,* to which we alluded above.

A study made in Connecticut, the results of which were presented to the General Assembly of the state in 1947, followed the careers of 4,538 delinquent and 898 neglected children. The conclusions reached were as follows: (a) the central causal factor behind most child neglect and delinquency lies in family disorganization; (b) the causal factor in family disorganization most frequently is emotional instability of the parent; (c) a disorganized family life sets in motion processes harmful to children, who may react to them by becoming delinquent or by developing traits that lead to breakdown later on; (d) from disorganized families come other serious and costly social breakdowns such as mental disease, mental deficiency, crime, and divorce.[27]

Recent clinical and pragmatic evidence into the psychological mechanisms by which parental attitudes may foster delinquency has been discussed by Johnson and Robinson in connection with antisocial sexual delinquencies:

> The research on causes of "individual delinquency" carried out during the past 15 years by Szurek and one of us suggested a new approach to the problem of the sexual deviant. By means of intensive treatment of parents of the non-gang-member delinquent child, we demonstrated that repeated stealing, arson, and vandalism are stimulated by unconscious or rarely conscious antisocial impulses in the parents. The parents derive unconscious but real gratification from the enactment by the child or adolescent of impulses socially forbidden the parent. In very large numbers of the cases studied of serious consequence deficiencies in children of "good families," parental fostering of antisocial behavior was discovered in every instance in which an adequate search was possible or was permitted.[28]

---

[26] "Family Disorganization as a Causal Factor in Delinquency and Crime," *Federal Probation,* Vol. 12, No. 3 (September, 1948), pp. 13-17.

[27] For details see *Needs of Neglected and Delinquent Children,* Hartford, Connecticut, 1946; for a summary of the findings see Reginald Robinson, "Beneath the Surface of Juvenile Delinquency and Child Neglect," *Survey Midmonthly,* Vol. 83, No. 2 (February, 1947), pp. 41-52.

[28] Adelaide M. Johnson and David B. Robinson, "The Sexual Deviant (Sexual Psychopath)—Causes, Treatment, Prevention," *Journal A.M.A.,* Vol. 164, No. 14 (August 3, 1957), pp. 1559-1565.

The National Conference on Prevention and Control of Juvenile Delinquency (1946) issued a report on Home Responsibility that attempts to place the responsibility of wholesome childhood on the family unit without placing the element of blame, which is so often done. The conclusion of this report is that the family cannot do the job alone but must have the support of community agencies.[29]

Community Influences

The Press and Crime News

Aside from the home the child is confronted with what Clinard refers to as "secondary community influences."[30] We are all familiar with the roles played by school, church, playgrounds, and character-building agencies. We prefer to postpone discussion of these important agencies until a later chapter dealing with prevention. In this chapter we shall discuss the influences of the newspaper, radio, motion picture, television, comic book, and other phases of our culture that are often charged with adversely influencing the child. We shall start with the press.

The influence of the press on the increase of delinquency and crime has been widely debated for years. Certainly the amount of space devoted to crime news has greatly increased during the past quarter century and many papers, notably the tabloids, have made it their stock in trade. It may be that this practice, through suggestion, stimulates the very commission of crime.

Naturally journalists and editors take the positon that all news that reflects human interest is the stock in trade of the newspaper; that this sort of human interest material is of vital concern to the public, and that any restrictions placed on the press is an encroachment upon its freedom. The late Grove Patterson, one-time editor of the Toledo *Blade,* speaking before a crime conference many years ago, expressed the opinion generally held by newspaper people:

> I will say, simply and frankly, at the very start that in my judgment the newspapers will go on printing all the crime news that is available, in the future as in the past, and that furthermore they should continue to do so as a matter of good newspaper making and sound public policy. Only publicity will awaken our people to the prevalence, constancy, and importance of crime. Only publicity will arouse public opinion. Publicity, more than anything else, will stir laggard public officials to courageous action. This publicity, copious and unpleasant, frequently hated by every conservative element in the community, disliked and deplored by the newspapers themselves, offers the surest and most sweeping approach to the clean-up.[31]

---

[29] *Report* No. 16, (Washington, D.C.: U.S. Government Printing Office, 1947).

[30] Marshall Clinard, "Secondary Community Influences and Juvenile Delinquency," *The Annals,* Vol. 261 (January, 1949), pp. 42-54.

[31] *Proceedings,* Attorney General's Crime Conference, (Washington, D.C.: U.S. Government Printing Office, 1934, p. 84).

Policies concerning what news is "fit to print" and how it should be presented to the public can be formulated only by the news profession itself. The public does its bit by supporting the papers that meet with its approval. Censorship from without is fatal in a democracy.

But newspaper methods of playing up crime news do have a morbid effect on many constitutionally weak persons. So long as the public wants to read racy and banal crime news we can expect little constructive reform from the news fraternity. Yet, it is their responsibility, and it is gratifying to note that some papers (notably the *Christian Science Monitor*) play down crime news.

The newspaper editorial writer, F. Perry Olds, contends that many newspapers throughout the country (there are approximately 9,000) do not measure up to their civic responsibility so far as crime news is concerned, and feels it is a part of the job of those who understand correctional philosophy to enlighten editors and news writers.[32]

In any study concerning the effect of crime news on the public, sweeping generalizations must be avoided, although it might seem that the "war on crime" could be more effectively waged through the editorial page than through lurid headlines or detailed descriptions of sordid crimes.

There is no statistical method of arriving at the number of persons who enter criminal activity through what they read in the newspapers or magazines. No doubt there are many. The constant repetition of crime stories in the press can affect readers in two different and dangerous ways: it may affect some highly suggestible persons, among whom are many young people, to commit similar crimes; or it may create an indifference to law and order through the constant reiteration and exaggeration of the details of the crimes. Stable people, juveniles and adults alike, will be little affected by what they read. The unstable and many of the socially maladjusted may be somewhat affected, and it is from this suggestible and abnormal group that most of our delinquents come.

Arguments concerning the amount of crime news published by the press are somewhat futile. It is not the amount of the news but the space given to it, whether on the front page or tucked away on the inside, as well as the manner it is "dished up" by reporters and handled by city editors. Editor Paul Deland, of the *Christian Science Monitor*, takes the position that crime news *does* encourage delinquency and crime.[33] The American public knows the *Monitor* policy which, since its founding, "consistently refuses to sensationalize crime, to exploit admitted reader interest in crime, or to build circulation on any appeal to morbidity."[34] Says Mr. Deland:

---

[32] F. Perry Olds, "The Place of the Press in Crime Control," *Yearbook*, N.P.P.A. (1947), pp. 245-259.

[33] See his article "Crime News Encourages Delinquency and Crime," *Federal Probation*, Vol. 11, No. 2 (April-June, 1947), pp. 3 f.

[34] *Ibid.*, p. 5.

The wrong perspective of conditions given by selection, proportion and relative size of stories is obvious. But the writing up or glamorizing of a story is a mystery not so readily comprehended outside the newspaper office. "Dressing up" the story for circulation purposes is the trick that does most of the damage. What figures can be made to misrepresent by a clever manipulator is scarcely comparable to the damage a clever rewrite wordsmith can do in making a dull, drab, sordid crime story deceitfully glamorous, thrilling and important by a misuse of words and correlation of ideas.[35]

Some of his examples: "Lone Wolf Shouts in Court" spread across the front page in type 1½ inches high; another, the "Black Dahlia" case in California several years ago, in which the newspapers seized upon such a name to build up a sensation; and a third, "a debonair, nattily dressed young gunman who flashed an automatic in approved western style and laughed as he coolly leaped out of the window and made his escape leaving his victim, etc."[36] In the summer of 1957 the San Francisco papers, aided by the police, all but fastened a sadistic crime on an innocent man through lurid headlines to news stories.[37] The pseudo-sophisticated public generally assumes that all acquittals are due to "slick" lawyers, anyway. This is one effect of what Raymond Moley calls "trial by city desk." Pertinently he points out that it is not the amount of news that affects the outcome of the trial, but the manner in which it is presented to the public by the press:

When the attitude created by the newspaper account is hostile to the defendant, the tendency is to destroy the practical value of the presumption of innocence, if not before the court at least before the public whose attitude will have a very important relationship to the subsequent life of the defendant whether he be acquitted or not. Because of the twisted information gained from the newspaper accounts of some cases, thousands of persons probably still believe in the guilt of the defendants.[38]

Some newspapers, besides "trying the case in the paper," publish information that defeats the efforts of the police to round up the criminal. Some years ago the criminal court of Baltimore, Maryland, ordered that any of the following acts would be grounds for contempt charges: (1) making photographs of the accused without his consent; (2) the issuance by the police, the state's attorney, counsel for the defense, or any other person having official connection with a case, of any statement relative to the conduct of the accused, statements or admissions made by the accused, or other matter bearing upon the issues to be tried; (3) the issuance of any forecast as to the future course of action of either the prosecuting attorney

---

[35] *Ibid.*, p. 4.

[36] Cf. Agness Underwood, *Newspaperwoman* (New York: Harper, 1949), Chapter 6, "Cops and Crime," and Chapter 7, "A Gallery of Murderers."

[37] See "Trial By Headline," by Decca M. Treuhaft, *The Nation*, Vol. 185, No. 13 (October 26, 1957), pp. 279-282.

[38] *Our Criminal Courts* (New York: Putnam's, 1939), p. 93. Courtesy the publisher.

or the defense relative to the conduct of the trial; (4) the publication of any matter which may prevent a fair trial, improperly influence the court or the jury, or tend in any manner to interfere with the administration of justice; and (5) the publication of any matter obtained as a result of a violation of this rule.[39] Unfortunately, this ruling was declared unconstitutional by the Supreme Court in 1949.

But have not editors a responsibility to society that transcends pandering to the prejudices and emotion of their readers? The late Bruce Smith, prominent police consultant, stated the case of newspaper responsibility in so far as crime news is concerned: "So far, this great instrument [the press] has scarcely done its part. The most carefully formulated editorial policies, through which might be secured able discussions of law-enforcement problems, are often offset by a news policy deliberately designed to appeal to the prejudices of the unschooled and ignorant."[40] At best we can give the newspaper credit for dramatizing crime and arousing public opinion to the menace of large-scale criminality.

Most discussions of the relation of the press to crime are limited to the effect of journalistic practices on sensationalism, crime news that may stimulate some neurotics to the commission of crime, the effect of crime news and editorials on trials and sentences, and the like. But perhaps the most disastrous influence that newspapers—and, today, the movies and the radio as well—exert on the crime situation lies in the all but neglected fact that these agencies of publicity are the most powerful single force in preventing the reformation of criminals.

Save in the more backward states, most administrators of correctional institutions today sincerely believe that their purpose is to rehabilitate rather than to punish. But the public has been taught to fear convicts, though it has, apparently, little fear of the more numerous and more dangerous criminals who remain outside the institutions.

Many newspaper editors and reporters, movie newsreel producers, and radio commentators fully share the popular attitude toward criminals and prisons. They denounce parole, commutation, and probation and excoriate judges who try to deal with offenders for the ultimate good of society. They believe that a good prison is one that merely prevents escapes, even though in the process it almost inevitably also prevents any reformation of the inmates. Moreover, escapes and prison riots make the hottest kind of news and promote brisk circulation of papers. The quiet and unpublicized process of rehabilitating inmates has little or no news value whatever, and a reformed prisoner cannot compete in publicity value with an incorrigible criminal who makes a successful prison break.

Hence, all the popular agencies of news and communication pounce

---

[39] "Trial by Newspaper," *J. Crim. Law,* Vol. 32, No. 4 (November-December, 1941), p. 448.

[40] "Enforcement of the Criminal Law," *The Annals,* Vol. 217 (September, 1941), p. 13.

on any evidence of laxity in the prison, especially if this may lead to an escape. A warden whose administration has permitted an escape or a break is violently denounced as a public enemy, even though he may have an enviable record as a humane official and be operating a relatively superior inmate program. Newspapers delight in nothing more than charging a warden who introduces a slight touch of humanity into his program with running a country club or giving free hotel service to convicts.

It is no exaggeration to state that no single force does more to prevent and frustrate the rehabilitative process than our newspapers and other agencies of publicity. Until there is a complete shift of emphasis in news interest and editorial comment from a demand for a repressive penal program to an encouragement of rehabilitative measures, there is slight prospect that we shall have better correctional programs or reform any considerable percentage of prison inmates.

If the newspapers actually carried out their intended function of public education they would reverse their whole procedure. Instead of carping criticism they would enlist the services of a correctional expert to run a column in their paper orienting the public on the whole process of criminal and correctional procedure. So far as the authors know there has never been an instance when a syndicate has commissioned a competent person to write a column on the various phases of corrections. The public is catered to with Hollywood and New York gossip, the current entertainers and their personal lives, but one may look in vain for information on sentencing policies, parole, probation, rehabilitation, prison labor, and other phases of corrections.

The Motion Picture, Radio, and Television in Relation to Crime

The motion picture, radio, and television have all been attacked for their supposed contribution to delinquency and crime. The comic books have also been held responsible for much delinquency. We cannot help feeling that the motion picture, especially, has been made too much of a scapegoat, much as we may deplore its shortcomings as a medium of art and education.

Common-sense notions testify to the fact that the movies have a great influence on children. Pictures dealing with crime occasionally show that it is easy to live without working legitimately, that crime is exciting, even though it may not pay in the long run; that there are proper methods of carrying guns, of snuffing out those who stand in one's way, of evading the law, at least in the early parts of the escapade; of enjoying many of the good things of life, snappy cars, cleverly dressed women, luxurious hotels and apartments, rich food, hurry-up trips by transcontinental airplanes, and other glamorous materialistic proofs of affluence. Young girls find from many movies that love can be thrilling and even pleasingly dangerous, that clothes make the woman, that men seek girls who have easy virtue, who use finesse in make-up and in wearing their clothes well; that if a girl is astute, she can have the clothes and good times she craves.

On the other hand, movies portray also much of what is conventionally called the good and the beautiful. They depict tales of heroism, courage, integrity, and other homely but socially-approved virtues. All movies tend to reflect experiences of human beings in which the average person would like to indulge but most of which, for one reason or another, he cannot ever expect to attain. In other words, the movies, like athletic events, make it possible for the average person to enjoy many desired activities vicariously. That is one of the greatest contributions of the motion picture. And it must depict life as it is or as we would like to have it; both are essential in a humdrum world.

All censorship codes are doomed to fail in part, since there is so little agreement as to what we wish to portray on the screen. Some of the best literature is filled with powerful suggestions to "evil" and glamor. An idea transmitted to one person, if it is acted upon, might have a detrimental effect on society; the same idea, accepted by another person, might ultimately benefit the community. To the millions of Americans who attend the movies, the worst possible films have little appreciable effect, because there is a leavening process present in life that minimizes any momentary deleterious effect.

We must accept the motion picture within the framework of our culture. It is merely a part of the conditioning process and must be regarded as only one contributing factor in the development of the personality of the individual. In like manner, the church, the home, the school, the place of employment, the neighborhood, and all the rest of the institutions whose influences impinge upon the structure of the individual play their part in determining the behavior of the person. As Hennrich puts it, "Considering everything, the influence motion pictures exercise over children seems to be proportionate to the weakness of the family, school, church, and neighborhood contacting them."[41]

The Payne Fund Studies, the most comprehensive critique of the movies ever made, were published in 1933. This series "of twelve studies of the influence of motion pictures upon children and youth [was] made by the Committee of Educational Research of the Payne Fund at the request of the National Committee for the Study of Social Values in Motion Pictures, now the Motion Picture Research Council."[42] One of the studies was by Professors Herbert S. Blumer and Philip M. Hauser, who wrote on *Movies, Delinquency, and Crime*. By various methods, notably the questionnaire and personal history techniques, they attempted to ascertain from several hundred children serving time in reform schools and from nondelinquent

[41] K. J. Hennrich in a review of the Payne Fund Studies, *Commonweal*, Vol. 19, No. 2 (March 30, 1934), p. 90w.

[42] Herbert S. Blumer and Philip H. Hauser, *Movies, Delinquency and Crime* (New York: Macmillan, 1933). By permission of The Macmillan Company.

children: (1) the role of motion pictures in the lives of delinquents and criminals of both sexes; (2) the effects on the inmates of motion pictures shown at juvenile reform schools and reformatories; and (3) the effects of crime pictures on nondelinquent boys and girls.

They found that only 10 per cent of the 368 cases of male delinquents interviewed "believed motion pictures to have had some direct effect on their criminal careers. While other forces played on their lives, motion pictures were regarded by them as of noteworthy contributory influence."[43] The authors say that "many impulses and ideas of crime are aroused in the mind of the individual by motion pictures, without coming to immediate expression in criminal behavior. Ideas and impulses are checked, they are held within the mind for the given time, being confined, so to speak, to mere incipient activity. In the course of time they may pass away, without leaving any trace; but they may also work in subtle ways into a pattern of life."[44]

The report that 49 per cent of the male delinquents investigated stated that the movies gave them a desire to carry a gun; 28 per cent, that the movies taught them methods of stealing; 21 per cent that they learned ways to "fool the police"; 12 per cent that they were encouraged to pull an adventuresome job because they saw a similar crime depicted in the movies; 45 per cent that they got notions of "easy money" from the shows they saw; 26 per cent that the movies encouraged them "to get tough"; and 20 per cent that they were led to daydreaming concerning bandits and gangsters.[45]

Blumer and Hauser further state that the motion picture has a serious influence on young girls:

Twenty-five per cent of the sample of 252 delinquent girls studied, mainly from 14 to 18 years of age, stated they had engaged in sexual relations with men following the arousing of sex impulses by a passionate love picture. Forty-one per cent admit that going to wild parties, cabarets, etc., "like they do in the movies," "got them into trouble." More specifically, 38 per cent of them say that they were led, in their attempts to live a wild, gay, fast life such as presented in the movies, to stay away from school; 33 per cent were led to run away from home; 23 per cent were led to sexual delinquencies. In their efforts to enjoy clothes, automobiles, lives of luxury and ease as depicted on the screen, 27 per cent have been led on occasion to run away from home. In their efforts to achieve a life of luxury easily through means suggested, at least in part, by motion pictures, 18 per cent say that they have lived with a man and let him support them; 12 per cent that they have engaged in other forms of sexual delinquency; 8 per cent have been led to "golddig" men; 8 per cent have been led to gamble; and 4 per cent that they have engaged in shoplifting. Fifty-four per cent declared they have stayed away from school to go to movies; and 17 per cent that they have run away from home after conflict with their parents over frequent motion picture attendance.[46]

43 The text of this code is furnished free to libraries and individuals by the Association, 28 West 44th Street, New York 36, New York. See also Gerald D. Block, "The Motion Picture Production Code", The Annals, Vol. 254 (November, 1947), p. 142.

44 Ibid., p. 35.

44 Ibid., p. 37.

45 Ibid., p. 71.

46 Ibid., p. 111.

It is doubtless true that delinquents attend movies more frequently than non-delinquents, owing, perhaps, to their lack of constructive interests in a drab home or neighborhood. Preferences for particular kinds of pictures divide along sex lines rather than according to delinquent as against non-delinquent. Adventure is top choice of the boys, the romantic love theme of the girls, although adventure is a close second for delinquent girls also. Non-delinquent girls prefer the romantic *motif,* just as do their delinquent sisters, but show little interest in adventure.[47]

We know of no universal formula to protect our constitutionally weak children from over-stimulation. Restrictions set up in the home are probably the best cure but, naturally, this must be done wisely. Censorship from without is no cure. Parent groups can do much if they possess the sustained energy. They may secure advice and help from contacting the Parents' Film Council as well as the Children's Film Library of the Motion Picture Association of America. Although the motion picture industry has set up a code to govern the making of pictures, it has had very little effect on the production and dissemination of shoddy or otherwise questionable pictures.[48]

As to the effects of radio and television, we might make the same statements as those in regard to the motion picture. It depends on what type of child is listening or viewing their output. Just as many criminals might claim that their start in crime was due to reading Sherlock Holmes, we might find some who contend that the exploits of a gunman, bank robber, or murderer, seen or heard on those great purveyors of culture, the radio and television, prompted them.

Up until a decade or more ago radio was the chief entertainer of American youth, aside from the movies. With the greatly expanded increase in television production (of both sets and offerings) the radio has apparently lost its appeal to the younger generation, aside from disk jockey recordings.

The influence of television as a medium of mass entertainment and dissemination of culture is of some concern to all thinking citizens. As the Senate Subcommittee on Juvenile Delinquency states:

> The cumulative effect of crime and horror television programs on the personality development of American children has become a source of mounting concern to parents. Several generalizations can be made concerning many of the programs shown during children's viewing hours. It was found that life is cheap; death, suffering, sadism, and brutality are subjects of callous indifference; and that judges, lawyers, and law-enforcement officers are too often dishonest, incompetent, and stupid.

---

[47] Cf. Maude A. Merrill, *Problems of Child Delinquency* (Boston: Houghton, 1947), pp. 257-258.

[48] The text of this code is furnished free to libraries and individuals by the Association, 28 West 44th Street, New York 36, New York. See also Geoffrey Shurlock, "The Motion Picture Production Code," *The Annals,* Vol. 254 (November, 1947), p. 142.

On the potency of television offerings to children the committee further states:

> There is reason to believe that television crime programs are potentially much more injurious to children and young people than motion pictures, radio, or comic books. Attending a movie requires money and the physical effort of leaving the home, so an average child's exposure to films in the theater tends to be limited to a few hours a week. Comic books demand strong imaginary projections. Also, they must be sought out and purchased. But television, available at a flick of a knob and combining visual and audible aspects into a "live" story, has a greater impact upon its child audience.[49]

Few studies have been made that throw light on the adverse effect of television or radio thrillers on children aside from overstimulation, nervousness and lack of interest in studies among heavily-addicted devotees. States the subcommittee:

> The subcommittee is aware that no comprehensive, conclusive study has been made of the effects of television on children. . . . Members . . . shared the concern of a large segment of the thinking public for the implications of the impact of this visual presentation upon the ethical and cultural standards of the youth of America. It has been unable to gather proof of a direct causal relationship between the viewing of acts of crime and violence and the actual performance of criminal deeds. It has not, however, found irrefutable evidence that young people may not be negatively influenced in their present-day behavior by the saturated exposure they now receive to pictures and drama based on an underlying theme of lawlessness and crime which depict human violence.[50]

While many individuals, most of whom are parents, may deplore the output of the television, it is of interest to note that in a survey made in New Haven in 1954 of some 3,500 households, 69 per cent of those parents interviewed generally approved of the programs being offered their children.[51]

The subcommittee had very little to recommend to improve the diet that flows from the television. It suggests that each subscriber to the National Association of Radio & Television Broadcasters be required to send to the central office in Washington copies of complaints and criticisms directed to specific stations by parents or others; that revisions in the code defining what is considered in good taste to be made; that the "seal of good practice" be displayed at frequent intervals with a brief statement giving reasons for

---

[49] Report of the Subcommittee on Juvenile Delinquency, 85th Congress, 1st. Session (March, 1957). (Washington, D.C.: U.S. Government Printing Office), p. 219. For more complete treatment see the Subcommittee's Interim Report, "Television and Juvenile Delinquency" (Washington, D.C.: U.S. Government Printing Office, 1955).

[50] *Ibid.*

[51] Cited by Herbert A. Bloch and Frank T. Flynn, *Delinquency: The Juvenile Offender in America Today* (New York: Random House, 1956), p. 216.

having a code; and to develop a "balance" in programming. The committee ends its recommendations with the following:

A constant vigil is required in relation to any large and powerful influence upon society. This is vital in a democratic state. The power of the people to direct their own destiny is enhanced by the energy with which they control the negative forces around them. If children are to live in an environment that is conducive to constructive attitudes and actions, they must live in communities where the adults about them are similarly motivated. Sober, unbiased adults can perform a useful function by maintaining steady watch over the programs offered to children and by promptly reporting offensive materials to responsible sources. The subcommittee hopes that leaders and citizen groups may stimulate development of the listening-council movement.[52]

Eternal vigilance—not official censorship—is the only democratic recommendation for control. Certainly there is much room for improvement. We should like more artistic menus both for adults and for children. But we should hesitate to pass judgment on the roles of radio and television as precipitates to delinquency and crime on a serious scale.

The Comics

Comic books are obviously misnamed. With few exceptions they are definitely not humorous. In fact, their subject matter has been denounced for years and their widespread circulation without restraint by their producers and distributors vigorously condemned. But one is reminded of the parallel condemnation in an earlier generation of the old dime novel—the Deadeye Dick and James Brothers era—when moralists and many parents were harassed by children's appetites in this direction. Then, as now, there was considerable difference of opinion regarding their harmfulness.

There are many types of comic magazines, in addition to the daily strips in the newspaper. Some, by all standards of culture and decency (and humor) are dull, unwholesome, highly suggestive, and even pernicious. Others are tolerably respectable and educational. As stated above, the Subcommittee on Juvenile Delinquency found that television programs seem more harmful to children, on the aggregate, than do the so-called comics.[53] Children and adolescents are very fond of the comics. As Paula Elkisch states, children tend to identify with the aggressor since imitation and identification are among the most primitive of impulses. Emphasizing that comic books are basically written in picture language, she further states: "It is well known that children indulge in 'bad' language and yet are repelled by it . . . . The same must be true of the children's reactions to utterly crude, violent, and, frequently, socially stimulating picture language of the comics. Again and again the child is exposed to conflict. Different methods

---

[52] Ibid., p. 221.

[53] See supra, pp. 192-193.

aim at the same goal, the goal of primitiveness, enhancing *vice versa* their fascination over the child."[54]

Parent groups and community action have done much to reduce the sale of objectionable magazines during the past decade. The Comics Magazine Association of America, resigning to community pressure, appointed a code administrator who, with a staff of reviewers, inspected all comics before they were printed. The code provided for a ban on all horror and terror comic books but not on crime books. As a result the industry declined from a high of 110 million copies per year to 70 million. In 1955 some thirteen states enacted laws to control the sale of the comics.

Recently the comics battle has died down. Part of this is undoubtedly due to the increase of television sets. It is the opinion of some who are in tune with the leisure-time pursuits of children that there is a definite trend away from the comics to television.[55]

Other "Snares" Attractive to Juveniles and Young Adults

We have heard of the marijuana and Benzedrine habit among young adults. In addition to their use, there are still other snares, vastly more dangerous potentially than the movies, television and comic book. Among these might be mentioned salacious literature. Police officials, reformers, and others concerned with youth have long been disturbed by pornographic literature and pictures that periodically deluge the hanging-out places of adolescent youth. Aside from newsstands, small corner stores, filling stations, and the like, it is not infrequent to have such material peddled outside schools and motion picture theatres. Such literature is highly suggestive, with the art work leaving nothing to the imagination.

Police women of our larger cities are constantly on the alert for such "contraband." They report that it seems to be available in waves, appearing and disappearing as publicity against it is turned on and off. While it often claims to be "art" it is frankly lewd and sexually stimulating to adolescents. Nudity is the rule and deviate forms of sexual behavior are graphically presented. Beautiful girls are depicted in compromising positions and nothing is left undone to titillate the sexual cravings of those who purchase the pictures. There is no doubt that most of the purchasers of these pornographic exhibits are maladjusted individuals and as such, presumably, need community protection. The only recommendation advanced is alert control by specially trained police officers, either male or female, who can see to it

---

[54] Paula Elkisch, "The Child's Conflict about Comic Books," *American Journal of Psychotherapy*, Vol. 2, No. 3 (July, 1948), pp. 483-487.

[55] A statement on community action to control harmful comics may be found in Paul S. Deland's article, "Battling Crime Comics to Protect Youth," *Federal Probation*, Vol. 19, No. 3 (September, 1955), pp. 26f. See also the *Interim Report* of the Sub-Committee on Juvenile Delinquency, "Comic Books and Juvenile Delinquency," (Washington, D.C.: U.S. Government Printing Office, 1955).

that city ordinances against salacious literature and pictures are enforced. Over-zealous policemen, however, with little or no training, do not make good censors since, in the course of delinquency prevention, they often confiscate what passes universally as good literature.

There are many borderline books, magazines, and pictures that cause a great deal of trouble, including an occasional court action. Those who produce these things are shrewd and determined. Their control and prosecution call for alert and equally determined action on the part of public officials who can count on community organizations for cooperation.[56]

Another form of snare for the young adult is the taproom or the suburban dine-and-dance club. While not outwardly offensive, these places do serve as contact-points between streetwalkers or holdup men and their potential partners among the youth. They also provide easy access to gambling machines and "the book."

It is not our purpose to be moralistic about the vices of the day. We merely wish to point out that such gadgets as the automobile, the nightclub, the dine-and-dance and taproom, the pinball and slot machine, the game of chance, the pornographic picture—all represent a wide variety of temptation to modern youth that were relatively non-existent in the earlier days of our culture. Since most parents are unaware of these potential dangers, or are too busy with their own pleasure or occupations, the police power of the community must accept its responsibility of controlling them with sustained effort—not only spasmodically.

Concluding Remarks

Marshall Clinard, in discussing these "secondary community influences," points out that, in general, these influences—newspapers, the movies, radios, comics, and television—are largely administered outside the home and immediate neighborhood. Their appraisal and control require a much broader perspective than the family, the gang, or the immediate neighborhood. He warns that it is dangerous to attempt to differentiate between a world of juveniles and an adult world, since both groups secure their values within the social framework of our culture.[57]

We can appraise cultural influences, not from opinions, but from concrete studies. Some years ago one of the writers was challenged, when he questioned the role of religious instruction in preventing delinquency, by the statement that most of our national leaders placed great stress on such training as a crime deterrent. Worthy as such opinions are, they are merely opinions. What we need is facts and studies. The roots of maladjustment lie deep and must be attacked on all fronts.

---

[56] "Obscene and Pornographic Literature and Juvenile Delinquency," *Interim Report* of the Sub-Committee.

[57] Marshall Clinard, "Secondary Community Influences and Juvenile Delinquency," *The Annals,* Vol. 261 (January, 1949), pp. 42-54.

13

Emotional Disturbances
as Factors in Criminality

The importance . . . of psychoanalytic concepts . . . as a method of individual treatment lies in the fact that the center of attention in the individual's maladjustment is shifted to his own psychological history and thus the importance of environmental causation is correspondingly minimized. . . . Search for motivations and incapacities for adjustment . . . is directed to the psyche. (From "The Social Theory of Professional Social Work," in Harry E. Barnes and Howard Becker, Contemporary Social Theory, *New York: Appleton, Copyright 1940.)*

PHILIP KLEIN

The psychiatric and psychoanalytical approaches to the problem of crime and delinquency have within the past quarter century or more made a deep impression on the thinking of students whose professional lives are dedicated to these areas of maladjustment. More recently psychiatry, psychology, and social work have produced an offshoot discipline known as "orthopsychiatry." Dedicated to the prevention of emotional disturbances, it has also made great strides in educating the public in delinquency and emotional maladjustment of children and adults.

In appraising the psychiatric and psychoanalytical approaches we must be reminded that, just as the extravagant claims of the early endocrinologists and psychologists have been tempered in the light of more modern research, so must the more enthusiastic visions of Freud, Adler, Jung, Rank, and others be re-evaluated.

We find that the normal mind is characterized by adequate responses on every level of stimulation. When we come, however, to the psychological, emotional, and social levels, "normality" of response must be judged with reference to the group in which the individual moves. Responses are normal if they harmonize with the ideas and behavior patterns prevailing in the group, whether or not they are the responses most desirable in other groups. Sadly erroneous ideas may dominate in any given group, and the customs and mores of the group may not be in harmony with what is best for the well-being of society or the progress of the race as a whole. With respect to these emotional and social levels of response, mental normality is viewed primarily as a matter of relatively complete psychic integration and thorough adjustment to the social environment. The ideas entertained by the individual must be relatively consistent, and the individual must feel in reasonable rapport with the social habits of the community. The tasks and responsibilities of life should be agreeable. In short, the individual must respond with enthusiasm and efficiency to the realities of his social environment.

The causes of abnormality are, therefore, both physical and psychological. This is to be expected because the lower and more fundamental levels of stimulation and response are primarily physical in their nature. The abnormal types of behavior that are due to physical causes may be brought about by disease, toxic or poisonous products in the system, changes due to age, and the like. Defects in the endocrine glands prevent normal responses on the physiochemical level and may result in diseases that manifest themselves in pathological excitability, abnormal apathy, physical monstrosities, and so forth. A serious mental disorder known as paresis, caused by syphilis, affects motor-sensory responses and makes it impossible for the individual to control his reflexes in normal fashion. Senility may bring about bodily changes that affect the normal responses on all these lower levels.

From the psychological point of view, mental abnormality is a marked inability to face reality, with resulting mental conflicts. The life experiences of the individual, from earliest childhood onward, may have created mental reaction patterns ("complexes") which make it very difficult for him to meet adult responsibilities. Insecurity, rejection, or frustration in childhood, harsh living conditions, disappointments in love, professional failures, economic insecurity, and a large number of other unfortunate experiences may make it extremely difficult for some individuals to face the realities of life. These individuals, tortured by mental conflicts, tend to escape from this intolerable reality by creating a world of mental phantasy that is more in harmony with their wishes and desires; or go to the other extreme and become aggressive or even cruel in their behavior toward others.

Since few persons have a completely satisfactory set of experiences from birth onward, or find life entirely to their liking in adulthood, the great majority of normal human beings create for themselves a realm of fancy in which they realize aspirations denied them in actual life.[1] But in all these normal cases reality occupies the leading role in their life interests and

---

[1] James Thurber has given us their patron saint in "The Secret Life of Walter Mitty," originally published in his collection, *My World and Welcome to It.*

activities. With the definitely psychotic, reality is almost wholly abandoned, and the realm of phantasy becomes their main concern.

These considerations will once more emphasize that there is no sharp line of demarcation between the normal and abnormal mind, the so-called sane and the psychotic insane. It is wholly a matter of degree. Nor are all flights into fancy pathological or socially undesirable. Much of art, music, and literature grows out of such flights. In these realms, the line dividing phantasy from constructive imagination is very faint indeed. It is a task of mental hygiene to be on the alert to prevent too great a conflict between the individual mental equipment and the reality with which it must deal. Through various remedial measures the mind must be toughened, reality must be softened, or both results approached simultaneously.

The orthopsychiatric attack on emotional maladjustment emphasizes the problems of childhood. Aside from working with troubled children it is interested in educating parents and teachers in wholesome childhood; thus preventing frustration and aiding them to understand danger-points as they emerge in the life of the child. The bulk of American parents have taken their cue from the poets and romanticists in assuming that childhood is a period of unmitigated bliss and that what problems they have are inconsequential. Yet we know today that insecure or rejected children—children who crave love and affection and who do not receive it—become the emotionally unstable adolescents of tomorrow.

Most parents take their children as they come, hoping they will outgrow their poutings, stubbornness, meanness, and temper tantrums. Just as unwittingly many parents have a second child without any knowledge of such a phenomenon as "sibling rivalry." Our culture is full of bewildered and thoughtless parents, as well as bruised children who will grow into suspicious, pathologically ambitious, or sadistic adults. It is the purpose of the orthopsychiatrist (psychiatrist, psychologist, and social worker) to guide prospective parents concerning the potential problems that arise in bringing children into the world.

The importance of the orthopsychiatric approach to the causes of delinquency and crime cannot be emphasized too much since, as we have seen elsewhere, it is this mobilization of professional skills that we find in the child-guidance clinic. Taking the initial cue from Otto Rank's thesis that the very process of being born is a devastating experience, some psychiatrists and social workers contend that many of the ills of children and adults may be traced to their inability to make the adjustment to it. Yet this is life and the amazing thing is that the vast majority of babies make the necessary adjustment in a reasonably satisfactory manner. But this happens only when the parents accept the child, cuddle it, and make it feel that it "belongs." Many hospitals are now permitting the child to be placed alongside the mother from the moment of birth rather than keep it in a nursery staffed by overworked and harassed nurses who either do not know the problems of the just-born child or are too busy to care.

The process of growing up in an adult-centered world is a serious problem for the child, even though he is mentally and physically adequate. Reports on the care of infants enumerate many types of neglect that are the lot of the average child in the average home. Certainly the child has the right to be wanted. Millions of them are not wanted; at least at the time they are conceived or born. They are merely tolerated by their mothers, but as they become an integral part of their parents' lives they are accepted, outwardly at any rate. But the harm done to their personalities prior to the moment of acceptance may never be completely remedied. And, of course, there are many definitely rejected children whose childhood is unduly harsh. For example: (1) there is the child who is accepted on the surface but is inwardly rejected by the mother; (2) the child who is over-indulged because of a compensation on the part of the parent who really possesses a latent hate for him; and (3) the child who is rejected because he is not as good-looking, bright, or talented as a neighbor's or relative's child. A revealing study that points up this rejection is that by Dr. John Bowlby, of the Tavistock Clinic in London, entitled *Forty-Four Thieves*. A large proportion of his cases showed definite signs of instability due to denial of mother love.[2]

The orthopsychiatric approach has, in recent years, become somewhat the fashion since it is based on solid postulates. Much of the early pioneer work in this area was accomplished by Dr. William Healy, in his Chicago clinic and at the Judge Baker Guidance Center in Boston.[3] The psychoanalytic approach was developed through the years and we see it manifested in August Aichhorn's work, *Wayward Youth,*[4] and later in Alexander and Healy's *Roots of Crime.*[5] Aichhorn, in his correctional school in Austria, used individual therapy in effecting cures. He insisted that it is important to know what the child is thinking so as to divert his overt antisocial acts into corrective channels rather than to repress them. He views delinquency as symptomatic of a neurosis.

Regardless of the apparent reasons why the child has gone astray, whether they be socioeconomic or biological, no complete picture can be gained until the less obvious phases of the problem are assayed by the psychiatrist working in conjunction with the psychologist and the psychiatric social worker.[6]

---

[2] Dr. John Bowlby, *Forty-Four Thieves* (London: Baillière, Tindall & Cox, 1947), p. 1.

[3] See Dr. Healy's early works, *The Individual Delinquent* (Boston: Little, Brown, 1915), and *Mental Conflicts and Misconduct* (Boston: Little, Brown, 1917).

[4] (New York: Viking, 1925; also Meridian, paper), with an Introduction by Sigmund Freud.

[5] Franz Alexander and William Healy, *Roots of Crime* (New York: Knopf, 1935).

[6] Cf. O. Spurgeon English and Gerald H. J. Pearson, *Common Neuroses of Children and Adults* (New York: Norton, 1937), and *Emotional Problems of Living* (New York: Norton, 1945). See also: Kate Friedlander, *The Psychoanalytical Approach to Juvenile Delinquency* (New York: International Universities Press, 1947).

In non-technical language, we see the human individual in the light of his original nature as mainly aggressive and assertive. He starts at birth to want things, many of which he is denied either at the moment he craves them or later. He naturally makes a vehement protest. The young baby cries until he is either exhausted or satisfied. As the child grows older many of his basic desires come into sharp conflict with adult norms and other realities of life. That is, under the normal conditions of life, society obstructs much of the direct and spontaneous expression of human nature.

The child is socially compelled to live as a member of several groups of which the family is the first he encounters. Through long experience the specific culture in which the child is placed by birth has worked out gradually, often irrationally and despotically, a code of beliefs and practices with respect to social behavior. These are imposed, consciously as well as unconsciously, upon the child from the moment of birth, a few at a time, until he conforms. He learns by bitter experience that other individuals have the same desires and that there does not seem to be enough satisfaction to go around. In other words, there is a constant struggle between the latent or innate drives of human nature and the discipline that is supplied by the customs and folkways of the social group. The socializing process is functioning bit by bit and the child must adjust or be penalized.

The wise parent will attempt to understand this conflict and reduce it to a minimum. For example, there are many situations in the home in which the growing child may be permitted to express himself without being disciplined or penalized. Many of these situations *seem* to make a great deal of difference, but in reality they do not. Modern experts in child care point out that during the various periods of development certain elements of growth manifest themselves and are thus considered normal. Should these be ignored, accepted, or penalized? Too many children are penalized or admonished for doing just what they should be expected to do at the specific period of growth.

Conversely, children often fail to do what is hopefully expected of them. Toilet habits are an example of this. Too early toilet training has caused many children to grow up with manifestations of maladjustment. Rebellion, either overt or latent often begins during this period. Here the child is master; it is one of his strongest weapons that cause parental concern. The mother must not only yield to the child's own whims incidental to this process, but she must convey to the child by her nonchalance that his apparent eccentricity makes no difference to her. A child left free to adjust to social good taste in toilet habits in his own time is not likely to show signs of rebellion during the early years of his life; assuming, of course, that his parents are as consistent on other important matters such as the feeding routine.

Strict adherence to scheduling the baby is no longer the fashion. As Mowrer and Kluckhohn point out: "Responsively gratifying the infant's needs during these early months does not mean that a child is spoiled. The spoiled child shows just those traits we have predicted from scheduling care, and he is either one who has never met any conditions for being rewarded or who has been inconsistently rewarded. . . . If the rewards which the child receives bear no consistent relation to his behavior, education will not proceed efficiently in its substantive aspects and an apathetic and anxious or hostile individual is likely to result.[7]

It should be noted that young children exhibit feelings of ambivalence, especially toward their parents because it is they who are not only the love objects but also those who are the withholding agents of society. Parents should recognize and understand the roles they must play in dealing with their offspring. This is especially difficult in a home of two or more children because of the rivalry among siblings for parental affection and recognition. Much conflict can thus be avoided by understanding parents. In other words, prevention of emotional misery and abnormal aggression in the child is all-important. Cuddling and other manifestations of affection are extremely potent. The older behavioristic approach of child-rearing ignored this fundamental preventive. To those critics of cuddling who contend that there is danger of the child becoming spoiled, one may well answer that it is the insecure or frustrated child that is often the spoiled child. Obviously, the child must learn the lessons of life and be taught a sense of responsibility. But the secure child will penetrate and explore the new and the unknown in his own good time if he knows he has a haven to which to retreat. Being ambivalent by nature, he seeks both security and self-assertion. While in many situations the child may select his course without interference from parents, it is the practice of the wise parent to set flexible limits as guides in helping him make his choices. There will be innumerable frustrations but the child will learn life's limitations through doing, coupled with parental skill.

If the child has an understanding parent who is there when needed he will have little difficulty in growing into a socially mature adult. Of course, this presupposes that the parent, especially the mother, is not emotionally dependent upon the child. She must accept the maturation process in an intelligent manner. Many traditional shibboleths and outmoded but widely accepted attitudes concerning child care conspire to thwart discerning parents. These must be appraised by those parents who are determined to do a good job in rearing their children.

The gradual social weaning of the child is the life task of both parents and should be accepted honestly and coolly. First experiences with sexual differentiation or meeting new playmates, sharing playthings, the first day at school (a very serious adventure), the first day at camp, or the first day

---

[7] O. H. Mowrer and Clyde Kluckhohn, "Dynamic Theory of Personality," J. MacV. Hunt (ed.), *Personality and the Behavior Disorders*, Vol. 1 (New York: Ronald, 1944), pp. 89-90.

in a new neighborhood—all of these must be met by the growing child. The parent can make or break a child's emotional life by the way in which he aids him in these experiences. The teacher has a responsibility in this too, as does the camp counselor.

It is difficult for any child to grow up without manifesting signs that are symptomatic of emotional stress or strain. The best thought concerning these mechanisms in children is that when they begin they should be noted with calm objectivity. If they persist, some professional guidance should be sought by the parent. This may be obtained from a psychiatrist, a child-guidance clinic, a progressive pediatrician, or by perusal of some reputable work on child care. It may well be that certain of these mechanisms emerge from some conflict in the home such as bickering between parents or jealousy on the part of the child toward either one of the parents or a brother or sister.

It should be stated that emotionally disturbed children do not necessarily become delinquent. Their constant frustration often does result in behavior that may be labeled delinquent, but just as often it may result in neurotic behavior. In any case the child is obviously unhappy. He may misbehave for attention or recognition or he may withdraw from members of his groups and compensate by leading an isolated existence.

It may be argued that the bulk of parents do not have the time nor the social background to devote so much attention to their children's growth. This is, indeed, a just comment. But, somehow, a new orientation concerning the parent-child relationship must be developed. The spadework must be begun in the schools. Teacher training institutions must scrap their old-fashioned philosophy and adapt their programs to this new emphasis. Parents can learn about the implications of this approach from well-chosen syndicated articles in the daily newspapers and by participating in parent-teacher groups. Nursery schools and kindergartens can be expanded in even the smallest school systems and parental guidance should become an integral part of their programs. The radio, television, and motion picture can be utilized in disseminating information to parents. Obstetricians and pediatricians, as well as hospital administrators, can advance this new approach to prospective mothers.

The above discussion leads us to perhaps the most important aspect of the psychiatric approach to crime, namely, just how prominent are the roles of mental disease and emotional disturbance in crime causation? Some early students ascribed crime to emotional disorder. Albert Morel held that crime is a product of mental degeneracy; Henry Maudsley viewed criminals as a product of moral insanity; and Raffaele Garofalo believed that criminals suffer from inherited moral degeneracy.

But all of these views represented little better than guesswork. In 1916, however, Dr. Bernard Glueck carried on his pioneer study of 608 prisoners in Sing Sing prison. He found that 59 per cent of these were mentally retarded, mentally diseased, or mentally abnormal. The retarded numbered 28.1 per cent; 12 per cent were psychotic or mentally deteriorated; and 18.9 per cent were classified as psychopathic.[8] This study was the first that attempted to obtain a "mental census" of a prison population. Since then there have been several made with a wide variety of results.

A study was made by Dr. Ralph Banay of prisoners in Sing Sing prison for the decade 1930-1940, which showed 1 per cent psychotic, 20 per cent emotionally immature, and 17 per cent unclassified psychopathic.[9] But, in all such studies, we do not know how many of these prisoners, aside from the mentally retarded, developed their emotional disturbance as a result of their arrest or imprisonment. The phenomenon of prison psychoses is an involved one.

There is no satisfactory body of evidence on the extent to which mental disease or serious emotional disturbances account for the commission of the first offense, before one has been demoralized by prison life. On the basis of many studies of mental disease among prisoners, who are the only criminals whose mental and emotional states can be ascertained, it may safely be stated that about 25 per cent of the prison population give evidence of psychotic, psychoneurotic, neurotic, and emotionally disturbed traits. It may well be that this is not in excess of the proportion of these types in the general population. We may assume that there is less mental defect and mental disease among the vastly larger group of criminals who are not apprehended and convicted. If this is the case, it is obvious that the criminal class as a whole is certainly as intelligent and stable, mentally and emotionally, as the general population. This does not, of course, nullify the fact that criminal conduct in any specific case may be due to abnormal or emotional conditions. The truth here can only be ascertained by a careful psychiatric examination of each individual case.

Perhaps the most complete body of evidence relating to the mental, emotional, and personality traits of defendants is that assembled over a period of years by the Psychiatric Clinic of the Court of General Sessions in New York City as the result of the initiative of the psychiatrist, Menas R. Gregory, begun in 1932.[10]

One of the most important generalizations to be drawn from the tests as well as from the results of the clinic is that personality traits and emotional disturbances that may favor criminality do not have to involve overt psychoses or serious psychoneuroses. They may be only strong or stubborn

---

[8] Bernard Glueck, "Concerning Prisoners," *Mental Hygiene,* Vol. 2, No. 2 (April, 1918), p. 178.

[9] Ralph S. Banay, "Mental Health in Corrective Institutions," *Proceedings,* American Prison Association, 1941 (New York: The Association, 1941), pp. 377-387.

[10] For an analysis of the findings of this clinic from 1932-48, see Walter Bromberg, "Personality Factors in Crime," in E. T. Stromberg, Ed., *Crimes of Violence* (Boulder, Colorado: University of Colorado Press, 1950), p. 18.

emotional disturbances which, in connection with the *situation* at the time of the crime, suffice to impel to delinquent action.

This clinic has been operating for a quarter of a century and has long since proved its value in assisting the judges in intelligently disposing of cases. In 1957, 2,947 cases were examined. Dr. Emanuel Messinger is psychiatrist-in-charge of the clinic.

The conclusion is inescapable that definite sentencing to a fixed term on the basis of alleged personal responsibility does not accord with elementary fact, sense, or logic in the circumstances. The handling of a convicted criminal, the determination of the time he should serve, and the prospect of rehabilitation are matters that can only be decided by a behavior clinic that can take cognizance of all behavior and personality traits as well as obvious psychoses. Mental responsibility for criminal action is a far more complicated matter than the mere issue of sanity or insanity, however scientifically these may be determined. Emotional disorganization is far more important as a factor in crime causation.[11]

---

[11] See our discussion of "The Plea of Insanity," Chapter 16, pp. 253-266.

14

An Attempt to Synthesize— The "Multiple-Causation" Approach

The general trend in criminological research is, and has been, toward an eclectic theory—a processual analysis, if you will— which is the only sound approach to a defiant puzzle like crime causation. (From "Trends in Criminological Research," Federal Probation, *Vol. 6, No. 4 October-December, 1942.*)

JEREMIAH P. SHALLOO

As we stated earlier in this section, the writers have felt a responsibility to review the various approaches to the intricate problem of causation. But we cannot espouse *in toto* any one of them, although each is able to explain certain types of behavior or help us understand why certain persons have committed crimes. It is the purpose of this chapter to attempt a synthesis.

Some years ago the distinguished criminologist, Enrico Ferri, wrote concerning criminal causation:

> Crime is the result of manifold causes, which, although found always linked into an intricate network, can be detected, however, by means of careful study. The factors of crime can be divided into *individual* or *anthropological, physical* or *natural,* and *social.* The anthropological factors comprise age, sex, civil status, profession, domicile, social rank, instruction, education, and the organic and psychic constitution. The physical factors are: race, climate, the fertility and disposition of the soil, the relative length of day and night, the seasons, meteoric conditions, temperature. The social factors comprise the density of population, emigration, public opinion, customs and religion, public order, economic and industrial conditions, agriculture and industrial production, public administration of public safety, public instruction and education, public beneficence, and, in general, civil and penal legislation.

As if the above were not enough to add to the complexity of the problem, Ferri continues:

To these factors we could add many others without ever exhausting them, since they include all that the Universe contains, not omitting a word or a gesture. What we must add, however, is the fact that as a whole they determine the *law of criminal saturation:* "Just as in a given volume of water, at a given temperature, we find the solution of a fixed quantity of any chemical substance, not an atom more or less, so in a given social environment, in certain defined physical conditions of the individual, we find the commission of a fixed number of crimes."[1]

De Quiros, the Spanish criminologist, states: "Ferri's theory is the most scientific production of modern studies." Certainly Ferri was a determinist without qualification.

Another penetrating statement regarding causes of delinquency and criminal behavior was made by the outstanding British psychologist, Dr. Cyril Burt, many years after Ferri's remark. He wrote:

> Crime is assignable to no single universal source, nor yet to two or three: it springs from a wide variety, and usually from a multiplicity, of alternative and converging influences. So violent a reaction, as may easily be conceived, is almost everywhere the outcome of a concurrenc of subversive factors: it needs many coats of pitch to paint a thing thoroughly black. The nature of these factors, and of their varying combinations, differs greatly from one individual to another.
>
> Hitherto, the fund of possible explanations invoked by the criminologist has been much too narrow. Ordinarily he is content to trace . . . [it] to but four or five all-powerful causes—sometimes, indeed, to no more than one. Drink, epilepsy, a defective moral sense, some outstanding feature of heredity, or some characteristic of city life, is seized upon in isolation, and made accountable for all. With the same exclusive emphasis, some solitary panacea has correspondingly put forward. It is as if one should explain the Amazon in its flood by pointing to a rivulet in the distant Andes, which, as the tributary that is farthest from the final outflow, has the honour of being called the source. Dry up the rill, and the river still flows on. Its tributaries are countless, though all stream into one sea.[2]

Dr. Burt concludes that there are at least four grades of factors that can be identified with any specific case of crime and that might be gauged in importance in terms of their intensity in the specific case. These are: (1) the principal or most conspicuous influence (if any); (2) the chief cooperating factor or factors; (3) minor predisposing or aggravating conditions; and (4) conditions present but apparently inoperative.

This is frankly the "multiple-causation" theory of crime. After examining much of the research and conclusions of the behavioral scientists (including the geographic and economic determinists) over the past two hundred years, the only honest conclusion is that thus far no unitary cause for crime

---

[1] *Studies on Criminality in France from 1826 to 1878* (Rome: 1881), quoted by C. Bernaldo de Quiros, *Modern Theories of Criminality* (Boston: Little, Brown, 1911), pp. 20-21.

[2] Cyril Burt, *The Young Delinquent* 1st ed. (London: University of London Press, 1938), pp. 599-600.

has been found. Each discipline, through careful research, throws considerable light on the perplexing subject but each of their conclusions can be regarded only as segmental.

An explanation for white-collar crime will not fit juvenile gang warfare; a suitable explanation as to why a person persistently commits arson will not suffice to explain syndicated crime; to offer an explanation as to why a man kills his spouse in a jealous rage will not be helpful to explain why a young adult wantonly kills a small shopkeeeper for a paltry sum of money. The reasons why the Mad Bomber of New York City persisted in placing bombs in many public places a few years ago for a fancied wrong will not apply to the black marketeer or the extortionist. Nor can any one explanation of crime cover both the occasional or accidental criminal acts of persons and the persistent patterns of professional criminal activity. To build up a system of crime causation that will include all types of criminals and their acts is unrealistic and futile.

As we have demonstrated in the preceding chapters, the causes of crime are legion. In many cases—but not all—the central theme was *conflict* or *disorganization,* either cultural or emotional. The constitutionalists may explain crime by morphological aberrations, or defective biological stock; the psychiatrist may speak of the "rejected personality" and the sociologist of the "socially maladjusted individual." This point may be summed up as follows:

> Probably more than in any other country is there a distinct absence of commitment to a particular school or system of causation. Etiological research has been characterized by an eclecticism which has not hesitated to include any and all possibilities for understanding the genesis of criminal behavior. Thus we have a socio-psychiatric approach genetically formulated and pursued. If contemporary research methods were reduced to particular emphasis, it would appear that conflict is central to such an emphasis, whether labeled "mental" or "cultural," whether within the mind of the individual, or between the individual and conduct commands inherent in differential norms, as in the case of native-born children of foreign-born parents, wherein the definitions of the situation reveal opposing principles of conduct. Contributions of American criminologists are to be sought in their penetrating analyses of the role of conflict in the development and expression of criminal behavior.[3]

In short, the eclectic or "multiple causation" thesis is the most fruitful, though perhaps frustrating, position that can be taken. As Sheldon and Eleanor Glueck state in their *Unraveling Juvenile Delinquency:* "At the present stage of knowledge an eclectic approach to the study of the causal process in human motivation and behavior is obviously necessary."[4] Sheldon Glueck goes further:

---

[3] Harry Elmer Barnes and Howard Becker, *Contemporary Social Theory* (New York: Appleton, Copyright 1940), p. 713. See also Thorsten Sellin, *Culture Conflict and Crime* (New York: Social Science Research Council Report, 1938).

[4] (New York: Commonwealth Fund, 1950), p. 7.

There are evidently criminologists who find such a "theory" [referring to Sutherland's "differential-association" theory] helpful in their thinking. . . . The multiple factor approach is much more illuminating and much more in accord with the variety of original natures involved in crime, the variety in kind and intensity of human and physical environmental influences involved in crime, the variety in the behavior patterns of the acts and mental states and mechanisms embraced in the single legal concept of "crime." For this not only recognizes the evident fact of a wide variation in influences, weights and combinations of traits and factors in crime causation; it recognizes, too, that while there is a "core type" of offender, there is also a variety of subtypes or fringe types. . . . It recognizes that all behavior is conditioned by both biological and sociocultural influences. . . . [This] leads to the conclusion that a *wise eclecticism . . .* is still the only promising and sensible credo for the modern criminologist.[5]

Professor George B. Vold, in his *Theoretical Criminology,* a work avowedly concerned with an analysis of criminal causation, states in his conclusion:

Crime must be recognized clearly as not being a unitary phenomenon but as consisting of many kinds of behavior occurring under many different situations. No single theory therefore should be expected to provide the explanations for the many varieties of behavior involved. The problem calling for clearer thinking in the future than has been given it in the past is the systematic and realistic delineation of kinds or types of criminality actually occurring that need to be comprehended. Consistent and unitary theory then should be possible for each type, so there would be less confusion due to the utilization of a non-applicable theory to any particular type. In this manner, too, the absurdities of a too facile eclecticism may be corrected.[6]

We may "explain" some crime on this or that basis, but there remains much crime that cannot be explained by resorting to specific hypotheses. What is perhaps needed is more tolerance among the experts for one another's work. As Professor Ruth Cavan puts it:

Unfortunately certain psychiatrists and sociologists have disparaged each other's studies, a situation that has prevented a necessary joint approach to the study of crime. Actually, both the psychiatric-psychoanalytic and the sociological approaches agree on certain important points. Both agree that no child is a born criminal but that he adopts criminal behavior as a result of experiences that begin in early life. Thus both agree that there is no easy way to understand crime. . . . Both agree, also, that there is no quick and sure cure for crime . . .[7]

---

[5] "Theory and Fact in Criminology," *British Journal of Delinquency,* Vol. 7, No. 2 (October, 1956), p. 108.

[6] George B. Vold, *Theoretical Criminology,* (New York: Oxford University Press, 1958) pp. 313-314.

[7] Ruth S. Cavan, *Criminology,* 2nd edition (New York: Crowell, 1955), p. 702. For an attempt to synthesize the "group" and "individual" approaches, see S. Kirson Weinberg, "Theories of Criminality and Problems of Prediction," *J. Crim. Law,* Vol. 45, No. 4, (November-December, 1954), pp. 412-424.

Aside from exploring for causes, the theoretically-oriented experts have a responsibility to interpret their findings to those working in "the field"—police, crime prevention units, parent groups, and others who bear much of the brunt of public opinion in its demands to "get something accomplished" in ridding our communities of the criminal menace, whether it be a crime syndicate, a group of traditional criminals preying on the lives and property of society, or a rash of juvenile gang antisocial behavior. Some integration of theory and practice is sorely needed. The expert in his ivory tower can spin theories and write books and monographs on the etiology of crime, but unless his theories are interpreted, little will result. Some scholars go little beyond asking for more and more research. Some actionists pay little attention to the theorists.

If we are to come to grips with the problem of crime in our communities, understanding and tolerance must be nurtured and developed. Scholars may continue with research but by some means, at the moment somewhat obscure, they must volunteer to participate in community councils and suggest ways and means as to how their theories may be implemented through social action.

PART III

CRIMINAL JUSTICE IN OPERATION

15

The First Line of Defense—The Police System

Nowhere else in the world is there so great an anxiety to place the moral regulation of social affairs in the hands of the police, but nowhere are the police so incapable of carrying out such regulations. (American Police Systems, New York: Appleton, Copyright 1920.)

RAYMOND FOSDICK

The Dilemma of the Legal Powers of the Police

It is axiomatic in civic circles that the first line of defense against the criminal is the police. Society has great confidence in the local police despite perennial and justifiable criticism as to their philosophy and methods.[1] Students of criminal justice recognize that the efficiency of the legal machinery depends basically on the quality of the initial work done by the police. The crucial relationship of the police system to the restraint of criminals lends special interest and significance to the books and articles written in recent years on such topics as police training, the selection of capable men and women for the police force, and the separation of police systems from political control. Comparisons of efficiency are frequently drawn between local police and the Federal Bureau of Investigation. While the problems of these two systems of police investigation are essentially similar, in various respects they are different.[2] While the F.B.I. is not with-

---

[1] For a convincing indictment of the American police, see Albert Deutsch, *The Trouble With Cops* (New York: Crown, 1955). For a discussion of the principles to which police conduct should conform, see Don L. Kooken, *Ethics In Police Service* (Springfield, Ill.: C. C. Thomas, 1957).

[2] For an account of the work of the F.B.I., see Editors of *Look* Magazine, *The Story of the F.B.I.* (New York: Dutton, 1954); Don Whitehead, *The F.B.I. Story: A Report to the People* (New York: Random House, 1956). For a highly critical account, see Max Lowenthal, *The Federal Bureau of Investigation* (New York: Sloane, 1950).

out blemish, the intelligent citizen would no doubt be glad to settle for a local police system that compares favorably with the federal system, or that is as widely respected as many of our state constabularies.

Part of the criticism lodged against local police officers comes from lack of knowledge concerning their legal limitations in making arrests. In general, an arrest for a felony (a serious offense) may be made with or without a warrant; a police officer (or a citizen) may arrest without a warrant if he actually sees the felony committed or if he has reasonable grounds for believing the person is guilty of the offense.[3] For misdemeanors (less serious offenses) the police officer (or citizen) may arrest without a warrant if the offense has been committed in his presence. But if he has not seen the offense committed, he must have a warrant. The above procedure is commonplace practice, but the matter is regulated by statute in the various states. It is generally not realized that a police officer may not arrest on "suspicion"; nor may he insist on the arrested person's answering questions. It is in this area that so many illegal arrests are made. So far as illegal arrests and seizures are concerned, it would take a Philadelphia lawyer to interpret recent complicated rulings by the higher courts.

But arresting on "suspicion," known officially as "detention," is a widespread police practice. Detention, or arrest on suspicion, means that a police officer may stop a person whom he reasonably suspects of committing a crime, demand his name, address, or his reason for being at the particular place. Persons making "unsatisfactory replies" may be requested by the officer to accompany him to the precinct station for further questioning or investigation. This procedure is not known as an arrest in police circles and is justified by apologists for the police in the interest of public safety. Thousands of persons are questioned every day in this country and a large proportion of them are released without formal charges being made against them.

Whether or not detention is sound practice is a moot question. Obviously police officers should have some authority to "detain" suspicious characters. Yet the law is vague regarding this practice, which seems to conflict with two legal ideals: that persons are innocent until proved guilty, and that no person may be obliged to testify against himself. The police officer, if permitted by law, could detain anyone he pleased without any evidence whatsoever, and although the detained person has the constitutional right not to answer any question asked him, zealous officers have ways of dealing with those who refuse to talk.

At common law the police officer has no right to search or frisk the suspect before actually arresting him. Thus, a detained person or one arrested on suspicion may not be legally searched for a weapon. The person de-

---

[3] Rocco Tresolini, Richard W. Taylor, and Elliott B. Barnett, "Arrest without Warrant: Extent and Social Implications," *J. Crim. Law*, Vol. 46, No. 2 (July-August, 1955), pp. 187-198.

tained may legally resist and "may oppose force with force and kill if necessary to protect his own life or limb." This philosophy comes down from historical times but it is outmoded today since a person can no longer defend a prosecution for murder or felonious assault on the grounds that his resistance to arrest was legal.[4]

The History of the Police

The history of the police is interesting. The early colonial guardians of peace and security were known as watchmen, or "watch and ward." This office may be traced back to the constables and night watchmen who guarded English cities. Fosdick tells us that, as early as 1636, Boston had its watchmen in addition to its military guard. New York (New Amsterdam) and Philadelphia soon followed the example set by New England towns. By the year 1700 we find citizens serving as night watchmen or paying substitutes to fulfill their municipal duty. In New York, the watch were known as the *Ratelwacht,* because they carried a rattle. In most towns the watchmen carried lanterns.

The transition from the night-watch to the organized metropolitan police force was evolutionary. Following the lead of the city of London, which organized its uniformed force in 1829 at the insistence of Sir Robert Peel, the larger American cities began to give attention to the establishment of an organized force. In 1833, Philadelphia made abortive attempts to establish a citywide force when it created a short-lived unit of 24 police officers under provisions of the will of Stephen Girard. In 1844, New York created a successful day and night police system. This system is the basis of the nation's modern police organization.[5]

The early police in the United States were not very high-grade material. The various groups fought among themselves. A keen rivalry existed between the day and night shifts, such as was found in the early days between the rival volunteer fire companies. They wore nondescript clothing, and, as a writer in 1853 said: "And look at their style of dress, some with hats, some with caps, some with coats like Joseph's of old, parti-colored. If they were mustered together they would look like Falstaff's regiment."

---

[4] See Lester B. Orfield, *Criminal Procedure from Arrest to Trial* (New York: New York University Press, 1947), pp. 28-29. For further details, see: Rollin M. Perkins, *Elements of Police Science* (Chicago: Foundation Press, 1942), Chapter 12, "The Law of Arrest"; Hubert E. Dax and Brooke Tibbs, *Arrest, Search and Seizure* (Milwaukee: Hammersmith-Kortmeyer, 1950).

[5] There is controversy regarding the "first" police units. Consolidation of day and night watches under one administration occurred in New York City in 1844, in Chicago in 1851, in New Orleans and Cincinnati in 1852, in Boston and Philadelphia in 1854, and in Baltimore and Newark in 1857. Uniforms did not come into general use until 1855. See Donal E. J. MacNamara, "American Police Administration at Mid-Century," *Public Administrative Review,* Vol. 10, No. 3, pp. 181-189. See also, Quentin Reynolds, *Headquarters* (New York: Harper, 1955).

Those opposed to putting the police in uniforms considered it "un-Ameri-can," "un-democratic," "militaristic," "King's livery," "a badge of degrada-tion and servitude," or "an imitation of royalty."[6]

With the growth of cities and the incidental growth of crime, separate day and night police had to be developed. These gradually came under a uni-fied control of a chief or commissioner, usually appointed by the mayor. During the latter part of the nineteenth century many a conscientious and capable commissioner of police developed what was earlier a crude "rabble-in-arms" into a disciplined professional corps of guardians of the law and of society.

Organization and Duties of the Police

There are several types of police organization found in American cities of over 10,000 population. According to Donal E. J. MacNamara, noted au-thority on police work, these are: (1) control by a state-appointed ad-ministrator or board, as in Baltimore; St. Louis; Boston; Kansas City and St. Joseph, Missouri; Fall River, Massachusetts; Lewiston, Maine; and a number of New Hampshire cities; (2) integration into a department of public safety along with other protective services, as in Philadelphia; (3) respon-sible to a member of a city commission designated as police or public safety commissioner, as in Newark; Birmingham; and other commission-form cities; (4) control by an independent elected chief, as in Laredo, Texas; Santa Ana, California; West Palm Beach, Florida; and some other smaller cities; (5) responsible to a city legislative body, as in Atlanta; Charlotte, North Carolina; Jacksonville, Florida; Corpus Christi, Texas; and some 20 other cities; (6) responsible to a "strong mayor," as in New York and Chicago; (7) responsible to a city manager, elected by the city council or chosen through civil service examination, as in Cincinnati; Oak-land, California; Fort Worth, Texas; Lowell, Massachusetts; Miami, Flor-ida; and most manager cities; and (8) responsible to a board or commission under a "weak mayor" form, as in Los Angeles, San Francisco, and a score of other smaller places.[7]

In our larger cities, the police are usually divided into two groups: the uniformed patrolmen and the detectives. The former are usually employed in those duties dealing with public order and civic protection, whereas the detective division is "essentially a secret service for combatting the criminal

---

[6] Raymond B. Fosdick, *American Police Systems* (New York: Appleton, Copyright 1920), p. 70.

[7] MacNamara, "American Police Administration at Mid-Century," pp. 182-183. For additional information regarding police organizations, see: Bruce Smith, *Police Systems in the United States*, rev. ed. (New York: Harper, 1949); O. W. Wilson, *Police Administration* (New York: Macmillan, 1950); "What a City Should Expect from Police," *Life*, Vol. 43, No. 12 (September 16, 1957), pp. 70 ff, an article on Cin-cinnati's police system.

elements. In the larger cities, about two thirds of the officers are uniformed and serve as patrolmen or traffic officers. Usually the city is divided into precincts, although some of the larger metropolitan areas are made up of inspection districts, which are then divided into precincts. In the smaller towns, there is usually no such division; all patrolmen, as well as detectives, work out of police headquarters.

In all cities, the traditional unit of police service is the patrol beat or post, a geographic unit over which the patrolman walks. He is supposed to know the area of his beat thoroughly, not only the streets, alleys and vacant lots, but the inhabitants as well. He is often familiar with virtually everyone on his beat, although in some cities he is changed around so often that this is impossible. While the patrolman is getting accustomed to his beat, the criminal element in that neighborhood is also watching his habits. They soon learn his movements, the time of night when he rings in or reports to headquarters, and where he "hangs out" during the long hours he is supposed to be on duty. This information is highly important to the professional criminal, and the ease with which he attains it is one of the serious objections to the outmoded system of patrolling the city.

Above the patrolman is the sergeant, whose duty it is to supervise a number of officers who walk a beat. This is theoretically a check on the patrolman, but sergeants usually meet their patrolmen at designated times, so that there is no surprise and no real check-up. Above the sergeant is the lieutenant who has charge of the police precinct in the absence of the commanding officer. He is also expected to "keep watch" over patrolmen and sergeants. Above the lieutenant is the captain, in complete charge of the precinct and responsible for the crime conditions in that neighborhood. Over the entire system is the Chief of Police, the Commissioner, or Director. Frequently he has assisting him a deputy chief or an inspector. Usually the latter is in charge of administrative details, keeps the records, and assumes charge of the office when the chief is away.

In smaller towns and cities, there is no such differentiation as exists in a larger metropolitan area between uniformed patrolmen and detectives. We ordinarily think of the detective spending all of his time solving the large city's most serious and baffling murders, but in reality, most of his work deals with crimes involving loss of property—robberies, burglaries, larcenies, frauds, and swindles. Except in the larger cities, the detective personnel is, as a rule, no more intelligent or better trained than the average of the patrolmen in uniform. Many are chosen for detective work because of some conspicuous arrests or outstanding bravery in line of duty. Others are political appointees, and still others are employed because they have special aptitudes for investigation. They seldom have any specific training for their work; more likely they possess only an interest in sleuthing.

Special vice squads in our larger cities are sometimes coveted assignments for police but all too often they prove to be snares, as personnel are wide open to temptations which may lead to blackmail. There have been many instances in recent years in which police officers have been corrupted by vice squad duty. They come into contact with drug pushers, prostitutes, sex deviates, and the fringe of the criminal underworld. In addition, they are forced to lead irregular lives away from their families, which sometimes corrodes their morals. Some police authorities contend that vice should be handled by routine officers and not by special squads.

The police system in this country gradually became a salient element in partisan politics and political spoil. In many cities this condition still exists. In the past, those who applied for positions on the police force were often chosen not for intelligence or integrity, but for their ability to work with henchmen of the political organization. Such a system encourages corruption and favoritism, notoriously stimulated by the problems of vice. Even civil service has not eradicated special influence, since in many cities the civil service commission appointees are subservient to the political organization.

There are debatable issues involved in the reform of our methods of repressing crime, but the police system is not among them. It is a practical matter that admits of no evasion. Unless we can provide a police personnel more intelligent, expert, and incorruptible than we have had in the past, we face the dissolution of orderly social relations in our communities.

The chief criticisms of the American police, all of a serious nature, are: (1) subservience to political bosses through a system peculiar to American cities; (2) lack of professional training and clear differentiation between the patrolman and detective, and ignorance of the law and of the duties inherent in their jobs; and (3) brutality, ranging from such techniques of interrogation as protracted questioning, to "third degree" methods.

The Police and Corruption

The system of which the police are a part is honeycombed with corruption, graft, and partisan politics. This alliance is indigenous to this country and is one of the costs of democracy and party politics. With the growth of the great political machines, it was natural that the policeman on his beat should be enlisted to get out the votes and, if necessary, to promise the hesitant citizen certain concessions for supporting the party. Thus, the patrolman often became a tool of the dominant party.

When the bosses began to levy on certain dubious occupations operating in the large cities, in order to build up a "war chest" to keep the party in power, they offered police protection in return for the financial support of gamblers, prostitutes, and other obnoxious or illegal small-time operators. This system could not work without the support of the officer on his beat. It was his task—whether he liked it or not—to collect this "hush money" and turn it over to his superiors. As the system grew, it reached into every conceivable shady occupation, including the operation of punchboards, pinball machines, burlesque shows, pornography, dog races, and bootlegging. The next step in building up the system was to allow the policeman a certain small share of the hush money or graft, often referred to as "ice." His duty now was to collect the weekly sum, deduct his share, and pass on the remainder to be split between the higher-ups. After the sergeant, captain, and inspector took their share—the amount pyramided from the many patrolmen up to the top—the politicians controlling the organization pocketed the remainder. If the city was a large one, the illicit levy was huge.

Once the patrolman yielded to the temptation of playing this sordid but lucrative game, he was compromised. Yet, if he demurred, he was assigned to "Siberia," a beat "in the meadow," and was seldom cited for promotion. This was frankly admitted during the 1938 Jimmie Hines trial in New York when it was stated that police officers were shifted about if they interfered with the favored criminal groups. The rookie policeman might enter the system with good intentions, but if he was to succeed, he had to succumb to the system of graft. Consequently, not only vice and graft flourished, but serious crime as well. Criminals tended to hide away in gambling joints and bawdy houses where they were known and were reasonably certain the police would not molest them.

In the early days of the system, the political higher-ups were definitely in control of the situation. But, as the game began to be played for higher stakes, the powerful gambler or the leader of the syndicate that operated organized gambling or controlled prostitution soon became the master and dictated not only to the police but also to some of the higher political incumbents. These frequently included the mayor, some of the judges, the district attorney, and even state officers, up to and including the governor. In such a web of corruption, how did the obscure police officer on his beat count? He, in his small way, was enmeshed in the toils of a grim game so that, if he refused to play or if he played it half-heartedly, not only his job but his life was jeopardized.

The police seem to be largely impotent in the large cities where great gambling combines flourish—Chicago, Los Angeles, New York, Kansas City (Missouri), Miami, and other places. In fact, the same may be said regarding the smaller counterparts of these metropolitan centers. The pattern is the same in hundreds of localities. Sensational exposures were made in Brooklyn in 1950 that bear out this corrupt activity of police who are enmeshed in the tentacles of big-time gambling. Bookie Harry Gross told Judge Samuel S. Leibowitz that "a bookmaker can't operate without protection." He admitted paying one million dollars annually for police protection. It was pointed out that thirty cents on each dollar collected by the bookie went to the police.[8] The Kefauver Committee, in 1951, found evidence of this collusion in practically every city where hearings were held.[9]

Aside from hush money taken from illicit operators by the police, there is the malicious practice used by the police officer on his beat to "shake down" legitimate businessmen.[10] This is a widespread technique and many police officers believe it an ethical area in which they may operate as a perquisite of the profession. Included in these practices are "accepting" weekly sums of money for "helping the merchant count his money" on Saturday nights; accepting or expecting merchandise at wholesale prices or as outright gifts; attending all sorts of entertainments free of charge; and taking gratuities for services performed in the line of duty. Merchants who do not "play along" with the police in their neighborhoods are likely to find themselves in embarrassing situations, such as frequent visits from city inspectors that result in their being forced to make needless repairs, install new plumbing, or undergo complete renovation of their premises.

It is easy to suggest that the police be taken out of politics but it is hard to bring about such a sweeping reform. So long as we have political parties currying favors to the voters through patronage, Christmas baskets, the "fix," social privileges, and so forth, and so long as we have the organized well-oiled political machine that appears to be *necessary* in our democratic government, we shall never be able to eliminate graft from politics and divorce the police function from the politician.

Lack of Proper Training

The policemen of the past were appointed because they could fight, and because they were strong enough to cope with the gashouse gang or the hooligans of the community. They were poorly paid, and the mental requirements were low. If they looked good in a uniform, so much the better, for then they could be called upon to lead the innumerable parades that are so characteristic of metropolitan life. A police rookie was not expected to know anything about the law. He could pick that up by word of mouth. Gradually, manuals were developed for the benefit of the new recruit, but all too frequently he was not able to digest the various rules and regulations in these booklets.

In more recent years, a number of the larger cities have kept up-to-date in equipping their police, so that they are more capable of dealing with the armed criminal. Motorcycles, fast squad cars, teletype, radio, and other recent inventions have made a real contribution to law enforcement. But

---

[8] *New York Times* (October 8, 1950). For the story of the gambling empire of Harry Gross, see Norton Mockridge and Robert H. Prall, *The Big Fix* (New York: Holt, 1954).

[9] For a story of the temptations confronting a police officer, see Robert McAllister with Floyd Miller, *The Kind of Guy I Am* (New York: McGraw-Hill, 1957).

[10] See page 23 for description of this practice.

this streamlining of police methods has its drawbacks. Removing the patrol-man from his beat elicited apprehension among pedestrians and property owners. The American people like to believe they are protected against hoodlums and thieves at all times. In Cleveland, after a "wave" of assaults on women by street thugs, the director of police was forced to put the police back on their beats. There is much controversy in police circles over this matter.

The charge that the American police are inefficient is perhaps unfair in the light of the courageous efforts of many police administrators, but it is a charge that can easily be substantiated. A study made some years ago by Van Vechten showed that of 610,000 reported offenses, arrests were made in only 25 per cent of the cases, convictions obtained in only 5.5 per cent, and prison sentences meted in only 3.5 per cent.[11]

Let us examine the figures submitted by the F.B.I. showing the number of offenses cleared by arrests for 1956: for each 100 crimes against the person, police cleared 79 by arrest. In the individual classes of offenses, the police cleared by arrest 92 of each 100 murders, 87 of each 100 negligent manslaughters, 77 of each 100 rapes, and 78 of each 100 aggravated as-saults. Of crimes against property, for each 100, 24 were cleared by arrest. Clearances of robberies by arrest amounted to 42 of each 100, of burglaries, 31 of each 100, of auto thefts, 30 of each 100.[12]

But the police frequently show themselves to be inefficient by "running in" thousands of persons who are subsequently cleared. This charge is frequently made by students of the police problem. Calling this practice "trial and error" arrest, Hopkins shows that this is done to cover up police inefficiency: "All police reporters know that night after night in virtually every city, scores of individuals are 'run in' without being charged with any concrete offense, or even being considered especially connected with any known offense."[13]

Many of these are petty offenders or suspected of petty crimes. This flot-sam and jetsam found in all large cities consists generally of those who are maladjusted economically, socially, or mentally, and who need treatment by a social case worker or a psychiatrist. They are unable to compete favor-ably in our complex economic maelstrom. Guidance is called for rather than badgering and "pushing around." Thousands of these marginal men and women are not fortunate enough to be discharged after they are arrested. They are thrown into our county jails and houses of correction. Such a practice is futile and an added burden is thus placed on the taxpayers with few results. They continue in their habits and are arrested time and again with a stern but stupid admonition from the magistrates, most of whom are

11 C. C. Van Vechten, "Differential Criminal Case Mortality in Selected Jurisdic-tions," *Amer. Soc. Rev.,* Vol. 7, No. 6 (December, 1942), p. 837.
12 *Uniform Crime Reports,* Vol. 28, No. 1 (September 26, 1957), p. 47.
13 Ernest J. Hopkins, *Our Lawless Police* (New York: Viking, 1931), p. 70.

incapable of understanding the real social problem involved. It is easy to pound the gavel and boom out, "Thirty days in jail." As one of these socially bankrupt persons once put it, "We do a life sentence in short stretches." Millions of dollars are spent annually by our counties and municipalities with nothing constructive to show for the cost.

This vast army of socially inadequate, derelict, and transient types is not the criminal class. They need supervision—there is no doubt of that—but we must label them in some way other than "criminal." Probably a new and specialized monitor could be employed in the future to handle this problem. We have dealt more in detail with the derelict and socially unfit in an earlier chapter.

Aside from the cases of more or less false arrest, we find that many are arrested by inept police officers who fail to make a case for potential guilt in the preliminary hearing before the magistrate, so that the suspect is subsequently released. A third group is known as "squared cases": serious offenders arrested on suspicion of having committed a felony, whose offenses are changed to misdemeanors and then frequently dropped altogether. Commenting on this device, Hopkins states, with a word of caution:

> The third factor is the "squaring" of cases. . . . This . . . deals with felony cases only; and when a felony accusation is "squared," the prevailing device is the "changed charge"—the reduction, first, of the felony charge to a misdemeanor, before it is dropped altogether. Serious charges are seldom "squared" outright, except where the complaining witness disappears. . . .[14]

What is the remedy in this situation? The answer is easy enough but getting general action is slow and difficult. If the American police are to measure up to their responsibility, the citizen must become alert to the problem and demand thorough and continuing reform. This will call for professionally trained police and a complete break with the political machine. During the past few decades a corps of police specialists has emerged, due to the pioneer efforts of such men as the late August Vollmer of Berkeley, California, and the late Bruce Smith. Men at the top levels should be selected on a nationwide basis and rank-and-file rookies from carefully built examinations. Opportunity for promotion should be generous, based primarily on efficiency, dependability, and in-service training—not merely on seniority.

A second line of attack takes as its ultimate objective the elimination of thousands of the small, inefficient local police units that bedevil the entire field of law-enforcement. This can be realized by expanding over-all metropolitan areas such as New York, Los Angeles and Philadelphia, by strengthening the state constabulary, and by employing strong county or township forces.

---

[14] *Ibid.*

Police Cruelty—The Third Degree

The third general charge against American police is their cruelty in making arrests and in attempting to get confessions from suspects or evidence from witnesses. This charge has been repeated so often and substantiated by so many students of the problem that denials from police or their friends have little meaning.

It is assumed by the American people that the third degree is used only in "extreme" cases. They are all too ready to condone drastic police measures that are "really necessary" to eliminate or clear up crimes. But, of course, what is "necessary" is a purely subjective, arbitrary, and capricious question.

The term "third degree" is vague and emotive rather than descriptive. Ernest J. Hopkins makes this clear: "Do not ask whether the police use the third degree, but put the query rather, 'What methods do the police follow in getting a man to confess? How are felony suspects treated during the period after arrest and before the police bring them to court?' "[15]

According to the *Encyclopaedia Britannica* the expression "is believed to have been suggested by the third masonic degree which is conferred with considerable ceremony."[16] Hopkins verifies this:

> A man taking the third degree in a lodge might receive a little treatment with a slapstick. Applying it to a police interview was, to begin with, a bit of American slang. In 1910, Major Richard Sylvester of Washington, then president of the International Association of Chiefs of Police, explained that the "first degree" was the arrest, the "second degree" the transportation to some place of confinement, and the "third degree" the interrogation of the arrested man as to his guilt. All these steps, he argued, were essential parts of the arresting process.[17]

The reader may find much evidence concerning specific cases of third degree in the writings of Ernest Hopkins and Emmanuel Lavine[18] and more recently of Albert Deutsch.[19] Innumerable instruments are used to beat a suspect into submission; rubber hoses, clubs, electric shocks, fists, blackjacks, and kicks in various parts of the body, notably the groin. Not infrequently men must be hospitalized because of the torture to which they were subjected.

The technique is insidious because those who defend its use usually frame their own definition of severity, adequacy, and physical or mental

---

[15] *Ibid.,* p. 189.

[16] *Encyclopaedia Britannica,* 14th edition (Chicago: Encyclopaedia Britannica, 1929), Vol. 22, p. 135.

[17] Ernest J. Hopkins, *op. cit.,* p. 191.

[18] Emmanuel Lavine, *The Third Degree* (New York: Vanguard, 1930).

[19] *The Trouble with Cops* (New York: Crown, 1954), especially Chapter 5, "What Price Brutality?" For a grim novel in which police brutality plays a part see Seymour Shubin, *Anyone's My Name* (New York: Simon & Schuster, 1953; also Permabooks).

pain. Police argue that hardened criminals expect it and do not feel pain as acutely as other persons.

Apologists for the third degree claim such methods are used only when necessary, and imply that certain fundamental rights should be denied those whom the police reasonably consider guilty of serious offenses. It is difficult for some citizens to see what a dangerous philosophy there is in this implication.

The better American police officers and chiefs of police contend that they oppose third degree tactics, but if pressed, many will condone "protracted questioning," which is a definite phase of the technique. A few, however, categorically oppose this inquisitorial torture. For example, the late Chief August Vollmer, of Berkeley, California, in a letter written some years ago to the writers, stated: "For forty years I have consistently fought third-degree tactics including all protracted questioning of suspects."

Periodically stories of third degree methods are brought to light, often years after the specific brutalities. In one, Rudolph Sheeler, of Philadelphia, was arrested on suspicion in 1939 for killing a policeman three years earlier. Six of the city's policemen held the suspect incommunicado for forty days during which he was beaten, tortured, and subjected to protracted questioning so that it resulted in a confession. The officers then trumped up a case against the man that resulted in his being sentenced to life imprisonment. After serving twelve years his case was investigated gratuitously by Professor Louis Schwartz of the law school of the University of Pennsylvania and was found to have been a gross miscarriage of justice. A new and speedy trial was arranged and Sheeler was freed. After excoriating the cruel police tactics, Judge Gordon stated that those involved "revealed themselves as unfit to be entrusted with police power." The six officers were suspended and efforts were made to compensate Sheeler for the great wrong that had been done him.[20]

Another example of brutality was brought to light, again in Philadelphia, in 1957 when one Aaron ("Treetop") Turner was freed from a death penalty inflicted upon him in five trials extending over a twelve year period. This was a *cause celebre* in Pennsylvania over the years since Turner, accused of murdering a watchman in 1945, was defended by a court-appointed attorney, Edwin P. Rome, who refused to accept defeat. In freeing Turner the state Supreme Court called the affair a "near tragedy of errors" in which a confession "had been wrung from the appellant by coercive third degree police methods" over a twelve day period.[21]

In 1957 a Puerto Rican was released from the Norfolk, Massachusetts, state prison after serving three years of a life sentence for murder after it was established that he was innocent of the deed. It was revealed that a con-

---

[20] This case, along with those of others wrongfully convicted, is detailed in Jerome Frank, *Not Guilty* (New York: Doubleday, 1957), pp. 165-186.
[21] Philadelphia *Inquirer* (June 7, 1957).

fession had been extorted from him by the Springfield (Massachusetts) police in 1954. Knowing very little English at the time, he was severely beaten until he signed a confession. Indignant citizens of Springfield called for reforms in the city's police system.

A case that gained national notoriety occurred in Chicago in 1946 in connection with the brutal slaying of a six-year-old child whose remains were found dismembered in a sewer. The janitor of the apartment house in which the little girl lived was suspected. He was arrested and thoroughly beaten up and ordered held for the offense. Later it developed that a University of Chicago student, William Heirens, an infantile sex deviate, was the killer. The janitor sued the police for $100,000 and obtained a $20,000 settlement.

Third degree tactics are a violation of the Fifth and Sixth Amendments of the Constitution which safeguard the rights of the accused, first in the United States courts and then in the several states.[22] In addition, the Fourteenth Amendment specifically states: "nor shall any State deprive any person of life, liberty, or property without due process of law." Many states have gone further and have written into their Constitutions that the accused has certain rights, including "due process of law."

In a United States Supreme Court decision handed down on February 12, 1940, in a scathing denunciation of illegal tactics, Justice Black, reviewing a Florida case of four Negroes subjected to long and protracted questioning by public officials, reversed the decision of the state Supreme Court by invoking the due process clause of the Fourteenth Amendment.[23] On January 2, 1952 the Court unanimously upset the California conviction of a man for having narcotics in his possession.[24] Officers had entered his home without a search warrant. The suspect swallowed two capsules of narcotics and the police seized him, drove him to a hospital, strapped him to a table and forced a stomach pump down his throat in order to retrieve the capsules. On this evidence the man was convicted. The Supreme Court of California refused to hear the man's appeal but the U.S. Supreme Court reversed the conviction on the ground that the police officers' methods of obtaining evidence violated due process under the Fourteenth Amendment.

These decisions would seem to make *any* questioning in any star-chamber situation a violation of the guarantees afforded citizens.[25] Some students of

---

22 Cf. Francis Heller, *The Sixth Amendment to the Constitution of the United States* (Lawrence, Kansas: University of Kansas Press, 1951).

23 Chambers et al. v. Florida, 309 U.S. 227 (1940).

24 Rochin v. California, 342 U.S. 165 (1952).

25 "Star chamber": Originally, the Court of Star Chamber, chiefly a criminal court developed in 15th-century England from the King's Council, which was wont to hold judicial proceedings in the apartment in the royal palaces at Westminster that was so-called, conjecturally "because at the first all the roofe thereof was decked with images of starres gilted." From the abuse of this court under James I and Charles I (it was abolished by the Long Parliament in 1641), its name has become a byword for oppressive and arbitrary legal proceedings.

criminal procedure consider that it is impossible to get along without third degree methods, physical or psychological. Others contend that any questioning of a suspect should be in the presence of responsible persons who will safeguard the interests of the accused. One of the persons should be the suspect's attorney or someone who can see that his legal rights are not placed in jeopardy.

Some argue that if the criminal or suspect has the privilege of refusing to talk or to incriminate himself, the ascertainment of guilt is more difficult for the police. There are all too many well-meaning people in penal reform work who are ready to compromise with the police and grant them "certain prerogatives" in regard to questioning the suspect prior to arraignment. Of course many professional criminals will not talk, and many are further legally protected by writs of *habeas corpus* which demand that the police bring the arrested person before a magistrate for an immediate hearing.

It is not possible (or legal) to distinguish between suspects with known criminal records and the average citizens rounded up in a crime hunt. *Every* citizen, prior to the actual establishment of guilt, is guaranteed these rights by the Constitution—and that excludes every distinction.

The burden is on the police and their skill in collecting evidence against the suspect. Current practices permit the police to keep the suspect "for a reasonable time" in the police station prior to the hearing of the magistrate. What constitutes a reasonable time? The police feel justified in keeping the suspect incommunicado (or "on ice") until they succeed in breaking down his resistance by rigid questioning.[26]

It is sufficient, without going further, to note that: (1) the third degree is not restricted to any one section of the country; (2) in general, it is used more frequently against friendless and inconspicuous persons than against dangerous criminals, who are known to have powerful connections; (3) ingenious devices or techniques that leave no mark on the body of the suspect are used by the more adroit police officers; and (4) the third degree still persists despite its illegality and its frightfulness. Its secrecy makes it all the more insidious, since this feature hides it from the general public, which would certainly condemn it if the methods were well understood.

The use by the police of informers from among the criminal element— "stool pigeons"—poses a knotty ethical problem in police administration. It is a practice universal among American police and is stoutly defended by many authorities in police work.[27] Such informers are usually drawn from

---

[26] See pages 235-236 for further discussion.

[27] Henry Fielding, the great English dramatist who was responsible for "cleaning up" the rowdies infesting the streets of London, is credited with using "informers" as early as the 1750's. His "Bow Street Runners" were undercover thugs who were paid "piecework" for their unsavory services. This is related by Patrick Pringle, in *Hue and Cry* (New York: William Morrow, 1956).

small-fry criminal elements—drug addicts, pickpockets, prostitutes, panderers, and others whose criminal activities are well known to the police. In return for information regarding crimes or the whereabouts in a city of top-flight criminals, the police usually offer sums of money to the informer, or in many cases, immunity against arrest for their trifling offenses.

Using informers is "dirty business," many police officials admit; but, they counter, so is crime and vice. Some contend that law enforcement could not possibly operate without such a system. There are, of course, many hazards in informing. This is the risk the person must take if he decides to live the life of a stool-pigeon. In fact, even a reputable citizen endangers his life if he informs on a criminal, especially if it is a dangerous or much wanted criminal. A tragic example of this occurred in March, 1952, when a respectable young resident of Brooklyn, Arnold Schuster, was shot to death a few days after the newspapers had lauded him as the source of information leading to the arrest of the notorious bank robber, Willie Sutton.[28]

Another quite prevalent practice of the police is what is known as "entrapment," a technique used primarily by metropolitan vice squads. In their zeal to arrest persons suspected of immoral conduct, use or sale of narcotics, prostitution and other similar vices, police resort to "decoys" to trap the victim. Entrapment was once defined by former Supreme Court Justice Owen J. Roberts as "the conception and planning of an offense by an officer, and his procurement of its commission by one who would not have perpetrated it except for the trickery, persuasion or fraud of an officer."[29] Writes Professor Donnelly of Yale Law School concerning this technique:

> Clearly entrapment is a facet of a broader problem. Along with illegal search and seizures, wire tapping, false arrest, illegal detention and the third degree, it is a type of lawless law enforcement. They all spring from common motivations. Each is a substitute for skillful and scientific investigation. Each is condoned by the sinister sophism that the end, when dealing with known criminals or the "criminal classes," justifies the employment of illegal means.[30]

But as John B. Williams points out, the test of intent to commit a crime lies at the basis of "entrapment." If the police know or have strong reason to believe that a crime is about to be committed, entrapment apparently seems justified. On the other hand, officers are paid to catch criminals, but not to create them.[31]

---

[28] For further information regarding the use of informers, see Richard C. Donnelly, "Judicial Control of Informants, Spies, Stool Pigeons and Agents Provocateurs," *Yale Law Journal,* Vol. 60 (November, 1951), pp. 1091-1131.

[29] Quoted by Deutsch, *The Trouble with Cops,* p. 85.

[30] Donnelly, *op. cit.,* p. 1111.

[31] John B. Williams, "Entrapment—A Legal Limitation on Police Techniques," *J. Crim. Law,* Vol. 48, No. 3 (September-October, 1957), pp. 343-348.

Police Wiretapping

Private individuals as well as policemen use several modern techniques to gain information clandestinely: wiretapping, tape recorders—"bugging"— and telescopic cameras.

There is no denial among police officers or the F.B.I. that wiretapping is practiced. J. Edgar Hoover admitted on January 16, 1950 that his organization had tapped 170 telephones. At that time he justified the practice in the name of "internal security," adding, "I dare say the most violent critic of the F.B.I. would urge the use of wiretapping techniques if his child were kidnaped."[32]

This, of course, is no excuse for illegal methods being used, and Hoover would be the first to admit it. Kidnaping is charged with emotion and even hysteria. Doubtless the parents of the kidnaped child would condone almost any illegal tactics in order to have the child returned safe and unharmed. But no law-enforcement agent can defend such a position in the cold light of day.

The legal status of wiretapping in this country is in a confused state. Only four states, including New York, permit law enforcement officers to tap telephone wires and introduce evidence obtained in criminal prosecutions. The courts in about half the states permit wiretap evidence although their legislatures have not authorized wiretapping.

In 1934 Congress passed legislation permitting wiretapping by the F.B.I. "under certain conditions"—primarily to protect internal security. In the Coplon-Gubitchev espionage trial in New York some years ago, Judge Sylvester Ryan refused to admit testimony gathered by F.B.I. agents through wiretapping. He stated: "This is still the law. It contains no exemptions as to any individual and no exception as to investigation of any particular type of crime."[33]

The question of the use, in a Federal court, of wiretap evidence collected in a state where the technique is legal was decided by a Supreme Court decision in December, 1957. The Court held that no one was exempted from the 1934 statute and thus all wiretap evidence, no matter where obtained, or by whom, is inadmissible in a Federal court except in the rather rare cases permitted by the law.

Yet the police insist that they must be permitted to collect wiretap evidence if they are to combat crime efficiently. They contend that the technique is not a violation of the citizen's civil rights. Nowhere in the Constitution, they say, is privacy guaranteed. The opponents of wiretapping contend that this privacy is implicitly guaranteed.

Starting with the classic statement by Justice Oliver Wendell Holmes, who

---

[32] AP news dispatch, January 16, 1950.
[33] "FBI—Outside the Law?" *The Nation,* Vol. 170, No. 5 (February 4, 1950), p. 99. See also, Sam Dash, *The Eavesdroppers* (New Brunswick: Rutgers University Press, 1959).

referred to wiretapping as "dirty business," the exponents of civil rights have resisted it and other methods of surreptitiously gaining evidence by the police. The whole matter will probably be settled by rulings of the Supreme Court as cases come before it. At the moment it is regarded as repugnant by those concerned with the rights of citizenship, but by most police and other law-enforcement agents as necessary in detecting criminal activity. If widely legalized, of course, the major effect of wiretapping and other such practices might be to force criminals to change a few of their own methods; henceforth they would no doubt plan their forays against society on the street corner in whispering sessions.

Scientific Detection Methods—Criminalistics

The scientific detection of criminals and the investigation of crimes is known as "criminalistics." The term comes from Hans Gross (1847-1915), an Austrian lawyer who is given credit for founding modern police techniques. As an examining magistrate he had opportunity to witness the unreliability of human testimony. This led him to try to supplant such testimony by scientific means. He made no actual contributions to the science of crime detection himself, but gathered from the different branches of knowledge everything that could be of use in dealing with evidence.

The field of criminalistics is as much concerned with protecting the innocent from unjust conviction as with aiding in the conviction of the guilty. It is, in reality, a composite science dealing with police investigation of crimes, and calls into use all branches of knowledge, especially those dealing with physical phenomena, that will aid in reaching its final objectives.[34]

More famous than Gross was Alphonse Bertillon (1853-1914), the famous chief of the *Service d'Identité Judiciare* in Paris.[35] Bertillon developed a system of anthropometrics carefully worked out on the assumption that an individual's physical measurements are constant after maturity is attained. Certain of these measurements (height, span of arms, sitting height, length of head, width of right ear, length of left foot, length of left middle finger, length of left little finger, and length of left forearm) make identification of a criminal relatively simple. The Bertillon system also records photographs (front and profile), descriptions of hair and eye color, complexion, scars, tattoo marks and any asymmetrical anomalies. Bertillon made a signal contribution to the field of identification, and his carefully worked-out system deserves high praise.

---

[34] See: Charles E. O'Hara, *Fundamentals of Criminal Investigation* (Springfield, Ill.: C. C. Thomas, 1956); LeMoyne Snyder, *Homicide Investigation* (Springfield, Ill.: C. C. Thomas, 1951); Harry Soderman and J. O'Connell, *Modern Criminal Investigation* (New York: Funk, 1952).

[35] Henry Rhodes, *Alphonse B. Bertillon: Father of Scientific Detection* (New York: Abelard-Schuman, 1956).

Stimulated by these two pioneers, many outstanding personalities worked in the field of identification after the turn of the century. Among these may be mentioned Edmond Locard of Lyons, France, concerned with poroscopy, the examination of skin pores for identification; R. A. Reiss, formerly of Lausanne and later of Belgrade, working on criminal photographic methods; DeRechter of Brussels, in the identification of bullets, cartridges and traces of burglars' tools; Edmond Bayle of Paris, in the use of optical and photographic methods; Robert Heindl, Berlin, in metric photography and fingerprints; Hans Schneickert, Berlin, in the analysis of questioned documents; Siegfried Türkel, Vienna, in the examination of objects of art and antiques; and Van Ledden-Hulsebosch, Amsterdam, working with numerous optical methods of criminal investigation. In our own country Dr. Albert Osborn should be mentioned for his study, *Questioned Documents.*

These fields of special study represent merely a sampling of what is now undertaken by the first-class police laboratory, and by no means exhaust the areas in which science comes to the aid of efficient police systems. As stated above, modern police technique borrows from all the sciences. The three that are most frequently called into service are biology, chemistry, and physics. Most of the biological methods fall within the field of biochemistry and physiological chemistry, and include fingerprinting, examinations of the blood and hair, stains, dust particles, poisons, and post-mortem techniques. Physics provides all sorts of measurements and methods of comparison, as well as spectrography, photometry, and luminescence analysis.

Probably the most interesting techniques that have been developed in recent years deal with blood analysis, although many phases of this subject are not yet mastered. Stains that were once difficult to analyze because they were on a background of the same color are no longer troublesome: bloodstains on rust-colored plush or stains from semen on dark woolen material are now identified through the aid of ultraviolet rays. To determine whether blood is human or animal, the old methods of measuring and determining the form of the blood corpuscles have been displaced by serological methods. The species of animal from which the blood came is determined by mixing a water extract of the bloodstain with antiserums for the bloods of a variety of species of animals. Reactions, usually precipitates, occur only in mixtures of bloodstain extract and antiserums that are homologously related. Thus, extracts of pig blood will form precipitates in antiserum for pig blood only, and extracts of human bloodstains will form precipitates in antiserum for human blood only. This technique is very accurate and dependable, and the test can be made rapidly by a skilled technician.

In addition to the above techniques, most of which are relatively unknown to the layman, we find the universally accepted dactyloscopy, or identification by means of fingerprints, which has largely supplanted the older Bertillon system of identification. Bertillon was thought to have discovered this system, but he did not sponsor it, and was for a long time opposed to it. It owes its origin and development to the work of Francis Galton, Henry Faulds, William Herschel, and Edward Henry, supplemented by the Argentinian, Juan Vucetich. The thesis supporting this system is that the fingerprints of each person are absolutely unique and that not once in a trillion cases will two persons have identical fingerprints.[36]

A field that calls for caution is that of *forensic ballistics,* the identification of bullets fired from a gun. Much elaborate equipment is necessary for this investigation and where it is available and employed by experts there is little question of its value. But ballistics often intrigues charlatans who are sometimes employed by the police and make dubious analyses, sometimes doing great harm. Even alleged experts can make serious mistakes. Some years ago, in Cleveland, a man was almost convicted of murder on the expert evidence about the gun used in the crime. The gun in question and the bullet taken from the body of the dead man were submitted to an expert of national reputation, who reported that the gun examined fired the bullet. The Cleveland *News* investigated and found that the revolver in question had not been sold until a month after the murder. In the famous Sacco-Vanzetti trial in Massachusetts it was shown that most of the murder bullets were fired from a Steuer automatic, a foreign-made revolver. But since Sacco had a Colt pistol, the expert tried to prove that the shot came from his weapon. The owner of the Steuer, one of the notorious Morelli gang, was known by the prosecution to be in custody at the time and could have been produced, but this would have broken down the hysterical case against the accused men.[37]

Other techniques and methods essential in the identification of criminals and the analysis of clues include moulage, or the making of plaster casts for use in identification; microphotography for various obvious investigations; the use of chemical agents for detecting and analyzing various types of stains; methods of interpreting dust and perspiration stains, and of analyzing and identifying tire treads, handwriting, and forgeries.

Handwriting experts are often very slightly removed from pure fakery, even today. Much credence is given this type of analysis by the public generally, but results have thus far deserved no enthusiastic approval. Even the efforts of Locard, Schneickert, Osborn, Hamilton, and others to create an objective evaluation of the importance of specific peculiarities of handwriting leave much to be desired. The methods used presuppose the presence of long texts rather than a single signature.[38] A certain degree of subjectivity is unavoidable when the expert graphologist tries to find and build

---

36 Douglas G. Browne and A. S. Brock, *Fingerprints: Fifty Years of Crime Detection* (New York: Dutton, 1954); Frederick Cherill, *Fingerprints Never Lie* (New York: Macmillan, 1954).

37 See Arthur Warner, "A Sacco Revolver Expert Revealed," *The Nation,* Vol. 125, No. 3257 (December 7, 1927), pp. 625-626.

38 For a discussion of this matter by one of the experts, see Ordway Hilton, "Handwriting Identification Vs. Eye Witness Identification," *J. Crim. Law,* Vol. 45, No. 2 (July-August, 1954), pp. 207-212.

up a composite whole of those peculiarities of handwriting that characterize a writer.

Probably the most spectacular device in the modern police laboratory today is the lie detector. The public "knows all about it," but is totally ignorant of its limitations. The mechanical device for lie detection used is psychophysiological in its nature. It is attached to the arm and chest of the suspect, and makes a continuous record of his blood pressure, respiration, and any deliberate flexing of the biceps muscle. Significant questions are then asked him, and his emotional reactions are recorded.[39] It is contended that a guilty person will betray himself by his recorded emotions to the questions.

Another method of attempting to extract the truth from a suspect is "truth serum." Formerly scopolamine was used but today a barbiturate such as sodium amytal is usually employed.[40]

Much of the equipment we have mentioned, along with a great deal more that is in use in some of the larger crime laboratories of the country, is not available in the majority of police jurisdictions. It is expensive, and personnel with a knowledge of its uses and applications to crime detection is scarce. As it is now, only the larger cities can afford an adequately equipped and staffed crime-detection laboratory. Almost every county in the United States, however, has various amateurs in the fields of criminalistics. Good police systems can build up a staff of experts who will agree to serve on call for a modest consideration, or gratis, and thus make available a group who can materially assist in the detection of criminals and the analysis of clues.

The noted criminologist, Bernaldo de Quiros, many years ago, posed the question concerning the criminal's reaction to scientific methods of detection and apprehension in a statement which he labeled "The Criminal's Reply." It is worth recording:

> This intelligent attack is met by an equally intelligent defense on the part of the criminal world. The novel reflects the struggle. Sherlock Holmes, that clever detective, finds a worthy rival in Hornung's *Raffles* or in Leblanc's *Arsene Lupin*. . . . This is only fiction but the reality is not far distant. . . . Thus, as Ottolenghi pointed out, if the Bertillon system led professional criminals to abandon tattooing, dactyloscopy makes them commit a crime with gloves on to avoid the betraying dactylogram. . . . Will the skill of the thief ever be overcome?[41]

---

[39] For a standard work on lie detection, see Fred E. Inbau and John E. Reid, *Lie Detection and Criminal Investigation* (Baltimore: Williams & Wilkins, 1953). See also: Benjamin Burack, "A Critical Analysis of the Theory, Method, and Limitations of the Lie Detector," *J. Crim. Law,* Vol. 46, No. 3 (September-October, 1955), pp. 414-426; also Charles A. McInerney, "Routine Screening of Criminal Suspects by the Polygraph (Lie Detector) Technique," *ibid., J. Crim. Law,* Vol. 45, No. 6 (March-April, 1955), pp. 736-742.

[40] For limitations on the use of "truth serums," see John M. MacDonald, "Truth Serum," *J. Crim. Law,* Vol. 46, No. 2 (July-August, 1955), pp. 259-263.

[41] Bernaldo de Quiros, *Modern Theories of Criminality* (Boston: Little, Brown, 1911), pp. 231-232.

The *modus operandi* system of identifying a criminal is based on the assumption that the criminal leaves behind him his *trademark* in both his crime and his methods, that is, clues that definitely show what type of criminal the operator is. This system frequently promotes the actual identification of the operator because each has his own pattern of operation. As Professor Robert H. Gault describes the method:

> One [criminal] habitually steals clothing, another silver; one goes about it in the day-time, another at night; one enters by jimmying a rear window; another by means of an improvised door key, another by pretending to be a salesman; one proceeds to lock the occupants of the house into a closet; one smokes a cigarette furiously while at work and scatters ashes and stubs and match sticks over the floor; one smokes cigars; one leaves behind an apologetic note; one leaves a "thank-you" note; another a threatening inscription on the wall, and another imprisons the dog. Each thief will leave behind him several traces of his method of working, every one of which may be discovered by the officer who knows how to look for them.[42]

The important feature of the *modus operandi* technique is the classification of these clues, which are almost a signal, leading to the specific person who left them behind.

The development and expansion of criminalistics or police science shows an earnest attempt of the more intelligent police systems to utilize new discoveries that will be of service in the detection of crime and the apprehension of criminals. A warning may, however, check overenthusiasm in this field: the expert rouses too much confidence among detectives and policemen—and too little among trial lawyers. Much misplaced faith of many zealous detectives and police officers has been observed by students of detection. It is important to avoid sweeping generalizations regarding the guilt of individuals, on the basis of snap hunches by "amateur experts" or even by scientific findings, as gauged by this new equipment.

Rural Crime Control

The traditional organization of the law-enforcement agencies found in rural areas is quite different from that in the cities. The officer who is in charge of the apprehension of criminals throughout the county, including nonurban parts (if unincorporated), is the sheriff.

In the days of the Anglo-Saxons the sheriff, as the "King's Steward," held a dignified and enviable position in the county. But gradually his powers were curtailed so that today he has become little more than a custodian of the county jail. However, in those sections of the country where the county has real meaning, notably in the West and some sections of the South, the sheriff still has considerable prestige as an apprehender of criminals. He is charged by law to see that criminals within his bailiwick, the county, are apprehended; but he is personally little concerned with this

---

[42] Robert H. Gault, *Criminology* (Boston: Heath, 1932), p. 370.

phase of his job. As Bruce Smith, in his authoritative book, *Rural Crime Control,* states: "In New Jersey, 16 of the 21 sheriffs admitted that they did not take seriously their statutory duty to apprehend criminals. . . . The sheriff of a large and important rural county stated that 'in all my experience with the sheriff's office for the last twelve years, the sheriff has never been called upon to apprehend a criminal.' "[43]

The sheriff's deputies assist him in his duty, whether it is the care and custody of the inmates of the county jail or the investigation of crime. As a rule, the sheriff pays his own deputies out of the "fees" which, from time immemorial, are the perquisites of his office. One of the powers of the sheriff is to deputize any citizen for the purpose of running down a criminal who violates the peace of the county. The group thus deputized is called the *posse comitatus* (county's power). Refusal to join this group entails a severe penalty on a citizen if he is summoned. The *posse* is still a familiar incident in the community life of sparsely settled counties.

The officer of the village or small town, upon whose shoulders falls the responsibility of handling the crime situation, is the marshal, or as he is known in many jurisdictions, the constable. In colonial times the constable was usually elected at the town meeting, although occasionally he was appointed by the selectmen. He had a wide range of minor duties, such as summoning the electors to the town meetings, collecting taxes, settling claims against the town, supervising highway repair, and taking charge of the "night watch." At first, the constables organized the compulsory night watch. This duty fell upon all able-bodied citizens on regular assignment, but later this practice was abandoned, and the constable became the local policeman.

The rural judiciary that holds court to deal with petty crime consists of justices of the peace, counterparts of the magistrates in the larger cities. The constable now is usually an officer of the justice of the peace. All persons arrested within the "realm" of the justice of the peace are brought before him for the initial hearing, and he disposes of most of these cases. Justices of the peace are usually distributed by towns, townships, districts, or similar subdivisions of the county. In most states, 36 to be exact, they are elected; in three, they are appointed or elected, as the legislatures see fit; and in six states, they may be appointed by the governor.

Almost invariably all these officers controlling crime in rural areas are untrained. Often they are incompetent and ignorant. The system of which they are a part may have sufficed when crime was rare, but today they represent a "horse and buggy" method of dealing with crime and delinquency.

---

[43] Bruce Smith, *Rural Crime Control* (New York: Institute of Public Administration, 1933), p. 56. For an interesting account of law-enforcement methods in the colonial period of the South, see Jack Kenny Williams, "Catching the Criminal in Nineteenth Century South Carolina," *J. Crim. Law,* Vol. 46, No. 2 (July-August, 1955), pp. 264-271.

They are outmoded and should be abolished. Rural crime control should become one of the functions of a state constabulary through a system of centralized control.

There is one other county officer, the coroner. He must ascertain the cause of death, and determine responsibility in suspicious cases or where there is no physician's death certificate. His responsibility, however, is great, since he must differentiate between natural causes of death and death at the hands of some person, which constitutes homicide. Qualifications for the position of coroner are meager. In some states he must be a physician or have "some medical knowledge," or be "learned in the law."

So much progress has been made during the past 50 years in postmortem techniques that the highly trained skill of a medical examiner should be employed instead of the many charlatans holding down the position of coroner. As early as 1877 Massachusetts abolished the position of coroner, substituting a medical examiner appointed by the governor to a term of seven years. Unfortunately he is compensated only by fees, except the examiners in the Boston area who have fixed salaries. In New Hampshire the officers who replaced the coroners are *medical referees* and hold office for five years. Every autopsy authorized must be performed by the pathologist of the state laboratory of hygiene.

Bruce Smith summarizes the law-enforcement situation in rural areas today by saying that "the justice of the peace is a vigorous institution," compared to the sheriff-constable system and the absurd coroner system, but "this marked vitality should not be permitted to obscure the fact that judicial pronouncements delivered over a kitchen table by a judge who not only lacks professional training but often even the will to serve, fail to satisfy the requirements of an increasingly complex rural society."[44]

The Problem of Centralization

Most students of the police problem agree that a centralized system is the only complete answer. The traditional practice of organizing the police on a municipal, county, or town basis is outmoded, especially in populous areas. Glaring examples of jurisdictional confusion are found, especially in Boston, Chicago, and Los Angeles. In the latter city the problem is serious indeed since gansters and big moguls of crime capitalize on the divided authority of law-enforcement agencies and carry on their lawlessness with arrogance and impunity.[45] A police officer in one jurisdiction has no au-

---

[44] Smith, *Rural Crime Control* (New York: Institute of Public Administration, 1933), p. 277.

[45] This problem is pointed up in "A Survey of the Los Angeles County Facilities for Dealing with Crime and Delinquency," State Department of Corrections, Sacramento, June 28, 1940. In Los Angeles county there are 46 separate policing units, each with its own arresting limitations.

thority in another. He may pursue his suspect further than his own precincts by means of the so-called "hot pursuit" philosophy, but this is a dubious procedure even though it has been upheld by the Supreme Court as legal. The question is just how far he may pursue.

Consolidation is undoubtedly the answer, but it must go far beyond the conventional metropolitan system we now have and there must be state or regional control of the police function. While this form of control has not yet been established to any appreciable extent, the formation of state constabularies is a step in the right direction.

Let us, therefore, look at the development of the State Police, or the State Constabulary. Although there was a frontier police force, known as the Texas Rangers, supplementing the military forces in Texas as early as 1835, the modern state police organization dates back only to 1905. In that year, Governor Pennypacker of Pennsylvania created such a force to cope with the disturbed industrial situation, because the sheriff-constable system had completely broken down in its attempt to maintain order.

Such units were established in many of the states after World War I because of the ever-increasing problem of automobile traffic and its incidental wave of car thefts. There was some opposition from rural police officers at the outset to the introduction of the state police, but today there is a large amount of cooperation between the two systems. Standards are usually much higher in the state systems and, limited only by state boundaries, they are better equipped than local peace officers for running down criminals who get away from the scene of the crime in fast cars. Every state has a constabulary of some kind or quality. They vary as to the degree of police powers they possess, but a standardized system of functions may be developed and generally accepted.

The state police can devote much attention to coordinating the facilities for the patrolling of state highways, during summer traffic congestion and in winter blizzards and ice storms as well. They give a feeling of security to persons who live in the more isolated sections of the state. In general, the prestige of the state constabularies is relatively high.

The creation and development of state police systems has been a progressive step in the right direction, since there are so many advantages of centralization in law-enforcement. One obstacle is local pride. But far more potent is practical politics. State police have been remarkably free of graft, corruption and political patronage. It is this fact that recommends this system to the public.

16 ―――――――――――――――

The Defendant
Before the Bar of Justice

The law itself is on trial in every case as well as the cause be-
fore it.

JUSTICE HARLAN STONE

If we are to keep our democracy, there must be one command-
ment—Thou Shalt Not Ration Justice. (Federal Probation,
Vol. 20, No. 1, March 1956.)

JUSTICE LEARNED HAND

The Preliminary Hearing—The Magistrate's Court

After a police officer has made an arrest of a suspected
person he is expected to take him to a magistrate for a preliminary hearing.
The purpose of this hearing is to determine whether the evidence in the
possession of the officer is sufficient to prosecute. The defendant may make
a statement in his own behalf but the magistrate may not question him in
order to bring out additional evidence concerning the crime he is alleged
to have committed.

Before we discuss the disposition of a case before the magistrate it is
imperative that we examine the time element between the arrest and the
preliminary examination. The statutes of the various states dealing with
this interval are somewhat vague. It is implied that the suspect should be
taken before the magistrate without delay; yet it is notoriously true that
police keep many suspects "on ice" or incommunicado for days, weeks, or
even months. This is done in order to give the police a chance to question
the person (by fair means or foul) and thus build up a better case. Many
police precinct quarters maintain a number of "cold-storage" cells. It is not
unusual for the police to whisk an interesting suspect from one precinct to

another so the man's lawyer cannot find him. This is known as "sending a man around the loop." It is during this period that the third degree can be used with impunity.[1]

Our forefathers were wise in insisting upon impartial, democratic instruments by which the common man would be protected even though he were accused of committing a breach of the peace. The philosophy that a man is innocent until proved guilty is an example of this passion for safeguarding the rights of the individual. One such institution that was created for this specific purpose was the office of the justice of the peace, now often called a magistrate.

This dignitary may be traced far back in history. Edward III (1327-1377) made of the justice a sort of superior police officer who kept order among the villagers, prevented riots, assaults, and other crimes of violence. In course of time he became more a judicial officer and less a regulator of the peace.

In metropolitan centers there is usually a magistrate's court in each police precinct. It is this functionary before whom all arrested persons must be taken for the preliminary hearing. If the offense is serious, the magistrate must decide to his own satisfaction whether the crime was committed and also whether there is probable cause to suspect the accused. If, in his judgment, there is not sufficient evidence, the suspected person is discharged. If, however, the magistrate is convinced there is evidence that the person did commit the offense, he remands him to jail to await further action by the grand jury, or sees to it that the bail he sets for future appearance is deposited with him.

It is the business of the magistrates' courts to dispose of cases but it is also their business to see that no one guilty of an offense is discharged through political influence.

The accused may or may not be represented by counsel. Many professional criminals, because they are acquainted with the technicalities and procedures of the magistrate's hearing, manage to have their lawyers on hand to see that their legal rights are safeguarded. While the preliminary hearing is not a trial, evidence against the suspected person is submitted, and he has the opportunity of making a statement if he wishes.

The magistrate then has the responsibility of disposing of the case in one of the following ways: (1) discharge the suspect for lack of evidence or because the offense is innocuous; (2) release him on bail until the grand jury disposes of his case; (3) release him on his own recognizance, that is, take his pledge to appear later when wanted; and (4) remand him to jail pending the deliberations of the grand jury. If the offense is serious, bail

---

[1] For time intervals between arrest and arraignment, see Lester Orfield, *Criminal Procedure from Arrest to Appeal* (New York: New York University Press, 1947), p. 78. The Mallory case, decided by the Supreme Court in 1957, reinforces the right of a suspect to a prompt preliminary hearing. See Richard H. Rovere, "Letter from Washington," *The New Yorker,* Vol. 34, No. 9 (April 19, 1958), pp. 79 f.

may be denied. If the accused has a capable lawyer he may "shop" for bail from some accommodating judge.[2]

The main function of these lesser courts is the disposition of minor offenses. Day in and day out these lesser judges dispose of hundreds of small-fry offenders. The magistrate's headquarters, if not in the precinct station-house, are usually special quarters in some drab neighborhood. An hour or so before the magistrate arrives, dozens of nondescript persons mill about, some sitting on the hard seats provided for witnesses and spectators. There is usually no dignity. Newspaper reporters are on hand with the hope of getting a good story. The magistrate finally arrives and all stand at half-hearted attention. He disposes of his cases as rapidly as possible, frequently meting out justice in what may seem a careless and arrogant manner. He uses the slang of the street, injecting an occasional oath to demonstrate his authority. There is, in some instances, a great deal of whispering between him and the lawyers for the various defendants. Many cases are fixed, especially if those charged with crime have political influence.

Magistrate courts are not destined to inspire men and women or to impress them with the majesty of the law. In fact, many of them are sordid places. But they are necessary adjuncts to the criminal procedure as they dispose of thousands of minor cases that would otherwise clog the mills of justice in our metropolitan centers.

Few magistrates have had any training or formal higher education that might be of service to them in their important duties. They come from all walks of life and professions. Whether legal training is necessary, opinion is divided. Such leaders of the legal fraternity as Chief Justice William H. Taft and Dean Roscoe Pound of Harvard have considered legal training for magistrates advisable and necessary. Legal training of magistrates by itself will not insure much reform in this field of criminal procedure, because it is no guarantee against inefficiency or corruption; but this requirement would at least set some standard for the men who are appointed or elected.

It is objected that the salaries of magistrates are not high enough to attract first-rate lawyers. The pay is actually above what many lawyers make in private practice, but it is argued that only inefficient or shyster lawyers will apply for the position. That may be true, but just as we find able men in the medical service of the state and federal governments who are willing to give their best efforts for modest salaries, so we might find lawyers willing to forego the possibility of fairly lucrative private practice and serve on the magistrate's court.

The Bail-Bond Evil

The magistrates' courts daily dispose of hundreds of petty offenders in a big city, shunting many away from the criminal courts. The majority of

---

[2] *Ibid.*, p. 114.

the lesser judiciary do a reasonably satisfactory job in adjudicating these breaches of the law and sending the accusers and accused on their way tolerably well satisfied. But in many cases these courts are guilty of haphazard decisions, and there is an opportunity for shady dealings with certain individuals who live off the crumbs that fall from the magistrate's bench. The bail bond broker hangs around the court, eager for an opportunity to bail out a victim who finds he must have someone to present security (bail) for his later appearance before the criminal court.

Many bonds are *bona fide,* and many bondsmen are unimpeachable. But professional bondsmen are usually parties to a questionable, not to say corrupt, political system, and frequently have no appreciable assets with which to provide bail for those who need it. They offer what is called *straw bond,* that is, they present evidence that collateral exists, when in fact, this collateral is nonexistent or is insufficient for the purpose. In a study made by the Philadelphia Criminal Justice Association it was found that during a period of a little over two years more than a quarter of a million dollars in forfeited bail was uncollected.

Other irregularities in the bail-bond practice are the illegal reduction of bail by the magistrate (which in effect frees the accused), and the retention of cash bail deposited with the magistrate (which is keeping money that belongs to another person). In one instance a person who gave cash as bail found it difficult to redeem the cash. His importunities were rewarded by a check that was returned by the bank marked "not sufficient funds." The check was not made good until the matter was reported to the district attorney's office. A judge, commenting on cases of this kind, lashed out against this practice by stating that cash bail for a further hearing "is a trust fund, and its personal use by a magistrate is a crime. Its undue detention is a wrong to the citizen entitled to it and justly casts suspicion on the purposes of the magistrates."[3]

This bail-bond abuse *can* be regulated. The professional bondsman can be legally defined—for example, as one who enters bail in three or four cases per year—and then his activities can be regulated. He can be forced to take out a license; the amount of fees he can charge can be set by statute; and adequate records can be required. Provision can be made whereby police and magistrates' clerks will be prohibited from acting as agents for professional bondsmen.[4]

---

[3] *Joint Legislative Commission* (Pennsylvania, 1928), p. 93. Many irregular bail-bondsmen practices were discovered in the District of Columbia, see "Crime and Law Enforcement in the District of Columbia," 81st Congress, 2nd Session (Washington, D.C.: U.S. Government Printing Office, 1951), pp. 23-24.

[4] For a comprehensive study of the operations involving bail in magistrates' courts, see Caleb Foote, James P. Markle, and Edward A. Woolley, "Compelling Appearance in Court: Administration of Bail in Philadelphia," *University of Pennsylvania Law Review,* Vol. 102, No. 8 (June, 1954). Also, John W. Roberts and James S. Palermo, "A Study of the Administration of Bail in New York City," *ibid.,* Vol. 106, No. 5 (March, 1958), pp. 685-730.

Grand Jury Procedure—The Indictment or Dismissal

When the accused is held by the magistrate after the preliminary hearing, a copy of the record or transcript containing testimony is sent to the office of the prosecutor, usually called the "district attorney" in the larger cities. When the grand jury meets, the district attorney or one of his assistants presents this skeleton evidence of the crime to that body of 12 to 23 private citizens drawn from the local voting lists, in the same manner as the *petit jury* is selected. In theory, the grand jury deliberates on this evidence and passes judgment on it: if it decides the evidence submitted warrants an indictment of the accused it renders a "true bill" of indictment; if it decides that there is not enough grounds for this action, it ignores the case and returns a "no bill." Although the grand jury receives the credit for its deliberations it is, in reality (with rare exceptions), a rubber stamp for the district attorney. It is he who decides which cases should be prosecuted and which ones dismissed.

There is a marked tendency to eliminate the grand jury. Approximately half of the states have abolished it and permit the district attorney to present an "information" against the accused. The American Law Institute, in its model Code of Criminal Procedure, recommends that where the grand jury is required by the State constitution, waiver should be allowed.

One of the law's many delays is inherent in the grand jury system. The grand jury is not in continuous session—criminal procedure in any specific instance must wait for its session. This is another good reason for abolishing it. England eliminated the grand jury system in 1933.[5]

The Charge and the Plea

The charge or indictment rendered by the grand jury is, in due course, filed with the clerk of the court in anticipation of the trial. Criminal procedure requires that the indictment be drawn in a technical and stereotyped form that is a relic of olden times. The laborious detail reminds one of a strip of movie film, in which every move of the character's arm or leg requires a separate picture. Professor John L. Gillin cites a case of an indictment in which "the name of the defendant and his victim are each repeated nine times, the phrase 'in and upon' four times, the phrase 'then and there' five times, the phrase 'unlawfully, purposely and of premeditated malice' five times and the word 'said' and 'aforesaid' twenty-five times."[6] He reminds us that in England the indictment avoids all this redundancy and succinctly presents the necessary facts and no more.

---

[5] For a scholarly analysis of the historical attacks on the grand jury since the early colonial period, see Richard D. Younger, "The Grand Jury under Attack," *J. Crim. Law,* Vol. 46, No. 1 (May-June, 1955), pp. 26-49.

[6] John L. Gillin, *Criminology and Penology* (New York: Appleton, Copyright 1945), p. 286.

There are various methods by which the defense lawyer may attack an indictment. It is to guard against such attack that the indictment is so detailed and formal. The defense lawyer who takes advantage of a defect in an indictment does so by a written plea called a *demurrer,* or a *motion to quash.*

Here are a few examples of defective indictment that permitted new trials: a defendant was convicted on the charge of stealing $100, "lawful money." The conviction was set aside because the indictment did not read "lawful money of the United States"—the accused may have been carrying Mexican money around with him; in another case an accused person was indicted for stealing a pistol, but the make of the pistol was misspelled—the indictment spelled it "Smith & Weston" whereas it should have been "Smith & Wesson." In Georgia, a defendant was convicted under an indictment charging him with stealing a hog with a slit in its right ear and a clip out of the left. The appellate court granted a new trial because the evidence disclosed that the defendant had stolen a hog with a slit out of his left ear and a clip out of the right ear. In such cases in England the judge would simply have corrected the indictments with his pen and gone on with the trial.

Some states have simplified the indictment, but most states still cling to the technical instrument that reflects the judicial philosophy and phraseology of a bygone day.

After the defendant has been indicted by the grand jury he is arraigned in court. The indictment is read and he is asked how he wishes to plead. If he pleads guilty, he may be sentenced immediately by the judge. If he pleads not guilty, he is tried before the petit jury or before the judge without a jury.

Bargaining in regard to the plea is common. The district attorney asks the accused if he will plead guilty to a lesser offense in order to insure a lighter sentence. In the slang of the courtroom, the prisoner "dickers for a light rap" in return for his plea. Roscoe Pound discusses this much condemned power of the prosecutor to compromise his cases: "Ninety per cent of these 'convictions' are upon pleas of guilty, made on 'bargain days,' in the assured expectation of nominal punishment, as the cheapest way out."[7] These compromises of pleas are vigorously defended by many district attorneys and judges since they tend to clear the docket and save expense; it seems better to be sure of a conviction for a lesser offense than of none at all.

This sort of individualized treatment, however, does not tend to make justice more equitable or more efficient. Rather its purpose seems to be ministering to the convenience and prestige of the district attorney. Certainly it puts too much power in the hands of prosecutors—making them almost judges in their cases. One result of this practice of accepting pleas of guilty is that dangerous criminals are often treated as minor offenders. For example, in a case in which an armed thug robbed a truckload of silk

---

[7] Roscoe Pound, *Criminal Justice in America* (New York: Holt, 1929), p. 184.

valued at $25,000, which would ordinarily carry a severe penalty, the culprit was actually sentenced to one year in prison and fined one dollar—because he agreed to plead guilty to the charge of petit larceny rather than stand trial for the more serious offense.

A New York state law attempted to reduce the number of lesser pleas in the state's courts by requiring the district attorney to submit in writing his reasons for accepting a plea of guilty to a lesser charge than the original indictment called for. Reasons given are reported in a study made some years ago in New York City. The investigators divided the cases studied into seven categories: second offenders, weak proof, homicide, "any" felony, cases involving a small amount of money, those with "various reasons," and those with no apparent reason for a lesser plea. Cases of "second, third, and fourth offenders" numbered 241. The reasons were: small amount stolen, 55; sufficient punishment, 53; weak cases, 51; restitution effected, 21; full confession, 16; no property taken, 16; real guilt involved, 14; defendant drunk, 10; plea accepted covering two indictments, 8; other reasons, 24. The number of weak cases is unstable. Requiring district attorneys to set down in writing their reasons for resorting to this practice is at least one step in the right direction.[8]

The Prosecutor and the Lawyer for the Defense

The Prosecutor

When we think of the role the courts play in protecting society against the criminal element, we usually regard the judge as the chief protector. That he is, but it is his business to see that the accused, no matter how depraved he may seem to be, is also protected and gets his rights. The judge's powers are not so great as those of the prosecutor or district attorney. It is the prosecutor or district attorney, even more than the judge, whose task it is to rid society of the individuals who violate the laws.

The prosecutor is frequently a second-rate lawyer, but in a surprising number of cases, especially in some of our larger cities, he is well equipped, and makes an excellent record, so that the office may be a stepping-stone to a higher post, such as governor or judge of the superior or state supreme court.

Within the administrative unit in which the district attorney moves, the potentialities of his office are limited only by his skill, intelligence, and legal and political sagacity. Since the interests of the public are placed largely in his hands, he must run down and check evidence in order to see whether crimes have actually been committed, bring suspected criminals to book,

---

8 Ruth G. Weintraub and Rosalind Tough, "Lesser Pleas Considered," *J. Crim. Law,* Vol. 32, No. 5 (January-February, 1942), pp. 506-530. See also Donald J. Newman, "Pleading Guilty for Consideration: A Study of Bargain Justice," *J. Crim. Law,* Vol. 46, No. 6 (March-April, 1956), pp. 789-790.

start proceedings that come before the grand jury, bargain with those who are willing to plead guilty, decide whether or not to prosecute others, and, last but not least, convince trial juries so they will convict rather than acquit. The prosecutor's success is measured by the number of convictions he secures while he is in office, almost regardless of anything else.

He is often called on to serve as legal adviser for county and local officers and to represent the county in bringing or defending civil suits in which it is a party. His role as the public's representative in his locality takes on considerable proportions and makes him often a more important figure than the judge in the eyes of those who must stand before him.

Reforming elements concentrate on the prosecutor, and periodically, especially near election time, he crusades against small-fry gamblers, prostitutes, liquor law violators, and gamblers who have not kept up their payments for protection to the police, or to the small-time politicians and higher-ups in the organization, for it is these minor offenders who feel the iron heel of the prosecutor's authority. The prosecutor must be an astute politician indeed to curry favor both with the reformers and with the small-time criminals. He knows he cannot offend either group over too long a period of time. He makes some half-hearted attempts at cleaning up the community, but there are limits beyond which the average prosecutor does not go. Newspapers, especially those with editors possessed of reforming zeal, can make a prosecutor's life miserable; if he is to survive, he must have or develop a thick shell of indifference that permits him to survive public criticism with an amazing effrontery.

The prosecutor's unusual prerogatives permit him to dispose of cases with what seems pure caprice, through the device of *nolle prosequi* (literally "won't prosecute"), without giving reasons. Professor Sheldon Glueck, commenting on the queer ways of the prosecuting attorney, states:

> A recent sketch of the field, in which statistical recording is difficult, rightly points out that what goes on "behind the closed doors of the prosecutor's office" when he is deciding whether or not to prosecute may be of even more importance than those other acts of his that are at least susceptible of record and are done more or less in the open.[9]

When a case finally reaches the trial court, the prosecutor earnestly prepares for a real battle, not for justice, but for a conviction. His professional reputation is at stake. He must resort to all the oratory and psychological trickery he can mobilize. He is ethically no better and no worse than the defense lawyer in this judicial bout. The average trial, unfortunately, becomes more a show or contest than a struggle for justice. The judge acts as referee—to see that there is something like fair play. The jury sits in amazement, at times flattered at the compliments paid them by the lawyers, and at times incensed at the threats and insults exchanged by the lawyers in

---

[9] Sheldon Glueck, *Crime and Justice* (Boston: Little, Brown & Co., 1936), p. 147.

reckless fashion. During the court recess the two lawyers may often be seen slapping each other on the back in perfect amity. Here is a basic American institution in action, with tragic implications that most Americans do not grasp.

We know that ambitious district attorneys want to be successful in their profession and often resort to questionable practices. The real go-getting prosecutor is adept in all the tricks of the trade. Public pressure demands that he clean up the rackets and crime rings. Hence he finds himself in a serious dilemma. If he cannot get his evidence through normal channels, he may feel obliged to resort to high-handed methods. It is procuring the conviction that measures his success; the methods he employs are overlooked by the general public.

It takes courage, skill, and unflagging resourcefulness to be a conscientious, capable, and honest prosecutor. But never should the American people condone the use of high-handed tactics, illegal or extra-legal methods to intimidate, threaten, or bully suspected criminals and obtain a conviction. American legal procedure must never become a system of intimidation or brutality.[10]

The Lawyer for the Defense

The other party in the judicial tilt is the defendant's attorney. Whatever the public may believe, there are many honest criminal lawyers. In fact, many of those who have gained a national reputation have also been known for their humanity and compassion. Perhaps the best known and most highly admired criminal lawyer of his generation was Clarence Darrow.[11] As Professor John B. Waite, a distinguished authority on criminal law, has said: "The attorney who can save only the innocent from punishment will have few clients."[12]

The prisoner before the bar of justice is entitled to all the privileges guaranteed any citizen by the Constitution and Bill of Rights. If he is fortunate enough to have a good attorney, these rights will not be denied him. Unfortunately, few apart from those belonging to the organized criminal syndicates can raise enough money to employ the best legal talent. Most criminals are poor and friendless, and hence fail to procure lawyers who have the knowledge or the desire to protect their rights and defend them adequately in court.

---

[10] Cf. Bernard Botein, *The Prosecutor* (New York: Simon & Schuster, 1956); Harold Danforth and James D. Horan, *The D.A.'s Man* (New York: Crown, 1957).

[11] For accounts of Darrow's legal battles, see Irving Stone, *Clarence Darrow for the Defense* (New York: Doubleday, 1941); Quentin Reynolds, *Courtroom* (New York: Farrar, Straus, 1950), including an account of another outstanding criminal lawyer, Samuel S. Leibowitz; Arthur Weinberg, *Attorney for the Damned* (New York: Simon & Schuster, 1957); Clarence Darrow, *The Story of My Life* (New York: Universal Library-Grosset).

[12] *Criminal Law in Action* (New York: Harcourt, 1934), p. 173.

Because criminal law has been regarded as the "ugly duckling" of the legal profession for so long, we find many unscrupulous lawyers in the courtroom. Shrewd lawyers, practicing in criminal court, know just how near they can flirt with the unethical before any one particular judge. In the sensational trial of the eleven Communist leaders in New York City in 1949, several defense attorneys encroached on courtroom ethics time and again, and subsequently were held in contempt of court.

While an attempt is now being made to change the court trial from a battle of wits between prosecutor and defense attorney to a presentation of facts that will lead to a conviction or an acquittal on the merits of the case, there is still much room for improvement. Some defense attorneys stoop to unfair tactics in order to build up a reputation among criminals that will make them seem indispensable. Such unethical lawyers are in the class of lawyer-criminals. Found in the large cities, they are virtually in league with the more affluent criminals and advise them often before they commit their crimes.

Judge John C. Knox, chief judge of the United States District Court, Southern District of New York, states with authority that bombast and trickery are still used by trial lawyers, but he contends that such tactics win fewer cases than preparation and courtesy. The jurist, however, tells of a trial lawyer in a murder case in Texas some years ago, who terminated his emotional plea to the jury with tears in his eyes, singing "Home Sweet Home." He adds that this lawyer won an acquittal for his client and thus made himself more in demand as a criminal lawyer.[13]

The Public Defender and the Voluntary Defender

Assignment of counsel is important for the defendant financially unable to obtain a lawyer. If a criminal belongs to a crime syndicate he will have no trouble securing adequate counsel, and even the small-fry criminal is likely to be able to raise some funds to secure a lawyer. But if the hapless criminal is poor and has stumbled into his delinquency by accident or passion, or because he was used as a tool by someone more clever than he, he may find no one to defend him, or he may be exploited by some shyster who hangs about the courtroom looking for a retaining fee without much effort in protecting his client.

It is a sorry spectacle to see hundreds of poor, witless delinquents and petty criminals come before the court without adequate or honest and sincere counsel to represent them. In those cases where the court supplies counsel, the novice lawyer or the court hanger-on is likely to place the defendant's rights in jeopardy, for the system of assigned counsel is farcical in its operation and tragic in its consequence.

[13] John C. Knox, "The Trial Lawyer as Seen from the Bench," *New York Times Magazine*, December 4, 1949.

This state of affairs was discussed for years among leaders in the legal profession and in conferences of social workers. Out of these discussions the idea of *volunteer* counsel arose. There also developed a feeling in legal and social work circles that if the state supplies a prosecutor, it should also supply a *public defender* at public costs. This motion did not gain many friends, however, since it was obvious from the start that such an official would encroach upon the legal profession at too many points.

Nevertheless, both ideas began to flourish, and today we see in the United States both types of service, especially adaptable to poor and indigent cases brought before the court. The two methods are often confused in the minds of the public and both are confused with legal aid, which is the service that attempts to provide sympathetic and adequate legal advice to the poor, usually in matters outside the field of crime and delinquency. It concerns itself primarily with such questions of law as small claims for wages, disputes over board or rent bills, damage to personal property, and the like. The legal aid movements have, however, been responsible for the development of voluntary defenders in some places, notably in New York City.

The first Public Defender in this country was employed in 1914 by Los Angeles County; in 1915 a city ordinance extended his work. The service met with such success that in 1923 a statute was adopted that introduced the office in other counties throughout the entire state. There were many objections to the Public Defender in the early days, but experience has proved it an excellent social service that meets many problems arising from the preponderance of indigent cases in the courts. Now there are public defenders in many cities, including San Francisco and Oakland, Chicago, Bridgeport, New Haven, Hartford, Columbus, St. Louis, and Tulsa.[14]

The best known Voluntary Defender organization was founded in 1917 in New York City. The movement resulted from an inquiry instituted by the various bar associations. The original plan, to ask for the volunteer services of the attorney, was changed to employing expert attorneys to serve as defenders with adequate salaries from funds raised through subscriptions.

It was decided that the defender could do a much better piece of legal work if each case he accepted was thoroughly investigated by a social case worker. So this type of service now makes a complete investigation of the client's background and furnishes a rational solution to his legal problem.

Both the public defender and the volunteer defender systems have their champions, and each system has many advantages and some few disadvantages. Criticisms of the public defender include the possibility that he might become political like the prosecuting attorney; that collusion would be encouraged if the defender and the prosecutor belonged to the same political

---

14 See David Mars, "Public Defenders," *J. Crim. Law,* Vol. 46, No. 2 (July-August, 1955), pp. 199-210; Jacob K. Javits, "The Need for a Public Defender," *N.P.P.A. Journal,* Vol. 3, No. 3 (July, 1957), pp. 209-214.

party; and that the social work investigation, an integral part of the volunteer defender's program, would be less emphasized. The disadvantage of the voluntary defender plan is mainly financial. Too frequently the officers of the organization are handicapped by the difficulty of raising funds.

Courtroom Procedure

Rules of Evidence

Despite the persistence of the conventional explanation that courts were conceived and developed to protect the ordinary man and his rights, the fact remains that many of the decisions flowing from the judicial courts of America have tended to restrict the poor man and sustain special privilege. Economic and ecclesiastical powers aligned themselves against the masses in earlier days and maintained their ascendency by setting up a friendly judiciary.

The courts have, however, been the object of many penetrating attacks in the past and gradually have been obliged to yield in many particulars to the overwhelming logic of their critics. We need cite only such reformers of the law as Beccaria, Montesquieu, and Voltaire. Even the English jurist Blackstone (1723-1780), while hardly a radical, did condemn the glaring injustices in the unspeakable English criminal code of his day. The diverse interests of Jeremy Bentham, the political economist, embraced voluminous writings for the reform of both criminal and penal administration.

As time went on, the tendency of the courts has been to grant more rights and even some leniency to an accused person of no outstanding rank or status. This trend has not been even and consistent, but the benefit of the doubt has been gradually extended to the defendant in court procedure.

As we view court procedure, we see the rights of the accused protected at every turn, at least in theory. In fact, as the rules developed, the courts bent over backwards to such a degree that the severe criticisms made earlier against them were revived in reverse form. It was claimed that, instead of society being protected, the criminal was receiving special privileges in court. It was argued that the observance of these constitutional rights all too often ended in mistrials, delays, and miscarriage of justice. Although the primary objective of courtroom procedure is to dispense justice, critics insist that this worthy ideal is realized only in a fraction of the cases. They further insist that the rules of the game are defective and outmoded, and merely permit delay and inefficiency.

In theory, the accused is presumed to be innocent of the charge against him until he is proven guilty. This fundamental and venerable presumption places the burden of proof upon the prosecution for the state. It follows, therefore, that the defendant is entitled to due process of law, from which flow these rights. If he is acquitted, he may not be tried again for the same offense.

When a person is indicted, he at once learns the nature of the alleged crime and, to some extent, can infer the nature of the plans of the prosecutor. This permits him and his attorney to prepare the defense by establishing alibis, rounding up witnesses who will testify in his behalf, and in general developing the necessary strategy for the trial. But the prosecutor has no way of knowing in advance what type of defense the opposing attorney may employ, and he is placed in a somewhat difficult position. On the surface, this practice would seem to protect those who are undoubtedly guilty, but it merely follows the concept that it is the responsibility of the state to *prove* a person guilty; hence the burden must fall upon the representative of the state.

Since it is the obligation of the state to prove the defendant's guilt, it is incumbent on the prosecutor to submit his evidence to the judge and the jury. He calls his witnesses in one by one and asks them to tell the jury what they know about the crime and the accused's alleged participation in it. All testimony is presented under oath (or affirmation), and anyone testifying falsely is liable to the charge of perjury. Though much testimony in trials is perjured, few witnesses are actually tried for this offense. Each witness may later be cross-examined by the opposing attorney, and he tries either to break down the testimony completely or else cast doubt upon it in the minds of the jury.

The accused can refuse to testify against himself. He may testify in his own behalf if he so wishes, but it is specifically assumed that if he does not, no presumption of guilt is thereby created. This right of the defendant has been criticized by many persons, since he is fairly well protected against unfair tactics by the prosecutor through his attorney and the rules of evidence. If he is innocent, he should not object to testifying; if he is guilty, his silence tends to incriminate him before the jury. Whether this constitutional right of the accused should be abolished is a moot question.

Another reason for not calling the defendant to testify is that the prosecutor might conceivably introduce his past record, if he has one, and thus throw doubt on his testimony. While this frequently happens, it is argued that it is the business of the defense attorney to see that his client is not taken advantage of in this manner. In formal law the past record of a defendant has nothing to do with his guilt in the case before the court—the law is clear on that point. While the defendant should be protected in the courtroom by legal procedure, society, too, must be assured that a chronic criminal or sex offender should be identified so far as his past record is concerned, although he may in fact be innocent of the charge for which he is being tried.

In addition to protecting the accused by permitting him to refuse to testify against himself, the state extends this privilege to those close to the defense. These are privileged witnesses and include a wife or minor children of the defendant, his physician, lawyer, or minister, all in so intimate a relationship to the defendant that they have confidential or privileged communications with him.

There are other rules besides these that protect the rights of the accused. Evidence consists of testimony from all witnesses, whether testifying for the defense or the prosecution. In addition, material objects or documents purporting to have bearing on the case and usually referred to as exhibits may be considered as evidence. The judge rules on the admissibility of evidence, but if it is admitted, the jury alone decides on its weight in the case. Hearsay evidence is not permitted in court; neither can a witness express an opinion unless he is asked for it or is an expert in some specific field where his opinion carries weight.

There are two kinds of evidence—direct and circumstantial. Direct evidence flows from the witness's personal knowledge or observation, whereas circumstantial evidence consists of inferences from observed facts or testimony. While the latter is not considered to be as relevant as direct evidence, it is admissible, and, at times, is convincing and conclusive. However, it is often misleading and unreliable and lends itself to exploitation by a clever lawyer or prosecutor.

We have discussed elsewhere the evidence submitted by the expert.[15] While there are some charlatans among the experts, their evidence is, in general, scientific. Scientific evidence seeks to introduce into a trial all the relevant facts in order to get at the truth. But so-called legal evidence, much of it hearsay, circumstantial, and perjured, obscures the truth. By employing this type of conventional evidence the prosecutor and defense attorney succeed in perpetuating the traditional legal procedure by deliberately covering up pertinent facts.

Since the burden of proof rests with the prosecutor, it is his responsibility to open the trial. This he does with his remarks to the jury, stating briefly the charges against the defendant as they are contained in the bill of indictment and outlining the evidence he expects to introduce during the trial. After the completion of his opening statement, he calls his witnesses one by one. After he directly examines each, the defense attorney has the privilege of cross-examination, his purpose being to discredit the prosecutor's witnesses or to minimize the weight of their testimony. Defense witnesses are then introduced, and they in turn are cross-examined by the prosecutor. After this, both sides may produce additional witnesses. Upon the completion of taking the state's testimony against the accused the prosecutor advises the court that he rests his case.

The Role of the Trial Judge

No matter how crude our early courts were, a halo of mysticism fabricated from ritual has been thrown about them so that they have come down

---

[15] See pp. 227-230.

to us surrounded with a large element of sanctity. The legal trappings of the court such as the wearing of robes, and, in England, powdered wigs, contribute to make up the illusion of mystery, pomp, and circumstance. We in America have abandoned much of this "legal magic" but there still are many vestigial remains.

However magic the courts may be, they are composed of human beings in action. Modern scientific criticism finds many objections to them.[16] The courts of today have many shortcomings in meting out even and accurate justice.

The trial judge is an imposing figure and possesses many powers; yet in many respects he is greatly limited in his authority, since every act and every ruling he makes in the conduct of the trial is carefully scrutinized by counsel, who often are more learned in the law than the judge. Counsel quite frequently takes exceptions to the judge's rulings; these are useful in case of appeal. Not infrequently the higher, or appellate, courts reprimand a judge who has erred in his conduct of a trial or who has made rulings indefensible according to law. But if the judge has been technically correct with respect to the law, the higher courts take little cognizance of any obvious bias on his part. If one were to make an analysis of the prevailing bias of American justice, he might well find that in many cases judges lack an understanding of the broader principles of an enlightened social philosophy.[17]

While in theory the American trial judge is vested with considerable authority and independence, in practice he is frequently indebted to members of the political organization that was responsible for elevating him to the bench. Judges feel this dependence quite keenly and resent any reference to it. But it does persist and will continue so long as the office is elective. As Professor Waite says in his work on the criminal law: "They are dependent for attainment and retention of their positions upon influential lawyers, upon newspaper reporters and editors, upon gang chief politicians, and upon whatever other persons and forces determine elections."[18] This close tie-up between the judges and political organizations is discussed by Percival E. Jackson in his book, *Look at the Law,* in which he quotes Joseph H. Choate, leader of the American Bar Association: "There is one other abuse against which we can at least utter an indignant protest. I mean the toleration of judicial candidates who are willing or are permitted to pay for their nomination or to pay their party for their election. No matter what their personal or

---

[16] See Jerome Frank, *Courts on Trial* (Princeton: Princeton University Press, 1949); Fred Rodell, *Woe Unto You, Lawyers!* (New York: Reynal & Hitchcock, 1939); John Barker Waite, *Criminal Law In Action* (New York: Harcourt, Brace, 1934). For another approach to the role of the judge and his office, see Bernard Botein, *Trial Judge* (New York: Simon & Schuster, 1952).

[17] Cf. Professor Morris Cohen's introductory statement to Louis P. and Eleanore Levenson, *Lawless Judges* (New York: The Rand School Press, 1934), p. vi.

[18] Waite, *Criminal Law in Action,* p. 203.

professional qualifications in other respects may be, such a means of reaching office cannot but degrade the Bench."[19]

It is not unknown for aspirants to the bench in some states to put up as much as $25,000 for a nomination to a state supreme court judgeship. A revealing article by Howard Whitman has pointed out the extent to which judges owe their posts to large sums of money paid out to the politicians.[20] It may well be conceded that most judges who pay for nomination are not corrupt; most of them may deplore organized criminality and maintain an uncompromising attitude toward it. An eminent criminologist has asserted that there is not a single crime ring in the country that does not revolve around a corrupt judge. Even the honest judge finds it difficult to smash crime because he is indebted to politicians who are in league with the criminals. The corrupt judge does not care to eliminate crime in his domain because it is too lucrative a business, as far as his own fortunes are concerned.

As to the ability of trial judges, Whitman shows how utterly incompetent many judges are; many of them employ bright young secretaries to write their decisions. Professor Waite says: ". . . neither appellate courts nor legislators evince great confidence in the capacity of elected trial judges to use power wisely or fairly. And, oddly enough, the appellate court judges are rather more skeptical of their inferior colleagues' ability than are the legislators."[21]

Whitman calls attention to the Missouri Plan of selecting judges. It functions as follows: judges are nominated by a commission made up of outstanding citizens, including lawyers; three names from this list are submitted to the governor of the state, who selects one of them. After serving for one year, this person's name appears on the ballot with the question, "Should Judge X be retained in office? Yes or No." If he has been competent he is then elected by the people; if not, he is retired from the bench. This plan has the good features of both appointment and election.[22]

The Criminal Trial and Legal Verbiage

Dean John H. Wigmore of Northwestern University Law School is given credit (or blame) for coining the term "sporting theory" of justice to describe the court trial.[23] While it is assumed that the purpose of a criminal trial is to ascertain whether a defendant committed a specific crime and thus

---

[19] Quoted in Percival E. Jackson, *Look at the Law* (New York: Dutton, 1940), Chapter 10, "Judges are Corrupt," p. 294.

[20] Howard Whitman, "Behind the Black Robes," *Woman's Home Companion* (February, 1949), p. 32 ff.

[21] Waite, *op. cit.,* p. 207.

[22] Whitman, "Behind the Black Robes," p. 112. For further discussion of this plan, see Lawrence M. Hyde, "The Missouri Plan for Selection and Tenure of Judges," *J. Crim. Law,* Vol. 39, No. 3 (September-October, 1948), pp. 271-287.

[23] Thus stated by Roscoe Pound, *Criminal Justice in America* (New York: Holt, 1929), p. 163.

dispense justice, the fact is that the process is a legal game (or battle) between two lawyers, the prosecutor and the defense attorney, with the judge acting as referee. The purpose of the judge is to hold the contestants to the rules of the game.

Going further, Leonard Moore, author of a handbook for practicing lawyers which was sponsored by the Practicing Law Institute, refers to litigation as "warfare." He writes: "Litigation resembles warfare. Opposing counsel are charged with the responsibility of so conducting their campaign that ultimate victory will result." He continues by stating that a lawsuit, like a battle, requires "stratagems," "tactics," "skirmishes and a series of battles," and a "scouting party to discover the enemy's position and strength, sometimes to draw his fire."[24] If this can be said of civil cases, how much more accurate it is of criminal cases!

The "sporting theory" of justice makes public spectacles out of some of our nationally notorious criminal trials. Dean Roscoe Pound thinks that criminal trials stopped being primarily a show around 1900 (some years before Dean Wigmore coined the term "sporting theory" to describe the process). But, of course, we have had many such spectacles since the turn of the century. In fact, the umpire metaphor was used as recently as 1950 by Supreme Court Justice Felix Frankfurter to refer to the role of the criminal court judge.[25]

One reason the criminal trial, especially, is like a football game is its rules and technicalities. From the moment the defendant is indicted until he is convicted or acquitted, legal technicalities are legion. As Percival E. Jackson remarks: "Applications for mistrials in criminal cases, motions in arrest of judgment, reversals of convictions of notorious criminals because of technical errors in the conduct of the prosecution, are of common occurrence. . . . In matters of giving evidence upon a trial, an apotheosis of technique is reached. What is to be admitted in evidence upon a trial and what is to be rejected, is a subject of such abstruseness that volumes of law books are constantly being written and rewritten on the subject."[26] He names *Greenleaf on Evidence,* published in 1899—three volumes of over 2,000 pages, citing almost 15,000 precedents—and Dean Wigmore, whose five volume work of 5,500 pages cites 42,000 precedents.

The courts insist upon a technical "legalese" unintelligible to the layman. Federal Judge Jerome Frank asks: "But why are lawyers peculiarly infected with what has been called 'verbomania'? . . . Legal thinking, it is said, is affected by a 'belated scholasticism,' by a 'blighting medieval prepossession' . . . In no other field of human thought is that prepossession to be

24 Leonard Moore, "Modern Practice and Strategy," Series I, No. 4 (New York: Practicing Law Institute), p. 1; quoted by Frank, *Courts on Trial,* p. 8.
25 *Minutes of Evidence,* Royal Commission on Capital Punishment (London: July 21, 1950), p. 582.
26 Percival E. Jackson, *Look at the Law* (New York: Dutton, 1940), pp. 113 and 120.

found in a more exaggerated and persistent form."[27] Court procedure and legal language must be made simpler and more efficient—adapted to the social and economic conditions of today—so that delays will be minimized and justice will be expedited.

Charging the Jury and Sentencing the Convicted

After the evidence has been submitted to the jury and the opposing attorneys have examined and cross-examined the various witnesses and have presented the final arguments, the judge makes his charge to the jury. The purpose of the charge is to give the jury the legal aspects of the case and to instruct them how to analyze the evidence in arriving at their verdict. The judge must be most scrupulous in making this charge since he knows that if he exceeds his prerogatives, reversible error may be charged against him. Sometimes both the prosecutor, the defense attorney, or both, point out to the judge that he was subject to error in the charge and ask him if he cares to correct the errors they specify.

Callender gives a summary and criticism of the judge's charge to the jury:

> The judge begins by stating the nature of the offense with which the defendant is charged, and explains what evidence is necessary to justify a conviction. He then reviews the evidence for the prosecution and defense, and instructs the jury how to weigh it in arriving at a verdict. He tells them that the presumption is that the defendant is innocent, and that the burden is on the state to convince them "beyond a reasonable doubt" that he is guilty. In criminal cases a verdict of guilty should not be based upon a mere preponderance of proof. The jury should be convinced that upon all the evidence, there can be no reasonable doubt that the accused is guilty. . . . What constitutes reasonable doubt is difficult to define, and the probabilities are that few jurors understand the definition which the judge gives them. He tells them that it consists in a substantial doubt arising from the evidence or lack of evidence, such as an honest, sensible, and fair-minded person may with reason entertain consistent with a conscientious desire on his part to ascertain the truth. He tells them that the state is not required to establish guilt beyond any possibility of doubt, and they should not acquit for the mere reason that there is possibility of error, but only if they have a *reasonable* doubt. Assuming that the jury understands what he means, it is not unlikely that they will have some difficulty in applying the doctrine.[28]

After the jury deliberates, it returns its verdict and it is the task of the judge to pronounce sentence.

If the offense is a serious one, carrying the death penalty or life imprisonment, it is customary for the defendant's attorney to make a final plea to the

---

[27] Jerome Frank, *Law and the Modern Mind* (New York: Coward-McCann, 1930), p. 63. Copyright 1949 by Coward-McCann, Inc. See also Fred Rodell, *Woe Unto You, Lawyers!* (New York: Reynal, 1939), pp. 193, 197-198.

[28] By permission from: Clarence N. Callender, *American Courts, Their Organization and Procedure* (New York: McGraw-Hill Book Co., Copyright 1927), pp. 193-194.

judge for mercy. The defense attorney may bring into the courtroom members of the family of the convicted man, in an attempt to sway the judge toward an attitude of leniency. This is an old trick, and of course both judge and attorney accept it as part of the "show" that makes up the entire court procedure. After sentence has been pronounced, the trial is over except for a motion for a new trial. If this is denied by the judge, an appeal may be made to a higher court.

Although the judge cannot order the jury to return a verdict of guilty, he "may take the case away from the jury" and insist on a verdict in favor of the defendant. If a judge thinks, however, that the jury should have convicted the accused, he may still hold the defendant by means of a bond to keep the peace, especially in minor cases that are not worth the expense or trouble of a new trial on another count. In some jurisdictions, persons acquitted of certain crimes, but reputed to be potentially dangerous, may be required by the judge to post a bond. If the peace bond is fixed at an exorbitant amount, impossible for the person to raise, an indeterminate sentence in jail may result. Sometimes the sentencing jurist forgets the case and the victim is kept in jail for a year or more. During the year 1949, for instance, the criminal courts in Philadelphia remanded 36 acquitted persons to jail "to keep the peace." Their bonds ranged from $500 to $5,000. This practice has since been declared to be unconstitutional in Pennsylvania.

While peace bonds for acquitted persons are an outgrowth of the hoary custom of surety to keep the peace, the resort to such instruments to restrain persons acquitted of a crime from further violence or threats of revenge has long been questioned. To cope with this dilemma the only legal way is to arrest the person on another charge.

The Plea of Insanity

There is probably nothing more confusing in the realm of jurisprudence than the degree of moral responsibility of the offender, especially with the more repugnant crimes such as murder, rape, or wanton violence. The massive literature dealing with this problem ably reflects the confusion but has produced nothing toward even a working solution.

When it is realized that the courts, prior to the eighteenth century, clung to the concept of free will in the commission of a crime, it is all the more significant that the insane were the first to be relieved of the onus of responsibility for their criminal acts. Even the humanitarian Classical School of penology of the eighteenth century, as interpreted by Cesare Beccaria, did not excuse the "lunatic" from his acts.[29] This exemption did not come until the exponents of the Neo-Classical School modified the concept of free will held by their predecessors. This school, represented by the jurists Rossi, Garraud, and Joly, contended not only that children were unable to choose

---

[29] See discussion of the postulates of this school, pp. 322-323.

between right and wrong, but that the adult insane criminal was not liable for his crime.

Eventually the courts came to recognize that if a person was suffering from some obvious mental disorder, an essential element in the commission of a crime was lacking, that is, a sense of responsibility. The problem immediately arose as to how this condition could be appraised; or, in other words, how insanity could be detected in a criminal. That there *is* a degree of mental derangement sufficient to excuse a person from further process of law is no longer questioned seriously, but the remaining problem is very exasperatingly alive: where is the line drawn, and by what yardstick?

As early as 1724, in the Arnold case, a test was enunciated. Known as the "wild beast" test, it set the principle that if a person was to be exempted from punishment he must be totally deprived of his understanding and memory and know no more than an infant, a brute, or a wild beast what he is doing. Another test grew out of the case of one James Hadfield, who attempted to take the life of King George III when His Majesty entered the Drury Lane Theatre. The presiding judge decided that Hadfield was suffering from a delusion, although he knew at the time that he was actually firing a gun at the king. Out of this decision grew the "delusion" test.

Modern English practice and consequently American practice acquired a degree of authoritative precision through the outcome of a famous case in 1843 involving criminal aspects of insanity. One Daniel M'Naghten was acquitted of the murder of Sir Robert Peel's secretary, whom he mistook for the statesman. The verdict caused so much public turmoil that it became a subject for debate in the House of Lords. In consequence, the Lords submitted to the judges certain questions about persons supposed to be afflicted with insanity. The answers to these questions, in which the judges stated their position, have since been known as the M'Naghten Rules, and have dominated the legal thought of both countries ever since. In essence, they are as follows:

> 1. He is nevertheless punishable, according to the nature of the crime committed, if he knew at the time of committing such crime that he was acting contrary to law, by which expression we understand your Lordships to mean the law of the land.
> 2. The jury ought to be told in all cases that every man is to be presumed to be sane, and to possess a sufficient degree of reason to be responsible for his crimes, until the contrary be proved to their satisfaction; and that to establish a defense on the ground of insanity it must be clearly proved that, at the time of committing the act, the accused was labouring under such a defect of reason, from disease of the mind, as not to know the nature and the quality of the act he was doing, or if he did know it, that he did not know he was doing what was wrong.[30]

---

[30] This is taken from John Paton, "Lunacy and the Law," *The Penal Reformer* (London) (October, 1934), p. 12. See also Henry Weihofen, *Insanity as a Defense in Criminal Law* (New York: Commonwealth Fund, 1933), p. 51 for an analysis of the judges' answers.

The M'Naghten Rules are based on an extremely narrow conception of the nature of insanity, which has long since been abandoned by experts in the field of psychiatry. Yet these rules are still the standard by which the responsibility of a convicted person is judged in our courts. Dr. Winfred Overholser, distinguished psychiatrist, says:

> For some reason or other, it has appeared that many courts have seemed to take the attitude since then that the opinions of the judges of 1843 were the last word and that no further progress in psychology and psychiatry need be considered. The Lord Chancellor of England in 1862, for example, spoke of the "vicious principle of considering insanity as a disease" and objected to the "introduction of medical opinions and theories into the subject." Indeed, as recently as 1924 the Lord Chancellor, in vigorously opposing any modification of the M'Naghten Rules, referred to psychology as "a most dangerous science to apply to practical affairs."[31]

Yet in this country alienists, as early as 1800, were aware that the capacity to distinguish between right and wrong is an inadequate test of insanity. The first American judges to adopt this more liberalized point of view were Shaw of Massachusetts and Edmonds of New York who as early as 1843 ruled that insanity was proved if the homicide "had no power of control" or "if his moral or intellectual powers were so deficient that he had not sufficient will" or "if he did the act for an irresistible and uncontrolled impulse." Judge Edmonds went further in stating that "the law in its slow and cautious progress still lags far behind the advance of true knowledge."[32]

Yet in 1871 the New Hampshire law expressed the doctrine that the question or irresponsibility by reason of insanity is one of fact for the jury since there is no legal test that they must observe.

Professor Raymond Moley calls attention to the state of confusion of the courts today in dealing with cases in which insanity is used as a defense: "American courts have fallen into boundless confusion in attempting to apply the M'Naghten rules or in seeking to improve substitutes."[33] Twenty-nine states abide by the M'Naghten rules. Some of these states hew close to the "right and wrong" test, whereas others use the "knowledge of nature" and "equality" tests. In 1884, the "irresistible impulse" test was introduced, and today we find this test being used in 17 states and the District of Columbia. This test is based on the view that besides the intellectual judgment there must also exist the capability of doing what is considered right and of abstaining from what is considered wrong.

---

[31] Dr. Winfred Overholser, "Ten Years of Co-operative Effort," *J. Crim. Law,* Vol. 29, No. 1 (May-June, 1938), p. 23.

[32] Quotations from a stimulating pamphlet first published in 1876 and republished, *J. Crim. Law* (November-December, 1949), pp. 397-444. This was a speech delivered by Stanford Emerson Chaille.

[33] Raymond Moley, *Our Criminal Courts* (New York: Putnam's, 1930), p. 146.

In 1954 the Court of Appeals of the District of Columbia set aside the conviction by a lower court in the District of one Monte Durham for housebreaking.[34] In his decision, Judge David L. Bazelon made a clear and deliberate departure from the M'Naghten Rule, which he criticized as not taking "sufficient account of psychic realities," and laid down in its place what soon became known as the Durham Rule: "An accused is not criminally responsible if his unlawful act was the product of a mental disease or mental defect."

Under the Durham Rule the question of fact before the jury thus becomes: (a) does the accused suffer from a mental disease or defect, and (b) was the unlawful act committed as a result of that disease or defect? Since both parts of the test are obviously matters for expert testimony, this rule is a far cry from the M'Naghten, whereunder the jury had to decide a question of fact upon which no expert could be called to testify, namely, whether or not the accused's mind had the capacity to deal with moral problems. The Durham Rule also allows for the type of mentally diseased person who, while clearly perceiving the wrongness of an act and its consequences, is still unable to stop himself, and legal insanity no longer turns on the question of one or two unscientific symptoms.

In the face of expert testimony under this rule, even of testimony tending to show only a probable disturbance or a probable connection with the unlawful act, the prosecution is nearly helpless. To prove beyond a reasonable doubt that a man's illness is not the cause of his crime, the prosecution must ask the jury to conclude that his actions would have been the same whether or not he were sick. Since such speculation is necessarily fraught with reasonable doubt, the tendency through a number of decisions would undoubtedly be to consider crime of any sort as invariably a symptom of mental disease.

The Durham Rule is operative only in the very limited jurisdiction of the District of Columbia, where it has, as might be expected, greatly increased the number of insanity pleas, and acquittals (ironically, Durham himself rejected the Durham Rule, a year after it was formulated, by returning to the lower court and pleading guilty to a charge of petit larceny, evidently preferring a fixed and limited sentence to an indeterminate stay in a mental institution). The Rule has been raised as a defense elsewhere in a number of state and federal courts, and rejected by all, although several state judges have expressed a willingness to follow it were they not bound to M'Naghten by statute. Perhaps the Rule's most salutary effect will be to serve as the basis for wider and more realistic discussion of the relation of law to the criminal. As the Washington *Post* has commented, "There lurks behind the relatively narrow confines of the Durham Rule debate the . . . startling question of whether or not most crime is not [sic] essentially the conse-

---

[34] Durham v. United States, 214 F.2d 862-76.

quence of mental aberration." Serious entertainment of the question by the responsible public could lead to a valuable reappraisal of legal and corrective philosophy. In the course of such deliberations we will do well to consider the key position of the psychiatric expert in any new orientation of the system of criminal law.[35]

It is important to point out that psychiatrists rarely agree completely in their diagnoses, disregarding the problem of the fakers who pose as psychiatrists, a problem for which the profession is hardly responsible. The core of psychiatry and mental therapy is certainly sound, and is something of which medicine and the social sciences may alike be proud. No other sociomedical movement of our generation holds so much promise for the relief of human misery and the curative control of social ills.

The partial disrepute into which psychiatry has fallen in connection with criminology has been the result of the handicaps imposed on it by courtroom procedure and rules of evidence. In such an atmosphere the psychiatrist cannot function as a scientist. He can present his material only as answers to questions put by the attorneys, which means that he cannot present his report completely or coherently.[36] There is little possibility that psychiatry can be adapted to the procedure in the conventional jury trial. In the first place, the legal definition of insanity, by which the court is guided, is an intellectual and moral concept with no medical significance. The old test of the power to differentiate between right and wrong in an absolute and metaphysical sense is entirely foreign to the conceptions of the nature of the human personality held by dynamic psychology and psychiatry. There is a large number of discernible mental aberrations, each with its characteristic systematology. Many of these are wholly emotional disorders with little or no impairment of intellectual faculties.

In the second place, the psychiatrist as we mentioned above, is prevented by conventional courtroom procedure from making a complete report on the case, so that his scientific knowledge is frustrated by such limitations. Finally, the effect of psychiatric testimony is rendered all but worthless in jury trials because it deals with technicalities far beyond the mental grasp of the average jury. These three huge gaps leave little prospect of utilizing psychiatry successfully in the courtroom until the criminal law squares itself with the scientific notion of human behavior, accepts medical concep-

---

[35] A reasoned and readable analysis of the Durham case is Richard H. Rovere's in his regular "Letter from Washington," *The New Yorker,* Vol. 34, No. 9 (April 19, 1958), pp. 79, and 81-86, from which the above quotations were taken. See also a symposium on its significance, "Insanity and the Criminal Law—A Critique of Durham v. United States," *University of Chicago Law Review* (Winter, 1955), pp. 317-404.

[36] See Henry A. Davidson, "Psychiatrists in Administration of Criminal Justice," *J. Crim. Law,* Vol. 45, No. 1 (May-June, 1954), pp. 12-20; Henry Weihofen, "Eliminating the 'Battle of Experts,' " *N.P.P.A. Journal,* Vol. 1, No. 2, (October, 1955), pp. 105-112.

tions of insanity, adopts the slogan of making the treatment fit the criminal, abolishes jury trial, and hands the defendant over to a group of permanently paid experts solely interested in questions of medical fact and social protection.

The Briggs Law of Massachusetts, passed in 1921, marked an enormous advance by taking the psychiatrist out of the courtroom and ordering him to make an examination of the defendant and submit his report in a complete form prior to the trial. This law provides that when a person is indicted by a grand jury of that state for a capital offense, or is indicted and is known to have been indicted for any other offense more than once in the past, or to have been previously convicted of a felony, notice shall be given to the department of mental diseases. This department will then be called upon to examine the person to determine his mental condition and the existence of any mental disorder that would affect his criminal responsibility. The department then files a report with the clerk of court in which the trial is to be held. The examination must be made by two psychiatrists appointed for this purpose by the Department of Mental Diseases. One of the experts must be a member of the department.

What can be done when we have an intelligent law is well illustrated by the disposition of the case of one Dr. Thierry in Massachusetts in the spring of 1925, under the operation of the Briggs Law. Here an insane murderer was brought before the court and jury, the report of the examining psychiatrists was read, the judge ordered a verdict of "not guilty by reason of insanity," and within two hours from his appearance in the courtroom, the defendant was on his way to the State Hospital for the Criminal Insane at Bridgewater.[37]

To ascertain how this law was working, Dr. Winfred Overholser examined 5,159 defendants during the first fourteen years of the operation of the law and found that 15 per cent were mentally abnormal and that the insane ranged from 3 to 8 per cent.[38]

Contrast this with the delayed and unscientific disposition of the celebrated Loeb-Leopold case in Chicago in 1924.[39] In this case, the facts as to the mental state of the defendants were as clear as they were in the Massachusetts case. But scientific procedure was frustrated by the conventional

---

[37] See Winfred Overholser, "Psychiatry and the Courts in Massachusetts," *J. Crim. Law,* Vol. 19, No. 1 (May, 1928), pp. 75-83; Winfred Overholser, "Two Years' Experience with the Briggs Law (1928-1930) of Massachusetts," *ibid.,* Vol. 23, No. 3 (September-October, 1932), pp. 415-426.

[38] "The Briggs Law of Massachusetts: A Review and an Appraisal," *J. Crim. Law,* Vol. 25, No. 6 (March-April, 1935), pp. 859-883.

[39] The Loeb-Leopold case (Chicago 1924) has long been taken to be psychiatry's entering wedge in criminal court. Richard Loeb and Nathan Leopold, two wealthy, brilliant, but morally perverted youth brutally murdered a younger boy, their stated motive being to prove they could commit the "perfect crime." They were defended by the skilled criminal lawyer, Clarence Darrow. Darrow was convinced that the psychiatrist had something to offer criminal law. To this end he enlisted the services of

Clarence Darrow before Judge Caverly in the Loeb-Leopold trial in Chicago, 1924.
(Courtesy of Chicago "Herald" and "Examiner.")

some of the country's outstanding specialists in this field. Appearing at the trial for the defense were Dr. William A. White, at the time Superintendent of St. Elizabeths Hospital, Washington, D.C., Dr. William Healy, for many years Director of the Judge Baker Guidance Center of Boston, Dr. Bernard Glueck, who for many years was chief psychiatrist at Sing Sing prison, and Dr. Harold S. Hulbert, Chicago neurologist. All of these specialists were in agreement that the boys were abnormal mentally—that the crime was "the result of diseased motivation."

However, the prosecution had arrayed a group of experts who were to look upon the defendants as "normal" mentally. These were Drs. William O. Krohn, Hugh T. Patrick and Harold D. Singer who, according to Darrow, "represented a past age of psychiatric medicine, which considered no one insane unless he were a raving maniac." The psychiatrists for the prosecution insisted that "they found no evidence of mental disease in the boys, that they were not without emotional reactions" and that a "paranoiac personality did not necessarily mean a diseased mind." Quotations here and following, from Irving Stone, *Clarence Darrow for the Defense*, (New York: Doubleday, 1941), p. 410. Copyright 1941 by Irving Stone. Reprinted by permission of Doubleday & Co., Inc. Judge John R. Caverly, before whom the case was tried, admitted that the expert testimony presented at the trial had been "of extreme interest and a valuable contribution to criminology," but, in not imposing the death penalty on the defendants, he was moved rather by the fact of their youth. Accordingly, Loeb and Leopold were sentenced to life imprisonment in Joliet penitentiary. This case will always remain one of the most notorious, if not famous, in American annals. Loeb died in prison in a brawl. After several unsuccessful attempts, covering many years, Leopold was paroled in 1958. See John Bartlow Martin's series of articles on Leopold, "Murder on His Conscience," *Saturday Evening Post*, Vol. 227, Nos. 40-43 (April 2, 9, 16, 23, 1955); Meyer Levin, *Compulsion* (New York: Simon & Schuster, 1956); Arthur Weinberg (ed.), *Attorney for the Damned* (New York: Simon & Schuster, 1957); Nathan Leopold, *Life Plus Ninety-Nine Years* (New York: Doubleday, 1958).

attitude of the State's attorney who stated: "I insist that the question of sanity or insanity is a matter under the law for a jury. . . . . From the moment you hear evidence on insanity everything you do becomes of no effect under the law, and this becomes a mock trial."[40]

While our states still have a long way to go to free themselves from the dead hand of legislators who lived and enacted laws long before the advent and fruition of psychiatry and clinical psychology, there are wholesome signs here and there throughout the country that point to a modicum of real progress. The psychiatric approach to criminality is a potent one, based on scientific postulates. The real job is to see that psychiatric services are introduced into the court trial with dignity and not through the degrading techniques now in use in most courts.

The recent addition of courses in forensic psychiatry or "legal medicine" to the law curriculum indicates a step in attempting to reconcile the two disciplines; another is the setting up of the Isaac Ray lectures, whereby outstanding legal scholars annually discuss the problems in this field. Judge John Biggs in *The Guilty Mind* demonstrates that responsibility is, at best, a tenuous phenomenon arrived at only by penetrating examination and analysis by qualified persons.[41]

Appeals to Higher Courts

New trials in civil cases were granted as early as the fourteenth century but this practice in criminal cases did not begin until the latter part of the seventeenth century.[42] Federal and state statutory provisions for taking appeals from court decisions to higher courts for review afford the convicted person further opportunity to vindicate himself. Within a short time after conviction the defense must file the application for appeal with reasons.

Appeals are very common. Many are not granted, but a stay in executing the sentence pronounced by the judge is effected by the appeal even if it is denied. If the defendant has money, or if his attorney feels he has a good case, despite conviction by the lower court, strenuous efforts are made for a new trial or for a review by the appellate court. Under certain circumstances, the state, too, may appeal a case, although this is somewhat rare.

Although appeals to higher courts are numerous, the proportion to the number of criminal trials is small. The percentage of cases upheld by the

---

[40] As quoted in the Chicago *Tribune* (July 31, 1924).

[41] John Biggs, *The Guilty Mind: Psychiatry and the Law of Homicide* (New York: Harcourt, Brace, 1955). See also Manfred S. Guttmacher and Henry Weihofen, *Psychiatry and the Law* (New York: Norton, 1952); Henry Weihofen, *Mental Disorder as a Criminal Defense* (New York: Dennis, 1954); and *The Urge to Punish* (New York: Farrar, 1957).

[42] See Orfield, *Criminal Procedure from Arrest to Appeal*, Chapter 8, "Motions after Verdict."

appellate court differs in various states and at different times, but, generally, most trial court decisions are sustained.

When a convicted criminal is denied a new trial by the appellate court, he may appeal to the governor or pardon board for a pardon or commutation of the sentence. In rare instances his case may be taken to the United States Supreme Court. This court has shown great reluctance, however, to interfere with state court rulings and does so only when in its judgment the due process clause has been infringed or some untoward set of events conspired to make a fair trial impossible, such as a mob storming a courtroom during the trial. In the few cases that have reached the Supreme Court from the state courts, the moving spirit has often been a group of liberals who have not only raised the necessary funds to prepare the case but have also contended that some very pertinent social question was involved, such as racial discrimination and persecution of minority groups.

A Baltimore jurist, Joseph N. Ulman, some years ago illuminated an appellate court procedure with this statement:

> Appellate courts . . . do not reverse decisions simply because they disagree with them. Reversal must proceed from error of law and such error must be substantial. But if this account is to be veracious I must call attention to a fact familiar to every experienced lawyer, yet not apparent in the classical literature of the law, and probably not consciously admitted even to themselves by most appellate judges. Practically every decision of a lower court *can* be reversed. By that I mean practically every record contains some erroneous rulings (and) they can nearly always find some error if they want grounds for reversal.[43]

Judge Ulman described a case over which he presided. Of 830 rulings on evidence that he made, 615 were adverse to the defense. The defendant was acquitted, but the judge pointed out that if the defendant had been found guilty and his attorney had appealed, there would have been 615 opportunities for reversal before the appellate court.

The practice of safeguarding even the convicted man against shoddy or inefficient "justice" is admirable, but the cumbersome machinery by which he is protected is inefficient, absurd, and time-consuming. Lawyers themselves are beginning to realize that court procedure must be made more simple and efficient so that justice will be expedited. Professor Raymond Moley summarized recommended reforms that would eliminate many appeals:

> Simplification of the indictment or information; provisions to facilitate the amendment of defective indictments and the cutting of defects by order of the court.
> That there be no reversals on appeal for error except in the case of miscarriage of justice.
> Rules for speeding up proceedings on appeal.

---

[43] Joseph N. Ulman, *The Judge Takes the Stand* (New York: Knopf, 1933), pp. 265-266.

Extension of power of court to modify judgment, sentence, etc., on appeal, without necessitating the granting of a new trial.

Simplification of the charge to the jury, to lessen the possibility of error.[44]

Aside from the cumbersome procedure, with its inefficiency, technicality, and delay, the cost of criminal justice is terrific. The cost of stenographic help is high. The record of a long trial will cost thousands of dollars. To appeal a case to a higher court, the defense must submit from 10 to 20 copies of the transcript of the trial and often these must be printed. Many of the more famous trials in our history have run into hundreds of thousands of dollars.

Sentencing a Convicted Criminal

On the surface, it would seem that sentencing a person found guilty of a crime should be a very simple process, for in most states the penal code provides for penalties to be imposed for the various acts considered crimes, whether they are serious (felonies) or less serious (misdemeanors). Some penal codes, however, give the courts wide discretion in determining what penalty to impose, and there are vast differences between the states regarding the severity of the penalty for the same crime.

The penal code is the "blueprint" the judges use in sentencing convicted prisoners. Most of our penal codes have come down to us from common law, modified by changes in philosophic thinking and, in this country, by statutes. As we view them today they are shackled by much voodoo, myth, caprice, and traditional morality.

In a later chapter we describe the excellent reform of the penal codes of his day by Cesare Beccaria, father of classical penology. Denouncing the arbitrariness of the judges who discriminated against the poor man in favor of the rich, he enunciated a penal code that caught the imagination of all lovers of democracy. The codes he and his followers (like Jeremy Bentham in England) developed were quite simple. To each offense a penalty was attached. Every person committing a certain crime was dealt with uniformly. Simple though it was, such a code did not envisage mitigating circumstances, but rather appraised all criminals alike. Then, too, the degree of severity of the penalty was capriciously arrived at either by the ruler of the country (such as Frederick of Prussia) or by majority vote of the group considering the offense. The penal codes in operation in our states have made little progress since the days of Beccaria. Rose and Prell have pointed out that there is considerable discrepancy between the law and popular judgment as to how severe punishment should be in regard to certain offenses.[45]

---

[44] Raymond Moley, *Our Criminal Courts* (New York: Putnam's, 1930), pp. 99-106.

[45] See Arnold M. Rose and Arthur E. Prell, "Does the Punishment Fit the Crime? A Study in Social Valuation," *Amer. J. Soc.*, Vol. 61, No. 3, (November, 1955), pp. 247-259.

In the early days of prisons in this country all penalties were fixed. Armed robbery might call for a ten year penalty. The convicted man was obliged to serve out his ten year stretch unless pardoned by the governor of his state. In due time a modified indeterminate sentence was introduced, state by state, which requires the judge to pronounce both a minimum and a maximum sentence, the maximum being no more than specified by the penal code for the particular crime in question. The minimum sentence was not set by law, so the judge had great discretionary power. Still more power was given the judge by permitting him to sentence a prisoner on more than one count of the indictment if he was actually convicted on more than one count. The judge may then "pile on" the number of years to be served in prison, requiring the various penalties to be served consecutively rather than concurrently. In Philadelphia a young thug was convicted of armed robbery on seven counts. The judge sent him to prison on a sentence of 70 to 140 years—ten to twenty years on each count. Excessive sentences are quite common. A member of the British Royal Commission on capital punishment reported that he had heard, while in Washington, D.C., of a man who had been sentenced to one of our prisons for a term of 300 years.

The penal codes of our states are hopelessly outmoded. Their philosophy lags far behind social and psychiatric science so that it is impossible to construct one based along sound scientific lines. Let us take Pennsylvania as an example. It is not accurate to say there is a penal code, for penal statutes are scattered throughout the various titles, subdivisions, and chapters with no order, coherence, or consistency. There have been several brave attempts to codify this *potpourri,* but they have met with little success.

While the penal codes reflect the social thinking of a specific period, there is no justification for keeping laws on the statute books, or definitions of crimes with their corresponding penalties that were defined by legislators and criminal lawyers who have been dead 50 to 100 years. Attitudes toward behavior change due to more enlightenment flowing from the social sciences, psychology, and psychiatry. The penalties attached to crimes should not be set by legislators. Socially disapproved behavior may be defined as criminal but the penalties should be set by persons trained in the knowledge of human behavior, not by legislators or by jurists.

There are many persons who are convinced that the sentencing power should be shorn from our judges and handed over to a board of scientists, known as a diagnostic clinic. In fact, some enlightened judges have recommended such an innovation. They see their task as one of interpreting the law in the courtroom, and seeing to it that the rules of procedure are scrupulously adhered to. This is a big enough responsibility for them and it should cease after the jury has brought in its verdict.

The diagnostic clinic would be staffed by a group of persons skilled in the fields of human behavior. It should be composed of a psychiatrist, a social worker, a psychologist, and other persons noted for their knowledge of human behavior. They could, through tests and investigation, distinguish those who suffer from mental or emotional disorders, from those who are retarded mentally, or who, through obvious circumstances, were accidentally precipitated into crime. As Dr. Paul Schroeder, Professor of Psychiatry at the University of Illinois, states: "It should be no great task to set up a diagnostic clinic for administration of all persons sentenced. In this clinic, impartial, disinterested scientists would function under conditions which never exist in the courts."[46]

Although we may be a long way from establishing a diagnostic clinic, we may see the "pre-sentence" clinic in much wider use than it now is. This is merely a device whereby the judge asks for help in meting out sentence. He usually consults with his probation staff, which, upon thorough investigation of the convicted man's background, assets and liabilities, makes a recommendation. But many jurists resent even this guidance.

It is natural that the majority of the judiciary in this country oppose, in principle, sentencing boards with personnel trained in various fields that deal with the problem of human behavior; but all the opposition is not prejudice or the dying gasps of a vested interest. No doubt there are some judges who would be most happy to relinquish the sentencing power but who have misgivings about turning over this power to this type of board. But the trend toward a "pre-sentence" clinic and perhaps even a diagnostic clinic will continue if the capricious character of present practice in sentencing by judges is not eliminated. In juvenile courts, the probation officer, especially if he is a trained social case worker, and the psychiatrist are of great assistance in helping the court dispose of cases.

Federal Judge Theodore Levin of the Eastern District of Michigan, speaking of the effects of sentence on a man, his family, and deterrence asks: "What equipment does the court possess to recognize these factors, and thereupon to allocate to them their relative and competing degrees of importance? The judge is a lawyer . . . but neither his education nor experience has embraced a familiarity with the compulsions underlying law violations, nor the factors to be appraised in . . . the treatment for the criminal offender." He calls for a sentencing tribunal.[47]

The present system of sentencing is referred to by many as the "hunch" system. Discriminatory sentences by the same judge and differences in severity among judges within the same state for the same offense have caused considerable bitterness on the part of convicts thus affected. There is scarcely a penal institution in which the administration does not have difficulty with inmates who feel, with justification, that they received a "rotten deal" from some befuddled or vindictive judge. Professor Lester B. Orfield, in his cautious legal work dealing with criminal procedure, has the following to say about the lack of uniformity in sentencing:

---

[46] *The New York Times* Sunday *Magazine,* September 14, 1947.

[47] Judge Theodore Levin, "Sentencing the Criminal Offender," *Federal Probation,* Vol. 13, No. 1 (March, 1949), pp. 3 ff.

Studies of the sentencing process show that there is no uniformity of approach among trial judges. Some tend to give the maximum penalty and some the minimum. Some penalize certain types of crime severely, while others may do the opposite. Some place many defendants on probation while others seldom provide for probation. Some judges penalize certain races and nationalities severely while others may be overlenient with the same groups. Individual judges often have no consistent policy of their own as to the cases they pass on. Often there is no thorough and scientific analysis of the behavior problems of individual offenders. Judges are often pressed for time, and receive little assistance from social workers and probation officers.[48]

There are quaint stories of the pure caprice of sentences by judges, and they would be humorous if they were not so tragic to the unfortunate persons receiving the sentences. A judge doubles a particular sentence because the defendant unfortunately happens to have the same name as the jurist, and tells the defendant that he has disgraced his honorable name—this actually happened in a Philadelphia courtroom.

Sectional attitudes regarding certain crimes are often reflected in the vagaries of sentencing by the judges of those districts. For example, some federal judges in the "moonshine" hills of the southern states mete out much longer sentences for violating the federal laws against manufacturing and selling untaxed liquor than judges are likely to impose in New Jersey or New York. Then, too, judges within a particular jurisdiction show little uniformity in their sentencing.

The advisory pre-sentence clinic is slowly gaining acceptance throughout the country. Aside from many federal courts where it is employed, various county courts are finding it extremely valuable in meting out a more even type of justice than the capricious "hunch" system. A California statute requires that in every felony case, before judgment is pronounced, the court shall order an investigation by a probation officer. In Colorado, in any felony case in which the court has discretion as to the penalty, a pre-sentence investigation is mandatory. Other states requiring pre-sentence investigation, before judgment is pronounced, are Michigan, New York, Massachusetts, Rhode Island, and New Jersey.[49]

Going somewhat beyond the use of the pre-sentence clinic are the prediction tables that have been developed during the past quarter century.[50] The most articulate in espousing the use of such devices in sentencing are Drs. Eleanor and Sheldon Glueck. As the latter puts it: "Prediction tables are but one added instrument in aid of the individualizer of justice." Factors in the constitution, social background, earlier behavior of the offender

---

[48] Orfield, *Criminal Procedure from Arrest to Appeal,* p. 557.

[49] Will C. Turnbladh, "Probation and the Administration of Justice," *Prison World* (November-December, 1950), p. 11. See also Manfred S. Guttmacher, "The Status of Adult Court Psychiatric Clinics," *N.P.P.A. Journal,* Vol. 1, No. 2 (October, 1955), pp. 97-104.

[50] See pages 580-581 for our discussion of prediction tables.

may be explored and analyzed. By means of correlation tables "the degree of relationship between each of these . . . factors (could be) determined."[51] Such prediction tables would supplement the pre-sentence investigation and be a further aid in computing the length and type of sentence to be imposed in each individual case.

Pending that probably remote time when the judge may be stripped of his power—and responsibility—to sentence, the reliance on the pre-sentence investigation and the prediction table is extremely urgent if individuality of justice is desirable.

Indemnity for Those Wrongfully Convicted of Crimes

Since it is assumed that the state can do no wrong, the practice has long been for the law to be indifferent to the erroneous conviction and incarceration of innocent individuals. This is a patent injustice. The late Dean Wigmore has well said:

> Because we have persisted in the self-deceiving assumption that only guilty persons are convicted, we have been ashamed to put into our code of justice any law which *per se* admits that our justice may err. But let us be realists. Let us confess that, of course, it may and does err occasionally. And when the occasion is plainly seen, let us complete our justice by awarding compensation. This measure must appeal to all our instincts of manhood as the only honorable course, the least that we can do. To ignore such a claim is to make shameful an error which before was pardonable.[52]

Only a few states have made provisions to indemnify persons wrongfully accused and convicted, and the relief is spasmodic even there. Usually considerable influence is needed to bring about the passage of a special legislative act to render restitution to the victim of a miscarriage of justice of this sort. Those states that have statutes providing relief are California, North Dakota, and Wisconsin, but Professor Edwin Borchard, an authority on this subject, says, "by reason of their apparent novelty, such statutes have been narrowly construed and but little has been accomplished by them."[53] European countries have gone much further in this regard.

One need only read Professor Borchard's carefully documented work, *Convicting the Innocent,* to realize the tragic possibilities of this miscarriage of justice. He records the cases of 65 persons who were wrongfully

---

[51] Sheldon Glueck, "The Sentencing Problem," *Federal Probation,* Vol. 20, No. 4 (December, 1956), pp. 15-25. See also, *Guides for Sentencing* (New York: Advisory Council of Judges of the National Probation and Parole Ass'n., 1957). For an informative survey as well as bibliography on the subject, see "What's New in Sentencing?" *Correctional Research,* publication of the United Prison Ass'n. of Massachusetts, Boston, Bulletin No. 7, October, 1957.

[52] Quoted in the United States Senate by Senator Robert M. LaFollette, Jr.; see *Congressional Record* for July 8, 1940.

[53] Edwin Borchard, *Convicting the Innocent* (New Haven: Yale University Press, 1932), p. xxiv.

accused and convicted of crimes. The crimes involved were: murder, 29 cases; robbery, swindling, or larceny, 23 cases; forgery or counterfeiting, 5 cases; criminal assault, 4 cases; writing obscene letters, 2 cases; and one case each of accepting a bribe and prostitution. They were scattered all over the country. In only a few cases was some financial restitution made by the states. A much more recent work by Judge Jerome Frank relates 36 cases of persons wrongly convicted. Guilty verdicts were returned because (1) exonerating evidence was suppressed or incriminating evidence fabricated, (2) witnesses were mistaken or perjured, or (3) defendants could not afford competent counsel or adequate investigation.[54]

The legal bases for compensation to the innocent convict are: *first,* the right of eminent domain, that is, that the owner of private property taken for public use shall be compensated, for in these cases private liberty, a right at least as sacred as that of property, has been taken by the state; *second,* the theory supporting the workman's compensation act, that in the operation of any great undertaking, such as the management of a large industry, or the administration of criminal law, there are bound to be a number of accidents, and that the accidents must be compensated.[55]

It is not to be implied, however, that anyone acquitted of a crime—which assumes that he has been at least wrongfully accused—is to receive compensation. In order to minimize abuses that may flow from statutes of this sort, the experience of some of the European countries would be most helpful in drawing up a law that would care for only those subjected to the "grossest injustice" and who were "most deserving" to be indemnified. As Professor Borchard recommends:

> We would first, therefore, compel the unjustly convicted person claiming the right to relief to prove that he was innocent of the crime with which he was charged and not guilty of any other offense against the law. And here he must show one of two things—that the crime, if committed, was not committed by the accused, or that the crime was not committed at all. This at once eliminates from consideration a vast class of possible claimants.
>
> In the second place, the loss indemnified should be confined to the pecuniary injury, that is, loss of income, costs for defense and for securing his ultimate acquittal or pardon, and similar losses.[56]

---

[54] Jerome Frank, *Not Guilty* (New York: Doubleday, 1957).

[55] Borchard, *op. cit.,* pp. 387-392.

[56] Quoted in the United States Senate by Senator LaFollette, July 8, 1940. For further evidence that persons have been wrongfully convicted, see Erle Stanley Gardner, *The Court of Last Resort* (New York: William Sloane, 1952; also Pocket Books). A television program of the same name was offered for some months to acquaint the American public with the all-too-frequent miscarriage of justice in criminal cases. Pennsylvania has pardoned two men wrongfully imprisoned in recent years: Rudolph Sheeler, who served ten years though innocent, and was pardoned in 1951 when evidence showed he had not committed the crime of murder (see page 222); and William Greene, pardoned by Governor George Leader in 1957 when it was established that he had been convicted of murder on perjured testimony.

Although Professor Borchard is cognizant that the damage is to a man's reputation and the injury is to his moral standing, he holds that these cannot be adequately compensated. Furthermore, to make such compensation would work a serious hardship on the state. He suggests that in no case should a sum larger than $5,000 be awarded. Nevertheless, the State of New York in 1945 awarded Bertram Campbell the sum of $115,000 in compensation for being convicted erroneously and being obliged to serve more than three years in Sing Sing prison. He had been charged with forging checks and his conviction was brought about by means of mistaken identity at the trial. This case attracted considerable national attention.[57] More recently the State of New Jersey awarded a man $15,000 in a mistaken identity case. He had been convicted of forgery and had served time in prison. Later the real forger, who closely resembled the convicted man, was apprehended in Wisconsin for the same offense. It was then that the mistake was discovered.

The federal law enacted May 24, 1938, provides for "relief to persons erroneously convicted in courts of the United States"; any person who can prove that he was wrongfully convicted and sentenced for a crime may bring suit in the Court of Claims for damages of not more than $5,000 against the Federal Government.

But this act applies to federal courts only. There is a decided movement for similar state legislation, notably on the part of the American Civil Liberties Union. This is a matter that should concern every American citizen interested in fair play and in maintaining and safeguarding the civil rights of citizens. The least that the state can do is to make some restitution to innocent persons who suffer the ignominy of a prison sentence or other unjust penalty.

---

[57] For more details of this case as well as others in which mistaken identity played a part, see Reynolds, *Courtroom*, Chapter 8, "That's the Man."

17 ⸻

The Place of the
Jury in Criminal Justice

The jury system, praised because, in its origins, it was apparently a bulwark against an arbitrary tyrannical executive, is today the quintessence of governmental arbitrariness. The jury system almost completely wipes out the principle of "equality before the law" which the "supremacy of the law" and the "reign of law" symbolize—and does so, too, at the expense of justice, which requires fairness and competence in finding the facts in specific cases. If anywhere we have a "government of men," in the worst sense of the phrase, it is in the operation of the jury system. (Courts on Trial, *Princeton, N.J.: Princeton University Press, 1949.*)

JUDGE JEROME FRANK

History of the Jury

The remote origins of the jury system are to be found in the nature of the Roman *fiscus* or imperial treasury. During the ninth century this power was carried over by the Franks, to whom the royal lands were known as *fiscal* lands. The king's representatives would summon a group of leading citizens and extract from them an opinion or statement as to the taxable wealth of their community, the state of public order, and the prevalence of offenses against the king and his laws. This group was known as a *jurata* and its report to the king or his representative was called a *veredictum*. Here we have in embryo both the terminology and the procedure of the modern jury.

The first stage of its development as a juristic rather than an administrative instrument was instituted by the Normans who took the practice to England in 1066. Exactly a century later, at the famous Assize of Clarendon in 1166, the *grand jury* took definite form. A varying number of country gentlemen were summoned before the royal representative and asked what they knew of those persons accused of crime in their neighborhood. At least 12 of them had to agree as to the accuracy of the report in order to secure royal action. It was necessary for so large a number to be summoned because a single individual feared to accuse powerful violators of the king's peace. This was the origin of the *grand jury*.

But the action that did more than anything else to establish the trial or *petit* jury was Pope Innocent's condemnation of the *ordeal* in 1215. Prior to that date, guilt was established by primitive and medieval devices known as "trial by battle," the ordeal (torture), and compurgation. In *compurgation* the defendant swore that he was innocent and then kinsmen or neighbors, the *compurgators,* took an oath that they believed he was telling the truth in proclaiming himself innocent.

The choice of the number twelve unquestionably was based on Scriptural precedent. The following quotation from *Duncomb's Trials* (1665) gives a good idea of the mystical and theological attitude toward this magic number:

> As to the sanctity and foreordained character of the number twelve, and first as to their [the jury's] number twelve; and this number is no less esteemed by our law than by Holy Writ. If the Twelve Apostles on their twelve thrones must try us in our eternal state, good reason hath the law to appoint the number twelve to try our temporal. The tribes of Israel were twelve; the Patriarchs were twelve, and Solomon's officers were twelve (I Kings IV, 7). Therefore, not only matters of fact were tried by twelve, but in ancient times twelve judges were to try matters in law. In the Exchequer Chambers there were twelve counsellors of state for matters of state and he that appealed to the law must have eleven others with him who believe he says true and the law is so precise in their number of twelve that if the trial be by more or less than twelve it is a mistrial.

For a time the grand jury acted also as a trial jury but soon the two bodies were definitely separated in composition and function. By the end of the fifteenth century it had superseded the ordeal, which gave way grudgingly, and instead of a body of witnesses it became a group to hear witnesses. For a century or so a person might decline a jury trial and resort to trial by battle, or accept torture. For a considerable period, the only witnesses summoned were those for the prosecution, and the defendant was at a decided disadvantage. Gradually, however, he acquired the right to call upon witnesses to testify in his behalf. Along with this came the development of rules of evidence, of the right to challenge jurymen, and of improved methods of impaneling the jury. The rules of evidence and courtroom procedure prevailing today are a curious mosaic, embodying some semi-magical and religious elements dating from the twelfth century or earlier, in juxtaposition to such highly modern devices as the summoning of trained experts and psychiatrists to aid judges and juries in their decisions.

It should be evident that the jury is hardly the bulwark of liberty it has been popularly considered to be. There is no justification, for instance, for such a statement as the following—a statement embodying what the public is prone to accept traditionally:

> The jury system is the very corner-stone of American democracy. It lies at the very foundation of the freedom and liberty of a free people.[1]

According to the late federal Judge Jerome Frank, part of the glamor that clings to the institution in our country is due to the fact that in pre-Revolutionary times juries often stood up to judges controlled by the hostile British government; thus our Federal and state constitutions embodied provisions guaranteeing trial by jury.[2]

Far from being a rampart of human freedom or a safeguard of democracy, the jury system was, in its origins, one of the most potent and highly prized instruments of royal absolutism and monarchical oppression. Compared to other institutions of the time, trial by jury probably made a fairly respectable showing in the sixteenth century, when there were relatively few highly trained lawyers, and the men summoned for jury service represented the intelligent and highly respected upper classes. But the progress of medical knowledge, sociology, jurisprudence, and democracy since that time has made it completely out of date. Moreover, the average jury today is chosen from an altogether less educated class than that which furnished jurymen in the sixteenth century.

Methods of Selecting Juries

The selection of the jury panel is determined by lot, the names of a definite number of citizens being drawn from a collection of cards bearing the names of all the qualified voters of the county. At best, any such panel can only in rare instances include a better than average group of citizens. In the usual case, the panel is made up of farmers, shoemakers, barbers, plumbers, salesmen, hodcarriers, and day laborers, with a few professionals or businessmen sprinkled among them.

There are no universally acceptable criteria for jury service. They vary widely throughout the country. In some states jurors are supposed to be "of ordinary intelligence," in others, merely "well informed." In one state the potential juror must be able to read and write the English language; in another he must be a taxpayer and own real estate. The following various qualifications of a juror are taken from the statutes of several states: must

---

[1] Leo E. Sherry, "Juries in Criminal Cases," *Law Society Journal,* Vol. 11 (May, 1944), pp. 192-193.

[2] Jerome Frank, *Courts on Trial* (Princeton, N.J.: Princeton University Press, 1949), p. 109.

have qualifications of all electors; must pay taxes; must be a taxable person; must be on the assessment rolls; must *not* be a public official; must *not* be an employee of city, state, or Federal government in any capacity whatever; must *not* have previously served as a juror in the past year; must *not* be insane or an idiot; must *not* have been convicted of a felony in state or Federal court or of any crime involving moral turpitude. In many states, businessmen, professional men, editors, school teachers, and other members of the better educated groups are excused from jury service.

One thing is certain: if names of ignorant and incompetent citizens go into the machinery for jury selection, many of them will be called to serve. One device to improve the jury is the "key number" system in use in Los Angeles, Detroit, and Cleveland (where it originated). In Los Angeles, over two million names are available on the lists in 4,300 voting precincts. To spread the service over the entire body of citizens, subject to the individual tests, every fifth precinct was sampled, beginning with number one. For the second half-year, every fifth precinct was taken, beginning with number two. The voters whose names are thus selected are then invited to appear at the jury commissioner's office at the rate of 392 for each of four days each week for nine or ten months in each year. The commissioner and his assistant use formal tests to examine these candidates and class them as fit to serve, exempted, or clearly unsuitable. The system thus far is effective in spreading the service and providing such a surplus of jury material that only the best one-fourth need be accepted, and in preventing any citizen from being summoned a second time. The jurors thus selected in Los Angeles serve in both civil and criminal cases.

The conventional method of selecting the jury by lot starts by exempting many who are likely to be the more intelligent and able members of the panel. When those who remain are called, they are asked whether they have read or formed any opinion about the case in question. Those who answer in the affirmative are usually disqualified, though in all significant cases any honest and literate person has to say yes to this question. Hence the actual choice of jurymen is obviously limited, to a large extent, to the illiterate and the liars. An appraisal of the jury by a student of court trials is of interest:

> We commonly strive to assemble twelve persons colossally ignorant of all practical matters, fill their vacuous heads with law which they cannot comprehend, obfuscate their seldom intellects with testimony which they are incompetent to analyze or unable to remember, permit partisan lawyers to bewilder them with their meaningless sophistry, then lock them up until the most obstinate of their number coerce the others into submission or drive them into open revolt. The average citizen rebels at the very thought of being forced to participate in such proceedings, and regards jury duty as an irksome and humiliating task, to be avoided if possible.[3]

---

[3] Statement made by Hon. Benton S. Oppenheimer and quoted by Dillard D. Gardner, "Courts and Records," in *Popular Government* (August, 1937), p. 9.

Naturally, an attorney desires to obtain a jury that will be most favorable to his side, and in many instances he resorts to devious and even unethical methods to accomplish this. "Jury substitution" usually results from ignorance—a person who has been summoned to serve sends another in his place; but sometimes politicians supply the right type of jurors by manipulating names or by summons—another type of substitution. "Jury tampering," or "embracery," frequently results from lax methods of selecting juries.

Defense attorneys who are leaders in their field develop shrewd techniques in selecting jurors. For example, they are likely to challenge all prospective jurors who may possibly be prejudiced against the defendant because of party affiliation, religious belief, class membership, or nationality. Judge Jerome Frank quotes Goldstein's *Trial Techniques* as follows:

> Always demand a jury if you represent a plaintiff who is a "woman, child, or old man or an old woman, or an ignorant, illiterate or foreign-born person unable to read or write or speak English who would naturally excite the jury's sympathies," especially if the defendant is a large corporation, a prominent or wealthy person, an insurance company, railroad or bank. Then, he advises, seek the type of juror who "will most naturally respond to an emotional appeal."[4]

Clarence Darrow, often called the "ace of jury picking," said that he was always more concerned with the political, religious, economic, and moral beliefs of the jury and the defendant than with the actual circumstances of the crime committed. The generous provisions for challenging jurors without cause and the all but unlimited right of challenging for cause makes this maneuvering easy.

Interminable delays in selecting 12 jurors are often characteristic of American courts. The necessity of wasting many dreary hours waiting to be called to the box drives citizens to seek exemption and legal excuses. There are so many ways of evading jury service, and the inconveniences of serving throughout a trial are so notorious, that scores are excused for each one finally selected.

In Philadelphia some years ago, one judge, after exhausting all those who had been impaneled for service, invoked an old law and ordered the tipstaves into the city streets to round up pedestrians for possible jury duty. In the famous Sacco-Vanzetti trial in Massachusetts, after 500 talesmen had been called and only five jurors selected, the judge sent court deputies to round up more prospective jurors; some were hauled out of bed, others were accosted on the streets, and still others were pressed into service while attending a Masonic Lodge meeting. In the trial of the New York racketeer, Louis (Lepke) Buchalter, it took five weeks to select the jury.

---

[4] Frank, *Courts on Trial* (Princeton: Princeton University Press), p. 121.

The Typical Jury Ready for the Trial

The jury, after a period of bewilderment in the new and strange atmosphere of the courtroom, often settles down to a state of mental paralysis that makes it virtually impossible to concentrate on the testimony and the rulings of the court. Awakened at times from this stupor, the jurymen may suddenly pounce on some more or less irrelevant bit of testimony and forget or overlook the more significant facts brought out by the witnesses. The testimony is generally dull and confused, punctuated by explosive remarks of the opposing attorneys who are trying hard to tear it down by ridicule.

If any testimony is pertinent, intelligent, and damaging, and if there are a few competent and alert jurymen, the lawyer whose side seems likely to lose by this testimony tries to obscure its significance and divert the attention of the jurymen from it. Every form of oratorical appeal is used by the attorneys. The jury may even be threatened by mob reprisal if it does not render a certain type of verdict. Particularly in closing appeals, rhetorical gaudiness is utilized. If the evidence is strongly unfavorable to one party, the lawyer representing it is likely to ignore the testimony and appeal solely to the emotions of the jury. Clarence Darrow once revealed the fact that his courtroom strategy was to make the jury identify themselves with the defendant, on the theory that no man will want to hang or convict himself. He held that, if the trial lasted long enough for him to apply his technique, he was usually able to get an acquittal or a "hung jury" (disagreement).

The Hon. Julius H. Miner, judge of the Circuit Court of Cook County, Chicago, while not advocating the abolition of the jury, is alert to its faults. He makes the following statement:

> It is unfair and unsound to expose twelve naïve and credulous jurors to the mercy of sharp, high powered battle-scarred veteran advocates, fortified with their bags of oratorical and procedural tricks suited to wrest advantage from every circumstance or situation developed during the trial. Under professed "solemn obligation" to their respective clients, the expert defense lawyer in a criminal case pertinaciously hammers away at "reasonable doubt," "moral certainty" and "presumption of innocence" throughout the tedious individual examination of the *voir dire* until there is an unconditional mental surrender by every potential juror. These legal doctrines are ideal safeguards for the protection of the innocent, but how often are they perverted avenues of escape for the guilty?[5]

Perhaps the most amazing feature of the modern jury trial is that neither the district attorney nor the counsel for the defense is vitally concerned with the hard facts. As Judge Miner points out, the jury becomes utterly bewildered with the "machine gun rapidity of the 'objections,' 'exceptions,' 'sustained,' 'disregarded,' 'exhibits' and the hundred and one stereotyped

---

[5] Julius H. Miner, "The Jury Problem," an address delivered February 9, 1946. See *Criminal Justice,* Journal of the Chicago Crime Commission, No. 73 (May, 1946).

instructions upon the law in the understanding of which even our supreme courts cannot agree." Certainly before a group of trained experts whose business it is to ascertain innocence or guilt, the vaporings of high-priced counsel could have no more effect than the gyrations of a whirling dervish.

The technical rulings on law are usually more ineffective upon the jury than the testimony. The average juror is abjectly ignorant of even the most elementary law, and in a great many instances misses the significance of the judge's interpretations of it. Even in those cases where the rulings are simple, explicit, and direct the jury on occasion goes counter to them. If a juryman has really been impressed by testimony, he will very seldom be influenced by a subsequent ruling that it is irrelevant and must be excluded from consideration.

Witnesses in the Courtroom

In most criminal cases there are few eyewitnesses, and those who are in that unenviable position are rarely persons of great intelligence. Quite as likely as not they are among the undesirable citizens of the community who would not be believed on oath anywhere else. On the other hand, counsel may seduce them into making vague insinuations or even overt statements about things they do not know. Judge Frank quotes Sir William Eggleston, a noted Australian lawyer, as saying that "no witness can be expected to be more than 50 per cent correct, even if perfectly honest and free from preconception."[6] As Frank points out, "Aside from perjurers, there are innumerable biased witnesses, whose narratives, although honest, have been markedly affected by their prejudices for or against one of the parties."[7]

But, worse than the fallibility of witnesses, there is the coaching of the witnesses, all too frequently, by counsel. From the counsel's point of view, the best possible witness is one who knows nothing about the case. He may then be taught a coherent story to tell the jury. Callender has the following to say about coaching of witnesses:

> In the interviewing of witnesses, the lawyer is interested in several things besides the procuring of evidence. The character and personality of the witness are matters of great importance. An ignorant and stupid witness presents a very different problem from an educated or intelligent one. A candid and affable personality may be counted on to be convincing to a jury, where a crafty or subtle nature may have an opposite effect. The lawyer must take these matters into consideration and devise the best method of handling each type. It is necessary for him to instruct each one as to the way he must conduct himself on the stand. He tells him how he is to answer the judge if questioned by him; how he must treat the opposing lawyer and with what patience he must answer all questions; upon what

---

[6] Frank, *Courts on Trial*, p. 18.
[7] *Ibid.*, p. 19.

matters he is competent to testify, and how careful he must be to be accurate in everything he says. If the witness appears to be likely to take liberties with the truth, the lawyer must warn him of the consequences of perjury.[8]

Convictions for perjury or confessions of perjury in all sorts of cases, from the celebrated Mooney case in California to the tragic miscarriage of justice in the Sacco-Vanzetti case in Massachusetts, have demonstrated the frequency of this building up of impressive testimony by counsel and witnesses without the slightest factual basis. It is generally agreed by lawyers and jurists that an immense amount of court testimony is deliberately false. Many cases are won through perjured testimony.

The average witness fears being caught in the act of perjury far less than cross-examination by opposing counsel. In relatively few cases is obvious perjury challenged. Perjury is a felony in some states, and a misdemeanor in others. It is extremely difficult to convict a person of perjury, since the prosecution has to prove "beyond a reasonable doubt" that the accused deliberately gave false testimony and that what he said was material to the issue, that is, of such a nature as to influence the decision.

It is one of the marvels and injustices of our criminal procedure that in the case of a conviction of perjury, few as they are, the witness alone, instead of witness and counsel together, is compelled to suffer the penalty of the law. Professor John Barker Waite cites cases in which there was no doubt that the defense attorney and other public officials connived with the witnesses at perjury.[9]

It is imperative that unscrupulous lawyers who aid and abet perjury by witnesses should share the penalty and run the risk of disbarment proceedings. As it is now, in most cases where collusion is obvious, lawyers are neither penalized by law nor purged by their fellow professionals. Professor Waite, referring to a notorious case of perjury involving a well-known lawyer, states: "I cannot find that either the court or his fellow members of the bar . . . made any attempt to exclude him from the practice of law, or to rebuke his vicious attitude toward criminal justice."

Hence, the stark reality is essentially this: a number of individuals of average or less than average ability, who could not tell the truth if they wanted to, who usually have little of the truth to tell, who are not allowed to tell even all of that, and who may have been instructed by unscrupulous lawyers to fabricate voluminously and unblushingly, present this largely worthless, wholly worthless, or worse than worthless information to twelve men and women who are for the most part unconscious of what is being divulged to them, and would be incapable of an intelligent interpretation of the information if they had actually heard it.

---

[8] By permission from: Clarence N. Callender, *American Courts, Their Organization and Procedure* (New York: McGraw-Hill Book Co., Copyright 1927), p. 80.
[9] John Barker Waite, *Criminal Law In Action* (New York: Harcourt, 1934), p. 184.

The art of cross-examination plays an important role in the court trial. It assists reputable, sincere, and truthful witnesses in having their testimony accepted as valuable; but it discredits the testimony of evasive, arrogant, or untruthful witnesses. Some witnesses, however, are so brazen and self-possessed in their courtroom attitude, even under cross-examination, that they make the jury accept their testimony.

In court procedure no hearsay evidence may be introduced. But there are two exceptions to this rule—a previous confession, which must have been made voluntarily, and a dying declaration.

It is a rule of evidence that the ordinary witness cannot express an opinion. The expert witness who is called upon to testify from his scientific knowledge may express an opinion, but it is not unusual to find both the defense and the prosecution calling in experts in the same field who present conflicting views. A warning concerning the expert is sounded by Percival Jackson:

> It has become the custom to call "expert" witnesses to give "expert" testimony on almost every conceivable subject, and many such "experts" make it a practice to sell any believe-it-or-not opinion for which a litigant will pay. Often, even an honest expert indulges in wishful thinking and stretches the probabilities to favor the side that calls him. In consequence, it is never very difficult to get an expert to testify "con" to combat the testimony of an expert who has testified "pro."[10]

The Deliberations of the Jury and Its Verdict

There is little published material that gives us an idea of what actually transpires in a jury room after the judge has charged the jury. There have been dramas as well as stories that give an approximation of how the jury arrives at a verdict but, of course, these are fictional. An especially powerful motion picture, *Twelve Angry Men,* depicted the tense and even bitter hatred that is sometimes engendered among the jury in its deliberations.

A few years ago a study financed by the Fund for the Republic was attempted but it came to grief. A tape recorder was hidden in the jury room with the consent of the presiding jurist and the two attorneys involved in a civil case in Kansas. But the members of the jury were unaware of the presence of the microphone. News of this leaked out and caused a na-

---

[10] Percival Jackson, *Look at the Law* (New York: Dutton, 1940), p. 305. See examples of erroneous testimony by experts mentioned on page 229 *supra.* See also, Robert Graves, *They Hanged My Saintly Billy: The Life and Death of Dr. William Palmer* (New York: Doubleday, 1957). Mr. Graves, the noted English poet, has revived the story of a trial that occurred in England in 1857, showing how expert testimony, superimposed on a gullible jury and a determined Chief Justice and Attorney General, brought about a dubious conviction by which an innocent man was executed. As Erle Stanley Gardner states, "Wrapping their opinionated testimony in technical jargon, they [the experts] created such an aura of infallibility that the mystified jurors unhesitatingly let the "experts" do their thinking for them." Quoted from his review in the *New York Times Book Review* (Sunday, May 26, 1957).

tional furor. While this was not illegal, it was believed passionately by many articulate people that it was at least unethical.

Verdicts are sometimes reached by flipping a coin. Prayer is sometimes resorted to. In a sensational rape case in Boulder, Colorado some years ago the jurors were quite frank in stating that they prayed for help. Judge Jerome Frank gives the following instance of jury procedure in arriving at a verdict:

> The jury at first stood six for assault and battery, and, as a compromise, the six agreed to vote for manslaughter, and the vote then stood six for manslaughter, and six for murder in the second degree; it was then agreed to prepare 24 ballots,—12 for manslaughter and 12 for murder in the second degree—place them all in a hat, and each juror draw one ballot therefrom, and render a verdict either for manslaughter or murder in the second degree, as the majority should appear; the first drawing was a tie, but the second one resulted in eight ballots for murder in the second degree and four for manslaughter, and thereupon, according to the agreement, a verdict was rendered for murder in the second degree.[11]

A bizarre story comes from Bedford, Indiana. A mother admitted that she was so anxious to get through with a trial and "get back to her children" that she voted for the conviction of a defendant charged with rape although she later admitted that she did not believe there was enough evidence to prove the man's guilt.[12]

Even when a jury is reasonably alert in following the testimony, desirable results are likely to be offset by the presence upon the panel of a powerful and impressive personality or an unusually stubborn individual. There have been innumerable miscarriages of justice because the jury was converted to the point of view of a prejudiced but convincing orator, or because a juror was present who, through bias, bribery, or stupidity, held out against the judgment of his eleven colleagues.

Often any criticism of the jury system is met by the allegation that most verdicts are sound. But how does one know that even one particular verdict is correct? The reader is referred to Edwin M. Borchard's book, *Convicting the Innocent,* for clinical evidence of the caprice of juries.[13] The majority of our convicted murderers go to the chair vehemently protesting their innocence, and many who seem obviously guilty are freed. By the mathematical laws of chance, verdicts should be correct in 50 per cent of all cases, taking a sufficiently large number of cases and extending them over an adequate period of time. Would any person of reasonable intelligence, sanity, and literacy contend that more than half of our jury verdicts are accurate,

[11] Frank, *Courts on Trial,* p. 115.
[12] Philadelphia *Evening Bulletin,* September 3, 1956.
[13] Edwin M. Borchard, *Convicting the Innocent* (New Haven: Yale University Press, 1932). The reader is also referred to Edgar Lustgarten, *Verdict In Dispute* (New York: Scribner's, 1950). The trials of several murderers are reviewed and the peculiar reasoning of the jurors in reaching verdicts is shown.

or that the majority of those that are sound are such for any reason other than pure chance? The modern jury trial is one of the many anachronisms that clutter up our system of criminal jurisprudence and penal philosophy.

Some of the rational recommendations recently incorporated into our criminal procedure deal with the outmoded jury trial. In over half of the states the defendant may waive a jury trial and come directly before the judge. Data show that in relatively few cases is the jury actually used. A great many defendants plead guilty and a large number are found guilty or are acquitted before judges. The jury system has all but disappeared in England except in serious criminal cases. In Scotland it has never played an important role.

There are many advantages and practically no disadvantages in waiving jury trials. It saves the state much expense, it reduces the number of appeals and retrials, and it relieves the work of the prosecutor's office, for more time can be placed on careful preparation of cases. The trend away from jury trials is healthy and is being encouraged by many whose professions are closely linked with criminal justice.

If we retain the jury system, the least we can do is to see that each citizen takes his turn in serving; that no one be excused except for the most excellent of reasons or for mental unfitness; that the compensation be raised so that it will work less hardship than the present niggardly pay; that jury trial be made optional with the accused; and that provision be made for the impaneling of extra or alternate jurors to serve in case of disability or disqualification of any of the regular jurors in a trial. The idea of briefing the jury prior to the trial concerning courtroom procedure has also been suggested.[14]

Desirable Reforms in Criminal Procedure

Our study of criminal procedure constantly reminds us that it is deeply submerged in the institutional mores. Hence, we curb our impatience with the apathy of those to whom we must appeal for change. It is difficult, however, to be patient with the enemies of reform who represent vested interests, many of them members of the legal profession. Yet it is to them, or to those too timid that we must call for change in criminal procedure even though we are aware of the great culture lag that exists. Many of our professors of criminal law mildly state that "it would seem reform in this area is necessary" but terminate their concern at that point.

Obviously education and research are necessary. Education must stimulate crusading zeal and critical judgment, but the zeal must be armed with facts gathered from scientific research. Michael and Adler make the following provocative suggestion regarding change in criminal justice:

---

[14] Dick Lansden Johnson, "Pre-Trial Education of Jurors," *J. Crim. Law,* Vol. 38, No. 6 (March-April, 1948), pp. 620-627. See also "Judicial Administration and the Common Man," *The Annals,* Vol. 28 (entire issue, May, 1953).

We are assuming, for the moment, that it is desirable to administer the criminal law as efficiently as possible. . . . Since we lack scientific knowledge of the etiology of administrative efficiency, our attempts to increase the efficiency of criminal justice must either be directed by common sense knowledge or they are merely efforts at trial and error.[15]

Many suggestions have been offered by laymen, ranging all the way from police apprehension of the suspect, through the court routine, and thence to the sentencing of the guilty criminal. We have discussed some of these and many of the measures offered by commissions composed of members of the legal profession. Laudable as these efforts are, they do no more than readjust the details of the game.

A *Code of Criminal Procedure* has been prepared by the American Law Institute, which is a draft of suggestions regarding improvements that can be made by the various states in the criminal process.[16] Yet, as Professor Sheldon Glueck states, "It is predicated upon the assumption of the continuance of the existing criminal law and administrative practices."[17]

In our discussion of the jury system, for example, we noted that it is rapidly disappearing in criminal cases. A suggestion that the next step, based on a scientific analysis of court procedure, would be to adopt a system whereby experts would ascertain the defendant's guilt is often branded as a "crackpot" notion. Yet this must come eventually. As Professor Glueck says:

> While a legally trained judge can act as an impartial referee during a technical trial, rule upon the exclusion or inclusion of evidence, give a legally adequate charge to the jury and perform other such functions, his education and habits of mind have not necessarily qualified him, acting alone, for the specialized task of determining the treatment best suited to various types of offenders. Since this requires accommodation of legal and extra-legal disciplines, a reasonable conclusion is that the sentencing function should be entrusted to a tribunal. After ascertaining the defendant's guilt, this tribunal . . . would determine the appropriative sentence as well as its tentative duration.[18]

The pre-sentence clinic composed of experts in the fields of psychology, psychiatry, sociology, social work, and medicine has been suggested to supplant the judge so far as sentencing is concerned and a modification of this is already in operation in a few scattered jurisdictions. There is every reason to believe that this group, or one like it, will eventually be called upon to ascertain the guilt or innocence of the defendant. Lawyers and judges might well sit in on this clinic but only in an advisory or supervisory capacity, the

---

[15] Jerome Michael and Mortimer J. Adler, *Crime, Law and Social Science* (New York: Harcourt, 1933), pp. 316-317; see specifically Chapter 10, "Increasing the Efficiency of Criminal Justice by Common Sense."

[16] Published by the American Law Institute, Philadelphia, 1930.

[17] Sheldon Glueck, *Crime and Justice* (Boston: Little, Brown, 1936), p. 234.

[18] *Ibid.*, pp. 225, 226.

judge to interpret the law and the lawyer to see that the defendant's rights are safeguarded.

We have little hope, however, for the immediate adoption of many significant changes in court procedure. The late Chief Justice of the Supreme Court of New Jersey, Arthur T. Vanderbilt, after demonstrating several areas in which reforms were badly needed, contended that they could only result from the legal profession's measuring up to its responsibility.[19]

---

[19] Arthur T. Vanderbilt, *The Challenge of Law Reform* (Princeton, N.J.: Princeton University Press, 1955).

Book Two

PENAL AND CORRECTIONAL PROCEDURES

Corrections involves the use of every phase of the science of human engineering. It must be just, humane, and scientific, not only for ordinary human decency, but also for our own safety and security. It is essential to make society understand that true protection—lasting protection—can result only through changing attitudes, *not* chaining individuals, *especially since 97 out of every 100 persons sentenced to penal institutions are eventually released to go back to the community.*

JUDGE ANNA KROSS
COMMISSIONER OF CORRECTION, NEW YORK CITY

Part 1

ANCIENT AND MEDIEVAL CONCEPTS OF TREATING THE OFFENDER

18 ——————————————————

Primitive Treatment of the Offender—Corporal Punishment

I cannot help entertaining the hope that the time is not far distant, when the gallows, the pillory, the stocks, and the whipping post will be connected with the history of the rack and the stake as marks of barbarity of ages and countries, and as melancholy proofs of the feeble operation of reason, and religion, upon the human mind. (An Inquiry into the Effects of Public Punishments upon Criminals and upon Society, *Phila-delphia, 1787. Dr. Rush was a noted physician and signer of the Declaration of Independence.*)

BENJAMIN RUSH

Cruel punishments have an inevitable tendency to produce cruelty in people. (Quoted in Arthur Koestler, Reflections on Hanging, *New York: Macmillan, 1957.)*

SAMUEL ROMILLY

The Primitive Mind

Doctrines concerning the desirability and objectives of punishment have been closely associated with theories of crime and criminal responsibility. In primitive times, crime was mainly attributed to the influence of evil spirits, and the major purpose of punishment was to placate the gods.[1]

Later in the evolution of punishment more stress was laid on social revenge. Crime was then considered a willful act of a free moral agent. So-

---

[1] This, of course, oversimplifies an intricate matter. For a relatively recent work on this subject, see E. A. Hoebel, *The Law of Primitive Man* (Cambridge: Harvard University Press, 1954).

ciety, outraged at an act of voluntary perversity, indignantly retaliated. Many forms of allegedly willful crime were later identified with sin and were believed to be a challenge to God and orthodox religion. This dual theory of crime and punishment, based on perversity and sin, still dominates contemporary criminal jurisprudence, slightly refined by sophisticated modifications.

An additional purpose of punishment is to deter potential wrongdoers from the commission of similar or worse crimes. This is merely a derived rationalization of revenge. Though social revenge is the actual psychological basis for punishment today, the apologists for the punitive regime are likely to resort to the futile contention that punishment deters from crime. In this concept of *deterrence* there is a childlike faith in punishment. The person undergoing punishment becomes an example for all to see. This is the main reason why public punishments persisted for so long in civilized societies. This is the philosophy underlying the traditional criminal code. It is deeply rooted in the social fabric. Certain concessions have been made by law, from time to time, to children and to others who are irresponsible and come within the doctrine of limitation, notably the insane.

With the development of the biological sciences and the growth of new philosophies regarding life and social responsibility, other concepts of crime and criminals developed. These concepts sharply challenged the doctrine of free will, notably a shift from theories of revenge, retribution, expiation, and deterrence, to those of reforming the offender and protecting society. The emphasis became focused on the individual who committed the crime rather than on the crime itself.

It is plain that, however futile it may be, social revenge is the only honest, straightforward, and logical justification for punishing criminals. The claim for deterrence is belied by both history and logic. History shows that severe punishments have never reduced criminality to any marked degree. It is obvious to anyone familiar with the activities of criminals that the argument for deterrence cannot be logically squared with the doctrine of the free moral agent, upon which the whole notion of punishment is based. If a man is free to decide as to his conduct, and is not affected by his experiences, he cannot be deterred from crime by the administration of any punishment, however severe.

Members of primitive societies live in fear of the unknown and the supernatural; nothing stands between them and the powers of darkness except the well-beaten path of custom. To deviate from this in the least is to tread on the brink of disaster. Any transgression of the code exposes the offender to untold woe and renders his group liable to the vengeance of the gods, for responsibility is collective, crime and sin are thus identical. To supplement the individual's fear of violating the prescribed mode of conduct, based on the vengeance of the unseen powers, there is his certain knowledge that his group will spring upon him for rendering them liable to destruction by spiritual forces.

The punishment for breaking such basic taboos as those against endogamy, exogamy, witchcraft, or sorcery was carried out with great severity. The whole group, sometimes including neighboring clans, turned out to eliminate the offender. He might be hacked to pieces in the frenzy built up by the mob.

To punish or repress crimes of a private nature, primitive societies often had no uniform practice or well-developed organs; it was left to private agencies. In the early stage of primitive social organization the settlement of private wrongs was purely personal. A murderer might escape unscathed unless some of the immediate relatives of the deceased took up his cause. In such cases the aggrieved party settled with the offender directly, and the adjustment often produced a greater grievance.

Blood feud retaliation was usually regulated by customary rules of procedure. The basis of this procedure was the well-known *lex talionis:* the "eye-for-an-eye and tooth-for-a-tooth" principle. In some cultures, the literalness of this method of retaliation, with the punishment made exactly to fit the crime, is astonishing.[2]

The system of unrestricted blood feud was attended by many difficulties and limitations, and changes were gradually introduced to eliminate its more obvious defects. The most serious shortcoming in the system of clan retaliation was that it provided no satisfactory method of bringing a quarrel to an end. If a man from clan A injured a man from clan B, the matter was not settled by the revenge that was visited by clan B upon the offender from clan A or upon his entire clan, but simply gave clan A reason to turn to revenge itself upon clan B. This was because no clan recognized a wrong against another clan or the right of another clan to avenge an injury upon one of its own members. Therefore, an injury once committed started a perpetual vendetta that was likely to render life extremely precarious to members of both clans.

A very widely practiced and successful substitute for blood feud was compensation or composition, that is, restitution for injury through fines or money payments. This principle could not be applied to public crimes but only to individual wrongs.

The amount of compensation varied according to the nature of the crime and the age, rank, sex, or prestige of the injured party. "A free born man is worth more than a slave; a grown-up more than a child, a man more than a woman, and a person of rank more than a freeman," were the usual principles governing compensations. Moreover, a crime that might be settled by compensation if it were committed against an average man, would call for blood revenge if committed against a nobleman. From these differences in

---

2 For a recent popular account of an area where blood feuding still flourishes among a Stone-Age people in New Guinea, see Robert Ruark, "The Land That Time Forgot, Part II: Stronghold of the Kukukuku," *Saturday Evening Post,* Vol. 230, No. 39 (March 29, 1958), pp. 24-25, 77-79. This account also gives some interesting support to the geographical school of crime causation.

the amount of damages and the value of the victim, there grew up such a complicated system of regulations that the earliest codified law of many peoples, particularly that of the Anglo-Saxons, was in considerable part devoted to this subject of man-money, or *wergild*.

Our barbarian ancestors were wiser and more just than we are today, for they adopted the theory of restitution to the injured, whereas we have abandoned this practice, to the detriment of all concerned. Even where fines are imposed today, the state retains the proceeds, and the victim gets no compensation.

Another important check on private vengeance was the Right of Sanctuary. Certain places were set aside to which the accused might flee and temporarily escape punishment. Holy places, or cities of refuge were often set aside for purposes of sanctuary. As late as the thirteenth century a person could claim refuge in a church for a stipulated period of time.

With the eventual rise of the king's arbitrary authority in settling disputes we find much of the philosophy of private vengeance transmuted to public disposition of wrongdoers. Codification of tribal taboos began with the potentates of the Middle East. The famous code of Hammurabi, king of Babylon, has generally been given credit for being the oldest, but more recent discoveries indicate that others preceded it. The codes of Lipit-Ishtar and of Eshnunna, Sumerian rulers, certainly predate Hammurabi's code by over a century. These two codes were formulated about 1860 B.C., and Hammurabi's about 1750 B.C. All of these codes preceded the Mosaic code by well over a thousand years.

Along with the growth of the various agencies and devices for mitigating the principle of blood feud came the basic modern principle of public control of private wrongs. The court, that "impartial third party" that was lacking in primitive days, first appears in the elders, or council of the tribe or clan, to whom appeal might be taken, though at first neither party was obligated to abide by its decision. The court, in its earliest appearance, had peacemaking rather than judicial functions. It lacked the support of a dominant public authority, which would enable it to enforce its decrees.

This need was filled by the establishment of kingship and the firm central authority that went with it. Kingship, like all human institutions, grew slowly, but when it had thoroughly established itself as a strong central authority, a great change came over the conception of the nature of crime and the methods of treating it. The crimes formerly settled by blood feud now became an offense against the king's peace, an insult to his formal vanity, and a breach of the public tranquility. The crime became a matter for the public authorities to settle.

With the growth of this conception of crime as an offense against the public welfare, as exemplified in the majesty of the king, there was a corresponding decline in the principle of compensation, which ultimately became obsolete. Of course, revenge through blood feud and the like did not

Shamash, the sun-god, presents the legal code to Hammurabi; portion of stone shaft on which is chiseled the
famous Babylonian legal code. In the Louvre, Paris.

immediately disappear; it lingered on, subject to certain restrictions. For
instance, revenge might be taken at the moment of the commission of a
crime, or with the consent of the court, or in case of the murder of a close
relative or the violation of marriage laws. In time, it tended to disappear,
but vestiges still remain in the so-called unwritten law.

In the early days of public control of crime the punishment of those convicted, since crime now was viewed as a challenge to civil and ecclesiastical
power and an affront to the king's station, was practiced rigorously. The
most brutal forms of corporal punishment were meted out for crimes.

In the history of early criminal jurisprudence, certain points must be remembered: the crimes that were considered a danger to the public and that

were punished by the local group were those that exposed the group to spiritual or human enemies, particularly the former. Crimes against persons were not controlled by the tribe or the family but by the clan under the principle of blood feud. Under the unrestrained action of this principle, responsibility and retaliation for crimes were collective on both sides, and the intent of the offender was ignored. Worse than all else, the clan revenge failed to put an end to the affair and furnished the means of keeping up a perpetual feud between clans.

Important in understanding the evolution of punishment is the fact that most early punishments, outside of exile and composition, were some form of corporal punishment. This set the general pattern in punishment for crime until the end of the eighteenth century. Humanitarianism and the resulting reform of the criminal codes gradually substituted imprisonment for corporal punishment for all except capital crimes and those punished by fines.

Corporal Punishment

Flogging

Not until the time of the American Revolution did western civilization begin seriously to consider the substitution of imprisonment for corporal punishment.

The most widely employed form of corporal punishment, flogging, has a sordid history. It has been one of the most popular methods of punishing crimes and has been almost universally used to preserve family, domestic, military, and academic discipline.

The whipping post may still be seen in Delaware, where it is not yet a historical curiosity.[3] The laws of that state call for a given number of lashes for certain offenses to be administered by the warden in the state prison near Wilmington or in the county workhouses. However, in recent years the whipping post has lapsed into disuse. The last flogging took place in that state on June 16, 1952, the victim receiving 20 lashes for breaking and entering. Maryland, too, used the whipping post for "assault on wife" or wife beating from colonial days until its repeal in April 1953. The last flogging in Maryland was in Ann Arundel county in 1948.

The instruments and methods of flogging have varied greatly. Straps and whips with a single lash have dominated in schools, but short pieces of rubber hose are highly effective since they leave little traces of the flogging. Especially cruel forms of the lash have been most popular in flogging criminals. The "cat-o'-nine-tails" was a lash made of nine knotted thongs of rawhide attached to a handle. Even more fiendishly cruel was the Russian knout, constructed of a number of dried and hardened thongs of rawhide,

---

[3] For the history of this institution, see Robert G. Caldwell, *Red Hannah* (Philadelphia: University of Pennsylvania Press, 1947).

A whipping post in Delaware as it appeared in 1897. (Courtesy Robert G. Caldwell from his book "Red Hannah.")

interspersed with wires having hooks on their ends that would enter and tear the flesh of those being flogged. Death usually resulted from severe floggings with the knout.

Mutilation

Another type of corporal punishment was mutilation. This procedure was early employed in connection with the *lex talionis* (literally, "law of retaliation"). It directed that punishment be inflicted by a method that duplicated the injury originally inflicted. If a person cut off the hand of another, he lost his hand, and so forth. As mayhem or mutilation was a common crime in some primitive and early historic societies, the infliction of mutilation according to the principle of the *lex talionis* had a wide application. This type of punishment is called poetic justice, though it happened more often in real life than in fiction or poetry.

Another justification of mutilation as a punishment appears to have been the desire to prevent the repetition of a particular crime. Thus thieves and counterfeiters had their hands cut off, liars and perjurers had their tongues torn out, spies had their eyes gouged out, sex criminals had their genitals removed, and so forth.

Even more horrible was the infliction of extensive mutilation for the purpose of creating a deterrent example and producing a gruesome object-lesson to discourage other potential criminals. Canute, King of Denmark and of England, following Danish custom, prescribed almost incredible forms of mutilation in his English forest laws of 1016, and it was the regulation penalty for poaching game in the royal preserves under William the Conqueror and his son. The Danes practiced mutilation with greater severity than the Saxons. It was not uncommon for offenders to have their eyes plucked out, their noses, ears, and upper lips slit, their scalps torn away and, as one might expect, death inflicted in the most horrible manner.

Branding

The branding of prisoners of war, slaves, and criminals was common in late oriental and classical societies. The Romans branded criminals with some appropriate mark upon the forehead. This penalty was quite common in England but was abolished in 1829.

Branding was frequent in American colonial jurisprudence and criminal procedure. In the East Jersey codes of 1668 and 1675 it was ordered, for example, that the first burglary offense was to be punished by branding with a "T" on the hand, while the second offense was to be punished by branding "R" on the forehead. In the colony of Maryland blasphemy called for branding with the letter "B" on the forehead. The adulteress of Vermont, as well as in other parts of New England, was forced to wear the symbolic "scarlet letter."

Stocks and Pillory

The stocks and pillory were used as a method of corporal punishment in early modern times. The stocks held the prisoner, sitting down, with his feet and hands fastened in a locked frame; and the pillory held him, standing, with head and hands similarly locked in the frame. The pillory was not abolished in England until 1837.

When the pillory or stocks were employed and not accompanied by any other mode of punishment, their operation was chiefly psychological, as they exposed the culprit to public scorn. But if the prisoner or his offense were unpopular, he was likely to be pelted and stoned.

The victims might also be whipped or branded while held fast in these engines. Sometimes their ears were nailed to the beams of the pillory and when released they would be compelled to tear their ears loose or have them cut away carelessly by the officer in charge. The pillory was still in use in Delaware as late as 1905.

Other Sundry Punishments

Confinement in irons was a common and brutal form of punishment. A prisoner might be confined in his cell, both hands and feet fastened by heavy chains to the sides, ceiling, or floor. It was not uncommon for prisoners to be chained in a reclining position upon bars of iron and left in such a position for days or weeks.

Chaining prisoners together at work or in cage wagons at night was practiced for many years in southern prison camps of the United States. Handcuffing inmates to their cell doors is a practice in some contemporary prisons. Most American prisons still have irons, handcuffs, or chains to be used on occasion.

The ducking-stool was a corporal punishment for village scolds and gossips. It was a device in which the victim was strapped to a chair, fastened to a long lever, and then dipped in the water by an operator who manipulated the affair from the banks of the stream or pond while a crowd would jeer at the culprit.

Other punishments for gossips, termagants, and "viragoes" were the old-fashioned brank and the gag. The brank or dame's bridle was an iron framework placed on the head and enclosing it like a cage; it had in the front a plate of iron, sharpened or covered with spikes, that was placed in the mouth of the victim in such a way that if she moved her tongue it was bound to be injured.[4]

The Quakers, or members of the Society of Friends, were the only considerable religious group in Europe in early modern times who discerned any discrepancy between the Christian religion of love and the infliction of brutal corporal punishments upon their fellow men. They were greatly repelled by the disgraceful cruelties and shedding of blood and protested vigorously against them. Unfortunately, they were greatly outnumbered in Europe and could do little beyond raising their voices against the barbarities of the age. In America, however, they controlled for a time the destinies of two English colonies, West Jersey and Pennsylvania. Here they introduced for the first time a criminal code that dispensed with corporal punishment, except in the case of murder. To the Quakers, then, we owe the first successful and significant protest against the savagery of corporal punishment. The Quaker protest resulted in the substitution of imprisonment for corporal punishment.

---

[4] For an adequate survey of corporal punishments, see: William Andrews, *Bygone Punishments* (London: William Andrews, 1899); Alice Morse Earle, *Curious Punishments of Bygone Days* (Chicago: Herbert S. Stone, 1896).

19

The Transportation
of Criminals

Criminals have always been judged a fair subject of hazardous experiments, to which it would be unjust to expose the more valuable members of the State. If there be, therefore, any terrors in the prospect before the wretch who is banished they are no more than he has a right to expect; if the dangers of a foreign climate be considered as nearly equivalent to death, the devoted convict naturally reflects that his crimes have drawn on this punishment, and that offended justice . . . does not mean to seat him for life on a bed of roses. (*From* Introductory Discourse on Banishment, *1787.*)

RIGHT HON. WILLIAM EDEN

The Origin of Transportation

The antecedent of the practice of transporting criminals to faraway places was banishment. Sending an unwanted person into exile was a common practice among primitive peoples, but it was usually carried out with certain religious or ritualistic ceremonies not found in the modern systems of transportation. Banishment was often tantamount to death and even in medieval ages carried a certain humiliation. Shakespeare depicts his Romeo as quite distracted with grief when he was exiled from Verona to Mantua, scarcely 30 miles away. Exile as a device gradually merged into outlawry with the earlier religious element largely supplanted by a political motive.

In oriental and classical times, criminals were sent away to work on galleys or in the mines. They were used at onerous tasks that were most frequently shunned by free men. During the Middle Ages and down to the time of Elizabeth, many criminals sentenced to death as well as captured outlaws were pressed into galley service. The famous fictional character, Jean Valjean, in Victor Hugo's *Les Miserables,* has served a long sentence in the galleys and is returning with the telltale yellow card in his pocket when the story opens.

The plight of those unfortunate enough to become galley slaves was about as terrible as could be imagined. They were chained to their seats and compelled to work and live like beasts of burden upon the threat of the most severe floggings. A vivid and terrible picture of the life and condition of the French galley slave has been preserved in the narrative of a French Huguenot, Jean Marteilhes, who with many coreligionists became "galley slaves for the faith," after the revocation of the Edict of Nantes in 1685. His description is paraphrased by Major Arthur Griffiths, a British prison administrator, as follows:

> The prisoners reached the sea-port on foot, traversing a large part of France in scanty clothing, and chained by the neck in large gangs. Once drafted on board ship, and posted to his bench, he remained there always, unless taken to the hospital or the grave. Six slaves, chained to the same bench, tugged at each oar, which was some fifty feet in length; they were compelled to keep time with the others before and behind, or they would have been knocked senseless by the return stroke. They rowed naked to the waist, partly to save clothing, still more to offer their backs to the thongs of the *sous-comités* or quarter-master, who flogged freely, and backed up every order with each stroke. When not rowing they sat at their benches at night, and slept where they sat. In all naval engagements of those days the oars were shot at first, hence the galley-slaves suffered first and most, and were often decimated while the garrison and crew escaped untouched.[1]

By the close of the sixteenth century the obsolescence of the galley and the increase of criminals presented a real problem, particularly in England. Even the authorities of those days could not bring themselves to a thoroughgoing application of the death penalty then prescribed for the great majority of crimes. They decided to ship criminals overseas, motivated more at first by the desire to rid the country of its criminals than to provide the colonies with manpower, for the first legislation ordering deportation came before England had established any permanent colonies in America.

The economic argument for the transportation of criminals was that they provided cheap labor in new countries. Hence, landed gentry staking out large claims in virgin lands often resorted to forced labor. But after a few generations even this alluring fallacy was exploded, since transportation of men against their will has always been executed at a terrific social cost. It took over one hundred years for England to realize her mistake, and France, at least prior to her defeat by the Nazis, was blind to the degrading effects

---

[1] *Secrets of the Prison House,* Vol. 1 (London: 1894), pp. 164-165. Oliver Goldsmith translated Marteilhes' *Memoirs of a Protestant* in 1758. For additional descriptions of the galleys, see George Ives, *History of Penal Methods* (London: 1914), pp. 104-105.

of the practice of maintaining penal servitude in the jungles of the South American islands off the coast of Guiana.

The English Experience in America

The English practice of transporting criminals to the American colonies has been all but forgotten by Americans. Yet from 1597, when the first law authorizing deportation was passed, until its forced termination in 1776, due to the Revolutionary War, a vast number of criminals and debtors was shipped to these shores. How many were sent is not definitely known but estimates by reliable students vary between 15,000 and 100,000.[2] Margaret Wilson, in her devastating book, *The Crime of Punishment,* estimates that only about three to four hundred malefactors were sent here annually, most of them to Virginia and Maryland. She says that some ten thousand servants were sent from Bristol alone between 1654 and 1679; that Oliver Cromwell "barbadoed" so many people that the name of the island to which they were sent became a verb; that from the Bloody Assizes of 1685 eight hundred and forty-one rebels were sent to the West Indies. This English writer maintains that American historians have minimized this factor in our population. She reports that "one historian of Virginia says that there is no record of any Virginia offender ever having left an offspring!"[3]

Probably fewer dangerous criminals were sent to the colonies after the general introduction of Negro slave labor, although the more respectable poor element continued to arrive until the slave trade ceased. Benjamin Franklin is said to have remonstrated with the London government against sending felons to America. When he was told it was absolutely necessary he replied by asking the ministers if the same reasoning would not justify the Americans in sending their rattlesnakes to England.[4]

The separation of the 13 British colonies from the mother country put an end to the transportation of criminals to America. The cessation of deportation played temporary havoc in England. Criminals piled up in the jails where heretofore they had merely awaited the arrival of boats to carry them to the colonies. Something had to be done immediately. A few were sent to Africa, but they perished quickly when exposed to the tropical climate and to indigenous diseases. Then many criminals were placed in old boats or "hulks" anchored in the rivers and harbors of the British Isles.

---

[2] *Journal of Prison Discipline and Philanthropy* (October, 1859), p. 14, suggests about 2,000 annually. For a relatively recent work on the subject of transportation of criminals to America, see Abbott Emerson Smith, *Colonists in Bondage* (Chapel Hill: University of North Carolina Press, 1947).

[3] Margaret Wilson, *The Crime of Punishment* (New York: Harcourt, 1936), p. 96. See also Daniel Defoe's *Moll Flanders* (1722). This work describes in fictional form all the features of the system of transportation to America, and tells how a thief and adventuress was rehabilitated in Virginia.

[4] *Journal of Prison Discipline and Philanthropy* (October, 1859), p. 15.

Prisoners returning to hulks after laboring in the arsenal (England).

Some of these were transport ships converted into nautical prisons by erecting great hammock houses on their decks, plugging up the portholes, and dismantling their rigging.

The conditions in these hulks were what one might logically expect from the social ideals and penal methods of the age. In all British penal history, there is scarcely a more sordid, brutalizing and demoralizing episode than that associated with these prison ships. The introduction of the hulks was a measure of stark necessity, and they were designed for temporary use only. However they remained an integral part of the English prison system until 1858.

Brutal punishments, particularly severe floggings, were administered in the hulks. There was little work, and, what there was, was degrading. The more able-bodied were worked on the docks and fortifications. Even where disease and brutality were not rampant, moral degeneration frequently resulted from the promiscuous association of prisoners of all ages and degrees of criminality. Men and boys alike were kept in irons for days. One slight redeeming feature was the rationing of weak beer after particularly arduous work.[5]

In January, 1787, what was thought to be a happy solution of the problem of caring for English criminals was suggested in Parliament. Captain James Cook had discovered Australia in 1770, and now the question was asked: why not send criminals to this wild and new land to prepare the way for regular colonization?

The Transportation of English Felons to Australasia

A brave and loyal officer, Captain Arthur Phillip of the Royal Navy, was chosen to undertake the hazardous duty of commanding the first contingent of felons to the Australasian regions halfway around the earth. He set sail from Spithead, England, on May 13, 1787, in command of eleven vessels, two of them ships of war, with 16 officers, 197 marines, 45 wives and children of officers and men, 552 male and 190 female criminals, several of the latter with child. The voyage took eight months. They had supplies supposed to last for two years, but many important items were forgotten: there was no clothing for the females, many of whom were nearly naked; there was no medicine or ammunition for the marines. Even the marines who went along to guard the prisoners mutinied because of the terrible conditions.

The ships, known as "floating hells," were fitted out specifically for the unholy trade that became so profitable for the greedy ship chandlers. The ships' crews were a depraved lot. They were in the habit of drawing lots for the company of the female felons aboard. Many of the women were not

---

[5] For a recent work dealing with the hulks, see, W. Branch-Thompson, *The English Prison Hulks* (London: Christopher Johnson, Ltd.), 1957.

averse to this, for it took them away from the fetid compartments below ship.

Much of our knowledge of these conditions comes from the reports of Samuel Marsden, senior chaplain for New South Wales. According to Marsden's account,

> The prisoners consisted of the most abandoned persons of all nations: British, Dutch and Portuguese sailors, the polite swindler, and the audacious highwayman, with their female accomplices. They were shipped off in chains; during the passage outward a detachment of soldiers was constantly on guard; and the voyage was seldom accomplished without bloodshed. The secret plots, in which the prisoners were continually engaged, broke out into open mutiny whenever circumstances offered a chance of success; for this purpose a storm, a leak, or a feigned sickness, was readily taken advantage of. When signs of such disturbances showed themselves, the ringleaders were seized and tried in a summary way by court-martial; but the sailors often refused to enforce the sentence, so that it became necessary to compel obedience with loaded muskets.
>
> The hold of a convict ship presented a melancholy picture of human depravity. In the course of the voyage most of the felons survived the sense of shame; the sounds of ribaldry and boisterous mirth, mingled with catches from the popular songs of the day, issued unceasingly from the prisoners' deck; this uproar was ever and anon increased by more riotous disturbances; blows and bloodshed followed; and occasionally the monotony of the voyage was broken by mock-trials among the prisoners, to show that even in the most profligate and abandoned the principle of justice was not altogether destroyed. When a prisoner committed an offense against his fellows, a judge was appointed, advocates were assigned to the prosecutor and the accused, a jury was sworn to try according to the evidence, witnesses were examined, and the prisoner, being found guilty, was sentenced to an immediate and brutal punishment.[6]

The first port of call in the new land was Botany Bay, where they arrived January 18, 1788. Immediately to the north, Phillip found "the finest harbour in the world" which he named Sydney, after the Home Secretary. The unprecedented problems confronting this first commandant are authoritatively and graphically told by Charles Nordhoff and James N. Hall, of *Mutiny on the Bounty* fame, in the novel, *Botany Bay*.[7] By the close of the first year the colony was in deplorable condition. No fresh supplies were received from home, scurvy broke out, and the natives contracted smallpox.

Many years of disorder, confusion, and drift, not to mention extreme suffering and depravity, were to elapse before any semblance of order or decency was to emerge from the tragic penal experiment in the Antipodes. One administrator, Captain King, founded an orphange for female children

---

[6] *Journal of Prison Discipline and Philanthropy* (January, 1859), p. 16, quoting *Life of Samuel Marsden, Senior Chaplain of New South Wales* (London: 1858), p. 20.

[7] (Boston: Little, Brown, 1941; also Permabooks).

and trained them as wives, married them off, and provided homesteads at public expense. In 1808 Captain Bligh, of *Bounty* notoriety, was sent out as governor. The entire colony, both free settlers and criminals, rose in rebellion and locked him in his own house. He was deported forcibly by officers of the New South Wales Corps, a company raised in England. He

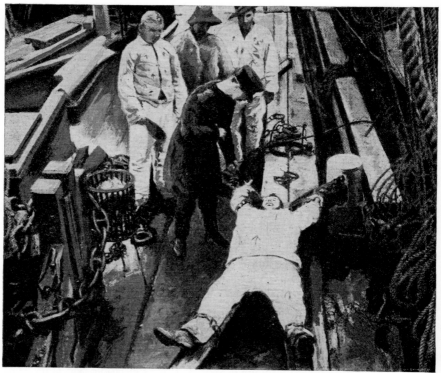

Method of Branding Prisoners on Convict Ship enroute to Australia penal colony.

was succeeded by Captain Lachlan Macquarie, an able and popular governor, who transformed the penal settlement into a colony, founded a bank, appointed an ex-prisoner to the bench, induced the government to send families of criminals out to the colony free of expense, and encouraged young girls who were inmates of English female reformatories to come out to be married off to the prisoners.[8]

Authorities estimate that 135,000 convicted criminals were sent out to Australasia between 1787 and 1875, when the practice was virtually aban-

---

[8] For a novel about an English female prisoner, see Janet Whitney, *Jennifer* (New York: Macmillan, 1941); for an account of the experiment from the standpoint of the Australian aborigine, see Eleanor Dark, *The Timeless Land* (New York: Macmillan, 1941). Also W. S. Hill-Reid, *John Grant's Journey* (London: Heinemann, 1957).

doned. There were several penal colonies in and around Australia. One was made in Van Diemen's Land, now Tasmania, in 1803, and prisoners were allotted to the settlers when free immigration began in 1816. Still another colony was Port Macquarie, in what was to be New South Wales. One of the most notorious penal colonies was at Norfolk Island, about a thousand miles east of Australia. Here, following 1840, Captain Alexander Maconochie carried out a reform program that helped to create the reformatory system.[9]

There was little or nothing to mitigate the sufferings of the prisoners. They arrived in Australia diseased and weakened from the effects of the long sea voyage passed under the worst conceivable conditions. They were put to work at the hardest tasks—clearing the land, mining, and lime burning among others. They were controlled and supervised by overseers who were often brutal and sadistic.

The English transportation system finally came to an end. Gold was discovered west of Sydney in 1851, and large numbers of colonists quickly came. Sheep raising and wheat growing were found profitable. The new settlers resented the prison camps. In 1840 orders had been issued that no more criminals be sent to New South Wales. The system of transportation theoretically came to an end in 1852, but as late as 1890 the home government was still paying for the support of prisoners in Australia.[10]

The French Penal Colonies

France proposed the British scheme of transportation as early as 1791, when it ordered life transportation of all persons convicted of a felony for a second time. Madagascar was selected as the site of the projected colony, but the destruction of the French navy by the English during the Napoleonic wars prevented the execution of this enterprise. In 1851, the proposal was revived during the reign of terror that succeeded the *coup d'état* of December 2. This illegal decree named French Guiana and Algiers as colonies to which prisoners might be sent; in 1854 the area was limited to Guiana.

In May, 1854, the legislation was made definite and legal. Transportation was to replace penal labor in the *bagnes* (hulks or shore prisons that had succeeded the French penal galleys). The hulks at Toulon were, however, used for another 70 years as a depot for prisoners awaiting passage to the colonies.

French criminals were sent not only to French Guiana but also to New Caledonia in the distant Pacific, the first boatload setting off for this island in 1864. New Caledonia, with an area of 8,500 square miles, is situated some 700 miles east of Australia in the South Pacific. It figured in World

---

[9] See pages 417-419 for an account of Maconochie's contribution to corrections.
[10] A good source for information on the Australian system is Eris O'Brien, *The Foundation of Australia* (Westport, Connecticut: Associated Booksellers, 1953).

Convicts praying in irons near Sweat Box on convict ship.

War II as one of the outposts controlled by the Free French forces, and American soldiers were stationed there.

The most degrading and incredibly brutal conditions were to be found in French Guiana and on the islands off the coast. Conditions there were surpassed only by those found in czarist and Soviet Siberian penal camps.

Americans were shocked in 1928 by a powerful and moving book that exposed for the first time the horrors of the French penal camps operated in Guiana.[11] The writer's story concerns a real person, whom she calls Michel, but who is now known by his real name, René Belbenoit, who wrote graphically of terrible experiences.

Cayenne is the capital of French Guiana. Ten miles offshore and 30 miles west of Cayenne are three islands, *Iles du Salut*, Safety Islands. To the condemned they were the Islands of Hell. The largest island is Isle Royale; the next, Isle Saint Joseph; the smallest is Devil's Island. The last is the most famous: Captain Alfred Dreyfus was imprisoned there in 1895-1899. His case stirred the entire world and brought a deserved notoriety to the penal colony. It is said that he was the first prisoner on the island, which is a rock about 1,200 yards in circumference, covered with coconut trees. The currents surrounding it are so treacherous that the only practicable method of reaching it is by a cable slung over from Royale, the island near by.

---

[11] Mrs. Blair Niles, *Condemned to Devil's Island* (London: Jonathan Cape, 1928).

The main penal camps were on the mainland. The islands were used only for the desperate cases, political prisoners, or those who tried to escape and were retaken. The camps were fever-infested, on the edges of the dank jungle, where escape was almost impossible. The men worked at making roads or cutting heavy timber from sunup to dark with no relief.[12]

Almost as many prisoners died in these camps each year as came out from France. It seemed to be the policy of the administration to kill them off. A few of these wretches escaped into the jungles and died; a very few got away, but more often they were reduced to starvation and actually killed and eaten by the others. Most of them were apprehended and returned. Many were betrayed by their comrades for the scant concessions offered to them by the guards. The nearby settlers simulated friendship, worked them unmercifully, and then turned them over for a mere pittance. Those who were recaptured after attempting to escape were sent off to the islands.

In short, the French penal colony was a more brutal duplicate of those formerly operated by the British in Australia. These vicious colonies were finally abolished during World War II by the Free French government.[13]

Russian Penal Transportation

By all odds the most brutalizing and degrading transportation system ever perpetrated, worse even than French Guiana, was in Siberia under the Czars. One need only read Dostoevksi's *House of the Dead,* Tolstoi's *Resurrection,* and other Russian classics to be sure of this.

Two types of exiles were most numerous in the Siberian system. The first embraced those who had forfeited all civil rights, and the second, those who retained their rights and citizenship but were undergoing long terms of penal servitude. Those belonging to the first category were considered dead. They lost their civil and political rights, their property was distributed to their heirs, and they could never hope to return home. If his wife and family did not wish to accompany a prisoner into exile, the wife could remarry. This class was distinguished in Siberia by having their heads half shaven. The second group were considered quasi-colonists. After serving their term, they could return to European Russia or remain in Siberia. Their lot was hard, but in many cases they could live outside the prison camps and work for the government.

The brutality of the Russian exile system started with the long grueling trek from European Russia to the various penal camps located in the min-

[12] For details see René Belbenoit, *Dry Guillotine* (New York: Dutton, 1938), p. 57.
[13] For details see: Clarence W. Hall, "The Man Who Conquered Devil's Island," *Reader's Digest,* Vol. 50, No. 299 (March, 1947), pp. 89-93; Aage Krarup-Nielsen, *Hell Beyond the Seas* (New York: Dutton, 1940); George J. Seaton, *Isle of the Damned* (New York: Farrar, 1951).

ing areas of remote Siberia. The elder George Kennan, in his definitive work, *Siberia and the Exile System,* graphically described the horrors of this forced migration. Between 1823 and 1877, a period of 54 years, the total number of exiles sent to Siberia amounted to 772,979.[14]

The prisoners worked from 7 o'clock in the morning until 5 in the evening, in winter and from 5 to 7 in summer. Much silver and gold was extracted from the Siberian mines by the prisoners, but little profit was obtained, since the system was very expensive, and much precious metal was mined clandestinely and smuggled out of the country. Kennan said that the expense of the Kara diggings alone ran to $250,000 annually.

The Siberian exile system was especially severe on the political prisoners, regardless of what type they might represent, or, for that matter, whether they were guilty of subversive activity or not. Periodically, a disciplinary lesson would be inaugurated by the authorities of the mine-prisons to "give the political convicts a lesson," and to "reduce the prison to order." This was usually done by depriving the prisoners of the few privileges they had such as their money, books, underclothing, bedding, and anything not furnished by the government to common criminals of the penal-servitude class. Later, the political prisoners would be separated into small groups and redistributed among the various prisons.

The czarist system of sending prisoners to Siberia has been continued by the Soviet Government on an even greater scale and, apparently, with just as much brutality. There is, however, one difference. Under the Czars a large portion of those exiled were criminals in the conventional sense. Under the Soviet rule, most of those sent to Siberia have been political prisoners. The first large-scale deportation was that of the kulaks, or rich farmers, who were disrupted when Stalin started to socialize and mechanize the Russian farms. Many of the kulaks were slaughtered, and others were sent off to Siberia. In the political purges of the mid-1930's many of those who were not executed were bundled off to Siberia. When the Russians occupied the Baltic States, Poland, and Czechoslovakia many of those suspected of hostility to the Soviet regime were deported to Siberia. It is alleged that even a considerable number of German prisoners of war were sent there.

There has been much controversy over the treatment of those sent to Siberia and herded into Soviet labor camps. David J. Dallin, and more recently Margarete Buber and Elinor Lipper, have written "horror books," describing the conditions as even worse than under the czars.[15] Soviet sympathizers have sought to discredit such tales. It is certain that the numbers sent today greatly exceed those sent during any comparable period of

---

[14] George Kennan, *Siberia and the Exile System,* 2 vols. (New York: Century, 1891).

[15] Margarete Buber, *Under Two Dictators* (New York: Dodd, Mead 1951); Elinor Lipper, *Eleven Years in Soviet Prison Camps* (Chicago: Regnery, 1951).

czarist rule. The precise figures are not available, but those sent since 1917 certainly run into the millions, perhaps ten million or more altogether. And, after making all qualifications necessary for the exaggerations of the bitterly anti-Soviet writers, it seems likely that the conditions in the Soviet labor camps in Siberia are as brutal and degrading as anything which was the rule under the Czars.

Conclusion

The practice of transporting criminals has been defended by a few eminent criminologists, especially in the past. Both Lombroso and Garofalo, well-known Italian students of the problem of crime, favored it. Lombroso held that it eliminated the hopeless and nonreformable types from the native criminal population and used the less serious offenders for colonizing. Garofalo saw in the practice a means of intimidating the criminal and of increasing the deterrent influence of punishment. But they both deplored the abuses incidental to the system wherever it has been resorted to.

Transportation, with its attenuated gestures of exile and outlawry, still remains in some countries, notably in Latin America, as one of the vestiges of an outmoded correctional philosophy. As may be seen from what we have reviewed, the annals of the practice are among the darkest stains on a supposedly civilized culture. It was one of the most repulsive phases of human activity in dealing with criminals.

Occasionally it is suggested that this type of penal treatment might be used by the United States. "Send criminals to Alaska, or some of our island outposts, anywhere in fact, to get rid of them, the farther away the better." That is the old shortsighted policy—out of sight, out of mind, and an observing penologist has noted that the cruelties connected with such a system were greatest in penal colonies farthest from the home country. Transportation has proved a ghastly failure wherever it has been tried.

20

Capital Punishment

Capital punishment, whilst pretending to support reverence for human life, does in fact, tend to destroy it. (Quoted in Arthur Koestler, Reflections on Hanging, *New York: Macmillan, 1957.)*

JOHN BRIGHT

Introductory Comment

The issue of capital punishment is slowly being resolved. The gradual enlightenment of succeeding generations, the difficulties in many trials of securing juries because of repugnance toward the death penalty, and crusading by various groups against it, have all assisted in the gradual loss of support for this vestige of a cruder and more retributive past.

Society has resorted to many different methods in executing criminals and other allegedly dangerous persons. Drowning, stoning to death, burning at the stake and beheading have all been used in the past. Of all the modern methods of administering the death penalty, hanging has been the most widely used. We read of hangings in the earliest historic literature and throughout the world even today it is still the most widely used. Our colonial ancestors used this method and, following the custom brought over from England, carried it on publicly. In due course these gruesome public spectacles were abolished by state laws. Here is an account of a public execution in Indiana in the late 1850's:

> The day of the execution was dark and cloudy. A slight, drizzling rain fell during the greater part of the morning, and the streets presented an unbroken surface of slop and mud. A large crowd had gathered from the country at an early hour—"arriving by the first trains from the east, west, north and south. Many came into town on horseback and in carriages and wagons," and the streets were thronged with a moving mass of human beings, eager to gratify curiosity.

A rope had been stretched across the enclosure in which the gallows stood, at a distance of forty or fifty feet from it, and guards, armed with muskets, were stationed inside of the rope to keep the crowd from pressing upon the place of execution. Outside of the stretched rope the streets were filled . . . The only incident that diverted the intense interest of this vast multitude was the appearance of a traveling auctioneer, who had selected that time to cry his goods and dispose of his wares.[1]

Here is the bill tendered the sheriff of Philadelphia county by the jailer for supervising the public execution of a prisoner in colonial times:

Jno. Reynolds' Bill   Octr., 11, 1782
      about
Execum of Negro Peter

      William Will, Esq. to John Reynolds Dr.

To making the Gallows & Putting it up	£1..10..0
To a Rope for Negroe Peter	0..10..0
To a Coffin for Ditto	1..10..0
To Cash paid Henry Byrns for Hanging Negroe Peter	3..10..0
To Liquer for the Constables	0..12..6
	£7..12..6

Not content with snuffing out the life of a malefactor, our ancestors quite frequently ordered, as part of the capital sentence, that the lifeless body of the victim be encased in the gibbet, an iron frame curiously wrought to fit the body, and exposed to public view for a period of time.[2] The purpose of this gruesome practice was to provide a warning to others. The movement for the abandonment of public executions began during the 1830's.

It is possible that the last public execution in this country took place in Owensboro, Kentucky, on August 14, 1936 when 22-year-old Ramsey Bethea was hanged. He was convicted of criminally assaulting and killing a 70-year-old woman. News dispatches stated that some 20,000 persons witnessed the execution.[3] Public sentiment was aroused, finally resulting in legislation that called for execution at the state penitentiary by means of electrocution. This was in 1938.[4]

Electrocution was first introduced in the Auburn, N.Y. prison, August 6, 1890, the initial victim being one William Kemmler. This now widespread

---

[1] "Public Executions," *Journal of Prison Discipline and Philanthropy* (July, 1859), pp. 117-123.

[2] For an account of the use of the gibbet in the American colonies, see Thorsten Sellin, "The Philadelphia Gibbet Iron," *J. Crim. Law,* Vol. 46, No. 1 (May-June, 1955), pp. 11-25. This specific gibbet was constructed for the hanging of one Thomas Wilkinson, convicted of piracy. However, he was later reprieved, largely through the petitioning of the patriot, Stephen Decatur, and other influential citizens. This was in 1780.

[3] See photograph on next page.

[4] For the story of hangings, see August Mencken, *By the Neck: A Book of Hangings* (New York: Hasting House, 1942).

**Last public execution in the United States
at Owensboro, Kentucky, August 14, 1936.**
(Courtesy Wide World Photos.)

method had been advocated as a humanitarian move. But, in reality, its original introduction was apparently the result of the effort of an electrical company to market its products.[5]

The first electrocution was generally condemned. The Utica *Globe* declared:

> . . . it is not improbable that the first will prove the last. . . . Manufactured lightning to take the place of the hangman's rope cannot be said to be satisfactory. . . . The men who witnessed the horrible scene Wednesday morning in the death chamber at Auburn prison never wish to be present at another such exhibition. Dr. Spitska, the celebrated expert, who was present, unhesitatingly pronounced the experiment a failure and declared it his belief that the law should be repealed and no more experiments made with electricity as a means of execution.[6]

It has been commonly held that death by electrocution is entirely painless, but a distinguished French scientist, L. G. V. Rota, disputes this contention. Labeling this punishment a form of torture, Rota contends that a condemned victim may be alive for several minutes after the current has passed through his body without a physician being certain whether death

---

[5] So stated by Nicola Tesla in *The New York World,* November 17, 1929.
[6] Quoted from *Correction,* New York Department of Correction (August, 1940).

has actually occurred or not. He adds that some persons have greater physiological resistance to the electric current than others, and that, no matter how weak the person, death cannot supervene instantly. Another attack on the pain of death in electrocution was made by Nicola Tesla, the electrical wizard.[7]

Lethal gas, used by ten states, is certainly painless. It is a relatively pleasant form of meeting death, and humanitarian sentiment would recommend it as a universal method of execution (until capital punishment is abolished). But even this method is revolting to many who are by profession somewhat inured to seeing men die at the hands of the state. The argument is not that the gas is painful to the condemned men but that the spectacle of gradual expiration is "torture to the spectators." There is reason to believe, however, that asphyxiation is less brutal than electrocution, and far less so than hanging.

The former warden of San Quentin prison, Clinton Duffy, a stern opponent of capital punishment, states that the operation of a gas chamber execution includes "funnels, rubber gloves, graduates, acid pumps, gas masks, cheese-cloth, steel chains, towels, soap, pliers, scissors, fuses and a mop; in addition, sodium-cyanide eggs, sulphuric acid, distilled water, and ammonia."[8]

We have been emphasizing the physical pain accompanying executions. Far worse than this agony is the mental torture experienced for weeks, months, and in many cases, years. A condemned man need not give up hope of reprieve until the last second, since in many instances a reprieve actually has been granted that late.[9] It is not uncommon for men and women to sit in the death house for six and eight years before their cases are settled one way or another.[10] This long mental torture is a social paradox. The ma-

---

[7] For an opposite view, see Robert G. Elliott, *Agent of Death* (New York: Dutton, 1940). Elliott, one time executioner for several eastern states, officiated at 387 executions. He maintains that electrocution is painless.

[8] From his series of articles, "San Quentin Is My Home," *Saturday Evening Post*, Vol. 222, Nos. 40-47, March 25-May 13, 1950 (above statement quoted from No. 42, April 8, p. 99). This series was later published in book form as *The San Quentin Story* (New York: Doubleday, 1950).

[9] An interesting bit of historical lore comes from the records of the Pennsylvania Colonial Council Minutes, dated June 6, 1737. Several persons were sentenced to be executed and their petition for a reprieve was turned down, except "The Board being inclined to spare the life of Isaac Bradford, on account of his Youth, yet that his Crime may leave a more lasting impression on him, It is Ordered that his Name be inserted in the Warrant, he be carried with the other Malefactors to the place of Execution, and there receive a Reprieve, to be in like manner signed by the President."

[10] Caryl Chessman in California, the "Fiesterville Boys" in Pennsylvania, and Vern Braasch and Melvin Sullivan in Utah, are typical. All waited in the death houses in their respective states for some eight or more years. The Pennsylvania men (the Darcy case) had their sentences commuted to life imprisonment in 1956 after an interminable legal battle. For the Utah case see Charles P. Larrowe, "Notches on a Chair: Utah Firing Squad," *The Nation*, Vol. 182, No. 15 (April 14, 1956), pp. 291 f. See also, David Lamson, *We Who Are About to Die* (New York: Scribner's, 1935); also, *Reprieve: The Testament of John Resko* (New York: Doubleday, 1956). Regarding the Chessman case, see his own book, *Trial by Ordeal* (Englewood Cliffs, N.J.: Prenitce-Hall, Inc., 1955).

Gas chamber at San Quentin, California, prison.

chinery of justice calls for the death penalty but at the same time insists on due process. Much of this could be eliminated if the death penalty were abolished.[11]

---

[11] The case that attracted most attention in this country illustrating the long and tortuous delay was the famous Sacco-Vanzetti case in Massachusetts in the 1920's. This is a case that should never be forgotten as it is now generally believed that two innocent men went to their death on a charge of murder. Nicola Sacco, a shoemaker, and Bartolomeo Vanzetti, avowed anarchists who had evaded the draft in World War

Just the opposite of this studied care for the condemned man took place in the San Quentin death house on April 7, 1956, in one of the most gory and brutalizing episodes ever to come from a modern scene of capital punishment:

> Husky Robert O. Pierce, 27, was carried kicking and screaming into the lethal gas chamber by five guards after his unsuccessful suicide attempt yesterday. Prison officials said he slashed his throat in a death cell just outside the gas chamber as the prison chaplain . . . was trying to comfort him. He [the chaplain] called the guards who grabbed Pierce and wrestled him into the lethal chamber as blood spurted from his throat.[12]

Another ghastly episode in "The San Quentin Story" occurred on March 15, 1957, when the convicted murderer, Burton Abbott, was actually undergoing execution in the gas chamber. At that very moment Governor Goodwin Knight was frantically trying to get a telephone call through to the prison to stay the execution "for another hour" so the condemned man's attorney could try again to save his client through court action. The trial of Abbott for the murder of a school girl of Berkeley was one of that state's most notorious and widely publicized. Reporters and feature writers from all over the country vied for stories of all the sordid and exciting details.

Yet most wardens attempt to be solicitous of the men—and women— almost hopelessly waiting in "death row." In that same San Quentin prison the Warden has advocated the installation of sun lamps for the use of the doomed inmates, who receive little sunlight. However, his recommendation got little support from his superiors.

I, were arrested on May 5, 1920 for the murder of a paymaster at South Braintree. Their trial was held in May-July 1921, and after many delays and nation-wide agitation, they were executed August 23, 1927. Perjured testimony, mistaken identity and a climate of hysteria regarding "undesirable aliens" are identified with this case. Wrote Judge Michael Musmanno, Pennsylvania jurist: "Each was placed in solitary confinement in a cell isolated from the rest of the world. The door to the cell was not an iron grating, but a thick oaken slab. The walls were solid, light-proof and sound-proof. The ceiling was low. The temperature was 90 degrees. The air was stifling. On July 10 Vanzetti wrote: 'Just think of it! They have persecuted us to death for seven long years. Now they feel positive that we will be executed on August 10 after midnight, and yet they transferred us here, just to deprive us for a month, of a little fresh air and sunlight, of some visits; just to inflict upon us thirty days more of solitary confinement, at the hottest of the summer, in a low, smoky, dreadful place before they burn us to death. And this is enlightened America.' " From a memorial speech by Judge Musmanno. See also his *After Twelve Years* (New York: Knopf, 1939). For other works on this case, see: Felix Frankfurter, "The Case of Sacco and Vanzetti," *Atlantic* (March, 1927), pp. 409-432, calling particular attention to the bias of the judge at the trial; Herbert B. Ehrmann, *The Untried Case* (New York: Vanguard, 1933); Osmond K. Fraenkel, *The Sacco-Vanzetti Case* (New York: Knopf, 1931), which covers the whole case including the report of the Lowell Commission to the governor which did not recommend clemency. The execution is described in Leo Sheridan, *I Killed for the Law* (Harrisburg, Pa.: Stackpole, 1939). For further proof that this case will long be a part of the American ethos, see Robert P. Weeks, *Commonwealth vs. Sacco and Vanzetti* (Englewood Cliffs, N.J.: Prentice-Hall, 1958).

[12] INS dispatch, The Philadelphia *Inquirer* (April 8, 1956).

Then, too, the poignant suffering experienced by the family and friends of those to be executed is another shocking phenomenon too little appreciated by the public. The infamy connected with the execution can never be eradicated from the memory of the survivors of the victim. Children, mothers, wives and other close relatives are suddenly confronted with an overwhelming life crisis that is out of proportion to the wrong committed by the criminal.

Extent of Capital Punishment Today

For many years there has been a dignified movement on the part of thousands of enlightened and humane individuals to eliminate the death penalty. Many countries have abolished it, although most of them with some qualification. This is eloquent proof that society can control its criminal element without resorting to this extreme type of punishment. Survival of the penalty is due either to lethargy or hysteria.

The status of capital punishment in Europe at present is: Belgium (in abeyance) since 1863; Portugal in 1867; the Netherlands in 1850; Switzerland in some cantons as early as 1874, in all by 1942; Norway in 1905; Sweden in 1921; Denmark in 1933; Lithuania in 1922; Italy abolished it in 1898, revived it under the Fascists in 1931, and again abolished it in 1944; Austria abolished it in 1919, but after the Nazis took over it was revived until 1950 when it was again abolished.

Most of the Latin American countries have no capital punishment. Brazil abolished it in 1891 but revived it in 1936 for "extreme cases of perversity only"; Ecuador abolished it in 1895; Colombia in 1910; Argentina in 1922; Costa Rica, Peru, Uruguay, and Venezuela all in 1926; Mexico in 1929. New Zealand abolished it in 1941 but unfortunately revived it in 1950. England has been in the throes of a long drawn-out debate on the issue following a recommendation by a Royal Commission for its abolition. Commons voted to abolish the penalty, but the House of Lords refused. Later a strange compromise was adopted in which only certain types of murderers could be executed.[13]

The following states have abolished capital punishment: Maine, Michigan, Minnesota, North Dakota, Rhode Island, Wisconsin and, in 1958, Delaware.[14] The methods of execution by the states are: 23 and the District of Columbia, electrocution; 8, hanging; 10, lethal gas; and Utah, by vic-

---

[13] Giles Playfair and Derrick Sington, *The Offenders: The Case against Legal Vengeance* (New York: Simon & Schuster, 1957), p. 248, state that only four countries in the world have abolished the death penalty for all crimes. These are Western Germany, Ecuador, Uruguay, and the tiny republic of San Marino. For details see Arthur Koestler, *Reflections on Hanging* (New York: Macmillan, 1957), pp. 54-55; also Appendix I, "The Experience of Foreign Countries," pp. 171-178.

[14] Qualifications are: Michigan, for treason; North Dakota, treason or murder by a lifer; Rhode Island, murder by a lifer.

Year	All offenses				Murder				Rape				Other offenses †		
	Total	White	Negro	Other	Total	White	Negro	Other	Total	White	Negro	Other	Total	White	Negro
All years...	3,568	1,617	1,911	40	(29) 3,096	(18) 1,542	(11) 1,516	38	411	41	368	2	(2) 61	(2) 34	27
Per Cent	100.0				86.8				11.5				1.7		
Per Cent	100.0	45.3	53.6	1.1	100.0	49.8	19.0	1.2	100.0	10.0	89.5	0.5	100.0	55.7	44.3
1957......	65	34	31	-	(1) 54	(1) 32	22	-	10	2	8	-	1	-	1
1956......	65	21	43	1	52	20	31	1	12	-	12	-	1	1	-
1955......	76	44	32	-	(1) 65	(1) 41	24	-	7	1	6	-	4	2	2
1954......	81	38	42	1	(2) 71	(1) 37	(1) 33	1	9	1	8	-	1	-	1
1953......	62	30	31	1	(1) 51	(1) 25	25	1	7	1	6	-	(2) 4	(2) 4	-
1952......	83	36	47	-	71	35	36	-	12	1	11	-	-	-	-
1951......	105	57	47	1	(1) 87	(1) 55	31	1	17	2	15	-	1	-	1
1950......	82	40	42	-	68	36	32	-	13	4	9	-	1	-	1
1949......	119	50	67	2	107	49	56	2	10	-	10	-	2	1	1
1948......	119	35	82	2	95	32	61	2	22	1	21	-	2	2	-
1947......	152	42	110	-	(2) 128	40	(1) 88	-	23	2	21	-	1	-	1
1946......	131	46	84	1	(1) 107	45	(1) 61	1	22	-	22	-	2	1	1
1945......	117	41	75	1	(1) 90	37	(1) 52	1	26	4	22	-	1	-	1
1944......	120	47	70	3	(3) 96	45	(3) 48	3	24	2	22	-	-	-	-
1943......	131	54	74	3	(3) 118	54	(2) 63	1	13	-	11	2	-	-	-
1942......	147	67	80	-	(1) 116	57	59	-	24	4	20	-	7	6	1
1941......	123	59	63	1	(1) 102	55	46	1	20	4	16	-	1	-	1
1940......	124	49	75	-	105	44	61	-	15	2	13	-	4	3	1
1939......	159	80	77	2	144	79	63	2	12	-	12	-	3	1	2
1938......	190	96	92	2	(2) 155	(2) 90	63	2	25	1	24	-	10	5	5
1937......	147	69	74	4	(1) 133	67	(1) 62	4	13	2	11	-	1	-	1
1936......	195	92	101	2	(1) 180	(1) 85	93	2	10	2	8	-	5	5	-
1935......	199	119	77	3	(4) 184	(3) 115	(1) 66	3	13	2	11	-	2	2	-
1934......	168	65	102	1	(1) 154	(1) 64	89	1	14	1	13	-	-	-	-
1933......	160	77	81	2	151	75	74	2	7	1	6	-	2	1	1
1932......	140	62	75	3	128	62	63	3	10	-	10	-	2	-	2
1931......	153	77	72	4	(1) 137	(1) 76	57	4	15	1	14	-	1	-	1
1930......	155	90	65	-	(1) 147	(1) 90	57	-	6	-	6	-	2	-	2

* From National Prisoner Statistics, Federal Bureau of Prisons, No. 18 (February, 1958), "Executions, 1957."

† 21 armed robbery, 17 kidnaping, 11 burglary, 8 espionage (6 in 1942 and 2 in 1953), 4 aggravated assault.

(The figures in parentheses show the number of female prisoners.)

tim's choice of hanging or shooting. Aside from first degree murder which is capital in all states having that penalty, rape and kidnaping are the most prevalent capital crimes, with a few states adding armed robbery, burglary, arson, and train wrecking.[15]

It is significant to note that for the past five years executions in this country have been fewer than one hundred for each year. The chart on page 313 shows considerable information on the extent of capital punishment as well as data on race, sex, and offenses, from 1930 through 1957.

One of the ironies of our current practice is the scrupulous medical care taken of a man or woman condemned to die, if he or she becomes sick. Letting the doomed person die a natural death would cheat the gallows or the electric chair or gas chamber. Much more attention is given to preserving the health and life of condemned persons than to improving the physical condition of prisoners serving a time sentence. It is also one of the quixotic features of capital punishment that a person who becomes insane after being sentenced to death cannot be executed in most states. Presumably, this legal anomaly exists because such a person would not be able to appreciate the lesson of being killed by the state if he is not in his right mind.

How Effective Is Capital Punishment?

The wisdom of capital punishment is still a subject for animated debate and has taken on renewed interest as it becomes increasingly apparent that those condemned men who have the funds can prolong their cases through the courts for years while those who do not have money go silently and promptly to their death.

The writers of this book do not consider the question of capital punishment as of prime significance, from the standpoint either of the conventional jurist or of the most radical scientific criminologist. The two reasons for this rather surprising statement are: *first,* it is a problem that relates to only a minute fraction of the criminal classes as a whole; and *second,* it involves a discussion of issues that are not supportable and have only a historical significance. Not a single assumption underlying the theory of capital punishment can be squared with the facts about human nature and social conduct that have been established through the progress of scientific and sociological thought in the last century and a half. In fact, the whole concept of capital punishment is scientifically and historically on a par with astrological medicine, the belief in witchcraft, or the rejection of biological evolution.

In its origins, the death penalty rested primarily upon the effort to placate the gods, lest their beneficent solicitude for the group be diverted as a result

---

[15] Cf. Leonard D. Savitz, "Capital Crimes as Defined in American Statutory Law," *J. Crim. Law,* Vol. 46, No. 3 (September-October, 1955), pp. 355-363.

of apparent group indifference to the violation of the social codes supposedly revealed by the gods. The complete blotting out of the culprit was a practical demonstration of group disapproval of the particular type of antisocial conduct involved in the case.

Later, with the rise of the metaphysical theories of human conduct, the individual came to be looked upon as a moral agent capable of free choice in every aspect of his conduct, irrespective of biological heredity or social environment. On these assumptions the criminal was inevitably regarded as a perverse free moral agent, who refused to think *right* and who had *wilfully* chosen to do wrong and outrage his social group and the gods. The theory of capital punishment that evolved in this period was revenge. The life of the individual who wilfully wronged the social group or brought serious loss to any one of its members must be forfeited.

More recently the exponents of capital punishment have added to this original and underlying notion of revenge that of deterrence: that the death penalty discourages criminal conduct on the part of those who are aware of the existence and horrors of this mode of treating criminals. They not only assume the overwhelming deterrent influence of punishment in general, but also contend that the death penalty is a far more powerful and effective deterrent than life imprisonment. The eminent dean of law, Dr. George W. Kirchwey, in a speech made in 1923, illustrated the absurdity of this claim:

> On June 21, 1877, ten men were hanged in Pennsylvania for murderous conspiracy. The New York *Herald* predicted the wholesome effect of the terrible lesson. "We may be certain," it said editorially, "that the pitiless severity of the law will deter the most wicked from anything like the imitation of these crimes." Yet, the night after this large scale execution, two of the witnesses at the trial of these men had been murdered and within two weeks five of the prosecutors had met the same fate.[16]

Many other such examples could be adduced to prove the futility of the claim that capital punishment has a decisive deterring effect upon our criminal population.

An interesting story is related by William Andrews in his *Bygone Punishments*. During the early part of the nineteenth century there were many minor offenses in England that called for the death penalty. It was assumed, of course, that such a severe penalty would serve as a deterrent. George Cruikshank claimed that, with the aid of his artistic skill, he was the means of putting an end to hanging for minor offenses. Here is his story, as related by Andrews:

> About the year 1817 or 1818 there were one-pound Bank of England notes in circulation, but unfortunately there were forged bank notes in circulation also; and the punishment for passing these was in some cases transportation for life, and in others death.

---

16 Julia Johnson, *Capital Punishment* (New York: Wilson, 1939), p. 79. Quoted from the *Prison Journal*, Vol. 3, No. 4 (October, 1923), p. 16.

. . . Returning home one night and seeing a number of persons looking up at Old Bailey . . . I saw several human beings hanging on the gibbet, two of whom were women. Upon inquiry I learned the women had [been hanged] for passing forged notes. I at once determined, if possible, to put a stop to this shocking destruction of life for merely obtaining a few shillings by fraud.

I went home and in ten minutes designed a sketch of this BANK-NOTE NOT TO BE IMITATED. William Hone saw the sketch and asked me what I intended to do with it. "Publish it," I replied. Then he said, "Will you let me have it?" I consented and it was published. Mr. Hone had seen many people hanging on the gibbet and when it [the sketch] appeared in the shop windows it caused a great sensation and the people gathered in such numbers that the Lord Mayor had to send the police [of that day] to disperse the crowd. The Bank Directors held a meeting immediately upon the subject and AFTER that, issued *no more* one-pound bank notes, and

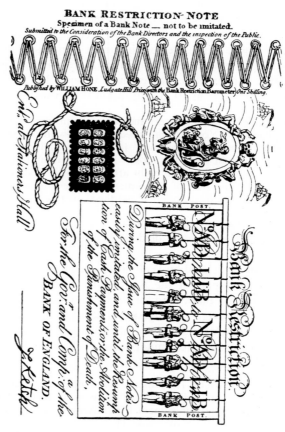

Engraving of Bank Note of England as protest against capital punishment 1818, from William Andrews, "Bygone Punishments," 1899, p. 30.

so there was *no more hanging for passing forged one-pound notes;* not only that but ultimately no hanging even for forgery.[17]

Since we usually associate capital punishment with the crime of murder, the first questions the criminologist asks are: "Why do people commit murder?" and "What types of persons actually kill?" Murderers fall into three main classes. First, there are those who suffer from serious physical, mental, and cultural deficiencies that make it possible for them to contemplate murder as a more or less natural form of conduct. Their point of view is so defective, judged by socially approved standards, or their sense of fellow-feeling and sympathy is so ill-developed, that the compunction against taking human life, which exists in the normal individual, is more or less absent in their case. The second type consists of those who are relatively normal physically, mentally, and culturally, but are subjected to intensely difficult or inciting emotional situations which lead them to commit murder, whereas under normal circumstances they would lead a law-abiding existence. The third group are the professional gunmen, who, in the matter of taking life, bear a close resemblance in their mental habits to members of the standing army. Their attitude toward the taking of human life is very much like that of the soldier on the battlefield; namely, it is taken as a matter of course, not involving any personal responsibility or depravity. The gangster's duty is to obey his superiors. It is obvious that the gunman is the victim of bad social habits that have deprived him of his sense of socially approved responsibility and the normal value placed on human life.

The next logical point at issue is whether the death penalty (under present conditions of police and court) acts as an effective deterrent to these various classes of murderers. It will be apparent at the outset that those who commit murder as a result of psychopathic compulsions or in fits of rage are relatively immune to the deterrent effect of the death penalty. No form of deterrence short of overt physical restraint before the deed, could serve to avert such murders. The same is true of those who commit murder as a result of defective personality or highly unfortunate social environment. Nor can the death penalty be supposed to act as an effective deterrent in the case of the professional gunman. He realizes that his chances of being apprehended for his crime are relatively slight; that the probability of his conviction after arrest is not more than 50 per cent; that he runs a fair chance of being released on a technicality in appeal, even if he is convicted; and, finally, that if he is sentenced to death, he is likely to have this sentence commuted to life imprisonment and may ultimately be pardoned and restored to a life of freedom.

Would the death penalty have a deterrent effect on those who commit murder to settle a deep-seated grudge? Probably not, for any fear would be

---

[17] Abridged from William Andrews, *Bygone Punishments* (London: 1899), pp. 32-34. See cut, page 316. For more grisly stories of hangings in England during this period, see Koestler, *Reflections on Hanging.*

outweighed by the strong pressure to commit murder and the consciousness of a large probability of escaping its application. Those in this group are most vehemently condemned by an outraged public, usually because such murderous acts are fantastic or brutal, being committed under terrific emotional stress. The triangle murders are cases in point; and so are those that terminate a long feud.

The death penalty, even if applied invariably to every apprehended and convicted murderer without any subsequent intervention of commutation or pardon, would not even seriously deter many of those whose personality types would respond to the operation of deterrent influences. The present defective police, detective, and court systems bring about a situation in which much less than half of those guilty of murder are even arrested, while a large proportion of those arrested are unjustly freed through the inefficiency of the court and jury system. There can be no doubt that a murderer would be more seriously deterred by the absolute certainty of relatively mild punishment than by a one-in-ten chance of having the death penalty inflicted on him. A study by Professor Sellin gives these figures:

> There were in the United States in 1930 probably some 10,000 homicides. How many of these were capital crimes it is impossible to say, but in that year there were 1,011 persons committed for life to state and federal prisons and reformatories in 45 states on the charge of homicide, most of them undoubtedly for capital offenses, at least in the 37 of these states where the death penalty exists. In the same year there were 155 executions. . . . Ten states with the death penalty did not execute a single prisoner. Sixty-nine of the executions were in six states: New York, Pennsylvania New Jersey, California, Arkansas, and North Carolina. Little is known of those executed.[18]

It is also quite obvious that the deterrent effect of capital punishment, even if it existed, is largely lost because of the mild modern methods of executing the death penalty. If one desired to make such penalty most effective as a deterring influence, it would seem to be desirable to make a great public spectacle of it and carry it out under the most brutal and degrading circumstances. But as a matter of fact, we know from history that even the most severe forms of capital punishment, publicly administered, have had slight or no deterrent influence. In England in the eighteenth century, hangings were public and almost like carnivals. Yet there was no evidence that the crime rate declined. Indeed, pocket-picking became so common in the crowds assembled to witness the public hangings of pick-pockets that hangings had to be made private.[19]

---

[18] Thorsten Sellin, "Common Sense and the Death Penalty," *Prison Journal,* Vol. 12, No. 4 (October, 1932), p. 12.

[19] The question of the deterring effect of a penalty has long perplexed students. For an analysis of the problem, see John C. Ball, "The Deterrence Concept in Criminology and Law," *J. Crim. Law,* Vol. 46, No. 3 (September-October, 1955), pp. 347-355.

In a study, by the Philadelphia Bar Association in 1951, of 215 persons pardoned after serving varying terms on sentence of life imprisonment for murder, it was found that only 7 were later arrested and of these only 1 was for murder.[20] It is generally accepted that lifers who have escaped the death penalty for murder make the best prisoners and after serving a relatively long prison term keep out of trouble.

One should also mention that capital punishment is more expensive than the high cost of maintaining prisoners, even for long terms, in institutions. Practical legislators who are allegedly looking after the taxpayer's money, might well ponder this seemingly paradoxical situation. But if we figure the cost of imprisonment per year at $1,000 and the average lifer's term as ten years, the total cost is but $10,000. Applied against this is the excessive cost of a court trial that deals with a possible capital offense, plus a *pro rata* charge for the construction of an expensive death house, plus added guard service in the prison. It is obvious, therefore, that the charge against capital punishment is extremely high.[21]

In states where the death penalty has been abolished, the rate of homicide is approximately the same as in states in the same area where the death penalty still persists. In other words, the presence of the death penalty appears to have nothing to do with the amount of homicide. As Professor George B. Vold states, life in Maine is just as secure as in New Hampshire or Vermont, yet the latter states have capital punishment, while Maine does not; similarly, Michigan (with no death penalty) has a homicide rate somewhat lower than the adjoining states of Ohio and Illinois (with death penalty).[22]

A Sane Criminological Perspective on Capital Punishment

The writers of this book do not advance any of the sentimental objections to capital punishment: that it is against the spirit of humanity, that it brutalizes the human intellect, that God alone has the right to take human life. Our position is that there may be convincing arguments for extending the practice of exterminating useless, defective, and dangerous human types. But, except for the immediate satisfaction of disposing of an occasional biological or social misfit, even these extreme eugenic measures are none too hopeful.

The final answer of the scientific criminologist to the exponent of capital punishment is that if we desire to get rid of crime we must adopt the same

---

[20] Summarized in the *Annual Report* of the District Attorney of Philadelphia for 1953.

[21] See Frank Hartung, "On Capital Punishment" (Detroit: Wayne University Mimeographing Department, 1951).

[22] George B. Vold, "Can the Death Penalty Prevent Crimes?" *Prison Journal*, Vol. 12, No. 4 (October, 1932).

scientific attitude that society has taken about the elimination of physical disease. It is absurd to think of punishing a person who is suffering from cancer or any other malignant disease. So is it likewise absurd to punish those who are socially ill to the degree that they commit socially disapproved acts. We must reduce, so far as possible, the unhealthy social environments that generate those bad habits that emerge in criminal conduct, and set up systems of treatment that will rehabilitate reformable convicts. The problem of capital punishment appears to be relatively unimportant in relation to the much broader and more fundamental series of problems that involve crime causation, criminal jurisprudence, and the rehabilitation of criminals.

We are thoroughly opposed to the continuation of capital punishment, and find it hard to concede that a person of any reasonable cultivation and possessed of even rudimentary knowledge of criminology can defend the perpetration of this relic of barbarism. At the same time, it is necessary to view the problem in proper perspective. Capital punishment is not the chief evil of contemporary jurisprudence. The number put to death annually is less than one hundred. The stupid, benighted treatment of some 200,000 inmates languishing in our correctional institutions and another million or more cooped up in county jails, most of which are not fit for human habitation, constitutes a situation infinitely more deplorable than the infliction of the death penalty upon a mere handful of persons who have, at least, presumably had a fair trial in the Anglo-Saxon tradition. Thus, anyone who attempts to represent the abolition of capital punishment as the core of the reform program in criminology is rendering a disservice to the cause of an enlightened criminology.

Capital punishment does have practical significance from one standpoint, however. Capital crimes, trials, appeals, and executions gather considerable publicity as they progress, and perhaps embody the public's ideals of justice more concretely than any other phase of criminal procedure. To persist in executing the worst offenders, then, is to keep alive a symbol of *lex talionis* where it will be most telling—to say, in effect, that the state believes in humanity toward offenders where it can afford to be humane, but that vengeance is the sure cure for the "really important" crimes. Thus the federal laws invoking capital punishment against narcotics peddlers probably indicate, not a greater disposition to execute people, but a desire on the part of lawmakers to draw attention to their exasperation with this type of offense, by calling down upon it the most attention-getting revenge known.

Perhaps, therefore, recent trends in public sentiment against capital punishment represent also a broader realization that correction is more important to society than is punishment. Contrary to what had been presumed to be true, that the majority of the American people are in favor of capital punishment, we find that in a poll conducted in 1958, fifty per cent of those asked for their opinion stated they were opposed to the death penalty, forty-two per cent were in favor, and eight per cent expressed no opinion. It was further revealed by the poll that opposition is strongest among lower economic groups with fifty-three per cent expressing such opposition; whereas, in the upper economic groups, forty-two per cent opposed the penalty.[23]

Encouraging as are the results of this sampling of American opinion, there is still far too much irrational resort to vengeance and retributive justice. Not until scientific concepts of corrections have thoroughly penetrated the educational and religious processes will the American people accept the thesis that brutal methods of dealing with those who flout the law are anachronistic and futile.

If such a time ever comes, those who commit murder or rape or other compulsive crimes may be subjected to a scientific analysis by those trained in the behavioral sciences in an attempt to ascertain why they are predisposed to such antisocial and dangerous acts. In this manner it may be of significant value in understanding and even preventing younger criminals from committing similar acts of violence. The case of Caryl Chessman, mentioned earlier, is relevant. Inured to a life of violent crime, this distorted genius (with an I.Q. of 172) developed, as a young boy, a strongly nihilistic philosophy of life. Efforts are being made to save him from the California gas chamber in the name of science if not humanity. Others who have been executed in the past might have made a significant contribution to science had their lives been spared. In this way they might have atoned for their wrongs in contributing to the understanding of persistent criminal behavior and thus, perhaps, save hundreds of future criminals and their victims from grief.[24]

---

[23] Poll conducted by Elmo Roper; Copyright, 1958, by National Newspaper Syndicate.

[24] For further information on capital punishment, see the symposium, "Murder and the Penalty of Death," *The Annals,* Vol. 284 (November, 1952). See also David Brion Davis, "The Movement to Abolish Capital Punishment in America, 1787-1861," *The American Hist. Review,* Vol. 63, No. 1 (October, 1957), pp. 23-46.

Part II

THE ERA OF REFORM: THE EMERGENCE OF THE CONCEPT OF IMPRISONMENT

21

Reforms in the Criminal Law

As has been well said of the men of the 18th century, they all grazed side by side in the same nourishing pasture. High purpose and moral obligation were the staples of their diet. Albert Schweitzer referred to this century's rationalism as the "greatest and most valuable manifestation of spiritual life the world has yet seen." (From "Introduction," Negley K. Teeters, The Cradle of the Penitentiary, *Philadelphia: Pennsylvania Prison Society, 1955.)*   EDMUND G. BURBANK

Reforms Abroad

It is generally conceded that the eighteenth century was remarkable primarily because it nurtured the noble concept of the "rights of man." The recognition of human dignity and human frailty abounded in many of the European countries but it was essentially in France that we witness its epitome in the writings of such scholars as Montesquieu, Diderot and Voltaire.

Montesquieu, in his *Persian Letters,* satirized the abuses in the existing criminal law, and Voltaire took an active part in the campaign for reform in criminal law and procedure. They both had considerable influence in stimulating the crusading zeal of an Italian, Cesare Beccaria (1738-1794). Beccaria's contribution to penal law is of special interest to us here; his published work is known best by the title *Crimes and Punishments* (1764).[1]

The essential principles that Beccaria recommended, from which rose the so-called "classical" school of criminology are: (1) The basis of all social action must be the utilitarian conception of the greatest happiness for the greatest number (a concept derived from the great English political philosophers). (2) Crime must be considered an injury to society, and

---

[1] See Marcello T. Maestro, *Voltaire and Beccaria as Reformers of Criminal Law* (New York: Colombia University Press, 1942), p. 54.

the only rational measure of crime is the extent of this injury. (3) Prevention of crime is more important than punishment for crimes; indeed, punishment is justifiable only on the supposition that it helps to prevent criminal conduct. In preventing crime it is necessary to improve and publish the laws, so that the nation may know what they are and be brought to support them, to reward virtue, and to improve education both as to legislation and life. (4) In criminal procedure secret accusations and torture should be abolished. There should be speedy trials. The accused should be treated humanely before trial and must have every right and facility to bring forward evidence in his behalf. Turning state's evidence should be done away with, as it amounts to no more than the "public authorization of treachery." (5) The purpose of punishment is to deter persons from the commission of crime and not to provide social revenge. Not severity, but certainty and expedition in punishment best secure this result. Punishment must be sure and swift and penalties determined strictly in accordance with the social damage wrought by the crime. Crimes against property should be punished solely by fines, or by imprisonment when the person is unable to pay the fine. Banishment is an excellent punishment for crimes against the state. There should be no capital punishment. Life imprisonment is a better deterrent. Capital punishment is irreparable and hence makes no provision for possible mistakes and the desirability of later rectification. (6) Imprisonment should be more widely employed but its mode of application should be greatly improved through providing better physical quarters and by separating and classifying the prisoners as to age, sex, and degree of criminality.

Beccaria's conclusions state his viewpoint with clarity and conciseness: "In order that every punishment may not be an act of violence committed by one man or by many against a single individual, it ought to be above all things public, speedy, necessary, and the least possible in the given circumstances, proportioned to its crime, dictated by the laws." Excepting only the modern psychiatric analysis of the criminal, with its substitution of the conception of treatment for punishment, one may safely say that Beccaria's treatise envisioned the major correctional advances made during the next century.

Extensive practical reform of criminal jurisprudence emerged from Beccaria's essay; in Austria under Maria Theresa and Joseph II, by Frederick the Great of Prussia, those carried out by Leopold of Tuscany, the criminal code of the French Revolution, the modification of the barbarous criminal code of England, and the reform of the criminal law in the infant United States after 1776, especially in Pennsylvania under the influence of the Quakers. Catherine the Great was much impressed by Beccaria's work and in 1767 invited him to St. Petersburg to aid her in drawing up a new code of laws. This trip, however, did not materialize. By 1800 the wave of reform had spread all over the Continent, influencing the social thinking of rulers and philosophers alike.

Perhaps the greatest leader in the reform of the criminal law in England at this time was the utilitarian, Jeremy Bentham (1748-1832). His interest in reform was a product of his humanitarianism and his desire to provide more scientific legislation. He was a prodigious writer on all aspects of criminal jurisprudence, penal administration, and many other economic, political, and social matters.

His *felicific calculus*—the idea that to achieve the most pleasure and the least pain is the main object of an intelligent man—is no longer taken seriously as the sole key to human behavior, but he applied his theory to the reform of the criminal law, recommending that penalties be fixed that would impose an amount of pain in excess of the pleasure that might be derived from the criminal act. This prospect of deriving more pain than pleasure from a crime would, Bentham believed, deter men from committing antisocial acts.

Thus Bentham, like Beccaria, looked upon punishment as deterrent and preventive. Pain should not be inflicted unnecessarily, but should "fit the crime." No one type of punishment suited all crimes, hence it was necessary to make a choice from among a wide variety of penalties, to vary and combine them. Restitution should be imposed according to the culprit's ability to pay. Infamy, he believed, was one of the most "salutary ingredients in penal pharmacy."[2]

Bentham's theories were implemented by four leaders in the practical movement that culminated in reforming the archaic English criminal code, with its 222 capital crimes. These four were Romilly, Mackintosh, Peel, and Buxton.

Sir Samuel Romilly (1757-1818), an able Whig lawyer, was the greatest leader in the direct and persistent agitation for the reform of the English criminal code. He devoted his main efforts to this enterprise in the generation preceding his death in 1818. His efforts led to the erection, in 1816, of the first modern English prison, the expensive medieval fortress known as the Millbank prison. Though the inertia and conservatism of his country at the time made it impossible for Romilly to achieve any sweeping results during his lifetime, he launched the program that was carried to success by his followers.

The two men who took up the work of Romilly were Sir James Mackintosh (1765-1832) and Sir Thomas Fowell Buxton (1786-1845). Mackintosh bore the brunt of the fight against the British criminal code after the death of Romilly. Sir Thomas Fowell Buxton married the sister of Elizabeth Fry, the well-known prison visitor and reformer. Like Bentham,

---

[2] For a further analysis of Bentham's contribution to penology see Gilbert Geis, "Pioneers in Criminology: Jeremy Bentham," *J. Crim. Law,* Vol. 46, No. 2 (July-August, 1955), pp. 159-171.

he was interested in humanitarianism, especially the reform of the criminal law and prison conditions.[3] The immediate legislative leadership in behalf of the reform of the criminal code devolved primarily upon Sir Robert Peel (1788-1850) who established the first modern police system. Thanks in large part to Peel's efforts, the program of Bentham, Romilly, Mackintosh, and others was formulated and ultimately engineered through Parliament in piecemeal installments.

As a result of the activity of these men and their supporters, the English criminal code was completely transformed between 1820 and 1861. In 1822, the death penalty was removed from some 100 petty offenses and by 1861 that penalty remained solely for murder, treason and piracy. This trend toward a less sanguinary concept of punishment had its counterpart earlier among the Quakers in America. It is undoubtedly true that correspondence between the reformers on both sides of the Atlantic was largely responsible for this more humanitarian attitude toward criminals.

Cultural and economic factors also influenced the development of a new humanitarian attitude toward man. The rise of mercantilism, the development of scientific improvements relating to navigation, and the discovery and colonization of new lands all contributed to an increase in commercial activity during the 16th, 17th, and 18th centuries. There arose concurrently a new attitude that dictated the preservation of human life to augment the available labor supply. In a stimulating work on the subject, Rusche and Kirchheimer have presented a clear idea of how changing social and economic systems fundamentally altered the ways of thinking and acting in relation to crime and penalties.[4]

The Quaker Influence in the American Colonies

The most important aspect of the reform of the criminal codes lay in the fact that these changes gradually substituted imprisonment for corporal punishment. Since it is generally conceded that Pennsylvania institutionalized imprisonment and its practical agency, the prison, the evolution of criminal law reform in that colony and state is especially relevant.[5]

From 1676 the American colonies came under a code of laws compiled at the direction of the Duke of York to include territories recently taken from the Dutch. Some of the component laws had been enacted as early

---

[3] For a picture of crime in eighteenth centry London, see Patrick Pringle, *Hue and Cry* (New York: Morrow, 1956).

[4] Georg Rusche and Otto Kirchheimer, *Punishment and the Social Structure* (New York: Columbia University Press, 1939).

[5] For the reforms in the Pennsylvania criminal codes that led to the rise of imprisonment, see H. E. Barnes, *The Evolution of Penology in Pennsylvania* (Indianapolis: Bobbs-Merrill, 1928), Chapters II-III; for the earlier West Jersey reforms, see H. E. Barnes, *A History of the Penal, Reformatory and Correctional Institutions of the State of New Jersey* (Trenton, N.J.; McCrellish and Quigley, 1918), Chap. I.

as 1664 and were referred to collectively as the Hampshire Code. Both codes, old and new, recognized many crimes as capital; lesser offenses brought corporal punishments—branding, whipping, the stocks, and the pillory—or heavy fines.

When William Penn landed at Chester with a royal charter to found a new colony, he brought with him a new legal corpus that had been agreed upon in England. The 61 chapters of this "Great Law," or "Great Body of Laws," were adopted by his first Assembly between December 4 and 7, 1682, and included the original Quaker criminal code. Actually, much of its substance had been incorporated into the constitution of West Jersey, also founded by Quakers; for example, the West Jersey laws for the first time in history prescribed imprisonment at hard labor for the great majority of serious crimes. The constitution had been drawn up in 1677 and confirmed in 1681, but its importance in the later history of criminology is not so great as that of the Pennsylvania laws of 1682.

The Great Law opened with a guarantee of liberty of conscience and freedom of worship to all who acknowledged the existence and attributes of the one true God. Gone were the endless enumerations of religious offenses that marred the criminal jurisprudence of the other colonies and of most European countries. In the main, crimes of violence were punished by imprionment at hard labor in the "house of correction." Neither murder nor manslaughter was treated in the Great Law, but both were included in a later supplement, which provided that manslaughter was punishable according to the nature and circumstances of the act, and premeditated murder, by death.

The wide reliance upon imprisonment as a punishment was epoch-making in criminal procedure. Excepting the laws in West Jersey, which are in any event derived from a common source, this Pennsylvania Code of 1682 is the first instance in criminal jurisprudence of prescribing imprisonment at hard labor as a punishment for a majority of the acts branded as serious crimes by the community.

Although there were many additions and some changes in details, this Quaker code of 1682 persisted as the basis of criminal procedure in the province until Penn's death in 1718. Through this period there was a noticeable change for the worse under various governors, however, and penalties gradually became more severe. Penn was suspended from power in 1692, and the Anglican Code was introduced in 1718, by which time influence from abroad and from other colonies crept into the Pennsylvania colony and all but nullified the milder Quaker philosophy.

The act of May 31, 1718, which once more brought the English criminal code into operation in Pennsylvania, resulted from the Quaker refusal to take an oath. Down to the beginning of the reign of George I in 1714 the Quakers had normally enjoyed the right to affirm their statements instead of swearing them, but in 1715 Parliament passed an act that forbade affirmation in criminal procedure, and this act was extended to the colonies. Governor Keith, appointed in 1716, advised the Pennsylvania Assembly that the best method of inducing the Crown to grant the right of affirmation was to adopt the English criminal code. This advice was accepted and the bill became law on May 31, 1718. By a strange coincidence, William Penn died the day before this *coup de grâce* was given his humane criminal jurisprudence.

Though the right of affirmation was guaranteed, this concession was more than offset by the complete adoption of the criminal code of England and of the other American colonies. The new code was even more rigorous than the Duke of York's laws. It listed 13 capital offenses. Larceny was the only felony not made a capital crime; it was punishable by restitution, fine, whipping and imprisonment. Counterfeiting was made a capital crime by the laws of September 21, 1756 and February 21, 1767 and in this form the sanguinary code remained until after the Revolution. The practice of whipping, branding, and mutilation for the punishment of lesser crimes, which was already growing up by 1718, was continued throughout the period.

The new Pennsylvania state constitution of September 28, 1776, directed a speedy reform of the criminal code. The absorption of attention and energy by the military struggle with England prevented any immediate action, but after the war ended, an act of September 15, 1786 first aimed to carry out the reforms.

There were two main reasons for reforming the provincial criminal code when Pennsylvania obtained her independence. The first was the feeling that the code of 1718 was not a native product, but rather the work of a foreign country forced upon the province by taking advantage of its religious scruples. This sort of reaction against English criminal jurisprudence was one of the first manifestations of the national spirit in 1776.

The second cause of reform was the growth of enlightenment and criticism abroad. The movement represented by Montesquieu, Voltaire, Diderot, Beccaria, Paine, Bentham, and others had affected the leaders of colonial thought, particularly in Pennsylvania, to such an extent that reform would probably have been inevitable without any strong patriotic impulses.

The revision of the Pennsylvania criminal code to conform with the ideals set forth in the Constitution of 1776 did not actually begin until a law was passed on September 15, 1786, which stipulated imprisonment at hard labor on the public streets and highways as the legal punishment for the majority of crimes hitherto punished by capital or lesser forms of corporal punishment. Further reforms were provided for in laws passed on April 5, 1790, and September 23, 1791, which reduced the list of capital crimes to murder in the first degree and ordered imprisonment at hard labor for all other serious crimes.

22

The Genesis and
Development of the Penitentiary

*While society in the United States gives the example of the
most extended liberty, the prisons of the same country offer
the spectacle of the most complete despotism.* (*From* On the
Penitentiary System in the United States, *1833.*)

GUSTAV DE BEAUMONT AND ALEXIS DE TOCQUEVILLE

Introductory Remarks

At the present time approximately 200,000 men and
women are locked up in more than 200 penitentiaries, reformatories, and
correctional institutions in the 48 states and the various facilities operated
by the Federal Bureau of Prisons.

Prior to 1800 there were no prisons, aside from county jails and houses
of correction which were used only for minor offenders or those awaiting
trial. The story of the advent and development of the penitentiary for the
treatment of serious offenders is a fascinating one.

Few realize that this country gave the modern prison system to the
world. Fewer still know that it had its origin in the state of Pennsylvania,
due primarily to the humanitarian zeal and ingenuity of the Quakers. The
first American penitentiary was the Walnut Street Jail in Philadelphia,
designated such by the Act of Assembly of April 5, 1790. This small structure was originally built to serve as a detention jail for the county of
Philadelphia, but following the act of 1790 it became a penitentiary-house
for all convicted felons of the state, aside from those sentenced to death.[1]

There have been prisons and dungeons for thousands of years, but prior
to the eighteenth century they were seldom used to incarcerate convicted

---

[1] See Negley K. Teeters, *The Cradle of the Penitentiary* (Philadelphia: Pennsylvania Prison Society, 1955).

felons.[2] Today we use the term "prison" to denote all places of restraint or detention of those either suspected or convicted of offenses contrary to law. The jail is the oldest of modern places of imprisonment and was used originally as a place of detention for those awaiting trial who were unable to obtain bail. Later petty offenders were sentenced to jail for short periods of time.[3] Debtors were also thrown into the common jail. The workhouse, or house of correction was the second prison to be conceived. This institution has always carried the connotation of punishment of petty offenders, vagrants and ne'er-do-wells through imprisonment at hard labor. We shall discuss this type later.

The "penitentiary" is the word applied to an institution designed to restrain for a long period of time convicted felons, or those guilty of serious offenses. The word is derived from the root words for "penitence" and "repentance," and it still denotes an ecclesiastical office concerned with the absolution of guilt. It is reputed to have been first used in connection with penal treatment by the English reformer, John Howard (1726-1790). "Penal," incidentally, is from the same Latin root, but belongs to the branch that has to do with punishment, pain, and revenge (cf. "penalty"); as we have seen, this word and its attendant concepts are not yet outdated by criminological practice. In our country we find at least one penitentiary in each of the forty-eight states (some of the larger ones have two or even more) and there are six in the federal system.[4] The term is also applied in most foreign countries to institutions where serious offenders are sent for long periods of imprisonment.

Another type of prison, developed during the years following 1870, is the "reformatory." It is used primarily for youthful offenders between the ages of 16 and 30.[5] Prisons set aside solely for women felons are also generally referred to as reformatories.[6] In more recent years the Federal Bureau of Prisons and some states have created another type of prison known as the "correctional institution." Persons are sent to these from penitentiaries when they have shown promise of rehabilitation, or they may be sent direct from the courts because their prognosis for reform

---

[2] It has been generally accepted that the concept of imprisonment as a penalty for convicted felons is relatively modern. However, research being carried on indicates that in certain countries and for certain offenses imprisonment was employed quite early. See Max Grünhut, *Penal Reform* (London: Oxford University Press, 1948), pp. 11-13. Marvin Wolfgang has found that many persons were sentenced to specific terms in the Florentine prison *Delle Stinche*, erected as early as 1300. Records indicate that there was a system of classification, a board of inspectors, recreational and work programs, that prisoners were segregated and kept separate, and that there were both definite and indefinite sentences. (In letter to one of the writers, dated May 22, 1958.) This prison is described in favorable terms by John Howard in his *State of Prisons*.

[3] See Chapter 24 for a discussion of county jails.

[4] About half of the states still refer to their state prisons as "penitentiaries."

[5] See Chapter 26 for discussion of the Reformatory.

[6] See Chapter 25 for discussion of women's prisons.

seems favorable, or because they are short-term offenders. All of the above are often referred to, even by experts, as prisons.[7]

The architecture of our prisons is so familiar that most of them can be recognized readily by any citizen. The prison plant is usually surrounded by massive walls or by high cyclone fences. Within the enclosures are long, forbidding cell blocks that house hundreds or thousands of inmates in small cubicles or crowded dormitories.

It is deeply imbedded in the social thinking of the public that all convicted criminals should be sent to prison. They look upon such persons as menaces and are relieved when a criminal is locked up. They hope that those held within the walls are securely guarded, because only then can society be protected. But the average citizen lives in a world of illusion, so far as his general notions of prisons and criminals are concerned. He has accepted the thesis that every person who commits a crime is dangerous and should be sent to prison to be punished as well as to deter others from committing offenses. He also fosters the illusion that the criminal will return to society properly chastened and willing to settle down, a contrite sinner, and become a law-abiding citizen. These views are woefully naïve; the purpose of penology is to correct such all too prevalent views.

The Workhouse, or House of Correction, Forerunner of the Penitentiary

After the breakup of the feudal system, pauperism and mendicancy abounded in all of Western Europe. Feudal barons disbanded their mercenary armies, and the soldiers, who had never worked at gainful occupations, wandered at large, congregating in the towns and cities. The suppression of the monasteries and the decline of the guilds also bred thousands of beggars and paupers who would not work. It was to meet this critical condition that the city of London established a workhouse in 1557, when an abandoned royal lodging was set aside for the purpose. The Bridewell, as it was called,[8] was used by the authorities to cope with the problem of caring for and disciplining the ever-increasing number of pillaging riffraff who were preying on society.

The philosophy of this establishment was that those sent there would be deterred from leading a life of wantonness and idleness by being forced to work at hard and disagreeable tasks.

> There was a spinning-room, a nail house, a cornmill, and a bakery. . . . Inmates made their beds and swept their rooms. The prisoners were paid for their labor. . . . Gradually the range of occupations increased. Women mended, men dredged sand and burned lime to make mortar. . . . By

---

[7] In addition there are prison camps, forestry camps, mobile units, large prison farms and special detention units in many states and the federal system.

[8] The word comes from a well nearby known as "St. Bride's well."

1579, twenty-five occupations were practiced in Bridewell including the making of pins, silk, lace, gloves, felts, and tennis balls. Besides this discipline of work other punishments were used such as whippings, restriction of diet, and torture.[9]

There was a strong economic motive in establishing the workhouse of this era. Rusche and Kirchheimer state that "the possibility of profits was a decisive motive for the institution of houses of correction."[10] They accumulate considerable evidence that humanitarian zeal was not the only reason for taking the vagrants and beggars off the streets and putting them to work. This may well be seen in reading the history of such workhouses as at Amsterdam, established in 1596, and at Ghent, Belgium, opened by the celebrated Jean Jacques Philippe Vilain (1717-1777) in 1773.[11] The businesslike methods developed at Ghent by the founder served as a pattern in later prison experimentation.

Because of the critical situation arising from the hundreds of able-bodied beggars and "sturdy rogues" swarming the countryside of Flanders, the Deputies called on Vilain to formulate a plan for relief. He did so by erecting his *"maison de force,"* or *"rasphuys."*[12] In Vilain's memoir to his deputies he quoted the Bible verses, "If any man will not work, neither let him eat," and "In the sweat of thy brow shall thou eat bread."

While Vilain may have been a stern disciplinarian, his innovations in correctional administration have earned for him the sobriquet of "father of penitentiary science." He provided for rudimentary classification of prisoners. Felons were separated from misdemeanants and vagrants. There was a distinct quarter for women and another for children. He urged that each criminal be sentenced for at least one year so he might be reformed by being taught a trade, but he was opposed to life imprisonment. He anticipated such services as adequate medical care for his prisoners, productive labor, individual cells, and a meaningful discipline without any semblance of cruelty.

The Life and Work of John Howard

One of the greatest prison reformers of all time, John Howard, deserves credit for suggesting the penitentiary system. He brought to the attention of the world the sordid conditions in jails, prisons, and hulks throughout Europe, and made recommendations that epitomized the philosophy under-

---

[9] Reprinted from Otto Kirchheimer and Georg Rusche, *Punishment and Social Structure* (New York: Columbia University Press, 1939), pp. 41-52.

[10] *Ibid.,* p. 48.

[11] For an account of the Amsterdam institution, see Thorsten Sellin, *Pioneering in Penology: The Amsterdam House of Correction* (Philadelphia: University of Pennsylvania Press, 1944).

[12] In English, *"rasp house,"* so called because the inmates chipped or rasped logwood, that is, grated it into powder for use in processing dyes.

John Howard—Great Prison Reformer (1726–1790).

lying the system of imprisonment later developed in America and England. His life and fortune were dedicated to the cause of humanity, more particularly to the incarcerated wretches of Europe.[13]

---

[13] For a recent biography of Howard, see D. L. Howard, *John Howard: Prison Reformer* (London: Christopher Johnson) 1958.

In 1773 Howard was appointed sheriff of Bedfordshire. In this capacity he saw abuses he had never dreamed existed: abuses that "though occurring amid the light of a brilliant and boasted civilization were fitted rather to cower, snake-like and slimy, in the jungles of darkest barbarism." As a result of his attacks on the current practices in penal management, two bills were passed by Parliament in the year 1774, correcting some of the abuses in the jails and improving the sanitary conditions in the prisons.

**Cells and corridor in Hospice of San Michele, Rome.
Note Pope Clement's inscription above window.**

But Howard was only beginning his work. In April, 1775, he left England and traveled on the Continent, where he spent the next year or so in visiting many of the prisons and workhouses. The two establishments that impressed him most were Vilain's *Maison de Force* and the Hospice of San Michel at Rome, erected in 1704. This latter institution was designed along monastic lines by Pope Clement XI for the treatment of wayward youth. Over the door of the establishment was the inscription: "It is insufficient to restrain the wicked by punishment unless you render them virtuous by corrective discipline." He provided for two types of delinquents: first, youths under 20 sentenced by the court for the commission of crimes; and

On the bank of the Tiber in Rome: "Ospizio di S. Michele" (Hospice of San Michele), first institution for delinquent boys, 1704. From an 18th century print by Alessandro Specchi. (Courtesy Ugo Donati, Lugano, Switzerland.)

second, incorrigible boys who could not be controlled by their parents. The young offenders worked in association in a central hall at tasks in spinning and weaving. Chained by one foot and under a strict rule of silence, they listened to the brothers of a religious order while they droned through the Scripture or religious tracts. The incorrigible boys were kept separated, day and night, in little cubicles or cells. Large signs, hung throughout the institution, admonished "Silence." Floggings were resorted to as penalties for "past mistakes" as well as for infraction of rules.

The Hospice of San Michele, designed by the architect, Carlo Fontana, is still standing and is used today as detention quarters for delinquent boys whose cases are in the process of being disposed of by the juvenile court of Rome. There are three tiers of cells on each side of the large central corridor, making a total of 60. This structure is historically important since it was the first establishment for delinquent boys in the world. The fact that it is still used for the same purpose, 250 years after it was erected, adds to its importance.[14]

It was the monastic philosophy of expiation dominating the Hospice that pleased John Howard most as he observed the program. Here, indeed, was personified true penance, out of which evolved the penitentiary concept.

---

[14] This institution was visited by one of the writers in 1958. He was escorted by the judge of the juvenile court, Guido Colucci, and the noted architect, Dr. Carlo Varetti.

In other parts of Europe, however, Howard found deplorable abuses existing in the prisons. After accumulating his data, he returned home and went to work on his famous book, *State of Prisons,* published in 1777. This classic in penology had tremendous influence on the course of penal reform both in Europe and America.

In 1778, Howard, with the aid of Sir William Blackstone and Sir William Eden, drafted the Penitentiary Act which was passed by Parliament in 1779.[15] This act established penitentiary houses throughout the realm. There were four principles laid down by the reformer: (1) secure and sanitary structures, (2) systematic inspection, (3) abolition of fees, and (4) a reformatory regime. The object of the act was "by sobriety, cleanliness and medical assistance by a regular series of labour, by solitary confinement during the intervals of work . . . to inure them to habits of industry, to guard them from pernicious company. . . ." Owing to official drift and the lack of funds, as well as to the war with the American colonies and with France, no penitentiaries were erected.

By 1794 a site for the much discussed penitentiary proposed by Howard had been selected, but again there were difficulties. The distinguished but erratic social reformer, Jeremy Bentham, prevailed upon the authorities to permit him to erect an ingeniously designed structure of his own conception. In 1799 a plot of ground was assigned him for his model institution. The plans were drawn up, but fortunately for penology this monstrosity was never built. Bentham called it a "Panopticon," or "inspection-house."

The plans for the Panopticon showed something like a huge tank, covered by a glass roof, with the cells on the outer circumference, facing the center, and an apartment for the keepers in the center. This made it theoretically possible for all cells to be under constant scrutiny. But this architectural sport was never adopted in England or anywhere else, except with modifications.[16]

The Penitentiary Emerges in America

After the Revolution a small group of Philadelphia citizens became concerned with the sad plight of convicted criminals. Spiritual descendants of William Penn who were impressed with the writings of John Howard,

---

[15] In a later trip to Russia, Howard succumbed to jail fever and died at Kherson, in the Ukraine, January 20, 1790.

[16] There have been three modified Panopticons built in this country: Virginia, in 1800, at the prison at Richmond, designed by the architect Latrobe with suggestions from Thomas Jefferson [Jefferson wrote in his memoirs that he had been influenced by an architect of Lyons, France, in designing the Virginia prison. See *The Writings of Thomas Jefferson,* Memorial Edition, Vol. I (Washington, D.C.: 1903), pp. 69-70. The Lyons architect was Pierre Gabriel Bugniet who, in 1785, designed the Roanne prison in Lyons. He died in 1806.]; the Western State Penitentiary of Pennsylvania at Pittsburgh, erected and designed in 1821 by William Strickland, a disciple of Latrobe; and at Stateville, Illinois between 1916 and 1925. See pictures on pages 358 and 367.

persuaded the legislature in 1790 to create the Walnut Street Jail mentioned earlier.[17]

In the same enactment lawmakers took cognizance of other suggestions enunciated by the reform group, and their action marks a turning-point in the field of correction. For the first time in America the principles of solitary confinement and labor, strongly suggested by John Howard, were actually put into effect. Beyond even this, the act of 1790 directed the separation of witnesses and debtors from the felons, and the proper segregation of the sexes; and ordered the erection of a block of cells in the yard of the prison for the complete segregation of the "more hardened offenders." This block became known as the "penitentiary house" and was the first of its kind in this country.

Walnut St. Jail, Philadelphia, an old print.

In this little prison we find the first classification of prison inmates, the first system of productive labor for prisoners in this country, a policy of firmness and kindness instead of punishments, and even a crude system of self-government. From 1793 to 1799 the establishment was actually administered by a woman, Mrs. Mary Weed, widow of the jailer who died in the yellow fever plague of 1793.

The decade from 1790 to 1799 was epochal in the correctional history of Pennsylvania. Students of penology came from abroad, as well as from states along the eastern seaboard, to examine the policies and administra-

---

[17] These citizens were banded together into a prison reform organization known as the Philadelphia Society for the Alleviation of the Miseries of Public Prisons which they founded in 1787. For a history of this oldest prison reform society in the world, see N. K. Teeters, *They Were In Prison* (Philadelphia: Winston, 1937). This society, now a case working agency, is still dedicated to correctional reform.

tion of this first penitentiary. But due to overcrowding, architectural limitations, public apathy, and the scarcity of productive tasks to give the
inmates, the system by 1800 began to break down. Anticipating this, the
Philadelphia reformers petitioned the legislature to erect a penitentiary
where the system of solitary confinement with hard labor could be more
effectively realized.

**First Balloon Ascension in the United States, from Walnut Street Jail, Philadelphia,
January 9, 1793. President Washington hands "first air mail letter" to pioneer balloon-
ist, Jean Pierre Blanchard.** (From a painting by Joseph Jackson. Courtesy Penn
Mutual Life Insurance Company, Philadelphia.)

It took the legislature several years to act; in the meantime, prison reformers in New York and other states were giving serious thought to the
problem of dealing with their criminals. Virginia erected her first prison
in 1800 and Massachusetts built one at Charlestown in 1805. But it is the
states of New York and Pennsylvania that progressed most in their correctional philosophy during the first quarter century of the 1800's.

The Pennsylvania and Auburn Systems of Prison Discipline

The Pennsylvania legislature finally appropriated money for the construction of two penitentiaries, one for each section of the state. They were
both established on the philosophy of separate confinement of prisoners,

one from the other, at hard labor. As there were no architectural models to guide them, the planners of the new system relied on two famous Philadelphia architects, William Strickland and John Haviland. The former was given the contract for the erection of the Western Penitentiary at Pittsburgh (Allegheny), the latter for the Eastern Penitentiary at Philadelphia.

John Haviland, architect, Eastern State Penitentiary at Philadelphia, 1821–1835. (Courtesy, The Metropolitan Museum of Art. Punnett Fund, 1938.)

The Pittsburgh prison, which was badly designed, does not enter into the controversy dealing with prison discipline. The Philadelphia establishment, with its unique philosophy and architecture, challenged the New York reformers who in the meantime had built a penitentiary in central New York State, at Auburn.

No correctional institution in the world is historically more famous than the prison at Philadelphia, known locally as "Cherry Hill" because its site had once been a cherry orchard.[18] When it was opened in 1829, it was not

---

[18] For a history of this institution see Negley K. Teeters and John D. Shearer, *The Prison at Philadelphia: Cherry Hill* (New York: Columbia University Press, 1957).

just another prison; it epitomized one of the most influential correctional philosophies ever conceived by man.

The architecture of this prison was deliberately adapted to the principle of separate confinement of prisoners. The outside cell blocks at the time were seven in number, all radiating from a common center, like the spokes of a wheel. To each individual cell was attached a small exercise yard to which each prisoner could repair twice daily for short periods of time. In the meantime he worked, ate, and slept in his cell, and saw no one but the officers of the institution and official visitors from the outside community. These latter were members of the Philadelphia prison reform society who, having initiated the system in Pennsylvania, took keen interest in its successful operation.

Eastern State Penitentiary. Note the exercise yards.

The occupations supplied the inmate, most of them adapted to the small confines of the cells, were of a handcraft nature; shoemaking, spinning, weaving, dyeing, dressing yarn, and the like. The prisoner, upon entering the prison, was blindfolded and led to his cell. Every precaution was taken to make it impossible for him to see either the plan of the prison or any other inmate.

The direct antithesis of the separate system of Pennsylvania was the program developed by the New York reformers, first at Auburn and later at the Sing Sing prison. The system in that state has been referred to as "congregate" or "silent." Its distinguishing feature was that the prisoners were permitted to work together in shops under a strict rule of silence (but not blindfolded), and at night were locked up in individual cells.

Facade of Auburn, N.Y. Prison with familiar figure of "Copper John" aloft.

Instead of using what is known as outside-cell construction similar to that employed at the Philadelphia prison, the designers of the Auburn prison used inside-cell architecture. Whereas the size of the cell at Philadelphia was unusually ample (so the inmate could use it as a workshop), the Auburn cell was excessively small, measuring only 7 feet long and 3 feet, 6 inches wide. Any light or air that might penetrate the cell could only enter by small and heavily barred niches built into the outer building-shell that surrounded the cell blocks. The cells were obviously too small for any degree of comfort, usually too damp in summer and too cold in winter, yet the Auburn type of construction served for years as the pattern for most state penitentiaries.

The system of prison discipline developed at Auburn was a clever compromise between the outmoded and repudiated congregate system of the county jails and the little understood system of separate confinement (usually referred to by its critics as solitary confinement). Great faith was placed on the rule of silence to prevent contamination. But this ideal could only be realized by the liberal use of the lash. It was in the Auburn prison that the "lock step" was devised for the purpose of making supervision

Old South Hall, Auburn Prison. Note the narrow cells of first important inside cell block. (Courtesy New York State Dept. of Correction.)

easier. This was a kind of slow motion shuffle of the prisoners. Each man placed one hand on the shoulder of the person in front of him and, with downcast eyes, or all facing the officer, and in strict silence, proceeded to shop or cell block. Conversation or even simulated communication between the inmates was regarded as the most heinous of offenses.

The prison was operated in its early days by Elam Lynds, a strict disciplinarian who was a firm advocate of flogging as the best means of maintaining discipline. Contemporary reports state that Lynds used a rawhide whip, but sometimes the floggings were applied with a "cat" made of six strands of wire. He is said to have inflicted as many as five hundred blows on a stripped prisoner at one time. He whipped the insane and those who had fits.[19]

Lynds contended that reformation could not be effected until the spirit of the criminal was broken. It was the purpose of prison discipline to accomplish this. He maintained that all prison inmates were cowards. A great deal of prison folklore has grown up around Elam Lynds, much of it from his tenure as warden of Sing Sing prison in 1825.[20]

---

[19] Stated by Margaret Wilson, *The Crime of Punishment* (New York: Harcourt, Brace, 1931), p. 224.

[20] For the best material on the early days of Sing Sing prison, see the historical novel by Lewis E. Lawes, *Cell 202* (New York: Farrar & Rinehart, 1935).

Prisoners in "lock step" (close formation) in Sing Sing prison in early days.

The Struggle between the Two Systems

The modern American prison is the direct descendant of the Auburn System with the additional bad feature of promiscuous intermingling of the inmates, a situation which the early administrators tried assiduously, even though cruelly, to prevent. Obsessed by the fear of contamination, they regarded it as a serious breach of prison discipline to permit inmates to communicate with one another.

In order, then, to understand the modern prison, it is of concern to review briefly the historic struggle between the advocates of the two systems, the Pennsylvania and the Auburn, which lasted for over fifty years. The chief protagonists of this epic controversy were the *Boston Prison Discipline Society,* headed by the zealous Rev. Louis Dwight, and the members of the *Philadelphia Society for Alleviating the Miseries of Public Prisons.*

Three of the main criticisms leveled at the Philadelphia prison by the Rev. Dwight and others are worthy of some comment. First, its expense in construction and operation; second, the mode of labor for the prisoners—handcraft as opposed to the use of power machinery that required workshops in which the inmates would be obliged to work together; and third, the tendency in its separate discipline to induce insanity or mental deterioration.

These were, of course, serious charges, but they were elusive of proof. So far as expense was concerned, taxpayers, then as now, were disturbed at the high cost of maintaining prisons. Any feasible work plan for

**Elam Lynds, Warden of early Auburn and Sing Sing
prisons.** (Courtesy of Russell Sage Foundation.)

prison inmates that would help defray the cost of maintenance was likely
to be more acceptable, even though the more laudable objective of imprisonment, that of reformation, might be partially nullified. The prison
movement was ushered in during the early days of the Industrial Revolution when power machinery was slowly supplanting hand labor. It seemed
absurd to prudent administrators as well as to legislators to maintain a
handcraft prison labor policy when the trend was definitely in the direction
of power machinery. But the Philadelphia zealots who believed implicitly
in reformation as the aim of imprisonment were never much impressed
with such an economic argument.

Charges that separate (or solitary) confinement tended to induce insanity were heatedly debated by the advocates of both systems. It must be
remembered that in those early days there were mental hospitals only for
those who could afford private treatment, that not too much was known
regarding mental diseases, and that mental disease and mental defect were

confused in diagnoses by some of the best physicians. Due to these factors as well as to the extreme partisanship of both groups, this charge loses much of its venom as viewed today.

The advocates of the Pennsylvania System contended that the Auburn System was not only cruel, but also a repudiation of the tenets of the new dispensation of imprisonment in the treatment of crime. As the Auburn System unfolded, and was adopted in the various states, the feature of contamination of prisoners became more obvious. In time, prisons operating under this system became just so many congregate institutions—the very situation the earlier reformers were so determined to avoid.

The Pennsylvania System had very little success outside the state of its origin, and even there it was consistently attacked because of lack of productive labor for prisoners, overcrowding, some cruelty, and expense. The following data indicate the extent of its adoption and the dates of its abandonment in other states: Maryland introduced solitary confinement in 1809 and abolished it in 1838; Massachusetts in 1811 until 1829; Maine from 1824 to 1827; New Jersey from 1820 to 1828, reintroduced it in 1833 and finally abandoned it completely in 1858; Rhode Island from 1838 to 1844. Pennsylvania did not give it up legally until 1913 but long before that year it was in all but name just another congregate prison.

But if the Pennsylvania System did not meet with hearty approval in its native country, its originators and promoters could take keen satisfaction in the admiration with which it was received by most distinguished penologists from abroad. Several European countries sent commissioners to study the merits of the two systems and in time practically all of them adopted

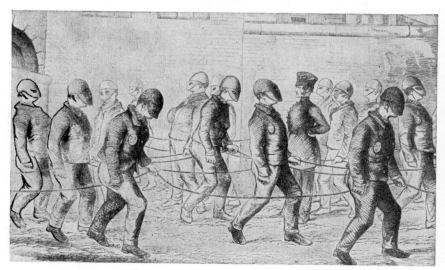

Prisoners exercising under separate system—Pentonville Prison in London, 1850.

Chapel services in a separate prison. While no longer in use, such chapels may still be seen in prisons in Europe.

the Pennsylvania System. Alert to the dangers of contamination of prisoners in the Auburn System, these astute students were especially impressed with the philosophy of strict separation of inmates. Wrote Gustave de Beaumont and Alexis de Tocqueville, the French commissioners:

> Whoever has studied the interior of prisons and the moral state of their inmates, has become convinced that communication between these persons renders their moral reformation impossible, and becomes even for them the inevitable cause of alarming corruption. . . . No salutary system can possibly exist without the separation of the criminals.[21]

These investigators saw the evils of the congregate prison; Americans, apparently, did not, since it was the Auburn System that spread throughout the country.

The Pennsylvania System, however, varied as it was applied in European countries. In some places it was reserved only for hardened criminals; in others, every prisoner was subjected to a period of strict solitary which, of course, indicated its use only as punishment; in still others, separation was applied primarily to maintain anonymity only, in order to minimize potential blackmail after release. Hoods or masks were used and the prisoners were exercised in the prison yards in congregate style. Various ingenious devices were adopted to keep prisoners from identifying each

---

[21] *On the Penitentiary System in the United States, etc.* (Philadelphia: 1833), p. 21.

other, though they often worked in close proximity. Contemporary engravings of London's Pentonville prison show inmates with their faces covered with visor-hoods as they shuffled about a circle which passed for exercise. In chapel prisoners were kept in coffin-like stalls while they listened to divine services.

Down to the present time the Pennsylvania System still dominates some of the prison systems of Europe, however modified in form they may be. In Belgium, France, and to a degree, in West Germany, vestiges of the system may still be noted.

Further Developments in the Penitentiary

We must not overlook the fact that the reformers and the more enlightened administrators were modifying the old regime in many particulars. Even before 1870, when the Reformatory ideal was being ardently considered, American prisons were trying out new concepts in the field of treatment, inadequate though we may regard them today. The new concept of reform (rather than expiation or penitence) was heralded far and wide as a panacea for criminality. But because legislatures, when urged to create these new institutions, were loath to overhaul the old order completely, the new form of rehabilitative treatment was developed to serve young men only, and institutions for adult criminals continued to operate under the Auburn plan.

But there were improvements in the penitentiary system that mitigated the monotonous and often brutal discipline that was its chief characteristic. These came slowly, some of them as a concession to prisoners to make discipline easier, some arising from the firm convictions of liberal administrators here and there throughout the country. Good wardens were in the minority in America during the first 100 years of the penitentiary regime, but they did exist. A concession to prisoners would be tried out. If it proved successful, it was copied in other institutions. The enlightened measures were usually harshly criticized by traditionalists. There have always been groups of people who criticize progressive penal administration, calling institutions where new ideas are adopted "country clubs."[22]

We do not know just when the first prison library was instituted, but the Philadelphia Prison Society furnished books to the prisoners in the renovated Walnut Street Jail in 1790. Sunday schools were introduced at an early date. Chaplains or "moral instructors" were provided by legislative act in most states. Compensation for overtime work was finally introduced and wages were paid prisoners. Pets and flowers were permitted in some prisons. Prisoners were eventually permitted to receive mail and write occasional letters to approved persons but this concession was generally restricted to first-grade or well-behaved prisoners and used as a device to

---

[22] See pages 458-459 for a discussion of this.

encourage obedience to the rules. Finally, musical instruments were permitted, which naturally resulted in the formation of bands and orchestras. There are few prisons in America today that do not boast a reasonably good band or orchestra. In many institutions inmates march to and from work and mess to the strains of stirring music furnished by their bands.

Other significant privileges included commutation of sentence, or "good time" remission, that is, reduction of the original time sentence for good behavior. Schools were created, and eventually outside teachers were sometimes employed, although in most prison schools today the teachers are inmates. Trained custodial officers and career wardens were brought in to replace untrained political administrative officers. The lock-step, the striped prison garb, and the silence rule finally passed away in all but a few of the more reactionary and repressive institutions. These, and a host of more recent changes in the administration of the prison, gradually brought about a new point of view concerning correctional treatment. The enlightened prison administrator accepts these today without any thought of criticism, but when they were first discussed many of them precipitated considerable heat among leaders in state and local political circles.

23 ————————————————————————

An Appraisal of
Conventional Imprisonment

There is something unkindly about the American prison. There is something corroding about it. It tends to harden all that come within the folds of its shadow. It takes kindly, well-intentioned people and makes them callous. . . . In some inexplicable manner the prison "gets" not only the prisoners but the prison guards as well. (From Osborne of Sing Sing, *Chapel Hill: University of North Carolina Press, 1933.)*

FRANK TANNENBAUM

The Transfer of Corporal Punishment to the Prison

One of the prime reasons the reformers of the early nineteenth century were determined to develop imprisonment was their repugnance to the corporal punishment that had dominated the penal philosophy of the preceding era. Yet as we survey the history of the traditional prison we cannot escape the stark reality that it has been responsible for more agony of mind and body than the branding-iron and the whipping-post.

Every form of corporal punishment of pre-prison days was transferred to the prison and more ingenious ones were invented to enforce discipline and to break the inmate's spirit. This was especially true in the Auburn-type prisons where the rule of silence was maintained by the fear of the lash and where prisoners were forced to hard-labor tasks, often beyond their endurance.

But the Philadelphia prison also stood condemned within five years after it was opened. According to the theory of separate confinement physical punishment was not necessary. If an inmate refused to work, or became noisy, or attacked a keeper, all that was necessary was to withdraw his work or books and let him suffer the remorse of "ennui, idleness, and loneliness." But, in 1834, the management was accused of "cruel and unusual" punishments such as the strait jacket, the gag or brank, the "tranquilizing chair," or excessive use of the dark cell on a bread and water diet. In the investigation of that year the management found it not too difficult to justify such penalties. And the majority report of the legislative committee exonerated the officials of any cruel motives.

Concepts of cruelty change through the years. No prison administrator would dare use a gag on a prisoner today. Strait jackets have all but disappeared in mental hospitals and certainly none has ever been seen in a prison since early times. But prison administrators, through the years, invented many bizarre punishments that compare in cruelty to those used in earlier times. Reports of commissions charged with investigating prison practices during the past century show conclusively that vicious forms of punishment and physical torture were used in many, if not all, institutions under various wardens.

Floggings have been prison practice down to our own times, and deaths have occurred due to over-severe whippings in southern prison camps and chain-gangs. Tying prisoners up by their hands and allowing them to hang suspended with their toes barely touching the floor or ground has been a common method of enforcing discipline.

Sweatboxes have been frequently used. They are relatively unventilated cells located, in the early days, on either side of a fire-place but in modern times, in a remote part of a prison camp where the heat of the sun can cause great suffering. For years this was a most frequently used punishment in chain-gangs. Cold baths and the pouring of cold water—known as douches—upon prisoners from a considerable height have been common methods of early prison punishment. The "water cure" had many variations. For years in the Ohio penitentiary at Columbus the most dreaded punishment was this type of cruelty. It consisted of a prisoner being strapped in a kind of tub and a strong fine stream of water turned on him. Blindfolded, he could not know when or where he would be hit. Usually the victim was released in a half drowned condition.

At the turn of the century the Ohio prison was notorious for its fantastic methods of punishment. News stories during 1890 told of a special cage being constructed to house four unruly prisoners. The story of this cage is graphically told by Al Jennings, train robber who wrote of his experiences after he was released and had turned evangelist. He tells how Warden S. G. Coffin had it built to house one Ira Maralatt, an insane murderer from Cleveland, and cashed in on the unusual exhibit by charging visitors 25 cents to view him. Manacled by the wrist the tragic figure remained in this cage until Coffin resigned and was succeeded by Warden W. N. Darby. This official realized the mental condition of Maralatt and was responsible for an operation on his brain which relieved pressure and restored him to his normal intelligence. Jennings, in his account, relates that Maralatt, when

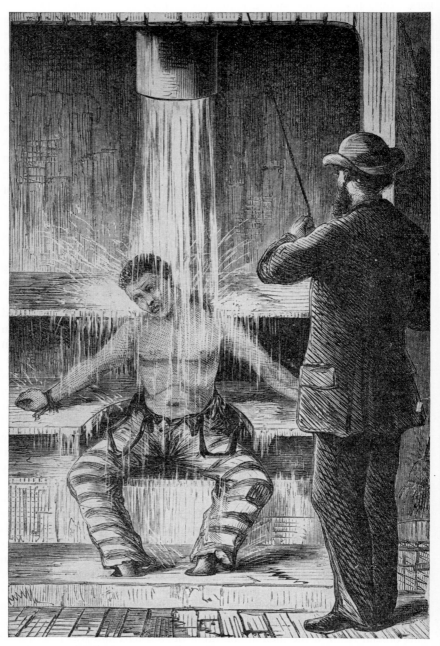

Use of the Douche as punishment device in Sing Sing, N.Y. prison in early days.

subdued by floggings or other inhuman treatment, was often sponged off by Sidney Porter, the short story writer who served a term in the Ohio prison at the same time and who, under the pen name of O. Henry, began his literary career while there.[1]

Unruly prisoners have often been condemned to wear an iron yoke or heavy collar. Within the first three months after the first prisoners were received at the Charlestown, Massachusetts prison (in 1805), the board of visitors voted that the warden should provide collars or rings to be worn by such prisoners "as shall discover a disposition to escape." The "Oregon boot," an excessively heavy leather footgear was placed on those who attempted to escape. Early guards carried "loaded" canes, a sort of elongated blackjack, for protection. Such a weapon could be used whenever the custodial officer thought it necessary. In the early Georgia prison, discipline was enforced by punishments of which we know nothing but their grim names: *cow skin, slue paddle,* and *wooden horse.*

Humiliation was one of the objectives of early prison treatment. One reason for prison stripes was to degrade the prisoners. This form of garb was introduced in New York prisons in 1815. Prison inmates in some states were obliged to wear garish clothing of varying hues. In Massachusetts the clothing was half red and half blue for first offenders; second timers were red, yellow and blue; and third termers a costume of yellow, black and blue. Conspicuous clothing makes escape more difficult. The shaving of heads or of just one half was and is still used to aid in preventing escapes, or, at least, in making recapture easy. Dogs have also been used by prison authorities, especially in the camps, to serve as a deterrent to escape.

The use of the dark or isolation cell—the hangover of the medieval dungeon—known in prison parlance as "Klondike," is probably the most universally used prison punishment in the history of American penology. The ingenuity of the twentieth century has devised the modern Klondike. A separate building, usually isolated from the general prison population (often it is in the basement), houses the recalcitrants for periods from one day to many weeks.[2] These cells are devoid of such necessary comforts as chairs, beds and lights. Food is restricted, quite frequently consisting only of the traditional bread and water diet.

The dark cell, being the "secret place" of the prison house, is the unhappy abode of hundreds of prison inmates every year in many institutions.

---

[1] See Alphonse Jennings, *Through the Shadows with O. Henry* (New York: H. F. Fly Company, 1912), Chapter 19, Jennings tells another interesting story about a safe cracker, Dick Price, who was the original Jimmie Valentine known in song and story for his exploits in opening safes.

[2] For data on use of dark cell or "segregation," see Negley K. Teeters, "A Limited Survey of Some Prison Practices and Policies," *Prison World* (now the *Amer. J. Corr.*), Vol. 14, No. 3 (May-June, 1952); also "Reformatory Practices and Procedures," *ibid.,* Vol. 15, No. 3 (May-June, 1953).

Some prisoners are kept in these gloomy places for months. What to do with a rebellious prisoner bedevils all wardens, but a sustained sojourn in a punishment cell is not the answer. The excessive use of Klondike is a grim example of what is known to students of corrections as "dead end" penology. Resorting to it for long periods of time is an illustration of total lack of imagination and outmoded prison administration, all too current in most of our prisons even today.

Not much different from the dark or isolation cell is the "segregation" block or ward. In this isolated part of the prison an inmate may be placed because he is "uncooperative," is considered dangerous or a bad influence, or for some other reason arrived at by the warden or his deputy in charge of custody.

Perhaps the most famous case known in prison annals is that of Jesse Pomeroy of Massachusetts. Convicted of a series of revolting sexual murders when he was but fourteen years of age, he was sentenced in 1874 to "solitary confinement for life." Pomeroy served approximately thirty-eight years in segregation where he had very few contacts and in 1932 was transferred to the state farm at Bridgewater where he died a few years later.[3]

A much more recent case which bids well to become a *cause célèbre* is that of Robert Stroud who has spent approximately the same period of time in "segregation" in the federal prisons of Leavenworth and Alcatraz. Stroud was first sent to prison when he was nineteen for killing a man in Alaska in 1909. While in the Leavenworth prison he killed a guard in the dining room for which he was sentenced to be hanged. This sentence was commuted to life by President Woodrow Wilson. While in prison, in a "segregated" cell, Stroud became an expert in diseases of birds and is alleged to have become a world-wide authority in his field. He became so much of a controversial figure that he was transferred to Alcatraz and his birds were taken from him.[4]

Twenty years ago harrowing tales came out of Texas relating how prisoners in the various prison camps subjected themselves to extreme mutilations. In 1935 the Houston *Press* told of several inmates cutting their legs with axes. This was done to escape the brutal beatings administered by the guards. One emaciated prisoner cut off his leg because he "was wholly unable to do the work assigned him" due to a previous appendix operation. He had been beaten before and had a terrible fear of the "bat." Another prisoner broke his leg. On being reproached for his self-mutilation his only reply was that "he could not bear the pain caused by the lash and that he had rather die than have another whipping." The Texas legislature abol-

---

[3] This famous case is written up by A. Warren Stearns, M.D. See "The Life and Crimes of Jesse Harding Pomeroy," *Journal of the Maine Medical Association,* Vol. 39, No. 4 (April, 1948), pp. 49 f.

[4] For details of this case see Thomas E. Gaddis, *Birdman of Alcatraz: The Story of Robert Stroud* (New York: Random House, 1955; and New American Library Signet Edition).

ished the bat in 1941 and in 1947 a new management of the state's penal farms emerged through which many reforms were initiated.

The practice of self-mutilation by prisoners is an old one. Records show that it was prevalent in the hulks at Chatham Dock in England and in French Guiana.[5] Here, in this country, four long-term prisoners in a Georgia prison camp at Dallas, known locally as "Little Alcatraz," broke their legs in order to escape work. A rash of self-mutilations in which 36 prisoners in the Rock Quarry camp in Georgia broke their legs was reported in the press in 1956; and in the Buford camp, same state, 30 inmates slashed their tendons in the same year, and in the same camp in May 1957 several inmates broke their legs with a sledge. Their reason was that they could not tolerate the working conditions of the camp. After an investigation of the 1956 affair, the authorities merely recommended a "tighter" policy.[6]

In February 1951, 31 prisoners of the Angola, Louisiana prison cut the tendons of their legs in order to escape work. The warden insisted this action was resorted to in order to "get rid of him." Again, in 1953, 17 inmates in the Louisiana prison slashed their arms.

A vicious regime involving cruel floggings in the Kilby, Alabama prison was exposed by a reporter, Hugh Sparrow. His account substantiates the charges made against Alabama's prisons by Haywood Patterson, one of the "Scottsboro Boys" whose bitter story will probably not be believed by many of its readers. Few more tragic books have been written.[7] These exposures prompted a clean-up of the state's prisons.

Both Louisiana and Alabama have since gone far in developing a more progressive and humane system of corrections. These reforms have come about by alert and courageous governors who, spurred on by an aroused public, urged their legislatures to act.[8]

Floggings have been all but eliminated in state prisons and reformatories but we still hear of them in isolated instances; the practice is still recorded on the statute books of some states even though orders often forbid wardens to resort to it. It is difficult to get precise data on the use of the whip and other forms of physical punishment since there is no clearing-house of information dealing with prison practices and where such disciplinary measures are used wardens are loath to admit it.

Floggings have been resorted to in many institutions in the recent past but it is generally agreed today that few institutions still persist in the

---

[5] For a psychological study of the practice of self-mutilation in prisons, see Rupert C. Koeninger, "What about Self-Mutilation?" *Prison World,* Vol. 13, No. 2 (March-April, 1951), p. 3 ff.

[6] New York *Times,* Tuesday, July 31, 1956.

[7] Haywood Patterson and Earl Conrad, *Scottsboro Boy* (New York: Doubleday, 1950).

[8] Regarding Louisiana, see "Our Prisons Need Not Fail," by Reed Cozart as told to Edward W. Stegg, *Saturday Evening Post,* Vol. 228, No. 16 (October 8, 1955).

practice. A former warden of the Canon City, Colorado, prison, was proud
of the fact that he used the "bat" and boasted of it for years. He was finally
removed from his post. A story from the Denver *Post* tells of the brutal
treatment of some of the inmates by this warden:

> One summer day . . . 50 guards and officials formed a ring around
> five Canon City, Colorado, prisoners in the penitentiary yard. The pris-
> oners were stripped and forced to keep their hands high above their heads.
> Occasionally a guard knocked down one of the convicts, picked him up,
> knocked him down again and kicked him in the head and groin. The sun
> was so hot that the man's bare feet began to blister on the pavement. Sweat
> from their bodies formed tiny pools, but whenever one of the men tried to
> stand in a pool to cool his seared feet, he was forced back onto the dry, hot
> surface of the pavement. After about a half hour of this the five men were
> pushed into the prison gymnasium. One convict who had booed at this per-
> formance was added to the group. There the six men were chained over
> the "gray mare" and lashed. The various officials took turns wielding the
> whip, a six-inch-wide leather strap. At intervals a wire noose was slipped
> around a convict's neck and he was dragged across the floor. When a con-
> vict fainted, he was revived with a bucket of water so none of the punish-
> ment would be wasted on him. After each had received about 45 lashes, the
> men were unchained, stretched out on the floor while a guard whipped the
> bare soles of their feet with a strap.[9]

The reason why the public does not hear more often of such repressive
measures is that the lone victim has no opportunity to be articulate, and
also because when intense physical torture is applied up to the acute danger
point, it is usually stopped in time to apply medical attention. It is only
when the administrative officers go too far, through ignorance or some oversight, and a casualty occurs, that the public learns of the cruelty.

The "Convict Bogey" and the Lock Psychosis

Some persons who are familiar with prison cruelty attribute it to the
deliberate viciousness and sadism of wardens, keepers, and guards. There
are such brutal officials, but it is the institutional and routine aspects of
prison administration and life, rather than the arbitrary personal sadism
and depravity, which accounts for most of the cruelty and personal demoralization that persists in our contemporary prison.

For a long time, cruel treatment of prisoners was considered highly
logical, and the warden performed his duty in making the life of the inmate
a harsh one. The aim of imprisonment was held to be a vigorous discipline
for the convicted offender in order to achieve social revenge for wrongdoing
and make him penitent, if not repentant. There was some hope, so the
theory ran, that rigid expiation might encourage the prisoner to reform and
thus avoid further punishment—the deterrent and purgative argument for

---

[9] Quotes and details of the Colorado prison at that time may be found in Harry
Elmer Barnes and Donald Wilson, "A Riot Is an Unnecessary Evil," *Life,* Vol. 33,
No. 21 (November 24, 1952).

punishment. So long as strict regimentation was regarded as the chief goal of imprisonment, cruel treatment was to be expected, and no prison official could fairly be blamed for his harshness.

But since the enunciation of the Declaration of Principles in 1870 which we shall discuss later (see page 425) all enlightened correctional authorities have repudiated punishment in behalf of curative treatment and rehabilitation. Many practical administrators have adopted this attitude in theory, and nearly all of them give lip service to this ideal. But the actual introduction of a prison regimen that is logically compatible with any extensive rehabilitative program has been either blocked or largely frustrated by what we refer to as the "convict bogey." By this we mean the public hatred and fear of criminals only—or at least mainly—*after* they have been convicted of crime. It is not the criminal but the prisoner that frightens us. The newspapers play this up by printing scare headlines when a jail or prison break occurs. Prisoners who escape are always referred to as "dangerous" or "desperate" or "killers." An interesting bit of folklore illustrates this pathological fear we have of escaped prisoners. It is from F. G. Pettigrove's account of a fire at the Richmond, Virginia prison in 1865:

> "About ten o'clock," writes one eye witness, "a cry of dismay rang all along the streets . . . and I saw a crowd of leaping, shouting demons, in parti-colored clothes, and with heads half-shaven. It was the convicts, from the penitentiary, who had overcome the guard, set fire to the prison, and were now at liberty. Many a heart which had kept its courage to this point, quailed at the sight. Fortunately the prisoners were too intent upon securing their freedom to do much damage."[10]

This tense fear of the escaped prisoner was dramatized a few years ago in the legitimate play, *Desperate Hours,* which faithfully paralleled the true story of a suburban family who were subjected to a harrowing experience when some escapees from a state prison entered their home and held them at pistol point for three days.

It is not the criminal, then, but the prisoner who frightens us. In Massachusetts, as early as 1815, "guards were exhorted to think of the prison as a volcano filled with burning lava, which, if not restrained, would destroy both friends and foes." As we have made clear earlier, the most serious and dangerous criminals are rarely convicted. By and large, it is only the petty and traditional criminals who ever get into prisons. Yet the very idea of freeing a prison inmate terrifies us. On the other hand, we are not moved to any deep fear of the thousands of dangerous criminals whom we know, from newspaper reports, to be about us every day. Indeed, if the criminal is notorious enough, we are likely to flock to a night club to get an admiring look at him.

---

10 In Charles R. Henderson, *Correction and Prevention,* Vol. 2: "Penal and Reformatory Institutions" (New York: Charities Publication, Russell Sage Foundation, 1910), p. 4.

We may well inquire why society fears the inmate rather than the criminal. The reasons are numerous and complex. The prisoner is the victim of the psychology of the primitive taboo transferred to our time and social surroundings. In primitive times, the violator of the law was regarded as one who had broken the rules laid down by the gods. Today he is the violator of the rules of the herd, which are still regarded as quasi-divine. The inmate bears the brand of the scapegoat and like the violator of taboos in an earlier time, he must be exiled from the group—in this case by being imprisoned. As a result of the operation of what the psychologists call the "projective mechanism," the convicted man bears the burden of our own sins and in his punishment expiates our own sense of guilt. The prisoner or the ex-prisoner is a marked man—a human dog to whom a bad name has been given. He is called a "convict" or an "ex-convict."

In Anglo-Saxon jurisprudence, a man is regarded as innocent until proved guilty. Hence a man must be arrested and formally convicted to bear the brand of a criminal. Imprisonment supplies the method of carrying out the revengeful spirit of outraged society, which thus secures satisfaction for the wrongs, real or alleged, that are brought upon it by the offender. Contemporary imprisonment offers a vicarious release of the sadistic traits of human beings under approved and seemingly respectable circumstances, whereas relatively few individuals would personally and individually find themselves able to carry out, or to admit themselves subject to, such obviously cruel impulses.[11]

The "convict bogey" logically leads to the jailing psychosis and to the idea that the prime purpose of the prison is to lock up and punish criminals. A few years ago when Sing Sing prison "went on the five day work week," there were cries of "coddling" prisoners. The more repressive the incarceration the more the uninformed public is satisfied. The problem of finding productive work for prisoners—one facing administrators constantly—is unknown to the man in the street. All he knows is the conventional folkway that prison inmates should be worked hard and long. Prisoners are socially branded exiles who must be safely kept from contact with society. It is dangerous to unlock them too soon and let them out until they have atoned for their crime; when that time arrives is not known, of course. And when they are released, we feel we must avoid them and keep up the sense of opprobrium attached to them.

The sporadic cruelty of sadistic officers is occasionally exposed—in some jurisdictions where a few intelligent persons know what is happening and call the attention of the public to the abuses—and such persons, either guards or wardens, are discharged. Yet, the writers of this book know of wardens who operate what is known as a "tight" prison with much routinized brutality, much of it psychological—for instance, a warden who holds

---

[11] For an analysis of this point of view see Paul Reiwald, *Society and Its Criminals* (New York: International Press, 1949).

Manual operation in controlling doors of cell block. (Courtesy New Jersey State Department of Institutions and Agencies.)

an exceptionally good reputation in and out of his state, makes his inmates face the wall when they appear before his disciplinary court, and of course, before they have been found guilty of a breach of the rules. Such wardens carry their uninformed public along with them by being "good fellows" and by ingratiating themselves to civic clubs, chambers of commerce, and similar organizations, and thus maintain themselves politically for years.

Even earnest administrators who sincerely believe in rehabilitation are afraid to introduce a wholehearted program that might improve treatment procedures. Such rehabilitative treatment requires flexibility and experimentation, but these increase escape risks and even the most enlightened warden realizes that his work will be judged by newspapers, politicians, and the public on the basis of how successful he is in preventing escapes.[12]

---

[12] In a book by Gladys A. Erickson, *Warden Ragen of Joliet* (New York: E. P. Dutton, 1957), the statement is made by a Catholic chaplain, "Warden Ragen has set the prison system in Illinois back twenty years." This startling statement is provoked by the fact that Illinois is not prepared to develop a system of smaller and diversified units so it relies on Warden Joseph Ragen to operate a "tight" prison at Stateville. See review of this book by Walter M. Wallack in *Amer. J. Corr.,* Vol. 19, No. 3 (May-June, 1957), p. 8.

Illinois State Prison, Stateville. Note Panopticon Features. Designed by C. Harrick Hammond, architect.

Since he makes his living out of his post, he does not relish the idea of losing it by any bold adventure in rehabilitation. In this manner, humane treatment programs are frustrated and a prison routine, wholly incompatible with the formally repudiated aim of punishment, survives and dominates the prison scene. In the most enlightened system of prison administration the officials live in constant fear lest their constructive daring be discovered. The better the system the greater the fear.

Not too many years ago one of the country's best known and progressive superintendents, Dr. Miriam Van Waters, was discharged from her position as head of the Massachusetts Reformatory for Women at Framingham by the then Commissioner of Corrections on charges that she was "coddling" her inmates. Her regime was simply compatible with some of the newer concepts held by modern correctional experts—something which the politician commissioner could not understand. Fortunately the board of inquiry appointed by the governor refused to take the charges seriously and Miss Van Waters was reinstated. This is one of the few cases where enlightened penologists dared to break with traditional retributive methods and found influential defenders.

The Custodial Staff

Books by students of the correctional problem and by prison inmates attest to the fact that most prison posts fall to a rather inferior grade of men, usually through the spoils system of politics. To make matters worse, salaries are low and the nature of the work is most unattractive to persons with any ambition.

Starting, then, with a pedestrian type as the basis of prison personnel, we must add the inevitable degrading and brutalizing aspects of the day-by-day job. As we have seen, the primary function of the warden is to act as an efficient jailer. To achieve this result he naturally resorts to the most rudimentary discipline and regimentation. He relies almost entirely upon rules and coercive procedure. This prevents nearly all sympathetic human contact with inmates of the prison, develops a purely mechanical spirit that brutalizes both the officials and the prisoners, and creates a fatal antipathy between them. This severe regimentation, so totally different from the life of freedom to which the prisoner will ultimately be restored, is the worst possible preparation for freedom and the responsibilities of citizenship.

Numbering, counting, checking, and locking up, constitutes the routine of prison personnel. The mental habits of the custodial staff revolve around the mania to keep prisoners either locked up or scrupulously accounted for. Considerations of reformation and humanity evaporate in the face of this inexorable and all-encompassing anxiety.

Then a crisis of unusual magnitude occurs as, for example, the great prison fire in the Ohio penitentiary in 1930 when 300 prisoners' lives were lost, partly through delay in releasing the inmates from their cells. This is perhaps the worst prison tragedy that has ever occurred.[13] Such an occurrence demands a reversal of all this locking and counting routine. It calls for independence and resourcefulness—strange attitudes to the jailers. In the Columbus fire, the locking-up crew made a poor rescue party. The inmates showed far more resourcefulness than the guards. The latter were temporarily stunned at the prospect of unlocking the cells and allowing the prisoners freedom in the yard. It was not that they wanted to see human beings burned or suffocated. They were merely paying the penalty for wrong-headed attitudes and stupid habits that could not be instantly thrown off in the time of emergency.

There may be found in the 100 or so correctional institutions in this country many efficient, progressive, and sympathetic wardens who actually understand the psychological handicaps of both the prisoner and the guard. In recent years there has been an effort made to staff prisons with a higher type of personnel, especially in the federal system and in some of the more penologically progressive states. Standards have been raised with examinations, with subsequent personal interviews, and later, in-service training courses.

The prison tends to warp the personalities of staff members because of the unnatural life they lead. They are scarcely more free than the prisoners they guard.

The guard's life is minutely charted for him when on duty. "A guard is required to watch a prisoner at least as closely as the prisoner watches him. Whenever an officer or prisoner comes within sight of a guard's post, such guard will immediately get up, if seated, and move around, so that the officer or prisoner may see that he is awake and on the alert. Guards must not read books or papers while on duty, or have them in his possession. When off duty, guards must not discuss penitentiary affairs or become a source of information pertaining thereto."[14]

There is also constant temptation for guards to be corrupted by the inmates. They have the power to give or withhold small favors, or to deliver messages to friends of the prisoners on the outside, or to render special services to "privileged" prisoners. It takes a person of heroic integrity not to yield to overtures from persistent inmates.

What has been stated above about the personnel of men's prisons applies to women's institutions as well. Few women who possess the finer sensitiveness of their sex hold positions in correctional institutions for long.

---

[13] For an account of this holocaust, see James M. Winning, *Behind These Walls* (New York: Macmillan, 1933).

[14] *Attorney General's Survey of Release Procedures*, Vol. 5, "Prisons," pp. 72-73.

The Prison Code

There are two distressing features of the modern prison: the harassed personnel, guards and wardens alike, and the conflict that seethes below the surface between prisoners and administrative officials. This latter is seldom discussed in meetings of prison people, either publicly or privately. Yet as we view the prison, we see frustration of both keepers and prisoners. The groups build up techniques and methods to block each other in everything they do. It is a devil-inspired game of conflict; in prison parlance it is called "the code."

On the one hand, guards and wardens make interminable rules in order to regiment and control. The natural reaction of the prisoners is to break these rules with as much cunning as they possess. The code permits no one to inform and those who do, are referred to as "rats."[15] So the guard must get his man, even though he may make a mistake by singling out an innocent person. This unfairness, practiced in almost all prisons, piles up further resentment among the inmates.

Collusion, or anything more than a casual contact between two or more inmates, is tantamount to plotting against the administration. The warden is under constant pressure from the public to have the prison operate without rioting or sit-down strikes. Hence, any indication that the inmates are ganging up, even though the action is only the normal desire to form a primary group, is frowned upon. When there is any sign of "ganging up," the men are separated and placed in different parts of the prison.

As we have seen, the guard is almost as much a prisoner as the inmate. Often he has little status outside the prison. His job within the institution supplies this status, so he must feed upon the situation, abnormal though it is, to obtain a degree of inflation.

The "social distance" existing between the staff and the inmates is admirably described by Weinberg in his analysis of the prison community:

> The officials, especially the guards, regard the convicts as "criminals after all," as "people who can't and shouldn't be trusted," and as "degenerates who must be put in their place at all times." "You can't be too easy with them," states one custodian. . . . "They're on the go to put one over on you. They don't think of us when they try to get over the wall." "There must be something wrong with every man here," states another, "else he wouldn't be here. They're scheming all the time, soon as you give them an inch. That's because there's something wrong with every one of 'em." Convicts are considered "born bad," as mentally, emotionally or morally deficient. Their only language, "the language they understand, is punishment." Attempts at rehabilitation usually are considered as futile. In exceptional cases, only in cases where the inmates are "not really convicts," reform does occur. Prisoners are "unfit, failures," and hard men without human feelings. They are considered calloused because they were unable

---

[15] For an analysis of the "rat" complex, see Richard McCleery, *The Strange Journey*, Chapel Hill, N.C.; Extension Bulletin, Vol. 32 (March, 1953).

"to make their way in life like honest folk." Hence they become recalcitrant, they must "be softened and broken" to get them "back in line." They are thought of as unintelligent and lazy, and, consequently, they stall in their work at every opportunity. Further, those with abilities usually have other undesirable qualities to offset their merits. . . .

Because they are unable to care for themselves, they must be held under leash. Because they are "wild" and uncontrollable, they require the sternest measures of discipline. Homosexuality is almost considered "natural" among inmates. As one custodian claimed, "It's in them. I couldn't believe it could happen till I saw it, and I had to give them both the hole." The "punk" and sex pervert are thus natural products of a degraded group and "prove" that the convicts are depraved and animalistic, for they resort to practices abhorred by conventional persons. Hence, "to act like an inmate" indicates disagreeable appearance. Convicts in their dress, speech, and walk are "different." They are enemies of and outcasts from society. Resultantly, they diverge from noncriminal persons in all the above-mentioned characteristics.[16]

It naturally follows that the "cons" counter "the repressive measures of the administration by condemning them and by intriguing against them." The prisoners say in effect: "Have nothing to do with a 'screw' [guard]; tell him nothing; take no favors from him." Many prison-wise "old-timers" even discourage newcomers from participating in rehabilitative programs, such as musical organizations, hobby shops, correspondence courses operated by the school, or discussion groups.

During the past fifteen years, sociologists have applied their knowledge to the prison community and have correctly described the conflict, hostility, push for status, and negativism that are so characteristic of the prison. There is a wealth of literature in the professional journals on these aspects of prison life.[17]

---

[16] S. Kirson Weinberg, "Aspects of the Prison's Social Structure," *Amer. J. Soc.,* Vol. 48, No. 5 (March, 1942), pp. 717-726; excerpt from p. 721; direct quotations are from guards and other officials.

[17] Hans Reimer, "Socialization in the Prison Community," *Proceedings,* American Prison Association, 1937, pp. 151-155; Donald Clemmer, "Leadership Phenomena in a Prison Community," *J. Crim. Law,* Vol. 28, No. 6 (March-April, 1938), pp. 851-872; Norman S. Hayner and Ellis Ash, "The Prison Community as a Social Group," *Amer. Soc. Rev.,* Vol. 4, No. 3 (June, 1939), pp. 362-369, and "The Prison As a Community," *ibid.,* Vol. 5, No. 4 (August, 1940), pp. 577-583; Donald Clemmer, *The Prison Community* (Boston: Christopher Press, 1940. Reissued 1958 by Rinehart, New York); Gresham M. Sykes, *The Society of Captives: A Study of a Maximum Security Prison* (Princeton, N.J.: Princeton University Press, 1958); N. A. Polanski, "The Prison as an Autocracy," *J. Crim. Law,* Vol. 33, No. 1 (May-June, 1942), pp. 16-22; Norman S. Hayner, "Washington State Correctional Institutions as Communities," *Social Forces,* Vol. 21, No. 3 (March, 1943), pp. 316-322; Ida Harper, "The Role of 'the Fringer' in a State Prison for Women," *ibid.,* Vol. 31, No. 1 (October, 1952), pp. 53-60; Paul B. Foreman, "Guide Theory for the Study of Informal Inmate Relations," *Southwestern Social Society Quarterly,* Vol. 34, No. 3 (December, 1953), pp. 34-46; Clarence Schrag, "Leadership among Prison Inmates," *Amer. Soc. Rev.,* Vol. 19, No. 1 (February, 1954), pp. 37-42; Lloyd McCorkle and Richard Korn, "Resocialization within Walls," *The Annals,* Vol. 293 (May, 1954), pp. 88-89; Mor-

Although progressive administrators have been attempting to break down this hostility, little cooperation has come from the inmates. But a few of the more vocal of these who can write are beginning to recognize the childishness of their attitude. The penal press carries editorials and articles calling attention to the sterility of this conflict between staff and prisoners.[18] In the *Atlantian,* a progressive magazine published in the federal penitentiary at Atlanta, Fred Ellis points out the absurdity of the "convict code." He states that it was justified in the old days when prisons exploited prisoners and treated them as cattle but he maintains that such enemy psychology defeats the prisoner today as well as minimizes the effectiveness of the rehabilitative effort of the staff. He calls upon the inmates to discard this outmoded code and to work with the administration to make the prison a bearable place.[19]

The psychology of prison cruelty, we may sum up, evolves from the custody or jailing psychosis of untrained and unimaginative officials who find it difficult to rise above the sterility that has gradually evolved in the practice and concept of imprisonment. While the professional staff are little affected by the insidious climate of the prison, the custodial staff must bear the brunt of inmate hostility. In-service training can be helpful but only after more care is used in the selection of line personnel. Fortunately, this is developing in some of our more progressive states.

The Prison Routine

Its Monotony

Let the reader imagine himself isolated from his family and friends and restricted to the one block on which his home is situated. Let him further imagine himself unable to cross the street, go uptown, use a telephone, send or receive mail more often than once a week, eat "boarding house" food with few delicacies, be restricted to one newspaper, and listen to no more than one or two radio programs daily. Then he may vaguely approximate what a prisoner must experience when he enters a correctional institution. Let him contemplate, if he can, a situation in which he rarely makes a decision himself, where he may never see a door-knob, not to mention grasp one; where one never opens a door himself or walks down stairs without permission. No matter how he may ponder this picture, he cannot imagine the restrictive situation that confronts the offender when he goes to prison.

---

ris G. Caldwell, "Group Dynamics in the Prison Community," *J. Crim. Law,* Vol. 46, No. 5 (January-February, 1955), pp. 648-657; Lloyd McCorkle, "Social Structure in a Prison," *The Welfare Reporter,* New Jersey Department of Institutions and Agencies (December, 1956), pp. 5 ff.

[18] See pages 519-521 for discussion of the penal press.

[19] "The Shift of Time," *The Atlantian,* Vol. 8, No. 4 (Winter, 1949-1950). This is a significant article, perhaps the first of its kind.

Caged men.

Most of the books written by those who have endured an extended prison experience comment on this lack of freedom, on the sheer monotony, and on the emphasis placed on the petty and arbitrary ways of doing things in prison. The mere business of tidying one's cell each morning becomes a disgusting ritual, the more so because in some institutions the procedures are the reverse of those found on the outside.

The reader is urged to become acquainted with some of the books written by ex-prisoners who vividly describe the deadly routine of prison life. A classic is Alexander Berkman's *Prison Memoirs of an Anarchist*.[20] Berkman served a "stretch" in the Western Penitentiary of Pennsylvania at Pittsburgh for the attempted slaying of the steel magnate, Henry Clay Frick, at the height of the bloody Homestead strike. A British work, *Museum,* by James L. Phelan, is well worth examination.[21] *Prison Days and Nights,* by an astute ex-prisoner, Victor Nelson,[22] gives us a bleak picture of the daily routine of the prison. He describes the insipid and unpalatable food, the "bucket brigade," the march to and from the prison shops, "yard-out,"

---

[20] Alexander Berkman, *Prison Memoirs of an Anarchist* (London: C. W. Daniel Co., 1926; first published in 1912).
[21] (New York: Morrow, 1937).
[22] (Boston: Little, Brown, 1932).

and the endless "count." While these books were written many years ago, the complaints registered by these men sound quite familiar today. The routine has not changed much.[23]

The Rules

The human animal, the prisoner, denied nearly every element of a normal expression of his personality by this insidiously monotonous routine, is still rendered "safe and secure" through an elaborate scheme of rules designed primarily to simplify the work of the guards. Let us look at the list of offenses for which prisoners may be punished in one state prison:

Altering clothing	Loud reading in cell
Bed not properly made	Malicious mischief
Clothing not in good order	Not out of bed promptly
Communication by signs	Not at door for count
Defacing property	Not wearing outside shirt
Dilatory	Not promptly out of cell
Dirty cells or furnishings	Out of place in shop or line
Disobedience	Profanity
Disturbance in cell house	Quarreling
Fighting	Shirking
Hands in pockets	Spitting on the floor
Hands or face not clean	Staring at visitors
Hair not combed	Stealing
Impertinence to visitors	Talking in chapel
Insolence to officers	Talking in dining room
Insolence to foremen	Talking at work
Insolence to fellow prisoners	Talking from cell to cell
Inattentive in line	Talking in corridor
Inattentive at work	Throwing away food
Laughing and fooling	

Some rules are inevitable in any situation where large groups are assembled. A few well-chosen rules are necessary in institutions but they should be the product of a slow evolutionary process and subject periodically to revision. Many transgressions against institutional routine cannot be anticipated; hence, when a specific type of overt behavior explodes, the administration has to insure against similar outbursts by making a rule.

It is plain that many rules cannot be abolished so long as inmates are not permitted to act as human beings with normal interests and motivations. In many prisons there are prohibitions against speaking in line, in the shops, and in the cells. The more repressive, the more rules there will be.

---

[23] Some more recent books by prisoners are: Paul Warren, *Next Time Is for Life* (New York: Dell, 1951); William Doyle and Scott O'Dell, *Man Alone* (Indianapolis: Bobbs-Merrill, 1953; and Bantam); Lew York, *Bighouse Banter* (New York: Vantage, 1954); Tom Runyon, *In For Life* (New York: Norton, 1954); Alfred Hassler, *Diary of a Self-Made Convict* (Chicago: Regnery, 1954); John Resko, *Reprieve* (New York: Doubleday, 1956). A work dealing with the problems of a woman in a modern prison is Helen Bryan, *Inside* (Boston: Houghton, 1953).

By the same token, the more advanced the institution, the fewer rules will be needed. Some institutions find it easier to handle men by changing the title of the rule book to "Helpful Hints to New Arrivals," with less emphasis on penalties and more on adjustment to the routine and purpose of the institution.

The Food

When the occasional outsider visits the average prison, he is usually guided to the kitchen and dining-room, where he is impressed by the shiny steam tables and other paraphernalia installed to mass-feed a prison population. He is shown a sample menu that he usually pronounces "pretty good"—for prisoners, at least. His guide takes him to the bakery and he smells the bread. Baking bread always has a pleasing aroma. This part of the tour is merely a window dressing or "Cook's Tour" set up for outsiders by the administration.

Assuming an adequate diet, in terms of calories, it is the method of cooking and the service that cause much of the discontent among prison inmates, so far as food is concerned. Many riots have been sparked by inadequate food or slovenly service. One reads of a "pork chop" riot; a "slumgullion" riot; a "bean" riot. For instance the riot at the Raiford, Florida, state prison on May 17, 1956 was referred to as the "beans rebellion."[24] Even too many pork chops get tiresome. There is an exclusion of the conviviality that generates a desirable psychic exuberance associated with good food.

In most prisons the food is served in cafeteria style with the men shuffling by the steam tables and going thence to a regularly designated seat, filling up the benches in strict order. To sit where one pleases is strictly against the rules in most prisons. In the older style prisons, where the food is served on the tables it is likely to be cold before the men file into the dining-room. The "plate" is frequently a large pan with small compartments to hold the various vegetables, very unattractive and cumbersome. In many prisons the inmates are obliged to eat with only a tablespoon as knives and forks are formidable weapons. Food is usually seasoned before it is served but in some prisons "community" salt boxes are on the tables, each man sticking his fingers in the box to procure a "pinch" of salt.

Inmates spend much of their time talking about the food. A piece of pie, a cup of coffee, some extra dessert—all loom large in their humdrum world. Fights among inmates are often precipitated over an extra cup of coffee. The dining room is, in fact, the place where many riots originate. The officials know this and take every precaution.

Inmates are usually allowed only a minimum of time to finish their meals—twenty minutes or less. In old-style prisons inmates all face the

---

[24] *Editor & Publisher,* Vol. 89, No. 42 (October 20, 1956).

front as they sit at the long narrow tables. In many prisons the wardens pride themselves on the little time consumed in feeding their prisoners. It is more important in "getting them out" of the dining room than in making the meal more attractive to the prisoners. Prisoners must not break line while entering or leaving. In one otherwise progressive prison, a guard is stationed in a specially constructed balcony well supplied with tear gas that the prisoners know will be turned on if rioting begins. In another prison, the dining-room is a large cage of iron bars. The men, after procuring their food from the steam tables, march into this iron-barred cage, and the gates are locked. No guards are inside; only inmate waiters. As the prisoners march out, eagle-eyed guards count the pieces of cutlery as they are dropped into a large tray.

Dining-room, Illinois Prison, Stateville. Note old-style feeding; all inmates face the same way, toward center.

Effects of Monotony and Routine on the Inmate

The prison regime brings into play a large number of disastrous influences that constitute a vicious circle. It would put the most severe strain upon even a thoroughly normal person, but it actually operates on those who are mostly physically or mentally abnormal or subnormal upon commitment. These emotionally unstable persons are denied in prison the assertion or enjoyment of the more important and basic human urges and impulses. Normal sociability is severely curtailed; self-assertion is practically denied; interesting or challenging work is rarely provided; play and recreation, if existent at all, are grotesquely inadequate for play space is limited and there is no variety of games. For instance, in many prisons where card playing is taboo, the inmates play dominoes, day after day. Cards are not permitted because they may be used for gambling. Yet gambling with dominoes is tolerated. In a survey made in 1952 "twenty-seven institutions permit card playing; twenty-six do not"; a few noted: "Under special conditions" or "farm group only."[25]

The natural oulet for the sex drive is totally denied in the modern prison, though sex cravings are abnormally strong because there is a blocking of the forms of emotional and intellectual expression that might drain off or sublimate such desires. In order to experience some vicarious sex catharsis many inmates plaster their cells with pin-up girls cut from magazines. Some wardens—and otherwise progressive ones, too—frown on the practice, while others humor the men by permitting it. There is no consistency.[26]

Nor can any sympathy be hoped for from the majority of those outside the walls, since the prison is society's mechanism for collective vengeance. Through it, as we have seen, the individual citizen gets a vicarious satisfaction, experiencing a pleasant indirect and symbolic release of the cruel and sadistic impulses that he could scarcely apply in direct contact with another individual. This perverse emotional motivation of the average citizen cannot be too greatly emphasized. It renders him unfavorably disposed toward any proposal for improving the situation, and this makes it very difficult to break into the vicious circle and bring about any significant and permanent advances and reforms.

At best, prison life tends to be deadly to the average human being. His vitality is sapped, his sensibilities are blunted, his "soul" is warped. Many become "stir simple," which is prison slang for the onset of some neurosis. Many men go to pieces under the strain and must be segregated. Modern corrections recognize the importance of prison psychoses as a legitimate area for investigation.[27]

Surly words between inmates over the slightest incident are magnified out of all proportion to their meaning or significance. Enmities, feuds, and

---

[25] Teeters, "A Limited Survey of Some Prison Practices and Policies," pp. 5 f.

[26] See two articles on this subject: Charles E. Smith, M. D., "Prison Pornography," and Stanley B. Zuckerman, "Sex Literature in Prison," both in the *Journal of Social Therapy* (hereafter, *"J. Soc. Therapy"*) (April, 1955).

[27] For details on the prison psychoses, see: Paul Nitsche and Karl Wilmanns, *The History of the Prison Psychoses* (Washington, D. C.: Nervous Disease Publishing Company, 1912); Arthur P. Noyes, *Modern Clinical Psychiatry* (Philadelphia: Saunders, 1939); or *New Horizons in Criminology*, 1st edition, 1943, pp. 608-615.

Typical inside cell block showing tiers. (Courtesy New Jersey State Department of Institutions and Agencies.)

gang warfare result from trivial situations and make the inmates' lives a virtual hell. The individual prisoner often develops unwarranted suspicions of his cell-mate, the guards, the warden or deputy, the chaplain, or anyone. During his spare time, usually at night, when lying on his cot, he wonders about his friends outside, or his wife or sweetheart, and imagines her unfaithfulness. Such thoughts drive him temporarily frantic. He lives a life of utter frustration. He may even try suicide. If the prisoner had any spirit when he entered the prison, it is often completely broken after months of this deadening routine. This is a serious indictment of the prison but it can be substantiated by any student of penology.

Such feelings of futility, shared by nearly all inmates of a penal institution, encourage conspiracies to escape. In the minds of many inmates escape looms large, even though carrying it into actual execution may never be realized. But it is exotic to contemplate. It furnishes the prisoner with excitement to spend hours dreaming and planning to escape. The public cannot blame men for attempting to escape from this sort of life. It is only surprising that more attempts are not made. To many behind prison walls, life is no longer meaningful and to die in an attempted escape holds no fear to many prisoners.

Recent prison history records many desperate escapes. The notorious Willie (The Actor) Sutton and several other inmates escaped through a sewer from the Eastern State Penitentiary at Philadelphia.[28] A Maryland prisoner tunneled his way out of the Baltimore prison, consuming about two years in hacking away at the dirt and stone; multiple prison breaks have occurred in recent years in the Delaware, Colorado, and West Virginia prisons. A Christmas Day tunneling attempt was frustrated at the Walla Walla, Washington, prison in 1957.

Most of those who escaped saw no hope in their future as they were doomed to serve fantastically long sentences. Death means little to men in despair. Merely piling on more years in duress is one of the elements of "dead end" penology. A very revealing autobiography entitled *In for Life* shows remarkable insight and relates a graphic story of the frustration and despair of a "double" lifer; a life sentence in Iowa and one in a federal prison under a detainer.[29] The writer, who died in prison in 1957, had made constructive use of the prison facilities under an understanding warden despite the regimentation and limitations of that institution. For years he edited the prison paper, *The Presidio*. His book points up the futility and unfairness of a long prison sentence.

There is a need for further sociological research in an attempt to verify the obvious frustrations of the prison community. The many books written by inmates, ex-inmates and trained observers strongly indicate that actual scientific research would clarify the situation and possibly convince legislators, judges and others in responsible public positions that it is futile to maintain an institution in which bitterness and human corrosion militate against rehabilitation.

Theoretically, a prison is intended to promote the reformation of the criminal, but in the light of its practical methods and actual operation probably nothing more ineffective or vicious could be devised as a method of protecting society from the depredations of the antisocial classes. Almost everything that could possibly contribute to the debasement and demoralization of the human personality characterizes present-day prisons and the contemporary methods of penal administration.

Our correctional institutions not only fail to reform; they are actually training schools in crime. A youngster sent to a juvenile institution by a judge "to teach him a lesson" and "to get some sense in his head," obtains his elementary education in criminal methods. His secondary school and his undergraduate collegiate career are passed in the state reformatory, where he has been sent "to learn a trade" to become a useful citizen. These reformatories are populated by late adolescents who have been small town

[28] For an account of Sutton's criminal career, see Quentin Reynolds, *I, Willie Sutton* (New York: Farrar, Straus, 1954).
[29] Runyon, *In for Life.*

nuisances and now have been changed by the contacts and routine of the institution into disillusioned "tough guys" familiar with all the ropes incidental to a life of crime. In addition they have become embittered against society. After discharge, the individual goes forth as a journeyman criminal, having won his spurs in antisocial conduct. If he is alert and well endowed by nature, he may avoid arrest and continue his criminal career with no important setback. If, however, he is lacking in intelligence and adroitness, or is a victim of hard luck, he moves on to state prison to start his graduate work in the field of crime in the prison seminars scattered among the long cellblocks and recreation yards under the greatest specialists available for his instruction in the ways of committing crime and avoiding arrest. He leaves prison not only a more competent criminal but even more an embittered man. Hans von Hentig, a German criminologist who spent many years in the United States, makes this indictment of the prison:

> The mammoth prison . . . breeds the professional criminal. It is the introduction to the organized underworld which receives here its material substructure and its ideological foundation.
> Idleness and overcrowding are but two sides of a dark picture. Even the best administered prison is the counterpart of a deeply disorganized society. The stratification in overprivileged and underprivileged is more strongly marked than in a revolution-ripe ancient regime.
> Prison life, however, means the survival of the most wicked; of the unfit. We may believe or not that crime outside pays. Here in prison deceit pays. We affirm that the honest and manly man is the best man. But who is the best prisoner? . . . In all times shrewd hypocrites, stool-pigeons and informers were regarded as "good prisoners." . . .
> All this shows that our prisons are unnaturally structured surroundings; they do not reflect the real world, its driving forces, its probable succession of fairness and attainment, fault and failure. In confinement there is no relationship of equal men; all human relations are unreal, fictitious, counterfeit.
> On the driest sand of a social Sahara we want non-swimmers to learn to swim. . . . We speak of delinquency areas, of degraded and degrading neighborhoods and of the ascendency of bad companionship. Like a burning-lens prison concentrates all these malignant forces and turns them on the convict. He cannot escape the powerful prison mores, and nowhere does this group give him the slightest support, as long as he does not descend to its muddy level.[30]

Such an appraisal as we have made of the prison in the preceding pages has been criticized by some administrators as unfair. They point to the "good" prisons throughout the country where, they claim, constructive therapy and progressive personnel are doing a fine rehabilitative job. We shall describe the so-called "New Penology" in a later chapter where we

---

[30] "The Limits of Penal Treatment," *J. Crim. Law,* Vol. 32, No. 4 (November-December, 1941), pp. 401-410; for a psychiatrist's analysis of the prison, see George J. Train, "Unrest in the Penitentiary," *ibid.,* Vol. 44, No. 3 (September-October, 1953).

Long outside cell block, Graterford, Pennsylvania prison.

contend that despite a better climate, a more enlightened program and a career staff, the results are none too flattering. Add to a term of imprisonment the frustrations the ex-prisoner must face in a free society and we see the bitter fruits of the system which augur little good to the control of crime.

Sex Deviation in Prison

There is probably no more delicate problem than that of sex in prison. Our mores still dictate against free discussion of sex, and prison administrators avoid a public airing of this problem situation, present in every prison, reformatory or children's institution, either for males or females. As one reviews the tables of contents of the various proceedings of the American Correctional Association and the many conferences on crime and correctional problems, he finds not one single article or speech devoted to this topic. It is mentioned only casually, if at all. Most textbooks dealing with corrections are also strangely silent on this subject. Sex is truly the pariah topic in the field of corrections.

Psychiatrists and psychopathologists have acquainted us with the intricate sexual behavior of the normal man and woman, yet we find jurists and administrators of correctional institutions flying in the face of these stark realities. Here we shall show the effects of prisons on the sex life of the incarcerated inmates, in the hope that readers will understand the wrongs we are daily inflicting upon convicted criminals.

Let us quote from a few well known persons who had insight enough to understand the problem of sex in prison. Oscar Wilde was afflicted with a sexual aberration that naturally focused his attention on the abnormal life of prison. His *Ballad of Reading Gaol* is a classic:

> With bars they blur the goodly sun
> And blur the goodly moon:
>> And they do well to hide their Hell
>> For in it things are done
> That Son of God, nor Son of Man,
> Ever should look upon.
>
> Each wretched cell in which we dwell
> Is a foul and dank latrine:
> And the fetid breath of living death
>> Chokes up each grated screen;
> And all, but Lust, is turned to dust
> In Humanity's Machine.
> . . . .
>
> The vilest deeds, like prison weeds
> Bloom well in prison air.
> It is only what is good in man
> That wastes and withers there.

One of the earliest books on prison life written in this country dealing with sexual cravings and aberrations of prison inmates is Alexander Berkman's *Prison Memoirs of an Anarchist*. More recently sex in prison was briefly discussed by Donald Clemmer in his *The Prison Community*.

We are also indebted to Joseph Fulling Fishman for making a comprehensive study in this field. His work, *Sex in Prison,* is authoritative.[31] Fishman has done a brilliant service in giving to the world an unvarnished account of the sex life in American prisons. His facts cannot be refuted. They are too well known by prison administrators and penologists alike, The only academic scientist to show any interest in this field of research was the late Alfred Kinsey. Sociologists and psychologists alike have avoided this important field of investigation. We include here a paragraph from a letter sent to the writers by Professor Kinsey relative to this subject:

> In regard to the incidence of the homosexual in penal institutions, I may indicate that we have never gathered histories from any male institution in which fewer than thirty-five per cent of the inmates were involved in homosexual relations while they were in the institution. We have never secured histories from any long-term institution in which fewer than sixty per cent of the men were engaged in such activity, and in one such institution we had over ninety per cent of the inmates admit such experience within the institution. Until prison authorities comprehend the magnitude of this problem, it is possible to be totally unrealistic in attempting to cope with it.[32]

---

[31] Joseph Fulling Fishman, *Sex in Prison* (New York: National Library Press, 1934). See also George Sylvester Viereck, *Men Into Beasts* (New York: Fawcett, 1952).

[32] In a letter to the authors dated November 20, 1950.

374 THE ERA OF REFORM

Victor Nelson, in his *Prison Days and Nights,* makes this illuminating observation:

> To the man dying of hunger and thirst it makes little difference that the only available food and water are tainted. Likewise it makes little or no difference to the average prisoner that the only available means of sexual satisfaction are abnormal. It is merely a matter of satisfying as best he can the hunger that besets him. I mean a hunger not only for sexual intercourse but a hunger for the voice, the touch, the laugh, the tears of Woman; a hunger for Woman herself.[33]

Deprived of all normal sexual relationships for a long period of time, it is almost axiomatic that men and women will develop abnormal liaisons in self-defense, heterosexual as the vast majority of people are, most of them try desperately when in prison to remain so. What percentage resorts to homosexuality cannot be ascertained, but it is high—far too high to be ignored or minimized. Most thwarted persons resort first to masturbation and there is no doubt that this practice is rife in prison. As it is indulged in secretly, its extent cannot be more than estimated. Fishman claims that men of very strong will who understand the ordeal they must experience while in prison pledge themselves that they will not succumb to self-abuse or to homosexuality. But those in this category are rare indeed. Sublimation of the sex drive is difficult under ideal conditions outside in the free community; it is much more difficult in prison with its world of phantasy.

If the inmate is to some degree cultured, he may turn to various avocations or hobbies to drain off his sexual hunger. Many go to extreme length in their lonely struggle to maintain their self-respect, if this is the word that should be used. They have carried over into a male community—the prison—the attitude of the free community that masturbation is degrading. They will do anything humanly possible to remain free from this taint. Many pace the floor of their cells, go through calisthenics, learn long passages of poetry—anything of a physical or mental nature to keep sexually normal. A man with remarkable self-control and a short sentence may succeed, but the majority of those immured will eventually succumb. When deprived of a normal sexual outlet the average man in prison will resort to masturbation.

Homosexual Tendencies in Prison

The true homosexual should not be a serious problem to prison administrators, although he will always be a nuisance. Special areas of the prison may be set apart for such deviants so that they cannot mingle with other inmates. This would entail a regime in which the group could work together, eat together, and sleep in separate cells. This represents the reality of the situation. Contrast this to a situation one of the writers witnessed in the Colorado state prison in 1948. Here the warden dressed his known

---

[33] Victor Nelson, *Prison Days and Nights* (Boston: Little, Brown, 1932), p. 143.

homosexuals in female clothes and permitted them to mingle with the rest of the population—at least at work. They marched into the mess hall in the regular line but were required to sit at a table alone. Such a method of dealing with these unfortunate persons is both cruel and senseless.

In the prison the young boy (*punk,* in prison parlance) is often the center of bitter enmities and even feuds that frequently precipitate brawls, sometimes of a serious nature. Inmates who go foraging for a "girl friend" usually have a knife on their person that is brought into use on the slightest pretext from a competitor. It is traditional for inmates of this type to carry knives fashioned from any stray bit of metal, including such harmless objects as spoons. They are usually honed down to razor-like keenness that makes them deadly indeed. It was in such a brawl that Richard (Dicky) Loeb was stabbed to death by another inmate of the Joliet, Illinois, Penitentiary some years ago.

A particularly revolting account of bestial sexual activity came to light a few years ago when Haywood Patterson, one of the nine Negro victims of race prejudice during the 1930's in Alabama, wrote a book in which he told his story. Patterson was sent to the Kilby prison for 75 years on an alleged rape charge. He eventually escaped. Here is one excerpt from his revealing book, *Scottsboro Boy:*

> I learned men were having men. . . . A fifteen-year-old had no chance.
> . . . Prisoner and warden were against him and he was quickly made into
> a woman. . . . "You go back there," said the warden, "and be good to
> that old man." The Negro boy was nothing to him. The boy returned to his
> cell. The old wolf beat him unmercifully. The other prisoners just looked
> on. They knew a young woman was being born. . . . After the boy was
> beaten up and lying on the floor the old wolf picked him up and brought
> him to his bunk. The covers went over the old wolf's double-decker bed.
> . . . The boy was being broken. . . . I saw men learn to love boys harder
> than they could any woman.[34]

Fishman graphically describes just how this vicious undercover activity works in a prison:

> This unfortunate condition is achieved not only through the negation of
> normal sex habits, but because of the constant talk concerning sex, enforced
> idleness, the loneliness of one's cell; and finally the relentless pressure of
> the "wolves" or "top men" housed among the normal inmates in the prison,
> who "spot" those among the younger prisoners whom they wish to make
> their "girls," and who "court" them with a persistence, a cunning, and a
> singleness of purpose which is almost incredible in its viciousness.
>
> They usually begin with a friendly offer to protect the newcomer, and to
> see that his life in prison is made as easy as possible for him. This offer is
> often gratefully accepted by the new inmate because he is not yet accustomed to prison life. The pressure of the older prisoners on the new and
> young arrivals can and often does make itself felt in a thousand subtle,
> irritating ways.

---

[34] Haywood Patterson and Earl Conrad, *Scottsboro Boy,* p. 82. Copyright 1950 by Earl Conrad. Reprinted by permission of Doubleday & Co., Inc.

The first advance is usually followed by the giving of small presents, such as a box of cigarettes purchased from the prison commissary. Unless the new prisoner has someone to "put him wise," assuming that he does not know the object of these advances, he gradually slips into a position of helpless dependency on his self-styled protector. When the final purpose of these attentions becomes known, and if the object of them resists, he is very often threatened with physical harm. This unfortunately is too often.[35]

Both Fishman and Victor Nelson give actual case histories of young boys who have succumbed to the wily techniques of such "top men."

This intramural sexual behavior, with its incidental but deadly warfare, exists in virtually all prisons and juvenile institutions. In any situation, in or out of prison, where men live together in fairly large numbers and are denied normal relationships with women, this type of perversion takes place: among sailors, lumbermen, transitory laborers, hoboes and tramps, and in the army. But it is only in prison that it is *absolutely impossible* for the individual to find some companionship among those of the opposite sex. And the development of a homosexual orientation in prison is directly the fault of society, for it is society that forces the convicted criminal to assume an abnormal existence.

Homosexual behavior is not restricted to male institutions, but is found in women's reformatories and in girls' correctional institutions as well. Many of the females sent to these places have not developed inhibitions and thus find the situation almost unbearable. They easily turn to the various forms of erotic sexual behavior and, as in the male institutions, debauch the more sensitive and feminine of their fellow prisoners.

It is particularly difficult for administrators to control this problem in female institutions, largely because the inmates have more freedom than male prisoners. Women's reformatories are usually of the cottage type with large campuses, where friendships between girls and women have few restraints. With little regimentation or rigorous supervision, "crushes" are more generally tolerated. But the same, subtle, beneath-the-surface struggles found in the men's prisons are present in the female institutions. The more feminine younger girls are in demand among the more masculine and older inmates.

It is assumed by some administrators that a vigorous program of work, physical activity, and recreation will stamp out this institutional practice. No doubt such a program does reduce it, but so long as healthy men and women are segregated according to sex, there will remain a fertile field for homosexual tendencies. The only solution is the abolition of the conditions that foster such behavior.

---

[35] Fishman, *Sex in Prison*, p. 84. For further descriptions of this type of debauchery, see Frank Tannenbaum, *Crime and the Community* (New York: Ginn, 1938), pp. 429-30; and more recently, Donald Webster Cory, "Homosexuality in Prison," *J. Soc. Therapy* (April, 1955).

This would unquestionably involve the practice of permitting inmates some opportunity of normal sex contact. There is nothing new nor daring in this suggestion, for this has been tried with success in several countries throughout the world, especially in Latin America. In some other countries, notably Sweden, a system of furloughs home meets the need for normal sex practice.[36]

A fact little known to the public is that even now, in this country, sexual visits are permitted by wardens who might be termed "over-indulgent." It is a strange anomaly that these visits are permitted in what may be termed "backward" prisons. While this is contrary to law, few persons actually object to the practice in the states where it is permitted.

The least that can be done, as matters stand now, is to see that every prisoner has *his own individual cell.* This is the rule in most female institutions, but there is a general laxity in this respect in most male prisons. No matter how good the intentions when an institution is built, the prison population invariably increases beyond the cell space. Thus, in most of our prisons today we find two men in a cell. This is regarded by most correction people as the worst possible housing arrangement. So far as preventing sexual irregularities is concerned, it is better to crowd three men into a cell originally built for one than to permit two men to live together. But even in this arrangement, two men may "gang up" on the third. The only answer is the individual cell.

Few persons realize that a basic need is some degree of privacy. It is so completely taken for granted that when it is mentioned as a right for inmates in penal institutions it carries little weight. Society should see to it that the prisoner has some privacy, at least at night. Since so many of our prisons were erected during the maximum-security and mass-treatment orgy of the nineteenth century, they do not lend themselves to such common decency. Many of the newer institutions neglect to provide enough individual cells, and instead, create the dormitory. In these, the prisoners get no privacy whatsoever, day or night.

The problem of sex is one of the most challenging that confronts the administrators of our prisons and one of the most baffling in a free society where large numbers of the same sex are thrown together for long periods of time.[37] The sex perversions learned in prison are normally correlated with the parallel or subsequent pathogenesis of many types of psychic abnormality and emotional instability that emerge in definite criminal compulsions.

It is alleged that many confirmed homosexuals become so because of an institutional experience. How far this is true we do not know. Justin K. Fuller, former medical director, U.S. Public Health Service, states:

---

[36] See our discussion of "conjugal visiting" and furloughs in Chapter 31.

[37] For further information on this problem, see Ben Karpman, "Sex Life in Prisons," *J. Crim. Law,* Vol. 38, No. 5 (January-February, 1948), pp. 475-86.

> When . . . a person is seduced—and many are—and indulges in such acts homosexual more or less regularly during incarceration, he is apt to suffer not only a permanent moral degeneration, but is likely to develop a preference for homosexual rather than heterosexual indulgence, and upon release is found to be a confirmed sex pervert.[38]

There has been no research in this area, although it is obvious that it is extremely necessary. Many who verbally chastise criminals have no compunctions about the sexual deterioration of men sent to prison, but they might be quite concerned were it established that these same persons returned from prison as confirmed homosexuals.

The Southern Chain Gang as a Penal Anachronism

The chain gang, so indigenous to the South, is in reality a prison road camp. In the era following the War between the States, the southern states leased their prisoners to private contractors. They were worked in construction gangs, on plantations, in turpentine camps, or sawmills. It was only natural that cruelty should develop in this type of open prison. Contractors tended to exploit the prisoners and the prisoners retaliated by shirking work and absconding. Shackles and chains were pressed into service as were the whip, the sweatbox, and other forms of corporal punishment. In time, southern prison camps became known as chain gangs.

An early document describing a penal camp in Florida is worth recording. Writing in 1891, Captain J. C. Powell, for eighteen years an overseer, called the system "The American Siberia." In 1876 much of Florida was a semitropical jungle. Turpentine extraction was profitable if labor was cheap enough, and labor had to be forced because the work was so onerous. Prisoners were worked in gangs, chained together in filthy bunkhouses, exposed to dysentery and scurvy. At first most of the prisoners were white men, "crackers," from the backwoods of the palmetto country. As time went on, however, the bulk of the gangs was composed of Negroes, for they were more amenable to the cruelty and indifference of the overseers. Here is an excerpt from Powell's book:

> It was not long before the camp was ravaged by every disease induced by starvation and exposure. The pestilential swamps were full of fever, and skin maladies, scurvy and pneumonia ran riot. Dysentery was most common, and reduced men to a point of emaciation difficult to describe or to credit. Every stopping-place was a shambles, and the line of scurvy was punctuated by grave-yards.
>
> The camp was at different times in charge of various captains, and under some of them the punishments were excessive. Hanging up by the thumbs was usually resorted to, and this led, one night, to a grisly tragedy. A Negro

---

[38] In a speech delivered February 7, 1948, at New York University before students enrolled in the "Field Course on Administration of Penal and Correctional Institutions."

convict was strung up for some infraction of the rules. Whipcords were fastened around his thumbs, the loose ends flung over a convenient limb and made taut until his toes swung clear of the ground. The scared convicts huddled about the camp-fire and watched their comrade as he writhed and yelled, expecting every moment that the cords would be unfastened and his agony ended.

But the captain had determined to make a salutary example, and he let the Negro hang. Meantime the poor wretch's anguish was a hideous thing to see.

They say his muscles knotted into cramps under the strain, his eyes started from his head, and sweat poured from his body in streams. An hour passed—then two. His shrieks had ceased and his struggles grown feeble, so they let him down and he fell to the ground like a log—dead.

Here was a study for an artist. Night in the palmetto woods, the flaming camp-fire outlining the circle of frightened convicts and the miserable barracks where they slept, the distorted corpse upon the ground, and the panic-stricken officer creeping away among the trees.[39]

Many articles and not a few books have been written on this benighted penal system and it would seem none has been overexaggerated. Probably the most sensational was Robert E. Burns' *I Am a Fugitive from a Georgia Chain-Gang,* which was adapted into a motion picture.[40] Burns gained notoriety when he escaped from a Georgia prison camp and was apprehended in New Jersey, from which state the governor refused to permit his extradition.

In more recent times prisoners are placed in the custody of the counties and engaged in various types of work for those political units. Private leasing has all but been abandoned throughout the country. The present system is highly decentralized and unsatisfactory, from a correctional standpoint, aside from the inherent cruelty so often associated with it. There is little attempt to rehabilitate the prisoner. He is mistreated in many units and is at best merely worked and guarded. The states vary in their persistence in clinging to the system, some making a serious attempt to abolish it completely, and others rationalizing it with amazing casuistry. For example, when Eugene Talmadge was governor of Georgia, in 1935, he defended the chain gang system because it kept men out of doors in God's open country where they could enjoy the singing of the birds, the beautiful sunrises and sunsets. He painted an idyllic scene in which the prisoners mingled on occasion with farmers and their families.[41] In his late term as governor in 1940 he told the citizens of Georgia that if they could not find workers he would get them a "convict."

Within the past quarter century some of the stories that have come out

---

[39] J. C. Powell, *The American Siberia* (Chicago: H. J. Smith & Company, 1891), pp. 13-14.

[40] Robert E. Burns, *I Am a Fugitive from a Georgia Chain-Gang* (New York: Vanguard, 1932).

[41] See *Proceedings,* American Correctional Association, 1935, p. 49.

of these camps have been almost unbelievable. Hitchhikers have been arrested and sent to a chain gang merely to supply the county with cheap labor. Sick prisoners have been beaten into insensibility and even unto death because the overseers have accused them of malingering. Men have been clandestinely buried in quicklime with no death certificate recorded so that their names and presence might be obliterated from the scrutiny of prying official eyes. Human degradation has fallen to lower depths in the chain gangs than any other form of penal treatment in America, closely approximating the conditions in the penal colonies of Siberia or Australia in the early days of the transportation system.

The administrative personnel of these camps is usually grossly underpaid and, as a rule, of low caliber.[42] As most of the prisoners are Negro and the guards white, the caste system aggravates the situation. There is little or no correctional treatment in most of these southern road camps. When one of the writers suggested in 1958 that a counselor be attached to each camp, the director of correction for that state ruefully replied that it would take at least ten years to obtain even this beginning of a treatment program.

Cage wagons in chain gang camp.

In the housing of prisoners in state and county camps one often finds anything from steel "circus wagon" cages and temporary shacks to more permanent barracks. Many chain gang camps turn back considerable money to the treasury by exploiting the prisoners' labor or by affording them a low standard of physical provision such as food, bedding, clothing, and medical care.

---

[42] In a southern state one of the writers visited in 1958 he found that guards worked 70 hours per week and starting pay was $175 per month.

Although there have been no official exposures of bad conditions in the chain gangs in recent years, it is obvious that some are maintained on a very low level of human decency. In 1949, Bayard Rustin of the Fellow-ship of Reconciliation Bureau published his own account of a chain gang in Orange County, North Carolina. In this he describes the rugged con-ditions he himself experienced through a thirty day period. He indicted the food, the personnel, the discipline and the types of punishment of the camp.

In 1956, William F. Baily, director of prisons in North Carolina, stated that the use of leg shackles in 84 of the 93 penal units was forbidden and added that leather leg cuffs were to be limited to nine camps for "problem cases." He stated: "The department looks forward to the day when the last vestige of an outmoded penology can be finally removed from the state prison system."[43]

This would indicate that in time the chain gang will be completely abolished. What is needed in all backward states, both North and South, is an enlightened correctional program with all that this implies. There is something definitely valuable in the southern correctional philosophy if it can be capitalized, administered and nurtured by progressive men and women. The segregation of prisoners into small units scattered throughout a state makes possible a real system of classification. What is needed, however, is an adequate treatment program for each of these small units with a pro-ductively efficient work program which is not the case in the road-building found in the southern states. California has blazed a trail now followed by some other states, in providing forestry camps for hundreds of men who otherwise would be housed in grim maximum-security prisons. A more humane philosophy of treatment, better-trained personnel, and a realistic work program can change the southern chain gang system into something possessing real merit.

Dead-End Penology—Alcatraz Prison

The year 1934 marked the beginning of a phase in which criminals were referred to as "mad dogs" and enlightened penologists as "sob sisters" or "cream puffers." In an editorial in the *Journal of Criminal Law, Crimi-nology, and Police Science,* we find a compilation of quotations from the speeches of J. Edgar Hoover that illustrates this argument between the "sentimentalists" and the advocates of the "treat 'em rough" school:

> Criminals are not just criminals. They are: "Scum from the boiling-pot of the underworld," "public rats," "lowest dregs of society," "scuttling rats in the ship of politics," "vermin in human form," "the slimy crew who feed upon crime," "desperadoes," "vermin spewed out of prison cells to continue their slaughter," "the octopus of the underworld." These "post graduates

---

43 From a news item in *Federal Probation,* Vol. 20, No. 1 (March, 1956), p. 74.

of outlawry" and "professors of crime" thrive "in the great fog of crime," and the "swamp and morasses of suffering" amidst the "appalling scourge of perjury" and the "oleanginous connivings of venal politicians," aided and abetted by "sentimental yammerheads," "moronic adults" of "asinine behavior," "maudlin sentiment," and "inherent criminal worship." Away with these "moo-cow sentimentalities" with their "mealy mouthings" and their "whining pleas for sympathy"; these "hoity-toity professors"—

This mosaic is from the speeches of J. Edgar Hoover of the Federal Bureau of Investigation, Department of Justice. He has announced a hearty contempt for "the cream-puff school of criminology whose daily efforts turn loose upon us the robber, the burglar, the arsonist, the killer, and the sex-degenerate." He condemns "these self-appointed ambassadors of the open cell block." He is horrified by the "ignorant blatherings of either ill-informed or selfishly-motivated persons" with their "blatant outcries."[44]

Alcatraz, San Francisco Bay; maximum security federal prison.

It was during this period of name-calling that the country was bedeviled by a hard core of desperate criminal gangs such as the Dillinger, Karpis, Barker, and Underhill aggregations. To the credit of the F.B.I. the gangs were liquidated although several brave men were shot down by gangsters.

To follow through with this war on crime Attorney General Homer S.

[44] *J. Crim. Law*, Vol. 28, No. 5 (January-February, 1938), p. 627. For an exchange of views concerning this matter between J. Edgar Hoover and Judge Harry W. Lindeman, of Essex County (Newark) New Jersey, President of the National Council of Juvenile Court Judges, see "The Mollycoddling Charge," N.P.P.A. *News* (May, 1957), p. 1.

Cummings set up Alcatraz prison in San Francisco Bay. This was in March, 1934. There were many misgivings among prison men about the advisability of creating such a bleak maximum-security facility.[45] Their protests went unheeded and through the years the Federal Bureau of Prisons has been obliged to maintain it although its excessive expense as well as its basic philosophy of correctional treatment have never been justified. Alcatraz is a monument to the thesis that some criminals cannot be reformed and should be repressed and disciplined by absolute inflexibility. But today not even the Federal Bureau defends Alcatraz. Yet it contends that a new and more modern maximum-security plant is the answer to those relatively few inmates that defy discipline and order. All of our states must handle this type of criminal without the benefit of new construction especially designed for the "rotten apples out of the barrel" as Director James V. Bennett calls them. Alcatraz is no longer justifiable, but to erect a super-maximum-security prison in its place is highly debatable.[46]

The Dilemma of Prison Unrest

Prison unrest, often exploding into costly rioting, is not a new phenomenon in the history of penology. While we read of an occasional riot abroad, they have occurred more frequently, through the years, in the United States.

There were riots and mass escapes in the old Simsbury, Connecticut, copper mine prison as early as 1774. There were riots, escapes, and assaults by prisoners on guards in the old Walnut Street Jail in Philadelphia between 1790 and 1835; also in the early prisons of Maine and Massachusetts. More recently there was a wave of riots during the years 1929-1932.

There have been riots in the super-prison, Alcatraz. In May 1946 that prison experienced its worst riot; at least it was played up with all the gusto that the public press could command. A few prisoners, armed with a pistol with 27 rounds of ammunition held off the guards and a detail of marines, flown in from west coast barracks, for a long period of time before they were subdued. This episode was blown up in the press and the public was informed periodically over the radio as to the progress of the "battle." The warden at the time, James A. Johnston, devotes an en-

---

[45] Among those opposed were former Commissioners of Correction A. Warren Stearns of Massachusetts, John Ellis of New Jersey, and Walter N. Thayer of New York State.

[46] A short bibliography on Alcatraz includes: Frederick R. Bechdalt, "The Rock," *Reader's Digest,* Vol. 28, No. 165 (January, 1936), p. 42; Anthony M. Turano, "America's Torture Chamber," *American Mercury* (September, 1938); E. E. Kirkpatrick, *Voices From Alcatraz* (San Antonio, Texas: Naylor, 1947); James A. Johnston, *Alcatraz Island Prison* (New York: Scribner's, 1949); Gaddis, *Birdman of Alcatraz.*

Prison in Southern Michigan at Jackson where serious riot occurred in 1952. This prison holds the dubious reputation for detaining the largest

tire chapter of a book on Alcatraz to the riot which he calls "The Battle of Alcatraz." He also draws a plan of the battle.[47]

But during the period 1950-1956 almost one hundred riots and serious disturbances swept the nation—a dozen on the average every year. Property destruction is estimated at more than ten million dollars—five million dollars in the Missouri prison riot of 1954 alone.[48] Few states escaped this demoralizing penal bankruptcy.

Many reasons have been advanced for this widespread unrest. These ranged from a general social unrest as reflected by the critical times in which we have been living, to the "bad apple" theory which places the blame on a "few disgruntled, hardened criminals" in each institution. It was alleged by some that the riots were sparked by tough, cynical young adults, a breed of youth peculiar to modern times, "the like of which we have never had" in previous eras. The "mass contagion" theory was also advanced which blames rioting on newspaper reports and radio bulletins that permeate most of our prisons and engender unrest.

The American Correctional Association published a report by correctional experts appointed to study the causes of the riots and, perhaps, suggest measures to prevent others in the future. This committee reported:

> The immediate causes . . . are usually only symptomatic of more basic causes. Bad food usually means inadequate budgets reflected in insufficient supplies, poor equipment, poor personnel and often, inept management. Mistreatment of prisoners, or lax discipline, usually has behind it untrained employees and unwise or inexperienced management. The fundamental causes of prison mal-administration may be categorized under a number of general heads:
>
> A. Inadequate financial support, and official and public indifference.
> B. Sub-standard personnel.
> C. Enforced idleness.
> D. Lack of professional leadership, and professional programs.
> E. Excessive size and overcrowding of institutions.
> F. Political domination and motivation of management.
> G. Unwise sentencing and parole practices.[49]

---

[47] Johnston, op. cit., pp. 221-237.

[48] So stated in a report issued by the Prison Association of New York in 1956. A complete list of riots and prison rebellions from 1855 to 1955 has been compiled by Vernon Fox in his revealing book dealing with the Michigan riot of 1952 in which he played a prominent mediating part: Violence Behind Bars (New York: Vantage Press, 1956).

[49]"Prison Riots and Disturbances," prepared by the Committee on Riots, American Correctional Association, 135 E. 15th St., New York City, May 1953. For further material on the riots of 1952-56, see: "Prison Riots—Why?" Prison Journal, Vol. 33, No. 1 (April, 1953), pp. 2-27; "Aftermath of Riot," Vol. 34, No. 1 (April, 1954), pp. 3-48; Harry Elmer Barnes and Donald Powell Wilson, "A Riot Is an Unnecessary Evil," Life, Vol. 33, No. 21 (November 24, 1952); Peg and Walter McGraw, Assignment: Prison Riots (New York: Holt, 1954), a synopsis of an NBC documentary of nine hour-long broadcasts: "U.S. Prisons: How Well Do They Protect Us?" Platform, published monthly by the Club and Educational Bureaus of Newsweek (September, 1952), pp. 1-21; John Bartlow Martin, Break Down the Walls (New York: Ballantine, 1954); Frank E. Hartung and Maurice Floch, "A Social-Psychological Analysis of Prison Riots," J. Crim. Law, Vol. 47, No. 1 (May-June, 1956), pp. 51-57.

But are these reasons only "symptomatic of basic causes" inherent in the prison? We quote from Albert G. Fraser's review of the monograph mentioned above:

> I make bold to suggest that until we make the prison a more human habitation, in a psychological sense, prison riots are inevitable for the very reasons which the monograph cites. Certainly, there must be qualified personnel, classification, full employment, smaller prisons, all the essential elements of a dynamic, positive program, and with all these, that which the committee fails even to suggest, a psychological setting in which "the prisoner's self-respect can be cultivated to the utmost." No authoritative statement on prison riots in this day and age can ignore or should fail to emphasize the feeling aspect of prison life.
>
> Should not one reasonably expect an authoritative statement on the subject of corrections today to include some reference to *individual help* and the process by which it is made available to the prisoner? The monograph sets down widely accepted theories of prison management. It resorts to a hackneyed, but still popular pastime, by placing the blame on "politicians" and an indifferent public for ills which afflict the prison. I suggest that we in the correctional field start to give an up-to-date interpretation to the challenging and too long neglected principles of the "Declaration of Principles."[50]

In short, riots of prisoners will continue so long as the individual's personality is ignored through the mass-treatment concept which insists on the "locking psychosis," regimentation and monotony of day by day existence in a setting where suspicion and fear and contempt dominate the personnel and in which the conflict between custody and treatment cannot be resolved and where treatment is eternally subservient to custody.

---

[50] *Prison Journal,* Vol. 33, No. 2 (October, 1953), pp. 23-24. See page 425 for further discussion of the American Correctional Association's Declaration of Principles.

24

The County Jail
and the Town Lockup

I am firmly convinced through my experience . . . that the more than 3,000 county jails and more than 2,000 city lockups of the United States are the most critical of the penal problems facing the American people if we would cut short the development of criminal careers. (From introduction of his report, "The Cook County Jail, December 6, 1954–August 1, 1957.")

JOSEPH D. LOHMAN
SHERIFF, COOK COUNTY, CHICAGO

The Historical Background of the Jail

Many are the harsh names that are hurled at the oldest and lowliest of all penal institutions, the county jails. They are denounced as "the cesspools of iniquity," "hot-beds of graft and corruption," "schools of crime," and "kindergartens of vice." These are all true indictments of the jail but they are not new. In colonial days, both in England and in the American colonies, jails were criticized for their promiscuity and corruption. Yet they are a highly important cog in the machinery of the correctional process.

There are over three thousand county jails in this country, in addition to some 10,000 city and town lockups. There are fewer workhouses, or "houses of correction," since only the larger cities maintain separate establishments for the maintenance of vagrants and derelicts, traditional occupants of such institutions. In a few states, two or more counties jointly maintain regional workhouses.

Of all the abodes for the criminal classes, the jail is the vilest from the standpoint of sanitation; the most absurd from a functional point of view; and the most inefficient in administration. In fact, the county jail

THE ERA OF REFORM

is an anachronism. If the American people had the will, the jail could be abolished. But jails are deeply rooted in local politics, and any suggestion that they be merged into a wider and more efficient system—under regional or statewide control—is indignantly resisted from local pride and suspicion.

The county jail is as important as the state prison. The majority of those sent to prison have already spent time in the jail awaiting trial. Thousands of others have served time in the jail for minor offenses. And, all too often, innocent persons have had jail experience. For instance, many of those acquitted by a judge or jury, have already been obliged to serve detention, sometimes for months and even years, in the jail.

In our contemporary crime situation the jail outweighs all other penal establishments in sheer bulk, for an infinitely greater number of persons are handled in the jails than in all other correctional institutions combined. It is difficult to estimate how many persons pass through our county jails each year, but it is a sizeable number. Louis N. Robinson, a close student of the jail problem, estimated that approximately one half million persons are committed to jails by the courts annually.[1] As early as 1930 it was stated by Hastings H. Hart, a pioneer in jail reform, that fully three million persons were committed to jails in a single year although he conceded that many were committed more than once.[2] If this were true in 1930 it is obvious that such a figure is conservative today.

A recent statement regarding the jail has been made by Myrl Alexander, assistant director of the Federal Bureau of Prisons and an indefatigable worker in jail reform. He writes:

> Far too many jails . . . are little more than the enforced meeting place for social derelicts who find there the greatest opportunity to infect the casual offender, the unsophisticated, the morally retarded, and the socially inadequate. Moreover, such jails are often unsuccessful in performing their basic mission of secure detention. In them, jailers' responsibilities are delegated to the most sophisticated and experienced criminals who proceed to prey upon the majority of other prisoners through tacitly approved kangaroo courts, "sanitary courts," and other devices and insidious methods concocted by those morally corrupt criminals schooled in the slimy culture of mankind's social backwash.[3]

Recognizing that most jails are not fit to house offenders before or after trial, the federal government has rated most of them as unfit for detention of its own prisoners. There are a few efficient jails scattered throughout

---

[1] Louis N. Robinson, Jails: *Care and Treatment of Misdemeanant Prisoners in the United States,* Philadelphia: Winston, 1944, p. 7.

[2] Report of the National Commission on Law Observance and Enforcement, No. 9 (Washington, D.C.: U.S. Government Printing Office, 1931), p. 329.

[3] Myrl E. Alexander, *Jail Administration* (Springfield, Illinois: Charles C. Thomas, 1957), p. 6. Reprinted by permission of Charles C. Thomas.

the country, but none measures up to any dignified standard as an institution for the scientific treatment of offenders.

The Colonial Jail

The jail (or "gaol") is an ancient institution, its origin lost in the hazy mists of antiquity. As a place of detention, however, its counterpart has existed since Biblical times. As we understand the jail in more modern times, we might accept the date of origin as 1166, when Henry II commanded the construction of jails at the Assize of Clarendon.

Jails were originally conceived as places for the detention of suspected or arrested offenders until they could be tried by the courts. This function still persists along with others not so logical. Later, another institution evolved in England known as the House of Correction (or Bridewell; see Chapter 22), which was a place of punishment. During the eighteenth century, these two types of institutions gradually merged and frequently were under the same roof and were administered by the same keeper. Hence, the jail was not only a place for the detention of suspects but also a penal institution for convicted petty offenders and vagrants. Herded together in small quarters were diverse types who had run afoul of the law, some of them possibly innocent.

When the early colonists came to America they brought with them institutions developed in the mother country. Hence, we see the jail set up as soon as numbers of people began to collect in any one place. There are a few of these early jails still in existence, although not in use. The jail at York Village, Maine, was erected in 1653; that at Williamsburg, Virginia, in 1701. Both are still standing and are open to the public as historical museums.

Colonial jails were alike in most respects. Nearby were the stocks and pillory, and in some places, the whipping post. Persons suspected of crimes were held in these houses of detention until the meeting of the court that was called "Quarter Sessions and General Gaol Delivery." In many places they are still officially known by this title.

There were no cells in these early jails; only small rooms in which were often herded 20 to 30 prisoners. There was no heat except that which the inmates could furnish for themselves by burning material in the fireplaces that were in each room. Food was sold by the jailer, or the inmates got it from friends or philanthropically inclined persons. The condition of most of these unfortunate individuals beggars description.

The Fee System in the Jail

The system of paying the jailer fees rather than a salary for maintaining prisoners is a relic of the dim historic past. The sheriff of the English county was, in the beginning, appointed by the Crown and his position was one of dignity. The position was a sinecure carried on with much pomp and little work. He farmed out the actual and more onerous duties of caring for the jail and its occupants to a keeper as a concession. The keeper's income came from fees exacted from the inmates or their friends. Each inmate paid so much for maintenance, for light, heat, and, in fact, for the privilege of being in jail. The more inmates in the jail and the longer they remained, the larger was the keeper's income.

This fee system continued in England without much criticism in official circles until the time of the great jail and prison reformer, John Howard. In 1773, Howard was named sheriff of Bedfordshire. Outraged with the fee system and the horrible condition of the prisoners, he determined to do something about it. But Howard found himself frustrated by a system that could not be eliminated overnight. After petitioning the county justices and eventually Parliament, he saw mitigated many of the grave injustices.

The fee system was brought to America with the colonists and took root in local governmental administration. Here it remained unchanged for many years; indeed, its stepchild is alive today: although the inmate no longer pays, the keeper and sheriff are paid a fee for every service by the county instead of a salary. The way the modern fee system works may be seen from the following schedule of fees in the state of Virginia in 1937:

For receiving a person in jail when first committed	$ .50
For keeping him and supporting him therein, for each day	1.00
But when there are as many as three and less than ten prisoners in jail, for each	.75
Where there are ten or more prisoners in jail, for each up to and including twenty-five	.60
For each prisoner in excess of twenty-five, up to and including fifty	.50
For each prisoner in excess of fifty	.25

The sheriff was here paid on a *per diem* basis for each prisoner out of the funds of the state, rather than by the county, as in most states. For the year ending June 30, 1942, jail costs in Virginia amounted to $471,099.53. Virginia is now on the list of states that has abolished the fee system.

Where the fee system is in operation, constables and deputies as well as sheriffs and justices of the peace are paid on a prorated basis. The more arrests, the more money paid these officials. These officers did not create this system; it was handed down to them and they cannot be blamed for playing it for all it is worth. Not until all officers identified with the jail are placed on a salary basis will this system be eliminated. As Alexander points out: "Personal and professional status requires businesslike management of the jail, an impossibility under the fee system."[4]

---

[4] Alexander, *Jail Administration,* p. 314. Reprinted by permission of Charles C. Thomas.

The Jail Today

As we have described the jail, we see it as a survival from pre-prison days. The original jail performed the function of keeping people in custody while they were being detained for trial, and until the 1830's and 1840's the only prisoners they held under long-term custody were debtors.

As state prisons and reformatories developed to take care of the great majority of those who had earlier been inflicted with corporal punishment, there were certain residual types in the varied criminal populations that did not seem to fit in anywhere. Some belonged in homes for dependents and feebleminded, but the tendency was to send them to the jail for winter—vagrants, dipsomaniacs, drug addicts, and those with mild mental disease. This naturally led to the demoralization of the jails through their utter promiscuity. And even now we do not keep out of our jails all types that should be sent to other specialized institutions. This practice has led to the permanent degradation of the jail population.

The jail, then, still serves the same functions as in colonial times. It is still a place of detention of those awaiting trial—its original function; a prison for the incarceration of misdemeanants and petty offenders; and a "parking place" for vagrants. Besides, most jails house those who are suffering from chronic alcoholism or from the effects of narcotics as well as prostitutes and panderers, shiftless derelicts, material witnesses, and

Acute overcrowding in county jail (Photograph taken in 1938).
(Courtesy "The Prison World.")

others of heterogeneous nature. The promiscuity and the utter lack of segregation of the contemporary jail and workhouse are two of their worse features. Especially serious, too, is the intermingling of first offenders, in most cases young boys and men, with older and more degraded types of depraved persons and adepts in petty crime and debauchery.

The report on Juvenile Detention of the Attorney General's National Conference on Prevention and Control of Juvenile Delinquency estimated that in 1946 over 40,000 children were detained in county jails. The problem of keeping children out of county jails is a serious and distressing one. One reason is because no alternative facilities are available in large areas of the country. Adequate detention quarters are not provided and thus, when a child is arrested for a delinquent act, or is a runaway from home, he is summarily thrust into jail.[5]

In addition to a lack of scientific classification and segregation, most jail administrators are inefficient and ignorant of correctional principles. Few of these jailers and sheriffs have a high degree of intelligence to bring to the crime problem. The county sheriff is like the county clerk and other such petty officials; he is in politics for what he can get out of the job. But we cannot blame the sheriff. Ever since the days of Howard, this functionary has run the jail for the benefits of the office. Part of these benefits have been financial rewards and political patronage, mostly through the fee system. So long as this system operates, this petty graft will attract only political hacks and untrained men.

Although the sheriff may have little scientific knowledge of penal methods, he is, in most cases, a more intelligent person than most of his underlings, be they deputy sheriffs or keepers of the jails. In the smaller counties, such a person may be little more than a mere *turnkey*. His job is to watch the jail and its inmates so that the prisoners will not walk off. He is also charged with feeding the inmates. His duties are, as a rule, very inefficiently dispatched, judged by the large number of escapes and the shoddy housing, sanitation, and food in most jails. In many jails there are not even full-time personnel.

Alexander lists a few of the jailers' problems:

> A sheriff has asked an inspector for information on maintenance and repair of cell-locking devices. Another wanted help and data to support his request to the county commissioners for more funds so that he might be able to feed his prisoners three decent meals a day. A jailer wanted a list of medications which he might keep in the jail to handle minor medical problems. What kind of gas should be kept in the jail for defensive purposes and when should it be used? Is there any written material on jail sanitation? What kind of mattress covers are recommended? Should jailers

---

[5] For further details, see Austin H. MacCormick, "Children in Our Jails," *The Annals*, Vol. 261 (January, 1949), pp. 150-157. MacCormick states that from 50,000 to 100,000 children are detained annually in jails, "most of which are unfit even for adults."

be taught *judo?* How much should jailers be paid? What's this business about civil rights? What about juveniles? women? addicts? escape artists? What a windmill![6]

Here are untrained men trying to cope with an almost impossible job.

In its architecture, the jail presents the greatest diversity of construction. Some places have structures erected over one hundred years ago; others have new and relatively efficient cagelike buildings. In more recent times some cities have gone in for absurd skyscraper or penthouse jails. In almost no case is modern architecture combined with humanitarian sentiment and scientific insight in the design of the jail.

In the bulk of our county jails there is no semblance of security. Escapes and just plain "walking off" are notoriously prevalent. It is recorded that a bank robber "sang his way out of jail" in Arkansas some years ago. He merely took his place among a visiting religious group and walked right past the jailer. In a Texas jail an inmate walked up to the trusty jailer and quietly said: "I'll go now." The obliging trusty opened the door and the prisoner walked out. The notorious John Dillinger escaped from the Crown Point, Indiana, jail in 1934 with the aid of a homemade wooden gun, after the female sheriff boasted that her institution was secure even for "tough babies" like Dillinger. It is estimated that some 3,000 persons escape or walk away from our county jails annually.

The modern jail is an expensive institution to maintain. It has been estimated that it costs the taxpayer over $50,000,000 annually to support the 3,000 county institutions. The per capita cost varies with the turnover in the population; larger jails in more populous centers cost less to operate than the smaller institutions where inmates are not so numerous. But it costs approximately $500 to $800 a year for each inmate. This is not the total social cost, for many of the persons who run afoul of the law are in poor circumstances financially, so that their incarceration works a serious hardship on their families. Consequently, thousands of such families must be supported by the state or city through its Department of Welfare.

The Kangaroo Court

An easy way of maintaining discipline in a jail is to institute the perverted form of inmate "self-government" called the "kangaroo court." The origin of the term is obscure. Some authorities claim it was imported from the Australian colonies, home of the kangaroo, where the practice was in operation. Others associate it with an activity of justices of the peace in the early days of the automobile whereby they maintained speedtraps to arrest unwary motorists who were "speeding." But the practice, like the fee system, began many years ago and has been abolished in all but the

---

[6] Alexander, *Jail Administration*, p. v. Reprinted by permission of Charles C. Thomas.

most reactionary jurisdictions. It can be traced to the jails of England in pre-colonial days. The kangaroo court was a thriving institution in Newgate Prison in 1818 when Thomas Fowell Buxton described how it worked.[7] This same "inferior" government existed in the Walnut Street Jail at Philadelphia and was described by a visitor, Robert Turnbull as a "secondary and inferior government among the criminals for their own convenience and comfort."[8]

In jails where it operates, the inmates elect a "judge" who presides over a "court." This judge is probably the most influential rascal in the establishment at the moment. As Buxton describes him and his court:

> Some one, generally the oldest and most dexterous thief, is appointed judge; a towel tied in knots is hung on each side of his head in imitation of a wig. He takes his seat with all form and decorum; and to call him anything but "my lord," is a high misdemeanour. A jury is then appointed, and regularly sworn, and the culprit is brought up.[9]

Obviously, the whole practice is a burlesque of the regular court of law.

Sanford Bates, former Director of the federal prisons, says of this institution, "the kangaroo court is a combination of blackmail, terrorization, and privilege." In jails where the kangaroo court flourishes, rules are drawn up that deal with such humdrum affairs as keeping the cells and bullpens clean, excessive noises and other nuisances under control and, in short, anything that comes within the scope of the personal conduct of the inmates. When a new inmate enters the establishment he is brought before the judge who presents him with a set of rules. Many of these rules are arbitrary and capricious. Here is a sample of such a set from a "sanitary court," supplied to the writers by the Federal Bureau of Prisons:

SANITARY COURT DECREE

This Sanitary Court is One Hundred Percent Strong and is also Iron Bounded and Copper Lined.

Every One who enters this Jail is Automatically a Member of the Court and is Fined $2.00.

Any One Denying Having Money on his Person Down Stairs will be Fined as the Judge Sees Fit.

(1) Every One must take a Bath on Entering the Jail and at Least twice a Week. If not Executed a Penalty of From 1 to 15 Lashes will be Imposed.

(2) No One is Allowed to make Unnecessary Noise after Lights are Turned Out at Night and Before Slams Open and During Meal Periods. If not Executed a Penalty of from 2-15. Or as Judge Sees Fit.

(3) No One is allowed to Resist Court. Penalty as Judge Sees Fit. All Fines Must Be Paid.

---

[7] Thomas Fowell Buxton, "An Inquiry Whether Crime and Misery are Produced or Prevented by Our Present System" (London: 1818), pp. 48-50.

[8] *A Visit to the Philadelphia Prison* (Philadelphia: 1796), p. 18.

[9] Buxton, "An Inquiry," p. 49.

(4) No One is Allowed to steal from his Fellowman. If so it will be a Penalty of from 50-100. Or as Judge Sees Fit.

(5) No Fighting Allowed. If so A Penalty of from 15-22 L. Or as Judge Sees Fit.

(6) No Malice Toward your Fellowman. 1-15 Lashes or as Judge Sees Fit.

(7) Every One Must fall in Line in Order at Meal Time. If Not A Penalty of from 2-25 Lashes or as Judge Sees Fit.

(8) No One is Allowed Above the Dead Line while Visitors are Present unless he is Called. Otherwise a Penalty from 1-15 L. Or as Judge Sees Fit.

(9) Loud Talking and Profane Language is Prohibited while Visitors are on Floor. If Found Guilty a Penalty of from 3-25 Lashes, Or as Judge Sees Fit.

(10) Every one must have on Clean Shirt Before Breakfast and Visiting Days which is Thursday and Sunday. A Penalty of from 2-20 L. Or as Judge Sees Fit.

(11) Hats and Caps must be Removed from your Head during Meal Hours and while Court is in Session. If found Guilty a Penalty of from 1-15 L. Or as Judge Sees Fit.

It is difficult for a new man who has not had time to establish status with his fellow inmates to avoid violating several of the rules. He is accordingly fined, the ultimate object being, not order, but to extract from the victim all the money he has on his person or lodged to his credit in the sheriff's office. If he refuses to pay, he is whipped. In many kangaroo courts there will be found hidden under some mattress a handmade lash. It is applied without fear of interference from the jailer, who often condones such action and shares in the final pay-off.

Gruesome and senseless tragedies have occurred in jails that tolerate the kangaroo court. A man was murdered in the Denver jail in 1938 and one in the Seattle jail in 1944.[10] But perhaps the most wanton killing took place in the San Mateo (California) jail at Redwood City in the summer of 1957. A vagrant, a pathological liar, was choked to death by two young thugs, 18 and 21 years of age, for no reason except that the man's "tall tales" irritated them. One of the killers, who had served time in a Youth Authority camp and was being held for peddling marijuana, had tattooed on his abdomen "Born to hate cops." Both killers were sentenced to life imprisonment for their act, the grand jury exonerated the officials, and a press statement remarked that the young criminals would be eligible for parole "after seven years."

The kangaroo court flourishes out of the sight of responsible citizens and is condoned by all too many administrators. While most states have outlawed this degenerate type of self-government, it flourishes undercover. This fact was brought out in a report on jails in California. The committee

---

[10] For details of the Denver case, see Courtney Ryley Cooper, *Designs in Scarlet* (Boston: Little, Brown, 1939), pp. 269-270; for the Seattle case, see *Congressional Record,* Vol. 91, No. 48 (May 14, 1945), pp. 2245 f.

making the investigation of jails were convinced that "inmate disciplinary authority" exists in 21 out of the state's 56 jails.[11] It might be added here that this high-handed form of "self-government" is also found in some penitentiaries. The trusty system often lends itself to this form of institutional blackmail.

Women in Jails

The treatment of women in jails is in harmony with the vulgarity and brutality characteristic of jail discipline and practice as a whole. The larger jails sometimes have a special apartment for female prisoners, set aside from the jail proper and staffed by women. But in most of the smaller institutions women prisoners are segregated in one of the tiers, not too far removed from the male inmates—and near enough to hear the ribaldry and derisive laughter from the men's side. Matrons are almost never furnished in the smaller establishments, and this makes a situation that is completely inexcusable—ministering to female needs by male jailers.[12]

Even in those larger jails that employ matrons to care for the female prisoners, few women of culture, refinement, or professional training can be persuaded to assume this onerous task, and administrators are forced to employ women of low caliber. As the positions are political, even if there were available women with some education and refinement to accept them, it is doubtful if they would be employed. The pay is usually meager and the hours long. Most jail matrons may have good intentions when they assume their duties, but lacking the insight necessary to care for the typically wretched female groups that infest our jails, they become intolerant, crotchety, and cruel. Granted that most of the females found in our jails are derelicts and prostitutes, it is all the more important that a high type of trained workers be placed in such positions.

Long range planning, so far as it affects the county jail, calls for the separation of inmates into various groups. This applies to women as well as to men. Women awaiting trial should be housed in a separate wing in a House of Detention and cared for by trained personnel. Female misdemeanants, drug addicts, alcoholics, and some other types should be taken from jails to receive scientific treatment in properly operated institutions for those specific ills. But always they should be rigorously segregated from male inmates and cared for by trained female attendants.

---

[11] "A Study of County Jails in California" (Sacramento: State Department of Correction, 1949), p. 10.

[12] In at least one New York county, however, where a regular matron is not provided, females detained overnight must be accompanied by a matron deputized for the purpose; any female citizen of good character and 21 years may be so deputized. The next day the female detained must be transferred to the jail of the neighboring county where there are staff matrons.

Town Lock-Ups

What we have stated concerning county jails thus far certainly also applies to city and town lock-ups where thousands of persons are detained because they cannot raise bail bonds. This country's town and city lock-ups, under the control of police, are altogether indecent for human habitation. This is an area that has been all but neglected by students of the crime problem. A revealing report on Chicago's lock-ups released by the John Howard Association of that city is the type of study that should be made in every large city and county seat of the nation. The most shocking conditions were found in Chicago's police lock-ups. The herding together of dozens of victims in cramped filthy cells, built originally for one or two persons, was the rule rather than the exception. Rules drawn up for the treatment of those detained for trial were ignored by custodians. On the basis of these conditions seven lock-ups were classified as "good," that is, approaching desirability, ten as "fair," and twenty-two as "poor." Owing to the widespread interest in the study, entitled "Held Without Bail," the shocking conditions were brought to the attention of the proper authorities and considerable reform measures were introduced. Here is a sample report of the conditions found by the investigators:

> Thirty-three lockups (studied) contain 336 cells, which housed 612 people on December 15, 1946. Eighteen of the 33 had repugnant odors. Some of the cell rooms were so malodorous as to affect the entire station. Twenty-six lockups had filthy walls which were also in need of major repairs. Two had as the only toilet facility a trough of constantly running water, which occasionally spread its contents over the entire floor area. Only one lock-up supplied toilet paper to prisoners, and none provided soap and paper towels for any of the men. The women's lockup in the first district was the only one having bathing facilities. Twenty-four lockups had wooden bunks and thirteen had metal. Only the first district supplied mattresses and blankets and these only for women. Twenty-five of the lockup keepers reported that their lockups were infested with vermin. Many reported that bedbugs, lice, cockroaches, mice and rats were so troublesome as to require the use of a gasoline torch to "burn them out." . . . With the exception of the women's lockups, prisoners are provided with two slices of bread, a slice of bologna, and a cup of black unsweetened coffee three times daily, if they desire.[13]

Recognizing the difficulties small towns, in particular, have in financing a lock-up, New Jersey has encouraged "regional" establishments whereby two or more small communities may pool their resources and operate one detention quarter. For instance, there are 565 municipalities but only 245

---

[13] "Held Without Bail" (Chicago: John Howard Association, 1949). See also, Sanford Bates, "Stepchildren of Penology," *Prison World*, Vol. 6, No. 2 (March-April, 1949), pp. 3-5. Another study dealing with this knotty problem is "Compelling Appearance in Court: Administration of Bail in Philadelphia," University of Pennsylvania *Law Review*, Vol. 102, No. 8, June 1954, pp. 1031-1079.

police lock-ups. Sanford Bates, former Commissioner of Institutions and Agencies, states that many police in small towns are anxious to maintain decent quarters. The department of Institutions and Agencies compiled a set of suggested regulations for the operation of these places, and many communities have availed themselves of this service.

Proposals for Jail Reform

Jail reform has come very slowly. John Howard found this institution, with its ancient characteristics of oppression and privilege, almost impossible to improve. Since his day many have denounced its evils, but very few improvements have been made. In America, the jail was accepted as a necessary adjunct of the penal system, and due largely to the famous jail architect John Haviland and his imitators hundreds of jails were built that were massive, secure (at least in looks), forbidding, ill-equipped, and ill-designed for health, not to mention comfort. Many of these structures are still in use, but they have been condemned by virtually all students of the jail problem.

Considerable impetus was given the movement toward genuine jail reform when the federal government, acting through the Bureau of Prisons, set up an inspection service of the nation's 3,500 jails. By means of this service, jails that would house federal prisoners awaiting trial must measure up to rigid standards in regard to security, sanitation, food and housing, medical attention, and other conditions compatible with common decency.

For the year ending June 30, 1950, 78.1 per cent of 3,115 local jails inspected were rated under 50 per cent and condemned as unfit for human habitation. Only 87 jails in the entire country received ratings of 60 per cent or over; only 8 jails received a rating of 80 per cent or over. Yet the Federal Bureau of Prisons was compelled to use 604 county jails for the detention of its accused.[14]

The Federal Bureau no longer "rates" jails but it does maintain a jail inspection program, with eight full-time trained inspectors. In 1956 an average of 3,027 federal prisoners were held daily in local jails. This service cost $2,600,000.[15]

The Director stated in his report for 1957:

> Our seven inspectors traveled nearly a hundred thousand miles this year, visited 47 states, and inspected a total of 599 jails. Twenty-four not previously approved were added to the approved list, and five were dropped. At the year end, 805 city and county jails were authorized for Federal use.

---

[14] From the Director's annual report, *Federal Prisons, 1950,* Report of the work of the Federal Bureau of Prisons (Washington, D.C.: U.S. Government Printing Office, 1950), Table 36, p. 92.

[15] Myrl Alexander, "Federal Jail Inspection," *Prison Journal,* Vol. 31, No. 2 (October, 1956), pp. 7-9.

Except for a few which were approved on only an emergency or temporary basis, all were rated "good" or "fair." None, we regret to say, was rated "excellent," our top rating. Jails we could not approve were rated "poor" or "bad." None of these will be inspected again unless an institution claims improvements have been made and asks for a reinspection. No jail could possibly be rated "excellent" unless it has eliminated to a substantial degree what we consider the major blight of local jails, inmate idleness. Three out of four of the jails inspected this year had no employment program worthy of the name; and in most jails what we term "inmate activities" were limited to some type of religious program, usually a weekly service provided by a local church group.[16]

In the vast majority of jails there are no rules concerning the management and discipline posted or even compiled. Well over half of the country's jails are never inspected by local boards of health or other official authorities except, in some instances, by local grand juries who are not supposed to know anything about jails. These bodies of citizens, however, do frequently condemn their local jails but rarely are their recommendations acted upon.

It is not to be inferred, however, that all jail administrators are deaf to jail reform. Many are genuinely concerned with the multifarious problems of their establishments and cooperate splendidly with the leaders in the reform movement to abolish, or at least radically change, the jail and all that it represents. The National Jail Committee of the American Correctional Association many years ago drew up and issued the following 14 points, or propositions, which represent phases of the over-all objective of jail reform. While issued in 1937, these objectives have as much meaning today as when they were first enunciated:

I. *Measures to Keep People Out of Jail*
   1. By law direct that the Courts adopt a more extended use of bail, recognizance, and other approved measures of release from custody.
   2. Secure a law providing for collection of fines by installment and for sufficient personnel to enforce it.
   3. Develop an approved probation system, not only to prevent people from getting into jail, but to supervise and guide offenders released from custody.

II. *Fundamental Changes in Jail Set-up*
   4. Abolish the locally controlled jail as a place for convicted prisoners.
   5. Place the jail and all its present functions wholly within the State Correction system and under centralized control.
   6. Reorganize the system to provide for secure and suitable detention places, properly staffed and equipped for segregation, classification of prisoners charged with law breaking.
   7. Establish regional farms and/or custodial centers for care, train-

---

[16] *Federal Prisons, 1957,* Report of the work of the Federal Bureau of Prisons (Washington, D.C.: U.S. Government Printing Office, 1957), p. 33.

ing, and needed treatment, with a regular work program under rigid discipline.

8. Eliminate the fee system in connection with arrest, trial, and custody of prisoners, and place all fee officers on fixed salary.

III. *Reform in Law and Court Action*

9. Simplify law and court procedure with regard to all arrested persons.

10. Adopt measures and reforms to shorten time spent in detention quarters by prisoners awaiting trial, witnesses, appeals, etc.

11. Secure an indeterminate sentence law with specified minimum sentence.

IV. *Standards and Records*

12. Fix minimum standards for custodians of prisoners and probation workers with merit system safeguards.

13. Establish a central state bureau of identification and record.

14. Create a uniform system of records and statistics for the whole correctional set-up, *jails included.*[17]

This is an ambitious program, and it will take many years to achieve. Here and there, however, some of the reforms mentioned are actually in operation. We consider some of the points in detail.

1. *The more extensive use of bail, recognizance, and other approved measures of release from custody and the collection of fines by installments.* Bail is in more or less general use, but it should be made easier to obtain and should be lowered in deserving cases. The vicious practices of the bail-bond racket should be promptly eliminated. Anyone who can raise bail stays out of jail now, regardless of the degree of criminality, except for those serious offenses that are denied this privilege by law. As now practiced, however, only the well-to-do can find such collateral; the friendless and impoverished go to jail because they cannot make bail-providing connections. The use of personal recognizance can undoubtedly be extended, especially in small communities where nearly everyone is known and will appear as directed by the court at the specified time. Recognizance, unlike bail, is essentially a form of credit with no security except the individual's personal integrity.

One of the strangest anomalies in present correctional practice is jailing a person who cannot or will not pay the fine imposed by the court. It is so easy for a magistrate or judge to pronounce the sentence: "Ten dollars and costs or 30 days in jail," without realizing that 30 days in jail is an expense to the taxpayer—and without realizing that he is sentencing the accused to be jailed for debt. If the accused cannot pay, the law becomes the creditor, exacting its pound of flesh from the hapless debtor to society. Nearly half of all the people in jail at any one time are there for inability to pay their fines. To remedy this situation, installment payment of fines has been sug-

---

[17] *Proceedings, American Correctional Association,* 1937, p. 320. See also Roberts J. Wright, "What! The County Jail Again?" *Federal Probation,* Vol. 11, No. 3 (July-September, 1947), pp. 17-20.

gested and in some places has been adopted. More recently the system has been successfully tried in Milwaukee (Wisconsin County) under provisions of the Huber Act. This act, passed in 1913, lay dormant until 1943 when it was implemented. The county saves thousands of dollars annually by permitting misdemeanants to continue at their work and to spend their nights at the jail. It is also working successfully in Santa Clara County in California.[18]

Aside from saving the taxpayer money and the prisoner an unwholesome jail experience and incidental demoralization to health, the installment system creates a wholesome effect on the family of the offender. If he was employed when he ran afoul of the law, there would be scarcely any interruption in his regular work; hence his family would not suffer economically. Another expedient with less to recommend it is the "week-end" sentence that is being resorted to in Rochester, N.Y., Camden, N.J., and other places throughout the country. This is usually employed in drunken driving, drunkenness, and other cases where it is assumed a jail sentence will "teach a lesson." The offender is permitted to serve any sentence imposed on him for an infraction of the law by appearing at the jail every Saturday afternoon after work and leaving Sunday evening until he has served the equivalent of his time sentence. This practice does not interfere with the violator's work and brings no financial hardship on him or his dependents. But if the purpose of the sentence is to teach the offender a lesson and to deter others, splitting the terms in this fashion makes the lesson pretty easy and greatly decreases any deterrent effect. In such cases, the offender is still subjected to the crudities and vulgarities of the jail and may be made a criminal as a result of vicious associations.

2. *Provide adequate detention quarters for the untried offender apart from those for any other custodial cases.* Since approximately five out of eight persons apprehended by the law and incarcerated in the county jail (or placed on bail) are finally acquitted of the charges, it is incumbent on society to guarantee decent and humane treatment of all persons under suspicion. But nearly all who await trial in county jails are herded together (often with vagrants and other short time offenders), and grudgingly supplied with minimum comforts. This is especially true in our smaller jails, where segregation of any sort is practically nonexistent.

In any movement to eliminate jails the first step, then, is to provide detention quarters for the untried. If this is accomplished, at least half of the present jail population will be removed from the dingy, unsanitary institutions where they are now housed. Aside from the difficulty of obtaining funds for such a House of Detention, the vested interests that control

---

[18] See Robert M. Yoder, "Wisconsin Throws Them Out of Jail," *Saturday Evening Post,* Vol. 228, No. 33 (February 4, 1956), pp. 25, 80-82; Gene Marine, "Part-Time Jail," *The Nation,* Vol. 185, No. 25 (December 21, 1957), p. 476. For an analysis of the law, see "Wisconsin's Huber Law In Action," a study made by the Wisconsin Service Ass'n. (Milwaukee, Spring, 1958).

New House of Detention for Young Adult Offenders, Brooklyn, N.Y. opened in 1957.

the patronage of the county jail usually oppose this innovation, despite the fact that untried prisoners do not belong in the jail as we find it today.[19] The larger cities can lead the way in erecting special detention facilities. The Brooklyn House of Detention, opened in 1957 under the able leadership of Judge Anna Kross, Commissioner of Correction for New York City, is serving as a model for other metropolitan areas. Erected originally for adults, it is at present used for adolescent offenders. Aside from a well-constructed building the Brooklyn facility has been geared to meet the personality needs of the older boy suddenly thrown into the toils of the law and held without bail for several weeks before he is brought to trial.

3. *Establish farms for misdemeanants in metropolitan areas with adequate provision for the treatment of contagious diseases, venereal diseases, and chronic alcoholism; such farms to be either under direct control of the state or an integral part of a state wide program for penal treatment. For the less populated counties, regional farms to be established to care for*

---

[19] See Louis Partnow, "Toward a Program for the Untried Adult in Detention," *Prison Journal,* Vol. 37, No. 1 (April, 1957), pp. 12-46; Louis Partnow, "Detention and the Untried and Probation," *Prison Journal,* Vol. 38, No. 1 (April, 1958), pp. 22-26.

this type of short-term offender, and adequate provision made for treatment of those suffering from disease.

James V. Bennett, Director of the Federal Bureau of Prisons, writing on "The Regional Work Farm," strongly advocates: the central control (by the state) of all correctional institutions for sentenced prisoners; the construction and establishment of a number of regional institutions for short-time offenders; and the transfer of sentenced prisoners from the county jails to the new institutions. He points out that since 1930 counties have been experiencing more and more difficulty in financing their local governments and that they have retrenched in education, road building, health programs and penal institutions. He comments:

> If we are to have decent correctional institutions with facilities for systematic study and treatment of the offenders committed to them; if we are to provide the prisoners with work, decent housing quarters, wholesome food, medical treatment and all the other things which they lack under the present system, it becomes necessary to consider the possibilities of operating the system on some basis other than that of independent county management. There is general agreement among those who have experimented with new means of treating the short-term prisoner that the construction and maintenance of regional institutions, largely devoted to farming, will provide a solution to the existing jail evils.[20]

Legislation should be passed in each state making it mandatory rather than discretionary for smaller counties to combine themselves in regional units and attack jointly the problem of adequate supervision and care of misdemeanants and other types that come within the jurisdiction of the lesser courts.

Some of the larger cities established farms at an early date, especially for vagrants and misdemeanants—Cleveland in 1905 and Kansas City, Missouri, in 1909. During the past 15 to 20 years other communities and counties have followed this trend: the St. Louis County Work Farm near Duluth, Minnesota, Berks County farm near Reading, Pennsylvania, and the detention roadcamps and farm operated by the sheriff of Los Angeles County.

Maryland operates a state farm for misdemeanants as well as long-term offenders near Hagerstown. The Indiana state farm near Greencastle has been in operation since 1915. The state law makes it mandatory for local judges to send to this institution misdemeanants sentenced for more than 30 days. They may send persons to the local jails for less than 30 days. This practice, however, does not solve the problem of the local jail. The problem of the jail, so far as misdemeanants are concerned, will not be solved until this class of law violator, as well as the vagrant, drug addict, and alcoholic, is sent to a regional or state farm, or to a large city house of

---

[20] James V. Bennett, "The Regional Work Farm," *Jail Association Journal* (March-April, 1939), pp. 17 f.

correction, on an indefinite basis, and while incarcerated be given the type of treatment he needs.

Other regional farms are in operation in Bridgewater, Massachusetts; at Lorton, Virginia, for the District of Columbia; at Vandalia, Illinois; and in work camps scattered over North Carolina, where the legislature permitted the state to take over the county prisoners (about 3,650 in all) and to use their labor under the control of the State Highway Commission. The counties, however, have not availed themselves of this opportunity to change the outmoded function of the jail.

The Berks County Prison near Reading, Pennsylvania has the advantage of a fine plant and fortunate location. (Courtesy "The Prison World.")

The instances mentioned are enough to indicate that a decided trend, albeit much too slow, is in the direction of differentiating the types found in the county jail into specialized institutions, or into segregated portions of a large jail farm, where they normally should be sent according to their specific type. We cannot be too optimistic at the moment, despite this healthy trend. Reports indicate that there is opposition in many political quarters against the movement. Many counties resent the break-up of their vested interest, with its political patronage, and many judges who are a part of the local county political organization lack the vision necessary to demand more activity in the direction of state control of misdemeanants, drug addicts, alcoholics, prostitutes, and other types that certainly do not belong in the traditional county jail.

Most important, however, are the program and personnel of the regional farm or house of correction. Rehabilitative work should be a fundamental part of the farm-institution program. Each man and woman sent to such an institution is a personal problem. It goes without saying that a program that does not include in its personnel trained experts in psychology, psychiatry, and social work, in addition to trained medical and dental specialists, cannot measure up to modern standards of therapy. Custodial officers should be of the highest type, selected only on the basis of merit, with no regard whatsoever to political expediency. Aside from the work program—physical, vocational, and institutional—an adequate program, including guidance, recreation, education, and religion should be maintained. Preparation for release should be considered most carefully by the trained members of the classification staff. Such preparation should envisage community preparation to such a degree that the released prisoner may adapt himself as easily as possible to the free outside world. This is vitally important, and it calls for a tangible plan to be agreed on before the inmate leaves the institution. Relationships with his family must be guaranteed where possible; a job or other type of helpful occupation must be provided and, from an ideal point of view, periodic check-up on the part of a paroling authority maintained. From a medical standpoint, adequate facilities for dealing with venereal diseases, alcoholism, and debilitating ailments should be provided, and then utilized in the best medical traditions. Drug addicts might well be sent to a regional farm to be segregated and to receive scientific treatment—but not to be punished.

Possibly wages can be paid for labor performed, to stimulate industry and thrift—one additional reason why such institutions should be under the control of the state and integrated with the prevailing statewide penal labor policy. Counties, especially the smaller ones, cannot hope to continue insisting on their jurisdiction over offenders against the state.

We are far from realizing the goals set forth in this section but the trend is definite. Realists are convinced that the county jail of the future will be shorn of its political authority and go the way of all outmoded institutions. In the meantime we must constantly inspect the county jails throughout the country. In one place we may find numerous reform measures being introduced by an intelligent administration; in another we shall see continued brutalities, filth, official indifference, lack of segregation, and other evils inherent in the county jail. Until we develop a uniform system of state-controlled regional farms for misdemeanants, effect a complete segregation of the sexes, separate those awaiting trial from others, and isolate and treat those individuals afflicted with various diseases so prevalent in jail populations, we must tolerate the jail. At the same time, we must demand cleanliness, sanitation, an adequate and nourishing diet, and a well-trained personnel as the very minimum for its administration.

Social case work in the county jail is also highly desirable. A few of the larger jails have such a service, but in the vast majority of them those in control fail to recognize its value. Organized case work may be introduced in jails either (1) through social workers attached to a private agency or (2) by direct employment of qualified and experienced case workers on the paid professional staff of the county institution or (3) by the appointment of internes in training as volunteers. To be of value, however, case work must be advantageous to both prisoner and administration.

Concluding Remarks

Obviously, if the jail is outmoded functionally, it should be abolished as a penal institution. As we pointed out, it was originally designed as a place of detention. Today detained persons awaiting trial should be housed in properly constructed places of detention. Approximately 40 per cent of those detained for trial are acquitted of their offenses as charged. All detainees are, under our system of jurisprudence, "presumed innocent." Yet in most detention facilities they are treated as if they were guilty.

All others sentenced to jail should be treated in institutions where some scientific program of treatment is in operation: drug addicts, alcoholics, prostitutes, and other such afflicted persons. Misdemeanants should also be given an analysis through investigations by social case workers and, if possible, kept in incarceration for an indefinite period of time. As mentioned earlier, if a county cannot afford an establishment specifically created to care for these types and their problems, two or more counties should be encouraged to work together to develop a regional farm or facility. The county can no longer justify the old-style jail.

The British people have abolished the county jail. Short-time offenders are sent to local prisons, all of which are under centralized control. Those awaiting trial are placed in remand centers.

In America many jail reformers have denounced the jail as an institution. Due to the deep-seated traditions surrounding the jail and the vested interests who profit financially and politically from it, there is little hope that it will be abolished in this generation. But its disappearance as we have known it for so long will eventually be consummated.[21]

---

[21] For further information, see American Correctional Association, *A Manual of Correctional Standards,* issued first in 1954, and revised in 1955 and 1956, Chapter 8, "Jails." This manual will be referred to in subsequent chapters as *"Manual, 1956."* See also, Martin E. Wyrick, "The Area of American Correction that Stood Still," *Amer. J. Corr.,* Vol. 20, No. 3 (May-June, 1958) pp. 7 f. For an appraisal of the House of Correction in modern corrections, see Edward J. Hendrick, "The House of Correction," *ibid.,* pp. 3 f.

25 ————————————————————

Correctional Treatment

for Women

*While I recognize in its fullest extent the validity of the claim
of every female to courtesy and protection from man, I reject
with indignation the requisition of homage to a sex which
should only be proffered to piety and virtue. Virtue, piety and
amiability are the sole ornaments of woman. Sex cannot oblit-
erate a crime or adorn a vice. I never will worship at the shrine
of vice and immorality, though their altars be decorated with
the image of a woman.* (*From* A Concise History of the Eastern Penitentiary of Pennsylvania, *Philadelphia: Neall & Mas-
sey, Printers, 1835, pp. 26-7.*)

THOMAS B. MCELWEE

Treatment of Females in Specialized Institutions

There has been the same evolutionary development of correctional concepts in dealing with convicted women as for men. For about a
hundred years after the penitentiary was developed, as well as prior to that
era, women were housed either in association with or contiguous to male
inmates. During the past half century special facilities have been created in
many of the states for the housing and treatment of women although there
are some states that have made little progress in this direction. Approximately twenty-two states still house their women in the men's prisons. While
most of these are the states with small populations, there are a few more
populous states that have as many as two hundred women in the men's
prisons. On the other hand there are small states that have separate facilities
for women although seldom do they have more than twenty inmates at a
time.

Any discussion of women's prison facilities must distinguish between the
institutions for long-term female felons and the reformatories which often
house both misdemeanants and felons. In some states, misdemeanants and

felons still serve their entire sentences in county jails or houses of correction; in other states, both types of offenders are found in reformatories, some of which are quite progressive. The national picture, therefore, is confused, which indicates that there is room for reform, especially in those states that still have no specialized establishments for women. Even though they may be segregated from the male prisoners, they deserve special facilities of their own.

In colonial days, women were housed in the same places as men—jails and workhouses—but in some cities separate apartments were set aside for them and their children. It was a common practice for mothers who had been arrested for minor offenses to bring their children with them.

There was no lack of humanitarian concern for these usually bedraggled females. Societies of women visitors, motivated by deep religious conviction, made it their objective to visit women in their segregated quarters. Such an organization was formed in Philadelphia as early as 1823, known as the Society of Women Friends. Here is part of a report the members made of a visit to the women's apartment of the Arch Street Prison, known locally as the Bridewell:

> The engagement was entered upon with feelings of weakness and fear, under a sense of the importance of keeping in view our blessed Redeemer's declaration, "Without me, ye can do nothing." The prisoners were collected and arranged: the Friends being accommodated on a raised seat facing the inmates. . . . Counsel and admonition were communicated as ability was afforded, and the stream of gospel at some seasons flowed toward these poor prisoners through ministering friends, who occasionally accompanied the visitors. Short texts of Scripture, printed in large characters and pasted on boards, were hung on walls in the prisoners' apartments. During these visits, while some of the prisoners evinced inattention and lightness, tears fell freely from the eyes of others. Their general condition was pitiable and affecting. Many of them were vagrants, taken out of the street in a state of intoxication, covered only by a few rags; some of them were committed for petty thefts, others for worse crimes, awaiting the sitting of the court.[1]

The routine in those early institutions where women were housed was likely to be monotonous and drab, but not cruel. Work, when provided, was mostly laundering and cooking with some sewing and perhaps spinning. Moralizing was quite prevalent, with feelings of guilt encouraged by the matrons. Rehabilitation was thought of only in terms of religious conversion.

Numerous scandals aroused public opinion so that women were gradually more fully segregated from men and a movement for special establishments for females began to crystallize. The 1866 Yearly Meeting of Friends in Indiana started the public demand for a separate female prison. The result was the creation of the Indiana Reformatory for Women and Girls

---

[1] Recorded in the *Journal of Prison Discipline and Philanthropy*, Philadelphia (April, 1845), pp. 111 f.

in 1873. In 1889 the name was changed to the Reform School for Girls and Women's Prison. In 1889 the two departments were separated, one becoming the Industrial School for Girls and the other the Women's Prison.[2] In recent years the girls' school has been moved to the country and the Women's Prison has become a reformatory.

Inside cell block in Massachusetts Reformatory for Women, Framingham (old photograph). Old type of prison for women.

In Massachusetts, in 1877 an institution for women was erected at Sherborn (Framingham) and called the Reformatory Prison for Women. It is still a prison in the sense that long term female felons are sent there, but "prison" was dropped officially from its name in 1911. In many respects it passes for a reformatory institution, especially in its philosophy of correctional treatment. This has been due almost exclusively to the enlightened superintendents that have administered the institution since its opening.

The first true reformatory for women was established at Bedford Hills (Westfield), New York, in 1901. It was for all female misdemeanants and felons (but not feeble-minded), at the discretion of the judge. In 1881 an institution had been opened at Hudson, as a house of refuge to care for women convicted of certain types of misdemeanors, primarily sexual. This Hudson institution is now the State Training School for Girls, but when it

---

2 Blake McKelvey, *American Prisons* (Chicago: University of Chicago Press, 1936), pp. 77 f.

was first opened, it put into operation such reformatory principles as indeterminate sentence, grades and marks, trade instruction, and wholesale farm labor.

The Reformatory at Bedford Hills made possible at that time the rehabilitative treatment of women convicted of more serious offenses. This Westfield State Farm lies among the rolling hills of Westchester County, not far from New York City. Sentences are indeterminate, the maximum three years and an arbitrary minimum of one year. Classification is by a cottage system, and progressive methods are employed. In 1933, the female felons housed in the Old Auburn prison were transferred to a new prison adjacent to the Westfield Reformatory, so that now two neighboring institutions care for the female adult delinquents of the Empire State. This makes possible a real reformatory for women convicted of minor offenses at the State Farm, and a prison nearby for the more serious offenders. The first superintendent of the Reformatory was Dr. Katherine Bement Davis (1860-1935), a pioneer in the field of corrections.

In the year 1911 Mrs. Jessie Hodder, an enlightened superintendent, took charge of the Massachusetts Reformatory for Women at Framingham. In 1932, Dr. Miriam Van Waters, who had previously had a wealth of experience in dealing with delinquent children in California, assumed charge. Under her guidance the Massachusetts institution became one of the most progressive in the country. Many of her reforms were questioned by the more conservative elements of the public, such as for example, her practical concept of indenturing girls into the community to work as a "proving ground" for release, and her scientific treatment of homosexuals. After some 15 years of distinguished service to the state she came under attack in 1948 by the commissioner of correction and was dismissed from her position. Fortunately the commonwealth of Massachusetts has a provision whereby an ousted official may demand a public hearing. Accordingly Dr. Van Waters defended her regime before an objective board that exonerated her completely of the charges and reinstated her.

The Van Waters case clearly demonstrates that a trained correctional administrator can effect reforms consistent with a scientific approach by carrying the public along such lines. It also refutes the mistaken notion that "we must educate the public before we can effect reform." Dr. Van Waters dared to be scientific in her treatment of the offender. The following words from the official report (1949) that cleared her, states:

> No institution can be static. Developments will occur despite individual administrators and despite rules and regulations and even statutes. Many bits of progress in the law have come about because the practices get ahead of the law and it has been necessary to change the law to catch up with what is being done. The whole history of equity is one of necessary developments made simply because the law was too static and too rigid. Even

statute law is not necessarily confined in its operation to the precise words as written.[3]

The chairman of the official board of Inquiry was the dean of Harvard Law School.

Life in the less progressive women's reformatories is still quite drab. The daily routine is usually highly regimented, the surroundings tawdry, the atmosphere highly charged with guilt feelings. Homosexuality abounds and repressive measures are pyramided one on the other to cope with the practice. Recreation is rarely provided and the program is poorly designed to be of real assistance to the offender.

Little authentic information regarding women's reformatories reaches the public, so that only rarely do we learn of the drabness and refined cruelty imposed on the inmates. Even the Federal Reformatory at Alderson, West Virginia, usually regarded as a relatively progressive institution, was indicted for its sheer monotony by an astute and intelligent woman, Helen Bryan, who was sentenced for refusing to testify before a Congressional committee.[4] Some years ago a newspaper woman, Virginia Kellogg, wrote an account of her voluntary commitment to a women's reformatory. Her report describes conditions that vary little from those experienced by Kate O'Hare in the Missouri prison following World War I. Here is what she found:

> Most state penitentiaries for women are ridden with political corruption.
> Women guards are on the payrolls of organized crime syndicates.
> Representatives of prostitution and shoplifting rings openly recruit the young inmates.
> Ignorant and sadistic matrons are appointed on the basis of political pull alone and cannot be fired by honest officials.
> Homosexuality and dope addiction run riot.[5]

The writer of this *exposé,* a journalist and screen writer, assisted in the preparation of the motion picture "Caged," which depicted life in a female prison. In her article she describes the tawdry clothing worn by the inmates, the unpalatable food served, the inefficient, untrained, and hardbitten staff officers, the cruel punishments inflicted for violation of petty

---

[3] See later analysis of this case in Thomas H. Eliot, "The Van Waters Case," *The Inter-University Case Program,* No. 22 (University, Alabama: University of Alabama Press, 1954).

[4] Helen Bryan, *Inside* (Boston: Houghton, Mifflin, 1953).

[5] Virginia Kellogg, "Inside Women's Prisons," *Collier's,* Vol. 125, No. 22 (June 3, 1950), pp. 15 ff. Kate O'Hare's book, *In Prison* (New York: Knopf, 1923), depicts the "old" prison, pointing up its brutality and exploitation of inmates. Other early books dealing with women's reformatories are: Mary B. Harris, *I Knew Them in Prison* (New York: Viking, 1936); Edna O'Brien, *So I Went to Prison* (New York: Stokes, 1938); Florence Monahan, *Women In Crime* (New York: Washburn, 1941). For a more recent book dealing with a girl's reformatory, see Creighton Brown Burnham, *Born Innocent* (Englewood Cliffs, N.J.: Prentice-Hall, 1958).

rules, the callous work program having no semblance of reformative value, and the utter despair of many inmates who are locked up for trivial offenses.

Her account, as well as the motion picture, were attacked by the American Correctional Association as being overdrawn and in no way affording an accurate picture of women's prisons. Certainly there are some excellent institutions for women in this country, but one must admit that in general women inmates are harshly treated and little understood in many state female institutions. We seldom read or hear of riots in women's institutions, but there was one in the New York prison (not the Reformatory) at Bedford Hills in June 1958. Guards from Sing Sing prison were rushed to the scene and some 80 female "trouble-makers" were segregated.

On the whole, however, women's reformatories are administered more progressively than men's institutions. Some of the latter, however, outstrip many of the female establishments in personnel, objectives, classification, general vision, and insight. Most of the men's prisons are severe looking so far as plant is concerned, but many make up for this handicap by well trained personnel and therapeutic programs. It is more difficult to secure and keep intelligent workers in most women's reformatories, especially in those far from the metropolitan areas. Their inaccessibility to cultural attractions such as theatres, for instance, makes it difficult to recruit staff. The California Reformatory for Women, formerly located in the mountains at Tehachapi, had to be moved to another site for this very reason. It is now located at Corona not far from the Chino prison for men. Here, in an attractive setting with modern and pleasant cottages and well-equipped workshops, a meaningful program can be carried on. Capable women will and do accept positions in such an atmosphere.

But, even at best, recruitment is a serious problem in women's institutions. The personnel in men's reformatories and prisons may bring their families to the vicinity and hence lead a reasonably normal life in their communities. This is difficult, if not impossible, for women staff members of the women's reformatories. Young college women who accept a post stay there for only a short time. They marry or seek greener pastures in the fields of social work or teaching.

The future of the women's reformatory is debatable. Many of the present inmates should never have been sent to a correctional institution. This is especially true of prostitutes, liquor violators, and drug addicts. Many of those committing such offenses need vocational guidance, psychiatric treatment, and casework therapy, all easily given without an institutional experience. When such an enlightened program can be carried out, the population of the women's reformatories will shrink almost to the vanishing point. Then, only the more "dangerous" felons will be sent there. Constitutional misfits will be sent to regional farms, to be kept occupied at various tasks compatible with their mediocre capacities, and to receive proper treatment for any obvious medical, psychiatric, and social ills they may be suffering.

Cottage-type women's reformatory. Federal Reformatory, Alderson, W.Va.

413

As we survey the country we are at once impressed with the lack of progress made in handling the problems of the female offender. There are still too many housed in county jails. Most of these are, of course, misdemeanants or short-term offenders. While they may be kept in separate wings of the institution, they are not too far removed from the noisy bedlam that so characterizes the county jail. All too frequently there are no women staff members available to minister to the peculiar physical or psychological problems of the female and their length of custody depends for as long as the judges see fit.

Whether the programs of our modern reformatories are of much service to the women sent to them is debatable, but we feel that there is little reformation, especially of sex offenders (in most cases prostitutes) and drug addicts. They are taught to sew, and most of them are put through a domestic science course, so that they may secure jobs as maids or "mothers' helpers" upon parole. But few prostitutes or promiscuous sex offenders, embezzlers, confidence women, or even shop-lifters are satisfied with a job that pays as little as housework.[6]

Babies in Female Institutions

A problem peculiar to women's reformatories that seldom comes to the attention of the public concerns the disposition of babies born of their inmates. Hundreds of girls and women enter reform schools or reformatories each year who give birth to babies while under sentence. There are also mothers in jails and houses of correction in many of our cities. In one of the few studies of this question, by Dean Shepard and Eugene Zemans, we find the following facts: in 1950 there were 364 babies born in 37 institutions—15 girls' reform schools, 14 reformatories, and 8 prisons that house women. It was found that in some states laws do not permit their institutions to take pregnant girls, and other institutions do not have facilities to care for prenatal and postnatal mothers. Of the total of 364 babies, 148 were born in hospitals of eight correctional schools and the remainder in county, state, or municipal hospitals outside the institutions. The term "illegitimate" is stamped on the birth certificates of such babies in Iowa, Minnesota, New Jersey, Ohio, and West Virginia if the mother is not married. Twelve other states reported that the word is not stamped on the certificate. In Illinois, Montana, and New Mexico the words "father's name unknown" are written, and in Wisconsin, the answer to the question "Is mother married?" is entered.

The newborn babies are "disposed of" in a variety of ways. Relatives are permitted first claim if they wish the babies. If not, state and county

---

[6] For a description of good procedures in women's institutions, see: *Manual, 1956,* Chapter 7; Henrietta Addition, "Institutional Treatment of Women Offenders," *N.P.P.A. Journal,* Vol. 3, No. 1 (January, 1957), pp. 21-30.

welfare departments, orphanages, or other agencies are pressed into service to handle the problem. Shepard and Zemans asked the superintendents of the female institutions two significant questions: "Should every prospective inmate mother be given the right to bear her child outside the institution?" and "Should the baby be separated from the mother at birth, remain with the mother only during her stay in the hospital, or remain with the mother in the institution up to two years, assuming the mother is competent and the institution has proper facilities?"

There was unanimous agreement that girls should have the right to bear their children outside the institution. There was little agreement regarding the second question. Some thought the child should be separated from the mother after the period of hospitalization; others felt provision should be made for the child to remain with the mother indefinitely or to be returned to her upon release.

This unique but highly important study points up several phases of this problem of the female sex offender who is sent to an institution. The stigma of being born in a prison is tragic and to have "illegitimate" placed on the birth certificate is compounding the tragedy. The study makes the following recommendation: "The entire matter should be examined on a statewide basis in order to determine whether or not this problem, particularly as it relates to female juvenile delinquents, is concerned, is adequately and humanely met on a case to case basis."[7]

Some European countries, notably Yugoslavia and Spain, have solved this problem by not sending pregnant women or women with small children to the regular female prison. They have special "mothers' prisons" which are especially designed as to architecture, staff and program, to meet the needs of mothers and babies. The children are never separated from the mothers while sentences are being served. These institutions are much more like nursing homes than prisons.

The experience of female reformatories in this country parallels the same destiny of male reformatories; high expectations and poor results, mainly for the same reasons—the jailing psychosis, poor staff, and inadequate appropriations. There is altogether too much moralizing in most of our state reformatories for women.

Modern, progressive reformatories, besides those already mentioned include those located in Clinton, New Jersey; East Lyme, Connecticut; Muncy, Pennsylvania; Dwight, Illinois; Goochland, Virginia; Taycheedah, Wisconsin; and Wetumpka, Alabama.

As a rule each girl coming to one of these progressive institutions is carefully studied on admission. A complete physical examination shows her physical abilities and limitations. The psychiatrist reports on mental disease or psychopathic tendencies. The psychologist reports on the in-

---

[7] Dean Shepard and Eugene Zemans, *Prison Babies* (Chicago: John Howard Association, 1950).

mate's mental capacity. Finally, a social history, scholastic, and industrial record are obtained from the various ex-employers and social agencies in the girl's home town.

Such an institution maintains an up-to-date classification clinic. From the data mentioned above and personal conferences with the new inmate, the clinic decides how best the girl can be prepared for life outside the institution while she is serving sentence. Assignments to the proper cottage and to suitable industrial training are made according to her physical condition, age, mentality, and previous training.

Reclassification provides for changes of assignment, necessary adjustments, and a review of progress made. A parole classification completely reviews each case and the parole officers present their plans for outside placement.[8]

---

[8] See Katherine Sullivan, *Girls on Parole* (Boston: Houghton, Mifflin, 1956).

26

The Rise of the Reformatory

The gathering of a group of offenders under one roof . . . creates a milieu through the common unit of selection—the commission of a crime. Naturally, then, crime will be the principal interest of the members of this milieu, their common tie, their first and chief topic of conversation. Here is an atmosphere in which crime is something to be admired. Such a milieu will go far toward solidifying delinquent behavior patterns already acquired. (Paraphrased by the National Commission on Law Observance and Enforcement, Report, Vol. I, "Causes of Crime," 1931, p. 87.)

WILLIAM HEALY

The Basis of the Reformatory

Several significant ingredients were characteristic of the reformative movement which emerged in the United States to check the senseless struggle between the advocates of the two prevailing systems of penal philosophy. These were indeterminate sentences, parole, and trade training. When the implementation of the movement was finally consummated in the Elmira, New York, institution in 1876, these concepts represented the program's core.

While several enlightened theorists and practical prison men had for years urged a reformative rather than a punitive program for treating criminals, their thesis was actually slow in crystallizing.

Historically the reformatory movement first saw light in an experiment carried on by Captain Alexander Maconochie (1787-1860) at one of the Australasian penal colonies after 1840. This astute penal administrator and his progressive correctional concepts have been badly neglected by historians and only recently, through the determined efforts of Justice John

Barry of Melbourne, has the program of this correctional genius been correctly appraised.[1]

Maconochie, a captain in the Royal Navy and a noted geographer, was placed in charge of one of the worst of the British penal colonies, located about a thousand miles off the coast of Australia on Norfolk Island. It was to this colony that the "twice condemned" criminals were sent—those transported from England for their crime, who had committed further crimes on the mainland.

Maconochie immediately eliminated the old flat-time sentence and introduced commutation through good behavior and industry. He installed a "mark system" by which each prisoner was debited with a certain number of marks based on the seriousness of his offense. These he had to redeem through work and good conduct. The more rapidly this was done, the speedier his release. Maconochie himself put it vividly: "When a man keeps the key of his own prison, he is soon persuaded to fit it to the lock." Justice Barry thus describes Norfolk Island prior to Maconochie's arrival:

> It is a natural paradise, but for almost seventy years, except during Maconochie's regime . . . it was defiled by the most appalling brutalities. These were perpetrated under the guise of penal discipline. The gallows stood permanently ready as a visible reminder to unhappy wretches of the fate that might soon be theirs; for years at a time men worked and ate and slept in irons; the lash and the cat were in habitual use, and the gag, solitary confinement and the pepper mill were constantly employed as punishments calculated to subjugate creatures made sub-human by deliberate policy. Any manifestation of resentment was classed as insolence, as was the contradiction by a prisoner of the evidence of an official. Subservience, the despot's substitute for respect, was insisted upon, and even the private soldier was entitled to require outward marks of servility from the convicts. When Dr. Ullathorne, a Catholic priest, went to the island in 1834, to bring the consolations of religion to men sentenced to death for their part in a rising against the authorities, the men who were reprieved wept with sorrow that they had to go on living, and those doomed to die fell on their knees and thanked God for the release that was to be theirs.[2]

The "apparatus" (as Maconochie called his correctional scheme), whereby he planned to put his theories into practice rested on these five postulates:

---

[1] John Barry, *Alexander Maconochie of Norfolk Island* (Melbourne: Oxford University Press, 1958). Beatrice and Sidney Webb, in their *English Prisons under Local Government* (London: Longmans, 1922), p. 165, observe that Maconochie is not even included in the *Dictionary of National Biography*, and quote W. L. Clay, *The Prison Chaplain* (London: 1864), p. 254, who writes that he was called "a noble-hearted old man . . . who passed from maleficent neglect to beneficent death." The painting of Maconochie is reproduced herein through the courtesy of his great-grandson, Kenneth J. Maconochie of London, and the good offices of Justice Barry.

[2] John Vincent Barry, "Alexander Maconochie," *J. Crim. Law,* Vol. 47, No. 2 (July-August, 1956), pp. 145-161.

Painting of Captain Alexander Maconochie done in 1836 by Edward Villiers Rippingille, Associate of the Royal Academy, London. (Courtesy, Kenneth J. Maconochie, great-grandson of Alexander Maconochie of London.)

1. Sentences should not be for a period of time, but for the performance of a determined and specified quantity of labor; in brief, time sentences should be abolished, and task sentences substituted;

2. The quantity of labor a prisoner must perform should be expressed in a number of "marks" which he must earn, by improvement of conduct, frugality of living, and habits of industry, before he can be released;

3. While in prison he should earn everything he receives; all sustenance and indulgences should be added to his debt of marks:

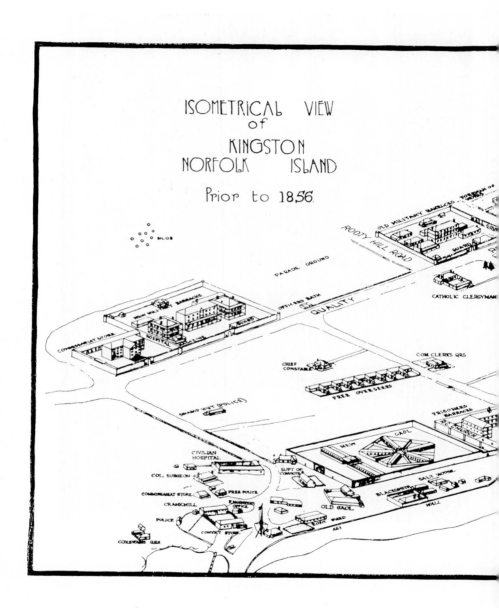

ISOMETRICAL VIEW
of
KINGSTON
NORFOLK ISLAND

Prior to 1856.

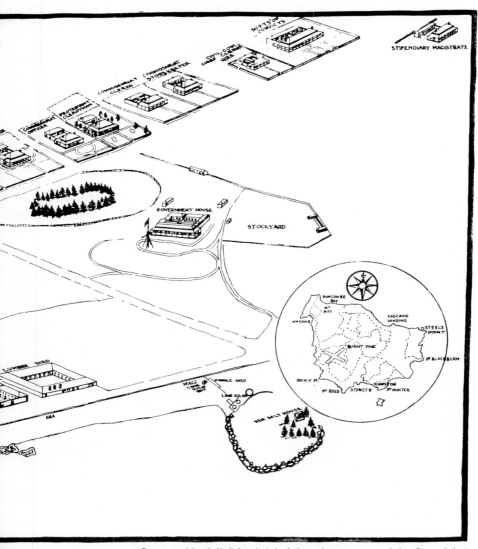

SUPT'S CONVICTS

STIPENDIARY MAGISTRATE

CIVIL OFFR

COMMISSARIAT CLERK

COMMISSARIAT STOREKEEPER

PROTESTANT CLERGYMAN

GOVERNMENT HOUSE

STOCKYARD

LUMBER YARD

SEA

WALL

SHINGLE SHED

LIME KILN

NEW SALT HOUSE

DUNCOMBE BAY

MT PITT

ANSON B

CASCADE LANDING

STEELS POINT

BURNT PINE

PT BLACKBURN

ROCKY PT

PT ROSS

SYDNEY B

KINGSTON

PT HUNTER

(Courtesy Norfolk Island Administration, owners of the Copyright.)

421

4. When qualified by discipline to do so he should work in association with a small number of other prisoners, forming a group of six or seven, and the whole group should be answerable for the conduct and labor of each member of it;

5. In the final stage, a prisoner, while still obliged to earn his daily tally of marks, should be given a proprietary interest in his own labor and be subject to a less rigorous discipline in order to prepare him for release into society.[3]

The effects of this man's ingenious system in this depraved colony may be judged by his well-founded boast: "I found Norfolk Island a hell, but left it an orderly and well-regulated community." But this brave pioneer's enlightenment was only a momentary gleam in the long night of savagery and benighted bureaucracy that characterized the Australian transportation system. Not only was his progressive correctional philosophy not encouraged, but it was actually undermined and repudiated by his superiors. He was recalled home after only a brief attempt to transform this brutalized colony into a semblance of order and hope. But his ideas did persist back in England as we shall point out below.

Prison compound on Norfolk Island, off coast of Australia.

During Maconochie's brief sojourn on Norfolk Island, the concept of the indeterminate sentence was being popularized by the writings of several reformers in England.[4] It was the theory of these people that the purpose of imprisonment was to prepare a man for release; hence the length of

---

[3] Quoted by John V. Barry in "Captain Alexander Maconochie," *The Victorian Historical Magazine,* Vol. 27, No. 2 (June, 1957), p. 5.

[4] Notably Richard Whately (1787-1863), George Combe (1788-1858), and the brothers Frederick Hill (1803-1896) and Matthew Davenport Hill (1790-1872).

sentence should not be fixed but rather be determined by the rehabilitation of the man. The logical supplement of the indeterminate sentence,
parole, was most effectively advocated by the French publicist, Bonneville
de Marsangy. In 1846 he had defined *conditional liberation,* or what later
became known as parole, as a "sort of middle term between an absolute
pardon and the execution of the entire sentence; the right conceded by
the judiciary to release provisionally, after a sufficient period of expiatory
suffering, a convict who appears to be reformed, reserving the right to
return him to prison, if there is against him any well-founded complaint."

The third ingredient envisaged in the reformatory movement was trade
or industrial training. While productive labor was used by the Philadelphia
reformers in the Walnut Street Jail during the decade of the 1790's, its
use as a rehabilitative device was elaborated and further emphasized by
two prison administrators in different countries during the 1830's. These
were Colonel Manuel Montesinos in his prison at Valencia, Spain, and
Georg M. Obermaier, a pioneer who headed an institution at Kaiserslautern, in Bavaria. Both of these administrators were truly pioneers in practical correctional philosophy. Wrote the former at that early date: "The
prison receives only the man; the crime remains at the gate."[5]

It was Sir Walter Crofton of Ireland who adapted Maconochie's mark
system to his prisons and whose practical vision stimulated the American
administrators. They were casting about to break the deadlock that shackled the penitentiary movement. Crofton referred to his system as the "Intermediate System." It was developed on the principle that criminals can
be reformed, but only through employment *in a free community* where
they are subjected to ordinary temptations. It was referred to as a "fil-
terer between the prison and the public, by which the reformed will be
separated from the unreformed, the former to be advanced toward personal liberty and restoration to society, and the latter to be returned to
further penal treatment."[6] For example, after four years of strict confinement in a conventional prison, a prisoner serving ten years was eligible
for removal to an intermediate prison. After four months' detention there,
if he received an offer of employment and his conduct was satisfactory, he
was granted a conditional pardon, known as a "ticket-of-leave." Since
his license was revocable for irregularity of conduct any time within the
original sentence of ten years, the person was under formal supervision
until his sentence expired. This was the first extensive use of what became
known in our own country as the parole system.

The Irish System combined the basic ideal for reformation; the determination of the time to be served by conduct in prison, promotion to

---

[5] José Rico de Estasen, *El Coronel Montesinos, Un Español de Prestigio Europeo,*
(Madrid: 1948).

[6] Frederick A. Packard, *Journal of Prison Discipline & Philanthropy* (January,
1858), pp. 16-25. Dr. Packard describes the Crofton system based on some of the
written articles by the Irish prison administrator.

The parent reformatory at Elmira, N.Y., about 1910.

release through graded stages of increasing freedom and responsibility, stress on the importance of teaching industrious habits to achieve self-reliance, and the use of a rudimentary parole system.

The American Pattern—Elmira

In 1863 Gaylord Hubbell, the enlightened warden of Sing Sing prison, visited Ireland and returned enthusiastic over graded institutions and the Irish System. He recommended its adoption in New York State. Interest in the Irish system, as well as numerous gropings for better prison administration in the United States, ushered in a wave of enthusiasm with unified leadership, galvanized reform principles, and a will to translate theory into action.

Prison administrators and others interested in correctional reform were summoned to meet in Cincinnati in the autumn of 1870. The Prison Congress initiated there not only expounded all the leading reform ideas of the era, but also brought into being the National Prison Association.[7]

It was felt that all men of good will throughout the world should join in formulating an ideal plan for dealing with criminals. The struggle between the Pennsylvania and Auburn systems had discouraged correctional people and all were ripe for some new dispensation. This first prison association met, listened to some excellent papers, and adopted a Declaration of Principles. This document has become famous in correctional circles. Its 37 paragraphs advocated a philosophy of reformation as opposed to the doctrine of punishment, progressive classification of prisoners based on a mark system, the indeterminate sentence, and the cultivation of the inmate's self-respect.

These enlightened principles anticipated virtually all of the New Penology which we describe in a later chapter. They have never been put into complete operation anywhere, and only incompletely in the most progressive correctional institutions. Therefore, it is not inaccurate to state that even the most enlightened practice of our day has not yet fully caught up with the theory expounded by the more progressive of the correctional leaders of that day. This fact is both a challenge to current practice and an effective answer to those who contend that enlightened penologists are sentimentalists and dreamers in their "novel and untried vagaries."[8]

---

[7] This organization is now known as the American Correctional Association and has in its membership prison administrators, parole and probation workers, prison chaplains, members of agencies for discharged prisoners, and others interested in the field of corrections.

[8] For an enumeration of the principles laid down at the Cincinnati Congress in 1870, see *Proceedings,* American Correctional Association, 1956, pp. xiii-xix. Justice John Barry states that "most of the ideas in the 1870 Cincinnati Declaration of Principles were taken from Maconochie's writings, the language sometimes lifted bodily." (Letter to the authors dated May 2, 1958.)

Those who urged the adoption of more humane and progressive methods at the Congress were, however, able to secure them only for the treatment of young first offenders in reformatories. The year before the Cincinnati meeting a law had been passed in New York State authorizing the creation of a "reformatory" institution to be erected at Elmira. Financial difficulties delayed construction, but the institution was finally opened in 1876 with the veteran prison administrator, Zebulon R. Brockway, its first superintendent.

Since Elmira became the parent reformatory plant, the new system began to be known as the Elmira System. Many other states followed the example of New York and soon glowing reports from the superintendents of the new establishments filled the journals of prison administration.

The Traditional Reformatory Program in its Early Years

There was little uniformity in the age limits set by the states for admission to their new establishments, except that the minimum was rarely less than sixteen. The maximum was usually thirty although a few finally sent men beyond that limit. It is sometimes as high as forty.

At the Elmira Reformatory the inmates were divided into three grades. At entry, each inmate was placed in the second grade. At the end of six months of good conduct, he could be promoted to the first grade, and six months of continued good conduct entitled him to parole. Misbehavior was punished by demotion to third grade, where a month's good conduct was required for restoration to the second grade. Incorrigible inmates were obliged to serve the maximum sentence.

Zebulon Brockway, first superintendent of Elmira, published his own conception of the essentials of the "Elmira system," but his views have been little, if at all, quoted directly since then. As Brockway saw it, the system must include:

> 1. The material structural establishment itself. . . . The general plan and arrangements should be those of the Auburn System plan, modified and modernized; and 10 per cent of the cells might well be constructed like those of the Pennsylvania System structures. The whole should be supplied with suitable modern sanitary appliances and with abundance of natural and artificial light.
>
> 2. Clothing—not degradingly distinctive but uniform, yet fitly representing the respective grades or standing of the prisoners. . . . Scrupulous cleanliness should be maintained and the prisoners appropriately groomed.
>
> 3. A liberal prison dietary designed to promote vigor. Deprivation of food, by a general regulation, is deprecated. . . .
>
> 4. All the modern appliances for scientific physical culture; a gymnasium completely equipped with baths and apparatus; and facilities for field athletics.
>
> 5. Facilities for manual training sufficient for about one-third of the population. . . . This special manual training covers, in addition to other

exercises in other departments, mechanical and freehand drawing; sloyd [manual training] in wood and metals; cardboard constructive form work; clay modeling; cabinet making; clipping and filing; and iron molding.

6. Trade instruction based on the needs and capacities of individual prisoners. (Where a thousand prisoners are involved, thirty-six trades may be usefully taught.)

7. A regimental military organization with a band of music, swords for officers and dummy guns for the rank and file of prisoners.

8. School of letters with a curriculum that reaches from an adaptation of the kindergarten . . . up to the usual high school course; and, in addition, special classes in college subjects. . . .

9. A well-selected library for circulation, consultation and, for occasional semi-social use.

10. The weekly institutional newspaper, in lieu of all outside newspapers, edited and printed by the prisoners under due censorship.

11. Recreating and diverting entertainments for the mass of the population, provided in the great auditorium; not any vaudeville or minstrel shows, but entertainments of such a class as the middle cultured people of a community would enjoy. . . .

12. Religious opportunities . . . adapted to the hereditary, habitual, and preferable denominational predilection of the individual prisoners.

13. Definitely planned, carefully directed, emotional occasions; not summoned, primarily, for either instruction, diversion, nor, specifically, for a common religious impression, but, figuratively, for a kind of irrigation.[9]

The reformatory movement was watched in correctional circles the world over for the two decades following the opening of the Elmira institution. Sir Evelyn Ruggles-Brise, director of the English prisons, visited Elmira in 1897 and returned home to launch his system of Borstal training for young offenders which is recognized today as far surpassing anything we have here for that purpose. Major Arthur Griffiths, inspector of Her Majesty's Prisons, writing in 1894, described the Elmira regime as one of extreme mildness, "where most of the comforts of a first-class boarding-school, ample diet, military music, the study of Plato, and instruction in interesting handcrafts are utilized in the process of amendment."[10]

Before the turn of the century, however, the enthusiasm of the 1870's had been dampened. Brockway himself anticipated some of the difficulties as early as 1887:

> Brockway at this time looks back over the seventeen years intervening and recalls that at Cincinnati in 1870 he had had an experience similar to that of the disciples on the Mount of Transfiguration. He had felt himself strengthened by a "mysterious, almighty, spiritual force. . . . I was going to have a grand success . . . but it did not work. . . . I found that there was a commonplace work of education to do with these persons I hoped to inspire. . . . That did not suffice. The industrial training of pris-

---

9 Zebulon Brockway, *Fifty Years of Prison Service,* pp. 419-423.
10 *Secrets of the Prison House,* Vol. I (London: Chapman & Hall, Ltd., 1894), p. 6.

oners was taken up, and that is drudgery. . . . " Thus he warns younger
men that hard work must follow inspiration.[11]

Disillusionment

It has often been pointed out that one important contribution of the
reformatory movement was that it tended to lessen the obsession with the
Auburn or Pennsylvania systems and reduced the controversy between the
partisans of each. It is true that its adoption distracted discussion from
the alleged virtues of the two systems by concentrating attention more
and more on the correctional ideals enunciated at the Cincinnati conference. But in actual practice in prison administration there was no change.
The Auburn system, by 1870, had dominated the entire country except
in Pennsylvania. But even in that state the cherished separate system was
slowly being dissipated by increased prison population.

The Auburn system continued to dominate the prison picture in both
building and administration and still does today. The only exceptions are
the federal prisons and a few states that have experimented in new types
of correctional institutions. Hence, the reformatory movement has had
little effect upon either the construction or the administration of prisons
for adults. This fact must be borne in mind when appraising the prison
picture in the United States.

There have been many explanations offered as to why the reformatory
movement proved a relative failure and disappointment. First and fore-
most, the ideals supporting the movement were actually never put into
effect; it never had a chance to succeed. One need only read the Declaration of Principles to see how far short of actual implementation was
its correctional philosophy in 1900, or today.

The most important explanation for the failure of the reformatory is
that it was blocked at the outset by the ever-present jailing psychosis
with the frenzied preoccupation of counting and double-checking inmates.
No rehabilitative program, however excellent in theory, can ever be a
success, so long as safe and secure custody is regarded as the prime
purpose of institutional activity, and administrators are judged to be efficient or incompetent almost wholly by their success in preventing escapes.
The public seems willing to accept a high price for such success: brutal
and repressive programs, not to say complete failure to rehabilitate inmates.

One cannot, of course, escape the necessity of security in any penal
situation. But so long as it completely dominates the institutional setting
it is impossible to operate a worth-while rehabilitative program. Whether
in a prison for adults, a reformatory, or even a cottage institution for

---

[11] Blake McKelvey, *American Prisons* (Chicago: University of Chicago Press,
1936), p. 71n.

juveniles, the factors and trends that explain the failure of the reform movement have almost invariably stemmed from the locking psychosis and its resulting regimentation. It is this fact that made it impossible to apply in practice the ideals of 1870—even at Elmira.

In no respect was this more fatally disruptive of the ideals underlying the reformatory movement than with regard to the architecture and physical setting that has operated down to the present time. The Elmira plant itself was originally built as a maximum-security prison for hardened adult criminals, and then the reformatory was set up within these bad structural surroundings. Most other American reformatories copied the Elmira prison architecture as well as the Elmira system of discipline. Indeed, the first three reformatories—at Elmira, at Concord, Massachusetts, and at Huntingdon, Pennsylvania—were all opened in structures that had been originally constructed as maximum-security prisons for adult felons. The Michigan Reformatory at Ionia was also a tight-custody cellular structure. Even when the Rahway Reformatory in New Jersey was built in 1901, it was designed along maximum-security lines and many thousands of dollars were wasted on a fantastic and useless dome. The tendency to build reformatories as nothing more than prisons for youth continued unabated into the twentieth century. The Camp Hill Reformatory of Pennsylvania, opened in 1941 near Harrisburg, was constructed with four maximum-security cell houses. When it was opened the papers of the state compared it to a "private school . . . run like a military academy." The blocks were named for Indian tribes. But its inmates and graduates would agree that it is a "tough" prison. It had been opened scarcely five months when it experienced a hunger strike and several prison breaks; in one of these a boy was shot to death. In the spring of 1942 one boy escaped in a fog and another was brought down with a shotgun fired by a guard. In recent years it has made some strides in living down its bad reputation.

Not even the enlightened Federal Bureau of Prisons dared to repudiate the conventional prison pattern in reformatory architecture. Both the El Reno Reformatory, completed in 1934, and the Chillicothe Reformatory, opened earlier, provided large maximum-security cell blocks with guard towers manned by expert marksmen.

It was impossible to develop true reformatory ideals in so forbidding an atmosphere as that provided by steel cages and high, gloomy walls. Many factors contributed to the emasculation of the reformatory program, but none was more important than this handicap of a completely penal and punitive setting for the operation of an enlightened system of rehabilitation. Even when architects, after long and fatal delays, went to the opposite extreme as at Annandale, New Jersey, in 1929, and erected an open unit, or cottage-type of institution, the morbid fear of escapes prevented the administration from fully exploiting the advantages of such a structure.

Annandale, N.J., Reformatory, 1929. First reformatory to have structure that conformed to reformatory ideals. (Courtesy New Jersey Department of Institutions and Agencies.)

Although the concepts of the reformers of the 1870's were enlightened, it was inevitable that the administration of the reformatories tended to become repressive just as in the adult prisons. The same contamination, inmate-feuding, sexual frustration, and other abnormalities of prison life were present. Even Brockway, who has often been called the greatest warden America has produced, resorted to corporal punishment. He was severely criticized for his frequent use of the strap, and, at times, even used a paddle with nails in it. Reformative discipline tended to degenerate almost everywhere into the same system of drab, repressive, and uninspiring regimentation that dominates the conventional prison.

We noted in our dicussion of the Irish System that the methods of grading and promoting inmates, as a leading means of encouraging good behavior, personal reformation, and securing early release, was one of the cornerstones of the system that ushered the American reformatory movement into being. It was actually the chief psychological motive to the improvement of conduct of the inmates. Yet, under the domination of the resulting system of regimentation and punitive discipline, this vital and crucial grading and promotion system tended to become perfunctory and almost totally ineffective in stimulating efforts at reformation.

Next to the system of grading and promotion, the reformers of the 1870's relied on teaching "good habits of industry" to promote the rehabilitation of inmates, giving them thorough vocational instruction that would provide them with a trade so they could live by non-criminal methods after release. But the industrial system in reformatories, as in the prisons, was hampered by the commercialization of institutional industry and by the erratic and senseless legislation limiting or destroying institutional industry during some 65 years following the opening of Elmira.

Vocational education has been more thoroughly worked out and applied by reformatories than any other single aspect of the reformatory complex of the 1870's. But no reformatory has ever provided anything like a thorough and ideal system of vocational instruction with the introduction of modern tests, and some institutions have made only the most perfunctory beginnings in this field. In some reformatories, the authorities have been chiefly concerned with making money out of reformatory shops. Even in those institutions where administrators thoroughly espoused the ideal of vocational instruction, made every possible effort to set up such a program, and completely subordinated profits to vocational education, they were rarely able to get adequate appropriations to maintain a satisfactory program of vocational education.

Many young adults who find themselves in a reformatory are emotionally unsuited for vocational or trade training. Modern tests show many persons manually unfit for trade training. Few reformatories employ such tests and many administrators know nothing about them.

Even a perfect system of vocational education would not suffice to constitute the core of a rehabilitative program. Adequate social and political education would be necessary also, but there has been little of it even in the best reformatories. There have been some excellent reformatory schools, but the curriculum has been chiefly limited in such cases to routine instruction in general academic education. It is true that reformatory administrators cannot be regarded as uniquely neglectful in this respect, for the failure to provide sufficient social and political education is also flagrantly true of our public school system in general.

It has frequently been stated that perhaps the greatest contribution of the reformatory movement was the introduction of the indeterminate sentence. It is certainly true that the reformatory propaganda did more than anything else to introduce the indeterminate sentence into penological thought and, slowly and gradually, into legal practice. But this basic conception of the reformatory has also tended to fall by the wayside. The indeterminate sentence has been far from universal in the sentencing policy employed with respect to those committed to reformatories. For example, in one very prominent mid-western reformatory, at the time of an official investigation, some 90 inmates were serving flat sentences of ten years or more, and 154 were serving minimum sentences of ten years. It was found that one inmate of the Concord Reformatory in Massachusetts had already served 28 years at the time of an official inspection. Moreover, since 1876 some form of limited indeterminate sentence has made considerable headway with respect to those sentenced to the conventional prison. The average time served by all prisoners in the penal institutions of the United States is only a little over two years, which is approximately the same as that served by reformatory inmates. Hence, there is little difference today between the inmates of prisons and the inmates of reformatories with respect to either the form of sentencing employed or the time actually served.

The "Siamese twin" of the indeterminate sentence in the reformatory complex is the parole system. The reformers of 1870 envisaged the universal use of the indeterminate sentence in committing reformatory inmates to an institution. They also took it for granted that, when these inmates were released, the period between their discharge and the expiration of their maximum sentence would be covered by adequate and efficient parole supervision. But the parole provisions that have actually operated have been even more deficient than the indeterminate sentence. Only the federal prison system and about a dozen states have anything that might be regarded as an adequately manned and professionally trained parole service. Hence, it is obvious that only a small minority of those discharged from reformatories have received effective parole supervision. The latter was regarded as an indispensable item in the rehabilitative plan of reformatories, and the failure to provide for it has been just one more reason why the reformatory has proved such a bitter disappointment as it has worked out in practice during the last 80 years.

The net result of all the defects and failures here briefly reviewed has been that the reformatory has degenerated into a system that provides conventional prisons for young offenders. It has become almost a junior prison, in which the program is often more repressive than in many adult prisons.

The reformatory system as envisaged by the progressive penologists of the 1870's has not failed, for the very evident reason that it was never actually installed and tested out. What has failed is the junior prison, which was set up and called a "reformatory." Since the junior prison had most of the traits of the conventional adult prison, it has been just as much of a failure as the latter.

Despite all this, the reformatory movement was undoubtedly an asset in the history of American penology. It galvanized progressive penological thought and has had a long-range, if spotted and gradual, beneficial effect upon both penal thought and practice. It would be unfortunate if the failure of what went under the name of "reformatory" should lead to the discrediting of the reformatory idea and its abandonment. The lesson that should be learned is that the reformatory conception is pre-eminently worth saving and trying. But saving the reformatory conception will require both a physical plant and an institutional program entirely compatible with the reformatory ideals of the 1870's, to which must be added provisions for a plan of inmate self-government.

The Future of the Reformatory

Some progressive penologists have raised the question whether there is any real justification for a reformatory, namely, a special institution to treat delinquents of the age-group between juveniles and adults. They suggest that our newer medium-security institutions for adults will serve the purpose. Nevertheless, it is the opinion of others that there is a real need for the reformatory. Young delinquents between the ages of 18 and 25 present very definite personality and adjustment problems that require specialized understanding, consideration, and treatment not to be found in "good" prisons for adults. Hence we must have institutions that are deliberately designed to serve this very particular and important need. But the matter of the age-group is not the only consideration. The reformatory is a natural and inevitable product of the theory and practice of classifying delinquents. Many young adults do not require institutional treatment; some states have demonstrated that with their forestry camps. But other prisoners may prove too dangerous to be remanded to a reformatory. It is to the classification system that we must turn in order to determine the type of young adult delinquent most satisfactory for reformatory treatment.

For those persons committed to a reformatory who present relatively slight custodial and disciplinary risks, an open unit of some type, such as a forestry camp or cottage institution, with or without a fence, would make for greater or less freedom for the inmates. If there is no fence, the authorities are usually so fearful of possible escapes that they gradually limit the use of the facilities and privileges that the open type of plant presents. The strictly open type of facility, such as a camp, provides both for maximum freedom and flexibility of program. If fenced in, it becomes a safe facility for all types of rehabilitative programs. These two types provide for the needs of every conceivable candidate for reformatory treatment.

Since classification with its screening procedures is the keystone of progressive correctional thought and practice, the reformatory system must include it. This calls for a well-staffed diagnostic center and clinic, to which all convicted delinquents would be sent for careful examination and study prior to their distribution among the appropriate facilities operating within the correctional system of the state. This diagnostic clinic would have the responsibility not only for determining which types should be consigned to reformatory treatment but also for determining which should be sent to either the close custody or the open facilities.

Adequate and well-trained staff within the institution itself is of signal importance. This fact was well stated many years ago by the Osborne Association:

> The real solution of the reformatory program lies in personnel first, and afterwards in program. In these institutions there is need not only of high-minded and especially trained superintendents, but also staffs that are composed of intelligent, trained men who are capable of exerting a stimulating and beneficial influence whenever they come in contact with the individual inmates. . . . There is no place in the reformatory for . . . the traditional guard.[12]

The system of grading and promoting inmates on the basis of good behavior, determined in a progressive and sensible fashion, which was the core of the Irish prison system and was intended to be introduced very completely in the Elmira reformatory system, has not been at all outmoded. It should be revived and applied unremittingly in any complete reformatory program. No other single device is likely to do more to promote rehabilitative efforts from the inmates. But this practice must never be allowed to fall into a perfunctory type of execution, which could virtually destroy its beneficial impact upon inmates.

---

[12] Osborne Association, *Handbook,* 1929, p. xxix.

Inmate education was originally regarded as essential in reformatory treatment, and all our experience since 1870 has only served to emphasize the correctness of this attitude. Vocational instruction is surely the most important aspect of reformatory education, and there should be adequate appropriations to supply all needed equipment and trained personnel, even though the operations do not provide one cent of profit from the sale of products. The profits here are the greater ones, even in terms of money values, which will come to the taxpaying public as a result of enabling former criminals to earn their living without any further depredations upon society. Counseling and guidance, as well as the practice of group therapy, should also be integral parts of the training program. Adequate social and political education should be included, far beyond that provided in the general run of public schools.

We cannot expect the reformatory system to be successful without the establishment of self-government plans that will train inmates in those responsibilities of citizenship they obviously do not possess when admitted.

Nothing could be more important in the disciplinary and rehabilitative program of a reformatory than elaborate provision for every kind of healthy recreation practicable within an institution setting. This is particularly true for those of the reformatory age group. Youths of this age possess a super-abundance of energy, especially of sexual desire and impetus. The Kinsey Report has recently shown that the human male reaches the apex of sexual desire and capacity in precisely this reformatory age period. One of the main reasons for severe disciplinary problems in a reformatory has been that the normal outlets of sex impulses are completely blocked and there has been no adequate provision for sublimating them through other forms of activity. Homosexuality and other forms of aberrations, as well as serious forms of incorrigibility and restlessness, are a natural result of all this. Strenuous physical exercise and exciting recreation can do more than anything else to distract attention from sexual desires and drain off unsatisfied sexual impulses. Recreation is also very helpful in developing physical strength and prowess, as well as in teaching the principles of sportsmanship and fair play that are highly desirable qualities in any good citizen. Also highly important is the fact that prison games are more likely to distract the attention of the inmates from their usual obsession with escape or release.

The net result of a good reformatory program, in its totality, should be to concentrate the attention of the inmates upon the opportunities for reform and training rather than upon the prospect of release or escape. We have already seen how such a program in the federal institution at Seagoville has served to hold in secure custody what are usually regarded as the most dangerous types of criminals, though Seagoville is an open and unwalled institution.

The original reformatory complex included as an indispensable item the

YOUTH AUTHORITY PROVISIONS

	Model Act	California	Minnesota	Wisconsin	Massachusetts	Texas
Name	Youth Correction Authority	Youth Authority	Youth Conservation Commission	Div. of Child Welf. and Youth Service in DPW	Youth Service Board	Youth Devel. Council
Stat. Ref.	1940 Adopted American Law Institute	1941 ch. 937; Welf. & Insts. Code Div. 2.5, ch. 1	1947 ch. 595	1947 ch. 546, 560 (1949 ch. 376)	1948 ch. 310	1949 ch. 538
Admin. Orgn.	3 man full time bd. apptd. by gov., 9 yr. staggered terms, bd. selects chmn.	3 man full time bd. apptd. by gov.; 2 from list of advisory panel (6 pres. of quasi-public prof. assns.) 4 yr. staggered terms, gov. selects chmn.	6 apptd. by gov. (3 pub. officials ex-off.; dir. of div. of insts., chmn. of parole bd., juv. ct. judge) 4 yr. staggered terms, gov. selects chmn.	Div. of DPW. Bd. may appt. citizen committee with gov.'s. appvl. to advise on programs and problems, DPW dir. selects div. head	3 man full time bd. apptd. by gov. from list of adv. committee on children & youth (15 apptd. by gov.), 6 yr. staggered terms, gov. selects chmn.	6 apptd. by gov. and 8 state officers ex-off.; DPW dir. is chmn.
Age and Juris.	16-21 from crim. or juv. ct.	Discretionary, 16-21 from crim. or juv. ct.	Minors convicted of crime, sentenced to less than life. Juv. ct. commitments to tr. schools	Minors sentenced to less than life, juvs. not released on probation	Juv. ct.; children convicted of crime after juv. ct. waiver (under 18)	Juv. ct.
Type of Commitment	Indeterminate	Indeterminate to 25, subject to max.	Indeterminate to max.	Definite (fixed by ct.), 1 yr. min.; or indeterminate to max.	Indeterminate to 21	Indeterminate to 21
Prob.	Granted by auth.	By ct.	By ct.	By ct.	By ct.	By ct.
Admin. of Facilities	Auth. is mainly diagnostic; permitted to admin. facilities	Manages insts.	Manages state tr. schools	May be given management of treatment facilities	Manages insts.; may use other facilities	Manages insts.
Duration of Control	To 25; juv. ct. commitments to 21; if over 18, 3 yrs. Unlimited control on appvl. of ct. may continue where release is dangerous	To 25 for felons; 23 for misdems. Control may continue by ct. order but only to max. for offense. Juv. ct. commitments to 21	To 25, or max. for offense, if shorter. To 21 for delinquents. Control may be trans. to another agency after 25 where release is dangerous	Unlimited, with appvl. of ct. Jury trial on dangerousness may be demanded. Juv. ct. commitments to 21	To 21; unlimited with appvl. of ct. Boy 14-18 convicted of crime may be committed to bd. to 23, or punished as adult	To 21
Disch. Parole	At any time	At any time	At any time	2 yr. min. for felons, except with appvl. of ct.	At any time	At any time

Chart from article by Sol Rubin, "Changing Youth Correction Authority Concepts," *Focus*, Vol. 29, No. 3 (May, 1950), pp. 77-82.

indeterminate sentence. This was a sound idea. All experience since 1870 has confirmed the wisdom of the indeterminate sentence and there cannot be any true reformatory system that does not provide for it in a thorough-going fashion. Crude maximum and minimum sentence laws are no answer to this need, and it is equally obvious that the indeterminate sentence is a farce unless linked up with a really effective parole system.

All this may seem an ambitious program but, in reality, aside from the dual physical plant required and the establishment of inmate self-government plans, there is nothing in it that was not known and recommended in 1870. As we remarked at the outset, all that is needed to institute the reformatory and to make it as much of a success as any human agency can well be under the circumstances is to catch up with the ideals of 1870. These, of course, need to be extended and implemented by the specialized professional knowledge we have gained since that time in certain fields, such as physiology, psychology, psychiatry, sociology, and community work.

To set up literal reformatories and to make them work reasonably well would be a most important advance step in correctional practice. But the vanguard of correctional thought today would indicate that progressive and workable reformatories are only a step toward a much more daring and experimental future in dealing with those who are now sent to reformatories. In the not too remote future we shall surely handle more and more of these types outside of any institution whatever. We will make greater use of probation, as well as of correctional camps.

The California Youth Authority, introduced in 1941, is the first major move in this direction.[13] Under this system, now adopted by Wisconsin, Minnesota, and Massachusetts, and Texas, as well as California, the judge does not sentence the youthful offender to any institution but turns him over to the personnel of the Youth Authority, which has the power to send the person to any institution, even to a state prison, or to place him on probation immediately after receiving the recommendations of a centralized diagnostic clinic. Upon receiving a youth remanded to its custody, the Youth Authority sends the youthful offender to a diagnostic clinic where trained personnel examine his case from every angle, both of personal history and scientific scrutiny. The clinic recommends the proper disposition of the case, and the Youth Authority has complete power to determine future action. The expansion of forestry camps has introduced a hopeful type of re-training for those capable to respond to such treatment.

There is little doubt that this Youth Authority innovation is the most important advance in correctional thought since the 1870's; in time it may be expected to supplement or supplant the older system.

---

[13] See chart indicating variations in the states regarding disposition of young adult cases.

Although New York State has not yet adopted a Youth Authority, it has installed a centralized diagnostic center. It is set up in a special block at the Elmira reformatory to which young adult offenders are sent from the courts for diagnosis and disposition.[14]

New Jersey maintains a diagnostic clinic, independent of, but available to, all other institutions. This center, located at Menlo Park, was opened in 1949 and serves both juveniles and adults.

At present, this clinic is concerned only with those cases referred to it for analysis by courts and by private and public institutions. It does not take any powers from the courts, since its role is purely advisory. It merely places at the disposal of judges and administrators an analysis of the client's personality in terms of significant facts about his life history and environmental conditions, his potentialities, liabilities, and assets.

New Jersey has also established a form of "short term treatment" for certain types of youthful offenders who, in the judgment of the classification and diagnostic authorities, do not need conventional reformatory restraints. It is the assumption that many young offenders can be "processed" for community supervision, on a type of probation, by adopting new techniques outside the traditional prison but within the framework of an informal group experience.

A group-centered project was inaugurated in June, 1950, on the Charles Lindbergh estate, near Hopewell, without any restraints whatsoever. Admissions to this type of center are voluntary. When the court is confronted by a youthful offender deemed suitable for this short-term specialized treatment, execution of sentence is suspended on condition that he agrees to attend this farm home for a period ranging from one to six months. Persons extended this treatment are generally those who are not good probation risks but who are not necessarily repeaters. The program consists of the following:

1. Group interaction sessions; in other words, a form of group therapy.
2. Individual counseling.
3. Self-government.
4. Individualized projects such as building a radio, writing a short story, overhauling a gas engine, etc. Each boy is expected to select and carry through to completion some such project.
5. A well-rounded program of work, recreation, and specially directed educational activities.

The youths "invited" to attend this farm establishment are those who find it difficult to settle down in their environment without getting into trouble by "borrowing" automobiles, going on all night joyrides, or engaging in sexual brawls.

---

[14] For details concerning the function and work of this center, see *Reception Center, Elmira, N. Y.* (Albany, N.Y.: State Department of Correction, 1958).

This center is, in many ways, similar to the California Youth Authority camps. The young adults sent to it work at truck gardening and janitorial service in a nearby state institution but the main objective of the project is to attempt to develop some degree of insight in the boys. A recent study of this facility, made by a disinterested group of social scientists, indicated that the recidivist rate was 19 per cent, compared with 43 per cent from the neighboring Annandale reformatory. This, of course, can be partially explained by the type of boys sent to the two institutions.[15]

Despite these laudable experiments, courageously attempted here and there throughout the country, we shall continue to have reformatories for many years to come. But during this period it is not only desirable but mandatory that reformatories should be reformative establishments in fact as well as in name.

---

[15] Nathaniel H. Siegel and H. Ashley Weeks, "Factors Contributing to Success and Failure at the Highfields Project," *Federal Probation,* Vol. 21, No. 3 (September, 1957), pp. 52-56. See also an article by the director, Albert Elias, "Highfields: A New Slant in the Treatment of Youthful Offenders," *N.P.P.A. Journal,* Vol. 2, No. 2 (April, 1956), pp. 163-167. Also, Lloyd McCorkle and others, *The Highfields Story* (New York: Henry Holt, 1958).

Part III

THE REHABILITATIVE PROCESS WITHIN THE FRAMEWORK OF IMPRISONMENT

27

The Emergence of the Concept of Corrections

Some seem to forget that the prisoner is a rational being of like feelings and passion with themselves. Some think that he is placed there to be perpetually tormented and punished. Some prescribe a certain time as necessary to his cure. One will not allow the light of heaven, or the refreshment of the breeze, the comforts of society, or even the voice of his keeper; while another considers a seclusion from his friends and connections, as a ground for accusation of inhumanity. These opinions have not been overlooked by the inspectors; they have adopted that plan which upon full consideration is deemed best, though not perfect; and the effects of which have not hitherto disappointed their hopes. (From An Account of the Gaol and Penitentiary House of Phila. and the Interior Management Thereof, *1792; remarks on the philosophy of the Board of Inspectors of the Walnut Street Jail, Philadelphia.)*

CALEB LOWNES

Change in Terms

Just as prison people refer to prisoners as "inmates" rather than as "convicts," so we find the word "corrections" slowly displacing "penal treatment" or "penology." Prison guards are now usually referred to as "custodial" or "correctional" officers. While these new terms have come into professional usage only within the past decade, we note references as early as 1935 to the "New Prison" and the "New Penology." Change comes slowly in such an area as the treatment of offenders. Yet most of the current philosophy found in progressive correctional circles was solemnly expressed in 1870 at Cincinnati by the prison administrators who met to establish the National Prison Association. No blueprint of progressive penology has ever been compiled since this Declaration of Principles was drawn up almost a hundred years ago.

But even prior to 1870 administrators with vision were contributing to progressive correctional treatment. Names like Amos Pilsbury (Wethersfield, Connecticut), Gideon Haynes (Charlestown, Massachusetts), and Gaylord Hubbell (Sing Sing, in New York) shine out in the history of corrections in this country since it was they and a few others who were heroically pioneering the New Penology.[1]

No great cataclysmic changes have actually occurred in any era, unless it was during the years following 1790 when the concept of imprisonment was translated into action by the Philadelphia reformers when they developed the "penitentiary" as a substitute for corporal punishment. One might also point, however, to the important effects on prison administration and the field of corrections of the renovation and expansion of the Federal Bureau of Prisons during the 1930's.

In the following chapter we shall also call attention to the conception and development of the diagnostic and classification clinic, following the 1920's and the more recent innovation of group therapy and counseling, as examples of milestones in the slow process of rehabilitation of criminals within prison walls.

We shall focus our attention in this section on several of the changes and advances made during the past fifty years; in these may be traced a philosophy in which there is a modicum of hope, however nebulous. If one were to make strict comparisons between the best of our modern correctional programs and the general run of prisons during the decade of 1830-1840 or even during a later decade, such as 1900-1910, he would be obliged to admit that considerable progress has been made. The trend has been toward considering inmates as individuals, rather as interchangeable parts of a mass. While there is still much mass treatment, individualization through classification, and diagnosis and treatment by means of clinical procedures, are widely accepted today.[2]

In this chapter we shall deal with (a) modern architectural plant, (b) modern provisions for physical needs such as food and housing, (c) personnel, (d) provisions for health and recreation, and (e) discipline. Other modern correctional practices and policies will be dealt with in appropriate chapters.

Modern Architectural Plant

Not much progress has been made in prison construction concepts during the past fifty years because most states have been shackled by the

---

[1] Nor can we ignore such names as Enoch C. Wines (1806-1879) and Franklin B. Sanborn (1831-1917) who, with Zebulon R. Brockway, was referred to as the "Big Three" of the correctional field during the last half of the nineteenth century.

[2] It can be cogently argued that individualized treatment in a prison is a contradiction in concepts.

massive prisons erected during the first fifty years of the prison regime. Perhaps the first prison architect to breathe new air into the construction of correctional institutions was Alfred Hopkins, who designed several establishments during the 1930's, including the federal prison at Lewisburg.[3]

But it has taken many years for architects and prison designers to free themselves from the older era when it was assumed that all inmates should be incarcerated in maximum-security institutions surrounded by an impressive wall.[4] Nearly all prisons and reformatories are still erected for maximum-security, although not over 20 per cent of those sent to prison require this. The walls vary from 30 to 50 feet in height and are several feet thick—and they are exceedingly expensive. Most of them are equipped with wall towers manned by guards with machine guns in constant readiness for possible escapes and riots. Many of the modern prisons are equipped with floodlights for use after dark and, in a few cases where wire takes the place of a wall or is strung along, concertina fashion, on top, the wire is electrified.

The Federal Bureau of Prisons and a few state systems in recent years have introduced certain modifications in the old maximum-security layout, substituting for it, so far as physical equipment is concerned, a medium or minimum-security prison. Institutions of this type usually consist of a group of dormitory buildings laid out in the form of a quadrangle without a wall or fence. The Lorton, Virginia, Reformatory for the District of Columbia, built in 1916, was the first unwalled institution of its kind in the United States. The federal prison opened at Terre Haute, Indiana, in 1940 is the first unwalled penitentiary in this country. It is designed on the telephone-pole style which was borrowed from the much older prison at Fresnes, France, erected in 1898.

In recent years, in typical American style, telephone-pole prisons have been erected with long fantastic corridors so that inmates must walk great distances to and from assignments without getting out of doors, even for recreation. There is noted in recent years a decided shift from these curious, deceptively "efficient" prisons to other designs.[5]

Another type of prison is the Norfolk, Massachusetts institution, built between 1927-1930 under the supervision of one of the country's most astute students of prison administration, Howard B. Gill. It was designed

---

[3] See his book, *Prisons and Prison Buildings* (New York: Architectural Book Company, 1930). Hopkins stated that with John Haviland, designer of the Eastern Penitentiary at Philadelphia and the New Jersey prison at Trenton, "progress in prison architecture stopped."

[4] For a history of prison construction, see Federal Bureau of Prisons, *Handbook of Correctional Institutional Design and Construction* (Washington, D.C.: U.S. Government Printing Office, 1949); also *New Horizons in Criminology*, 2nd Ed. (Englewood Cliffs, N.J.: Prentice-Hall, Inc., 1951) Chap. 24.

[5] California, after erecting telephone-pole prisons at Soledad and Vacaville, is shifting to small unit facilities.

The new federal mixed-custody, telephone-pole penitentiary at Terre Haute, Indiana. Note absence of wall.

443

Norfolk, Mass., "Community" Prison. Medium-security.

as a community prison and, although surrounded by a wall, consists of several small buildings, each housing 50 inmates. The idea for this institution was that, the security of the inmates being assured by the wall, life within could be as normal as possible with the treatment program approximating a "therapeutic community."

There are some correctional people with progressive ideas who insist that a wall is necessary. They contend that an unwalled institution places too great a strain of temptation on well-behaved inmates. Others counter that the presence of a wall merely challenges many to attempt to escape. Certainly if the problem of security is settled by a wall or an electrified fence with guard towers, life within the institution can be more normal with the main objective of rehabilitation unhampered by fear of escapes.

Perhaps the best example of a fence-enclosed institution is the California Institution for Men located at Chino, near Los Angeles. Built in 1941, it was originally designed for maximum-security inmates. This idea was changed so that today it really functions as a minimum-security institution, but with many medium and maximum custody risks sent to it. This establishment encloses the largest area of any institution in the country surrounded by a fence—1,250 acres. The guard towers originally erected have never been used. They are disintegrating, much to the pride of its first superintendent, Kenyon Scudder, who protested their erection from the start. In addition, there are no armed guards on the grounds. The Chino institution has been regarded as one of the few promising institutions in the country. But during recent years increased population with a corresponding increase of housing facilities has begun to nullify its conception.

Federal Minimum-Security Correctional Institution, Seagoville, near Dallas, Texas.

Other establishments that have broken from the past in architecture are those located at Seagoville, Texas, a minimum-security federal prison, a complex of two units each housing and programming 600 inmates at Soledad, California, and the New York institution located at Wallkill. The Seagoville institution was formerly built for females on the cottage plan but was later adapted for the housing of males. The Wallkill prison was designed by Alfred Hopkins in attractive Gothic style. The Hopkins firm has long accepted the challenge of new architectural concepts for prisons. Under the leadership of Hopkins' successor, Clarence B. Litchfield, ingenious new establishments are on the drawing-boards that substantiate Alfred Hopkins' conviction that detention quarters and prisons can be both functional and artistic. The institution being built at Enfield, Connecticut for minimum-security prisoners is a case in point. A new maximum-security institution is being built not far from the "campus-like" facility.

But even a progressive maximum-medium security prison is still likely to have a good strong wall, tool-proof steel bars, and modern gadgets for detecting such contraband as materials out of which guns and knives might be fashioned, illicit letters, censored pictures, and narcotics.

All too many states are still handicapped by century-old prisons. And although a rational and constructive program of corrections can be realized in such old plants, it is difficult. The staff is decidedly handicapped by the atmosphere of inside cell blocks, unsanitary plumbing, lack of recreational facilities, makeshift schoolrooms, chapel and shops. Custody and security have always been the objectives of prison planning, not reform or rehabilitation. It is little less than amazing to view today what is being done by progressive administrations in some of the old prison plants, such as Cherry Hill at Philadelphia, erected in 1829, San Quentin in California (1852), or the Iowa state prison at Fort Madison (1839), to mention a few.[6]

Much confusion exists in correctional circles concerning maximum, medium, and minimum custody of prisoners. In all too many prisons are to be found hundreds of reasonably good risks who are locked up in old fashioned cell blocks of steel and stone. On the other hand, many dangerous criminals are enjoying comparative freedom on prison farms and in

---

[6] To demonstrate how an old austere prison can be humane and constructive, see Hal Burton, "Inside Dannemora Prison," *Saturday Evening Post,* Vol. 229, No. 26 (December 29, 1956), pp. 11 f. This prison, long known as New York State's "Siberia," has been greatly improved without sacrificing custody. See also a series of articles in the Los Angeles *Times* by Norris Leap, entitled "The Folsom Story," March 21-April 5, 1957. These articles trace the grim history of this maximum-security prison from its opening down to modern times, now under the wardenship of Robert A. Heinze. The story is replete with rioting, murder, and petty graft behind the walls; in short, it presents a conventional picture of the traditional old-style prison. In recent years realistic prison men, under the direction of the Director of the California Department of Corrections, Richard A. McGee, and Warden Heinze have developed a humane institution without sacrificing the fundamentals of custody.

Proposed maximum-security federal prison designed to replace Alcatraz.

other minimum-security facilities. This maximum-medium-minimum complex may be acceptable to laymen and many jurists but it represents a snare and a delusion to understanding correctional people. It really oversimplifies the alleged purpose of imprisonment. Most lifers could respond to responsibility in minimum security units whereas many small-time criminals would abscond; socially inadequate first offenders may find a minimum security facility unbearable whereas an habituated four- or five-time loser may possibly find such a type of restraint constructive. The quicker the correctional field repudiates these terms and orients its philosophy to individualized treatment, based on an inmate's potential rather than on his crime or record, the more rehabilitation may result.

But the disheartening situation regarding the housing of prison inmates is that too many monolithic maximum-security installations are being conceived and constructed, even in these modern days of progressive corrections. Correctional leaders have been focusing attention on treatment in small installations but fail to raise their voices when legislators, architects, and prison construction people plan and build new correctional establishments. The dead hand of the past demands that massive piles of stone, concrete and steel with all the modern expensive security gadgets be built even though, as we stated above, only a bare twenty per cent of those sent to prison require maximum security. The frenzy for security and custody is costing the taxpayers millions of dollars in every state when new facilities are suggested. Even the proposed new federal prison designed to replace Alcatraz is labeled a "super-security" prison.[7]

---

[7] See the model in the *Handbook of Correctional Institution Design and Construction,* pp. 46-48; also photograph at top of this page.

Food and Housing

The New Prison has gone far to make the inmate more comfortable within the framework of discipline and routine. We have described poor food, the unattractive manner in which it is served, the poor facilities that pass as dining-rooms, and the inordinate haste that is demanded at meal-time in all too many prisons. An honest effort is being made in many institutions to counteract the objections and even evils of conventional feeding. All correctional administrators worthy of their hire recognize that good wholesome food with variety, served up in appetizing manner, is essential to keep prisoners from becoming disgruntled or rebellious. Yet some of the chief complaints of the prisoners in the riots of the 1950's dealt with poor quality and monotony of the food. Even the best wardens rarely have enough money to purchase good food. There were food riots in Sing Sing prison under the able and humane administration of the former warden, Lewis Lawes.

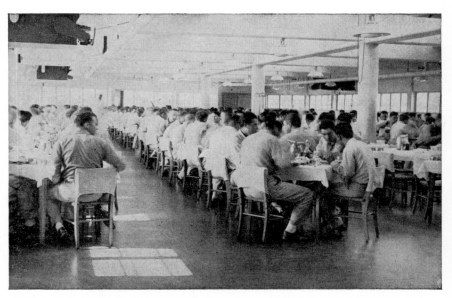

Dining-room, Federal Prison, Seagoville, Texas. Note tables for four, a relatively recent innovation in prisons.

In the modern prison meticulous care is given the preparation of food with the latest kitchen equipment and a trained steward in charge. The culinary staff are scrupulously examined for contagious diseases and charged to keep clean above all else. Those with venereal diseases are excluded from kitchen work. Constant use of soap and hot water as well as the mop is reflected in the immaculate and roomy kitchen. The warden in such a prison considers a clean kitchen of prime importance to the program. The dining-room is attractively furnished with modern tables and chairs, or benches. In such a prison the men may sit at the tables facing one another—which was never permitted in the "old" prison (unfortunately there are still many prisons where archaic seating arrangements may still be found). At the federal institution at Seagoville, Texas, mentioned earlier, the inmates sit at small tables for four. Few prisons have adopted this common-sense plan even yet. In a few prisons, plastic plates have been introduced to substitute for the old style tin compartmental tray.

In some of the more modern prisons the inmate orchestra often provides music during meals or records chosen by the inmates are played over an amplifying system. The prison band sometimes plays stirring music as the men march or walk to and from the dining-room.

Although little has been done in most prisons to see that every man gets the proper number of calories he needs for his work, those who need special diets are provided for through the advice of the medical department. Some prisons maintain a special diet kitchen. Many inmates, when they appear at the prison, show signs of malnutrition. The New Prison detects this and provides special menus for such persons until they are built up physically. In some states the food is checked by a dietician from the state department of home economics who visits institutions periodically.

In most of the new maximum-security prisons we still find plenty of the old inside-cell construction but the cells are more commodious and afford more light than earlier ones. Strong outside-cell construction should suf-

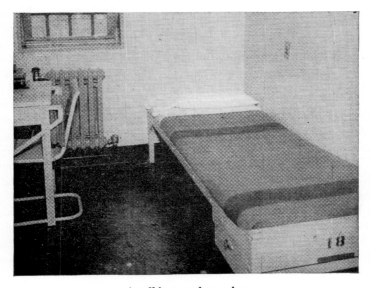

A cell in a modern prison.

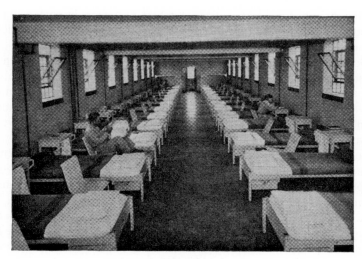

An open dormitory.

fice for any convicted criminal but the lock psychosis takes precedence
with legislators, architects and apologists of new prison construction.

Unfortunately, in many new prisons, especially in the federal system,
dormitories have been substituted for individual cells. This is usually done
for the sake of economy and then is rationalized by stating that certain
inmates need this type of housing. Yet most penologists consider the dormitory a mistake. There is no privacy. A prison inmate is under surveillance 24 hours per day. The individual cell at least permits him to be
alone for several hours but the dormitory eliminates any privacy whatsoever.

The sleeping quarters of a dormitory look not unlike a charity ward in
a hospital. If one goes into a dormitory in some prisons, when the men
have a few minutes off before meals, he will see many of them lying
on the springs of their beds with the mattresses folded back on itself, so
the bedding will not be "mussed up." In the evenings, during recreation
hour, when many of the younger men are playing checkers or listening to
the radio in the so-called "day rooms" at the end of the dormitory, the
older men are sitting on their bedding, waiting patiently for "lights out."

The New Prison provides the men with clean bedding and hot and cold
running water in their cells. Toilets may be in the cell or at the end of the
block, but this latter provision does not take care of night needs because
the men are *locked* in their cells at night. Instead of malodorous disinfectant, which is so liberally used in most prisons, a regime of soap and water,
with plenty of elbow grease, keeps the cell block and cells clean and
sanitary. There is little that is more nauseating than a prison smelling of
disinfectant.

Provisions for frequent bathing with plenty of hand (not laundry) soap are provided. Frequent changes of clothing, as well as bedding, are also an integral part of an intelligent administration. Freedom within the cell is allowed and is not, as in many prisons, a breach of discipline. That is, the inmate is permitted to decorate his cell walls, move his scant furniture around to suit his taste, sit or lie down when he wishes, and use magazines, books or musical instruments, so long as the last is not disturbing to the men in the rest of the block. Rules are few but strictly enforced. In short, wholesome housing accommodations that insure privacy are striven for in the really New Prison outside of dormitories.

Personnel

Most progressive administrators contend that personnel is more important than new plant and equipment. We may start with the warden. In the conventional prison the warden is appointed through politics with little attention to his qualifications for a responsible position. It is assumed by him and by those who appoint him that a tough policy is the only one the convict will "respect." Another type of warden found frequently is the ex-military man. It is reasoned that with his service experience he knows men and discipline and thus can operate a prison effectively. But of course this does not always follow.

Many good wardens have come up through the ranks. So have many poor ones. It is assumed that through their original training and subsequent experience they can operate a prison humanely and efficiently. In general, however, their concept of a prison is merely that of custody. They know that the public expects them to run a "tight" prison so rehabilitation or treatment is regarded by them as secondary. If a little treatment rubs off on the prisoner that is fine but routine and discipline have priority.

In the more progressive state systems, civil service has been introduced. In such states, a reasonably good warden may be employed. The personnel in states using some type of merit system are usually on a much higher level than in those states where such is not employed. The residence rule in states without civil service tends to stultify the system, since the right man for the post cannot always be secured from within the narrow confines of a state. The federal system has civil service rules for all its posts. However, civil service has drawbacks: frequently a mediocre man secures a toehold in the system by passing an examination in the lower brackets and gradually works himself up to a responsible position, though he is devoid of imagination and resourcefulness. In some state systems the personnel are hampered by the regulations of civil service, and a dead level of administrative efficiency has been reached. There is no opportunity for a man of vision to be appointed a warden in such systems unless he goes through the positions in the lower brackets. This policy is unfortunate, since many such persons might have a real contribution to make.

This is the situation in the federal system. Then, too, the methods used in the efficiency rating of the staff provoke much real or fancied cause for discontent. Nevertheless, it is merit that counts. How to find persons of ability without resorting to the mechanical civil service device is a real problem.

The truly modern correctional institution maintains a good in-service course of training for custodial officers. The objectives for such training are set forth by the Committee on Personnel Standards and Training of the American Correctional Association as follows:

1. To improve the capabilities of personnel for participation in the custody, classification, and treatment of prisoners,
2. To increase the effectiveness of personnel and thereby obtain greater efficiency and economy in operations,
3. To promote personnel capacity to recognize, understand, and solve the problems which occur in the correctional institution,
4. To prepare personnel for greater job satisfaction and broader career service.[8]

While in-service training courses are given in many institutions today, in most of them the courses deal almost exclusively with standard custody procedures. Often they vary only slightly from courses offered to state constabularies. Sharpshooting training is, of course, necessary even in a prison as is also judo. But there are many other phases of prison work that may be included in effective in-service courses.

In the really modern prison, training such as we have mentioned should always be on state time; not on the free time of the officers taking the course. Provision should be made for annual increments, tenure, vacation with pay, and adequate pension provisions. Only in offering such fringe benefits can we be assured of recruiting good men for correctional service and holding them against the temptations of other professional fields.

Provisions for Health and Recreation

We can do no better in opening our discussion of the topic of medical care of prisoners than to quote from the *Attorney General's Survey of Release Procedures:*

> The prison physician faces one of the most difficult problems in medical service. To him are sent the very essence of society's misfits, for his patients are not only the abnormal, the subnormal, and the maladjusted— the handicapped, the sick, the surgically unfit, the degenerate, the dissipated, the diseased, the psychotic, the psychopathic, the neurotic and the feebleminded, but they are also the socially undesirable—men and women whom society has cast out, who cannot or will not co-operate according to the rules of the game. To this beginning must be added all the complicat-

---

[8] "In-Service Training Standards for Prison Custodial Officers," (New York: American Correctional Association, 1951), p. 3. See also *Manual, 1956,* Chapter 13, "Selection and Training of Personnel." Howard B. Gill, "Training Prison Officers," *Amer. J. Corr.,* Vol. 20, No. 4 (July-August) p. 8 ff.

ing forces which are peculiar to a prison and which so materially affect the prison physician's patients. They begin with the mental worry and disgrace of imprisonment and they by no means end with the drab, dull existence which characterizes prison life. Not only does lack of stimulating work weaken the mental keenness of active minds, but lack of any work in most prisons today saps the moral fibre out of the most stalwart spirits. Crowded together in the narrow confines of a walled institution, housed in cages like animals where insufficiency of light and air and no privacy wilt the hardiest, fed on a monotonous diet, allowed only a restricted amount of exercise—the problem of keeping people well under such conditions is itself a problem.[9]

Caring for the health of the inmates through remedial service is only a part of the prison physician's responsibility. He must prevent or control contagious diseases, and see that all food, milk, and water are free from contamination, that garbage and sewage disposal is efficient, and that drugs are not making inroads on prisoner morale. It is his business, too, to handle all operations, many of them major in character and often on an emergency basis. In the New Prison he is furnished with adequate staff and facilities, including ample space, equipment of the latest design, anesthetics and sedatives, medicines of all kinds, and above all, independence from annoying bureaucratic restraints.

The attitude toward the "sick line" in the New Prison has changed from that of a routine administrative detail to one of careful contemplation. In the old prison, the sick line (which collects every morning at the dispensary) was all too often filled up with old-timers who regarded it as an opportunity to lay off work or to rest up for a spell. In the New Prison, the best medical knowledge is assiduously applied with a touch of realism. The *Survey,* quoted above, states: "The two most important conditions affecting the operation of the sick line have to do with (1) control of access to the sick line, and (2) facilities for professional attention by the doctor toward the inmates."[10]

In modern institutions, any inmate may report for sick call without getting the permission from a guard formerly required; the guard seldom knew whether a man was really ill or merely malingering, and his suspicions were often justified, but sometimes worked real hardship on the inmate. If the New Prison physician finds a man is a chronic malingerer, he reports the fact to the warden and may even suggest the type of discipline to be applied. The malingerer is a problem in all prisons. Ingenious devices, some medical in nature, are resorted to by the alert physician and the warden, to break up this traditional and annoying practice.

In the best prisons, excellent medical facilities are the rule. Devices for eye, ear, nose and throat treatment, X-ray equipment, electrocardiograph,

---

[9] *Survey of Release Procedures,* Vol. 5, "Prisons" (Washington, D.C.: U.S. Department of Justice, 1939), p. 159.

[10] *Ibid.,* p. 165.

Athletic Field, Federal Penitentiary, Terre Haute, Indiana.

and equipment for physiotherapy, basal metabolism, venereal treatment, and laboratory analysis are all supplied.

The first-class prison will also explore the possibilities of performing plastic surgery for the inmates when it is determined by classification clinics that physical anomalies have contributed to a criminal career. Obviously, inmates submitting to any form of plastic surgery will be apprized of its possibilities and limitations and their consent obtained.

The Federal Bureau of Prisons, in its *Handbook on Correctional Institution Design and Construction,* has outlined the salient features of adequate medical care as well as the physical features necessary for efficient functioning of the medical staff.[11]

What we have said about medical facilities and personnel applies also to the dental department. The teeth of most inmates are notoriously bad, and the prison dentist has a treatment job laid out for him. Modern and adequate equipment is furnished in the New Prison. Facilities for the fitting of eyeglasses are also available, as well as the services of a full- or part-time optometrist.

A well-rounded program in any prison, following in the tradition of the New Penology, must include a carefully thought-out recreational program supervised by a full-time trained worker. This initial requisite is a rarity in our correctional institutions. Recreation is important enough to be considered an integral part of the treatment program and, to most inmates, looms large in the daily life of the establishment.

Few of the early prisons made any preparation for recreation. Aside from the monotonous "yard-out" (the vernacular term for a breather) no constructive program was developed. During the brief respite, usually scheduled between closing time of the shops and the evening meal, yard-out consisted merely of walking up and down the available space in the prison enclosure. In early English prisons the inmates walked around in a circle, their faces encased in masks to prevent recognition. (See picture, page 344.)

In some prisons inmates were permitted to play baseball, pitch horseshoes or quoits (this was rare because of the danger of their use as spontaneous weapons), and play handball.

The New Prison attempts to provide recreational activities for all inmates who desire them. In some prisons, the authorities *require* the inmates to participate in some outdoor recreational pursuit. This is especially true in reformatories for older boys. In these, the wardens consider vigorous exercise as absolutely necessary and use it sometimes as a form of discipline. Many contend that the program should be so strenuous that the inmates go to bed worn out with recreational fatigue.

Other activities of a recreational nature often found in the better prisons

---

11 Chapter 15. See also *Manual, 1956,* Chapter 15, "Health and Medical Services."

A day- or play-room in a prison block.

are caring for flowers, growing a small garden, tending such pets as birds, rabbits, and squirrels. The Sing Sing birdhouse, where many tropical flowers have been collected, is famous, as are also the flower gardens at San Quentin and Stateville, Illinois. Just how much good is derived by the prisoners from the petunias that abound in the Illinois institution is debatable, however.

Hobbies are also encouraged in the New Prison. This activity, in many of its aspects, may be considered economic and thus falls to some degree within the purview of "prison labor."[12]

Punishments in the New Prison

In our chapter dealing with the conventional prison we described the punishments inflicted on unruly inmates. It is difficult to ascertain just what penalties are used today since wardens are loathe to publicize them. As we stated earlier, psychological badgering is constant in many modern-day prisons but there is no record of this type of punishment in the rule manuals. In an attempt to find out just what penalties were current practice, one of the writers conducted a survey a few years ago. Of the 58 state prisons (not reformatories) answering a questionnaire we find the following penalties:

[12] For further information see *Manual, 1956,* p. 341.

Deprivation of privileges: 56
Solitary confinement: 42
Other segregation such as special cell block: 3
Loss of "good time": 47 (2 by parole boards; 1 for escapees only)
Reduction in grade: 23 (some prisons do not have grades)
Closer custody: 1
Transfer to other prisons: 1; to "yard work," 1; to "special detail," 1
Reassignment to other work: 2
Loss of bonus money: 1
Removal from industrial job except in hardship cases: 1
Tightened handcuffs: 1 (in extreme cases)

Punishments listed on questionnaire but not answered by any:

Whipping or flogging
Ball and chain
Thumb racks
Handcuffing to cell door

The question of daily diet while in segregation punishment is another interesting one. There are many varieties of diet or feeding represented in the 58 institutions. In 20 prisons there is a bread and water diet, but in all cases it is apparently broken every two or three days with a regular meal. In a few prisons, the inmate in isolation is restricted to one full meal daily with no bread and water diet at all. In a few institutions, regular meals are given. In the federal prisons, as well as in some states, the "federal diet" is given. This consists of a thoroughly mixed up dish of meat and vegetables containing the normal amount of calories but most unpalatable and "messy." In one state, a special "loaf" diet is furnished. This is a hard baked loaf including vegetables ground up, not too dissimilar to the old "hardtack" army ration. In a few institutions, two regular meals are furnished daily. Here are a few items of general interest regarding segregation and diet practices:

Full meal at noon of third day
Solitary not more than 14 days, doctor checking each day
Omission of sweets and desserts
Restricted diet provided
Bread and water for five days, then regular meals
One full meal daily for three days, then regular diet
Monotonous diet of 2,100 calories
One meal every 36 hours, with limit of solitary, ten days

While most prisons answering the questionnaire did not state the limit of their punishment in solitary or segregation, a few stated that it was "unlimited." "Ten days" was mentioned in a few cases as the maximum. A reprimand was noted as a punishment in a few prisons. One specialized punishment for escapees only: "have their heads shaved."

It may be seen from the above that the century-old penalty of bread and water is still widely used and that some type of segregation, even the dark cell of solitude, is still favored by some prison administrators.[13]

The Prison as a Country-Club

On occasions some articulate critic of "prison laxity" labels our correctional institutions "country clubs" in which the inmates have practically unlicensed privileges and in which "they live like kings." There is nothing new about this unwarranted criticism of correctional treatment.

Perhaps the first to condemn "coddling" of prisoners was the "common-sense" British economist, Sydney Smith, who wrote in the *Edinburgh Review* as early as 1822:

> I would banish all the looms in Preston Gaol and substitute nothing but the treadwheel or the capstan, or some species of labour where the labourer could not see the results of his toil—where it is as monotonous, irksome and dull as possible—pulling and pushing, instead of reading and writing —no share in the profits—not a single shilling. There should be no tea or sugar, no assemblage of female felons round the washing-tub—nothing but beating hemp and pulling oakum and pounding bricks—no work but what was tedious, unusual and feminine.[14]

Thomas Carlyle, after visiting a London prison in 1850, wrote: "In my life I never saw so clean a building; probably no Duke of England lives in a mansion of such perfect and thorough cleanness." He lists the food, "bread, cocoa, soup, meat" which he tasted and found "superlative."[15]

McKelvey notes that as far back as 1891, "hoarse cries were heard against the 'collegiate and hotel prisons' " of that day.[16] Wrote an English visitor to the Tombs and Sing Sing prisons of New York in 1893:

> I was surprised on a visit to the Tombs Prison in New York City when prisoners took matches from their pockets and lighted cigars. . . . It was a surprise to me, too, to learn that if a prisoner had money, he could have any fare he chose to pay for, and friends might bring dainties to the prisoners. I saw one young fellow eating cake and raspberry jam, that his mother had brought him. . . . In Sing Sing I found a prisoner lying in bed, although it was eleven o'clock a.m., reading a newspaper and smoking his pipe and in another cell a convict was making tea over a lamp.[17]

---

[13] Negley K. Teeters, "A Limited Survey of Some Prison Practices and Policies," *Prison World,* Vol. 14, No. 3 (May-June, 1952), pp. 5 f.

[14] Quoted by Sidney and Beatrice Webb, *English Prisons under Local Government* (London: Longmans, Green, 1922), p. 88.

[15] Thomas Carlyle, *Latter Day Pamphlets: No. II, Model Prisons* (New York: Harper, 1850), p. 7.

[16] Blake McKelvey, *American Prisons* (Chicago: University of Chicago Press, 1936), p. 124.

[17] William Tallack, *Penological and Preventive Principles* (London: Wertheimer, Lea & Company, 1896), p. 105.

Austin MacCormick quoted the following tirade against the modern concepts of corrections in 1937:

> Punishment in all too many prisons has become a thing of the past. We are amazed to learn that there is no such thing as a life prisoner; that the average man who receives a life sentence for murder spends about ten years behind bars, and that the average sentence served by a person convicted of murder is only sixty-one months.
>
> Even then, he usually is granted the benefits of a private radio, of the daily newspaper, selected magazines, the latest motion-picture shows, orchestras, traveling-bands, hand-decorated cells, baseball, hand-ball, football, and basket-ball, and any other amusements which over-sympathetic and sob-sister wardens or prison boards may contrive to make his stay in prison more enjoyable. Many of our prisons today may well be classed as country-clubs. . . . I do not believe that the majesty of justice can appear in white and untrammeled garments so long as such disgraceful, sentimental convict-coddling is allowed to persist in our alleged penal institutions.[18]

No matter how many progressive practices and amenities are incorporated into the modern prison, it cannot be called a country club. Men incarcerated are deprived of that which they wish most—liberty and normal human contacts. Society and the law demand punishment for the wrong-doer but aside from incarceration for a period of time the law is strangely silent as to what constitutes punishment.[19] The privileges found in prison today are there, *not* for the prisoner's benefit, but for the sake of a society that must accept these inmates once again. They have been introduced gradually and hopefully for purposes of reformation, since it was found through bitter experience that to deny them and to insist on a program of repression merely added to society's burden. We know that there was a wave of mistrust when books (other than the Bible) were taken into the prisons of a century ago; when prisoners were permitted to receive and write letters or retain their musical instruments in a solitary cell.

The introduction of the radio in prisons is another example of cultural prison lag. The public enjoyed the radio years before the first cautious wardens allowed them within the walls of their institutions. Even now there are many restrictions on their use. It is not an unmitigated privilege as it is so regarded by a free society. Some prisons supply headphones whereas others require the inmates to purchase them from their own earnings. The prisoners are restricted to the programs selected by some official

---

[18] Quoted by Austin MacCormick in "Dead-End Penology," *Proceedings,* American Correctional Association, 1937, p. 16.

[19] See a stimulating article on the dilemma of "punishment versus treatment," Donald R. Cressey, "Rehabilitation Theory and Reality," *California Youth Authority Quarterly,* Parts I and II (Spring and Summer, 1957). See also a warden's refutation of the "country-club" charge; John C. Burke, "We've Got One of Those Coddling Prisons," *Federal Probation,* Vol. 20, No. 4 (December, 1956), pp. 34-36. Mr. Burke is warden of the Waupun, Wisconsin, state prison.

and only at those hours when the master receiver is on, prior to lights-out, which is seldom later than nine o'clock at night. We are, more recently, experiencing the same phenomenon with the universal diffusion of television sets. In one progressive prison the warden seriously contemplated receiving a gift of a television set, but on sober thought politely declined because of the fear of public disapproval. However, there are now many prisons that boast television sets.

Other innovations such as newspapers, magazines, pictures on cell walls —the little things the free person enjoys—are granted the prison inmate with apprehension by administrators because of the potential criticism from outsiders who charge that the inmates are being pampered.

Sometimes the reverse is true. A new warden, or a Commissioner of Correction, hoping to breathe new light into a benighted institution, attempts to introduce the daily newspaper, or some magazines (at the inmates' own expense) and is confronted by opposition by one or more of the administrative staff.

Temple University choir at Eastern Penitentiary, Philadelphia. (Courtesy George J. Roebas.)

An historic innovation was introduced into the Eastern Penitentiary in Philadelphia on May 20, 1957. For the first time in the history of the establishment (opened in 1829) the prisoners were allowed an evening "yard-out." The new privilege was celebrated with a mixed choir concert of Temple University students. Under the more enlightened dispensation of the warden and his staff there were few, if any, misgivings as to the ultimate success of the affair. The inmates responded normally and showed their appreciation. Thus evening "yard-out" in old Cherry Hill prison became the usual rather than the unusual.[20]

If society is to be protected from the men and women who are sent to prison, it is necessary that the regime set up should be, so far as is humanly possible, as normal as the outside free society. As an eminent British prison commissioner, Alexander Paterson, once wryly remarked: "The prison is a monastery in which the inmates do not choose to be monks." Imagine the personal disorganization that exists when a released prisoner tries to dodge traffic, something that he has not been obliged to do for years; of his embarrassment when a woman speaks to him and he attempts to tip his hat—something he never wore in prison; or of his confusion when he is confronted with the simple task of opening a door—he has not seen a doorknob for years. These are trivial to the free man but they are psychologically mountainous to the ex-prisoner. The prison denies the inmate these day-by-day little things of life. The more normal innovations for prison life, the better for society.

An Appraisal of Modern Corrections

In our earlier chapter we described the conventional prison. Thus far we have been discussing the emergence of the New Penology—now referred to as "corrections." During the period between 1935 and the present we have witnessed amazing paradoxes in this area. We have seen the expansion of the efficient Federal Bureau of Prisons and the development of modern concepts of corrections in several of the states, but in the first instance we have witnessed the sorry career of that nullification of progressive penal treatment—Alcatraz, the super-maximum-security prison in San Francisco Bay, maintained by the same progressive Federal Bureau of Prisons. We have seen the introduction of trained personnel in many of our systems, food and housing improved, recreation facilities expanded and medical facilities improved, but despite all this we have seen a rash of costly prison riots in which custodial officers were held hostage, shops and dining facilities destroyed, and bitterness and despair permeate both inmates and officials. We have heard and read of such progressive establishments at the California Institution for Men at Chino, the federal correctional institution at Seagoville, Texas, and the minimum-security institution at Wallkill, New York, but on the other hand we have read of brutality at the Cañon City, Colorado, and Kilby, Alabama prisons, the shambles in Louisiana, and the self-mutilation of inmates in Georgia.

Such paradoxes have been grimly reviewed by James V. Bennett, Director of the Federal Bureau of Prisons:

> Even our modern prison system is proceeding on a rather uncertain course because its administration is necessarily a series of compromises.

---

[20] See the "yard-out" photograph on preceding page.

On the one hand, prisons are expected to punish; on the other, they are supposed to reform. They are expected to discipline rigorously at the same time that they teach self-reliance. They are built to be operated like vast impersonal machines, yet they are expected to fit men to live normal community lives. They operate in accordance with a fixed autocratic routine, yet they are expected to develop individual initiative. All too frequently restrictive laws force prisoners into idleness despite the fact that one of their primary objectives is to teach men how to earn an honest living. They refuse a prisoner a voice in self-government, but they expect him to become a thinking citizen in a democratic society. To some, prisons are nothing but "country clubs" catering to the whims and fancies of the inmates. To others the prison atmosphere seems charged only with bitterness, rancor and an all-pervading sense of defeat. And so the whole parodoxical scheme continues, because our ideas and views regarding the function of correctional institutions in our society are confused, fuzzy and nebulous.[21]

Perhaps no more eloquent text for a sermon dealing with the tragedy of our concept of imprisonment has been uttered than that just quoted—and by a man who has worked intimately in the field of corrections for thirty or more years and who heads the largest and one of the most progressive correctional systems in the world.

We confess that we are skeptical that any adequate reconstruction of our correctional systems can emerge from the so-called New Prison. We look at it only as a patching-up process. Not even the inmates are impressed. As one looks through some of the magazines edited by prisoners he is struck with the fact that many of them are articulate in denouncing the system—and with cause. They call attention to the prison frustration, which is especially noticeable when the inmate wishes to make choices within the framework of the routine. He is denied all choice except in a few inconsequential matters once his "plan" has been agreed upon—in which he may have had some element of choice.

The jailing and lock psychosis still persists in all but a very few of our modern institutions. The conflict between custody and treatment is still unresolved in most of our establishments, even among the most progressive.[22] The writers have visited many prisons and have talked with many well-trained men who are doing a courageous job in what is generally called the treatment program. Their chief complaint is that much of what authority they believe they have is constantly whittled away by custody personnel who always take the position that "custody comes first." Inmates also have complained that certain decisions are made by the treatment staff regarding individual plans for certain prisoners only to have them

---

[21] *Federal Prisons, 1948,* A Report of the Work of the Federal Bureau of Prisons (Washington, D.C.: U.S. Government Printing Office, 1949), p. 3.

[22] This conflict is admirably reviewed in Vernon Fox, *Violence Behind Bars* (New York: Vantage, 1956). This is the story of the Michigan prison riot which occurred at the Jackson institution in 1952. For a caustic but at times sympathetic review of Mr. Fox's book, see *N.P.P.A. Journal,* Vol. 3, No. 3 (July, 1957). The review is by Roberts J. Wright, a prison administrator.

nullified by the deputy warden or captain of the guards. It is this *impasse* that makes the progressive prison more of a mockery than a real treatment center.

The daily routine in the modern prison must be unbroken; men must be counted and recounted. Charts in the deputy's office show at a glance where every man is at every moment. The mechanics of administration is a relentless strain on guards and other personnel, even the psychiatrist and chaplain. That is a part of the business for which they are paid. In the *better* prisons, the personnel is picked for efficiency in this respect!

The guards are constantly on the alert to ferret out potential plots that might lead to escape or riots. Guards cannot afford to converse with inmates, and inmates must be kept in small groups if they are allowed even that much freedom.[23]

The inmate-guard complex is still prevalent. The guard or custodial officer represents force. Above all else, the main purpose of the prison is to keep the prisoners from escaping. Even if the guards are college graduates, their duty is to serve as institutional police. In the best prisons, if they must accept this function, they can serve little in the treatment program. "Social distance" must be maintained. There is the same old "pushing around" so characteristic of the old prison, although some of this has been eliminated, especially in institutions where custodial officers are relieved of many of their onerous duties and drawn into the treatment program. The wise penal administrator capitalizes on the abilities of his officers and calls on them to be of service in assisting the inmate with his personal problems. Naturally, this can only be accomplished when the officials are acceptable to the prisoner on that level.

Contamination, which dismayed the early Quakers, still persists in the New Prison. Despite the smaller units developed in recent years—minimum, medium, and maximum security—this insidious process by which one prisoner adversely influences another still obtains. There is no complete answer to the problem. Much can be done by splitting up the inmates into smaller and smaller units, even to 100 or less, but the contaminating process would still persist, though in a lesser degree. Of course, contamination is found outside prison also, but more normal activities are available to minimize the influence of a confirmed law-breaker on a novice. In the prison, the man who wants to go straight finds every obstacle in his way. Denied normal contacts, he develops attitudes inimical to those for which

---

23 For an analysis of the difficulties confronting the custodial staff, see Lloyd McCorkle, "Social Structure In Prison," *The Welfare Reporter,* New Jersey Department of Institutions and Agencies (December, 1956), pp. 5 f. Mr. McCorkle is former warden of the New Jersey State Prison at Trenton. For a cogent rebuttal of this article, see Charles L. McKendrick, "Comments on 'Social Structure in Prison,'" *Correction,* New York Department of Correction, (September-October, 1957), pp. 3 ff. These two articles present the dilemma of staff-versus-inmates in the traditional prison and in the "new" prison.

society stands. He more often than not listens to a veteran convict with amusement at first, then admiration. Seldom does he fight back by denouncing such a point of view. He finds himself, after a while, saying to himself: "Perhaps he's right. I haven't had a square deal. Society is persecuting me and my family." The demoralization is gradual. And the New Prison cannot stop it. The inmate develops the prison code, the prison honor, the prison pattern. He often succumbs and hardly notices it. This degenerating process, with its nonsocial or even antisocial attitudes, is found in *all* prisons, regardless of their good intentions.

The same old prisoner feuds still persist. Ganging up by "prison wise" inmates on hapless individuals is still a characteristic of the prison. It was in such a situation that William Remington, convicted subversive, was brutally murdered by three inmates in the federal penitentiary at Lewisburg, Pennsylvania in 1955. Every institution has its quota of inmates who have been placed in segregation for "protective custody."

Psychoses, superinduced by the monotony and despair of imprisonment, are prevalent. These mental aberrations thrive in all prisons. Psychiatrists detailed to our new prisons are frankly worried that psychopathic behavior is prevalent among an ever larger group of inmates. Their program, worthy as it is, is one of diagnosis rather than of prevention and cure. Instead of wiping out the conditions that superinduce these psychoses, they are occupied only with those individuals whose overt behavior manifests obvious signs of having succumbed to the prison machine. Many of these persons would function reasonably well outside the prison under community supervision. Many are keen, intelligent men and women. Hundreds of their kind are walking the streets outside every day and get no worse, simply because they have more normal activities to which they attempt adjustment. Not so in prison, where conditions merely aggravate the incipient or latent predispositions to psychopathy.

Sexual perversions are prevalent in the modern prison. Masturbation is as rife in the progressive prison as in the old traditional regime. Sodomy is practiced as well. These manifestations are the direct result of the denial of normal contacts with the opposite sex that are a part of the society outside. Sublimation is much easier to attain outside prison than within.

On this charge alone, the prison should be abolished. Even the best wardens in the best prisons cannot cope with this problem of sex. We pointed out in our earlier chapter that it is found in all prisons, reformatories, and reform schools for boys and girls. It *cannot be eliminated in such an unwholesome environment.* No heterosexual person should be thrown into a situation that makes of him a deviate. If the prison is to persist, it should provide, at the very least, periodic visits home for an opportunity of meeting members of the opposite sex. The present regime is cruel in the extreme in this respect.

28

The Processing of the Convicted Offender by the Prison: Diagnosis and Classification

The progressive classification of prisoners, based on study of the individual, and administered on some well adjusted system, should be established in all prisons. (From Declaration of Principles, *1870.)*

NATIONAL PRISON ASSOCIATION

1. Admission of the Prisoner

A convicted offender is brought to the prison by a county official, either the sheriff or by one or more of his deputies. He is usually shackled. Official papers from the court and a document calling for the sheriff to "deliver the body" of the convicted man to the prison are presented to the receptionist of the institution. Thus the sheriff dispatches his responsibility and the offender is from that moment on in the keeping of the state prison.

The new prisoner is then taken to a room where initial preparations are made for his reception. He is registered and his papers are checked. Next he disrobes and bathes. His clothes and belongings are placed in a locker and he signs a statement verifying that he is the owner of these effects. He is then given a prison uniform, after which he is finger-printed and photographed. In most prisons the new inmate is "identified" by means of the old Bertillon method with such physical items as height, color of hair and eyes. Later additional data are added, such as date and place of birth, schooling, family, and work record. In due time he is given a thorough physical examination. He is also given a number by which he will be officially known so long as he remains in the custody of the state's prison system.

After preliminary and essential data are provided the prisoner is led to a cell on what is known as the quarantine or reception block (in some states there are regular reception, or "guidance" centers set apart from prisons) where he remains for a period ranging from a week to two or three months. During this period he is "diagnosed" and "classified." Before we describe these processes let us review what was originally meant by classification in our early prisons.

2. Historical Types of Classification

In the Walnut Street Jail in Philadelphia, following its renovation in 1790, segregation, or crude classification, began. The sexes were separated and children were provided separate quarters. It was not until the 1820's that separate institutions were provided for children in the establishment of "houses of refuge" in New York, Boston and Philadelphia.

In 1797 the management of the Walnut Street Jail actually initiated a classification that may be regarded as the first attempt to segregate prison inmates as to type.[1]

Later in the history of the penitentiary, many of the insane were sent to asylums that were established beginning with the 1840's. Much later, special facilities were created for the care and treatment of the mentally retarded but it was not until comparatively recently that prisons for the "defective delinquent" were established by some of the states.

A more recent movement toward classification, or segregation, is that of special detention quarters for the untried in our county jails. Brooklyn blazed the trail in this area with its House of Detention for Men opened in 1957.

Another type of segregation that eventually emerged in most correctional establishments is that of race. Negroes have been segregated from whites in most prisons where they appear in appreciable numbers. Negroes make up the bulk of the population in southern prison camps. Few northern prisons exercise the courage and social insight to break with the outmoded custom of racial segregation. What the Supreme Court decision will bring about in our prisons, we do not know. But eventually the practice will be discontinued.

Another method of classification was based primarily upon the degree of prospective reformation of the individual. It was this step that laid the basis for what is technically known as classification of prisoners today. In the larger penitentiaries the early practice was to segregate the first offender from the recidivist, the youth from the older offender, the sex deviate from other prisoners and the reformable from the "hopeless." A variation of this practice was that every prisoner was put in one grade,

---

[1] See N. K. Teeters, *The Cradle of the Penitentiary* (Philadelphia: The Pennsylvania Prison Society, 1955), pp. 58-60.

from which he could be promoted or demoted on the basis of his behavior. Often the various grades had different uniforms, or wore different insignia on their sleeves or braid on their trousers, by which they could be easily distinguished. Still another method of separating prisoners was the development of the *trusty system*.[2] It is obvious that such classification methods were not based on diagnosis but rather through a kind of "hunch" system or by the offense.

By 1920, many types of differentiation and classification of prisoners were recognized and recommended if not literally followed. Few institutions followed their programs of classification to the letter, because there was no unified control within the state, because there were not enough specialized institutions to care for the various types, or because the administration was apathetic. Classification then, is not a new idea; modern improvements and expansion of the procedure merely extend an old concept developed by pioneers in prison administration.

Today classification, or differentation, is a continuous process of individualizing correctional treatment starting with attempted diagnosis. Classification and re-classification is carried on day in and day out, the process being refined with the development of new concepts and techniques.[3]

The function of classification is to differentiate the various inmates in a state's correctional program in terms of their potentialities for rehabilitation, regardless of the offense or the sentence. Thus the avowed purposes of "classification" are diagnosis and treatment. These, however, are but the theoretical objectives: so long as the theories of punishment and retributive justice permeate our penal codes it is difficult not to classify prison inmates on the basis of offenses and sentences. This is actually what is done in most prisons and it is this imperative that generates the nullification of true re-socialization or rehabilitation from the very day of the reception of the prisoner.

3. A Centralized Authority Necessary

Although there has usually been a loosely set-up centralized bureau or agency in each state that controlled or supervised its correctional institutions, each facility has been traditionally independent and self-sufficient. The legislatures periodically pass laws relating to the discipline or the management of the prisons, but pay little attention to the actual management.

Because of the gradual trend in the past toward centralized authority, there is now a great variety of classification within the various states and more integration and coordination than formerly. Some centralized author-

---

[2] See Chapter 30, pp. 496-497.

[3] For an informative article on classification, see Robert G. Caldwell, "Classification: Key to Effective Institutional Correction," *Amer. J. Corr.*, Vol. 20, No. 2 (March-April, 1958), p. 10 ff.

ity is needed to control the state's classification philosophy and administration, usually referred to as a Department of Correction. With a strong centralized authority, as found now in many states, each institution becomes an integral unit in the whole system and a specific functional unit in the placement of offenders.[4]

The generally accepted facilities to implement a classification system are minimum, medium, and maximum custody. There are also many inmates who are mentally abnormal to a marked degree, including the pre-psychotic, the "psychopath," the defective delinquent, the epileptic, the aged, and others who need care in other types of facilities. In such a classification program it must be presupposed that there exists within the over-all system the machinery to transfer inmates from one type of institution to another when and if this seems expedient. Such transfer must be efficiently and harmoniously handled.

4. The Classification or Diagnostic Clinic

As we have mentioned, a few states have special reception or quarantine centers where diagnosis and initial treatment is handled apart from the prisons proper. Most states, however, set up diagnostic or classification clinics in each institution. In some instances the clinic is presided over by a special director; in others, the warden or his associate or deputy is in charge. These clinics are generally composed of the following specialists: the warden or deputy, the medical doctor, the psychologist, the chaplain, the psychiatrist (if there is one), the guidance counselor or psychotherapist, the social case worker, the director of industries, the educational director, the parole officer, and any other professional person whose duties bring him into direct contact with the prisoner's routine. The clinic personnel is essentially the same as what is referred to as the treatment staff, to differentiate it from the custodial staff composed of the warden, his deputy, the keepers, and the guards.

The clinic is confronted with the following questions: What in the prisoner's biological makeup or in his evironmental situation predisposed him to crime? What shall be done with him? Where shall he be sent? What kind of work program shall be provided him? What educational or vocational facilities will be of benefit to him? In short, what will be the nature of the treatment program for this specific inmate as can be determined in this initial processing? These are difficult questions to answer, but the clinic personnel does attempt to do so. If the state has a number of diversified units for the wide variety of cases that enter the correctional system every day the clinic will be more effective. It is obvious, therefore, that in a state where there are only a few giant correctional units and a

---

[4] For further details on classification, see *Manual, 1956,* Chapter 18, "Classification."

large and varied inmate population, the results of the clinical work will be inadequate; but where there are many small units, each serving a specific type of individual, the system will be of considerable value in the final objective of rehabilitation or preparation for release.

The *Handbook on Classification,* prepared in 1947 by the American Correctional Association, states:

> Classification implies not only a thorough analysis of the individual and the factors in his background and environment, which influenced his personal development but also a procedure by which this information can be utilized as the basis for a well-rounded, integrated program for him, looking toward his improvement as a social being. . . . Classification includes not only diagnosis but also the machinery by which a program fitted to an offender's needs is developed, placed in operation and modified as conditions require.[5]

It points out the advantages of classification as follows:

1. Proper segregation of different types of offenders
2. More adequate custodial supervision and control
3. Better discipline
4. Increased productivity of inmates
5. More effective organization of all training and treatment facilities
6. Greater continuity in the training and treatment program
7. Higher personnel morale
8. Better inmate attitude
9. Reduced failures of men released
10. Better guides in long range planning of building requirements
11. Classification reports have many values (for use for parole boards, by institutions in other states, etc.)[6]

After an inmate is finally classified and sent by the clinic and the centralized authority to some institution suited to his needs, he is theoretically not forgotten. After an interval of time, sometimes agreed on at the initial classification, the inmate comes before the clinic for *reclassification.* The purpose of this is to check on the implementation of the original recommendations. The case is reviewed and any new information concerning the man is introduced. This information may come from outside community agencies or from reports from the prison staff which, by this time, has come to know the man, his potentialities, and his weaknesses. The *Handbook on Classification,* quoted above, has this to say about reclassification:

> Reclassification guarantees that there will be neither forgotten men in prison nor "dead end" placements. Goals are established and attained. The inmate is provided an incentive to make progress, knowing that his efforts are officially recognized, that they will be brought to the attention of the

---

[5] *Handbook on Classification in Correctional Institutions,* 1947, p. 2. Another extremely important handbook dealing with phases of this subject is the *Handbook on Pre-Release in Correctional Institutions,* American Correctional Association, 1950.

[6] *Ibid.*

parole board and that consideration will be given to changes in his program so that he may gain the most from his period of confinement.[7]

This *reclassification* meeting is designed, especially in the more progressive systems, to allow the inmate to have his day before this tribunal. Before such a group he is encouraged to "beef" about his assignment and to state his wishes for a change. If the clinic is sincere, and many of them are, the members will permit a free expression and afford the man a genuinely friendly reception. But, unfortunately, prison people are human and, in addition, are bedeviled with their own problems. It is a rare clinic indeed, that follows through with sustained honesty what has been set down in its paper objectives.

The disciplinary function of the clinic is of the utmost importance and one that has met with considerable resistance from old-line custodial men in prison. As a preliminary statement to this function let us see what the *Handbook on Classification* has to say about prison discipline, since it points up the capricious techniques used in the traditional prison:

> Whether we like it or not, our emphasis in prison discipline has been largely negative. Despite our academic acceptance of the individualized approach to treatment in problems of human behavior, the fact remains that isolation, good time forfeiture, segregation, forfeiture of privileges, etc., remain the basic earmarks of our approach to those who do not adjust to the strict routines in our institutions. We have given lip service to, or weakly attempted to approach discipline in psychiatric, psychological, case work, or religious terms. But the arrest-trial-guilt-punishment philosophy has continued to heavily predominate our practice. . . . We must stop *assuming* that the organization of a classification program *automatically* extends throughout all phases of inmate life within the institution.[8]

Discipline cases can come to a sub-committee of the clinic, where a fair hearing can be given the inmate after he is confronted with the charges. The staff officer who made the complaint need not be present unless there are discrepancies in the stories. The board can then mete out such punishment as the case merits. The *Handbook,* written by prison men, points out some of the larger issues that can be handled by the clinic so far as discipline is concerned:

> The whole classification committee determines general policies of discipline and a sub-committee deals specifically with adjustment problems. It is concerned with causation not only within the individual, but within the whole fabric of the institution administration. It is concerned with the total atmosphere and tone of the institution. It is concerned with the adjustment of behavior problems and incipient violations before they actually occur as serious violations.[9]

---

[7] *Ibid.,* p. 60.
[8] *Ibid.,* p. 78.
[9] *Ibid.,* p. 79.

Many perplexing problems face the clinics. In most correctional systems where a fairly good classification program is in operation, the minimum security units, those to which are sent the best risks, the young offenders or first offenders who are obviously reformable, are likely to be prison farms with no wall about them. This walless feature is the actual reason for sending good risks there; it is not that the minimum risk inmates need farming in their plan of treatment. It does not necessarily follow that those who show the best evidence of adjustment should be sent to a farm to milk cows and tend chickens; neither does it follow that poor risks should be sent to the jute mill, laundry, or weaving room.

Smaller units, each housing not more than 500 inmates, is probably the answer to the classification question. In each of these smaller units, facilities should be provided to meet all the vocational needs of the inmates sent there by the clinic.

5. The Clinic Personnel

The trained personnel that make up the modern classification clinics did not all arrive in the prison system at the same time. In the early prisons there was a physician, a chaplain, a prison teacher, and sometimes a director of industries, possibly in the employ of the contractor who ran the prison factory, but who knew something of the individual inmate.

After World War I, which had popularized mass testing, trained clinical psychologists were placed in a few prisons to do mental testing. When New Jersey set up the first classification program in this country in 1918, the psychologist was asked to develop a state-wide system, so that this specialist controlled, if not dominated, the program in that state.[10]

The psychiatrist also entered the prison early, although his services even now are far from being universally accepted. He has a much harder case to present to the groups that control our prisons. He has been called upon for many years as an expert outside the prison in specific cases, but has not often been *persona grata* on the institutional staff. Psychiatrists find private practice much more remunerative than institutional work. This fact, together with the resistance to their services, explains in part why so few are found in our prisons.

The Federal Bureau of Prisons makes use of the psychiatrist as well as the psychologist. Aside from supplying special hospitals for mental cases, it assigns psychiatrists to many of the correctional institutions. In this system, the psychologists and psychiatrists are under the administration of the United States Public Health Service. Some of the more progressive states are also able to employ psychiatrists, but it is difficult to find the

---

[10] Diagnostic clinics were first developed in Buenos Aires by José Ingenieros in 1907 and by Louis Vervaeck in the Forest Prison in Brussels the same year.

funds necessary or the men trained to expand psychiatric treatment in all but a few correctional programs.

Psychiatrists, where they are available, usually sit on the classification board. In the federal system they are also on disciplinary committees, good-time forfeiture committees, and safety councils. In addition, some of them, coordinating their activities with the psychologists, have been engaged in recent years in interesting research with psychopaths, homosexuals, and other types of sexual or mental deviants. Research in group therapy, since the last war, has also entered the prison, which makes an extra type of therapeutic work accessible to inmate treatment.

Psychiatry can take a significant place in the prison only if there is a great change in attitudes regarding the function of the prison. The bulk of our correctional institutions still do not recognize its importance in the treatment program. There is some danger that only superficially trained men will take employment in prisons and an added hazard is that these men may become institutionalized due to the nature of the prison. Another criticism of the use of psychiatry in prisons is that it is rarely used in treatment. It is assumed a prison is progressive if the psychiatrist is doing research in *diagnosis*. A great deal of money and time is spent on diagnosis by members of this craft but precious little effective treatment.

The trained social worker is another professional expert who has entered the prison. We find, however, that mere sporadic and routine recording of data should not be his main function, important as this procedure is. All too often the so-called social worker is not a case worker. When we speak of social workers we mean individuals who understand case work technique, have graduated from an accredited graduate school of social work, or have had at least three years of training in an accredited social service agency. In addition, they should have had reasonable success in dealing with behavior problems or be able to show potential ability to handle such work and be recommended by reputable case workers.

Just what does social case work mean in prison? Because social workers are taking their places in public schools, hospitals, and other institutions, so they are being accepted in correctional work. As Edmund Burbank puts it: "Social case work is the art of helping individuals to find and use satisfying and constructive social relationships in the unfolding of life experiences in which they happen to find themselves."[11] All inmates have problems: on the assumption that inmates should be helped, social workers feel their philosophy and technique *can* operate in an authoritarian setting such as the prison is.

There are basic prerequisites for effective case work in prisons: First, it

---

[11] Edmund Burbank, "The Place of Social Casework Service in the Pre-Release Program in a Correctional Institution," *Prison Journal*, Vol. 29, No. 1 (April, 1949), pp. 33-39. See also Herbert H. Aptegar, *The Dynamics of Casework and Counseling* (Boston: Houghton, Mifflin, 1955).

must be accepted by the administration that there are constructive elements in a prison experience; second, that these elements can be translated into a sound correctional program; third, that the staff dedicate itself to the task of helping the inmate find maturity as a mark of social responsibility; fourth, that the administration provide the structure or climate in which the case worker can operate effectively.

The function of the case worker in prison, then, is to help the offender, within the limits of prison regulations, make constructive use of the many programs and social relationships that are a part of his prison experience, to help him find within himself, through interviews, the feelings and the will to assume adult responsibility in his strictly limited environment and to learn to live in terms not solely his own.

Sociologists have also gained admittance into the prison to help in treatment. Their role was described initially by Donald Clemmer in his book *The Prison Community*.[12] Aside from gathering social data from social agencies and interpreting it, sociologists can be of help in the analysis of the prison's structure as a "closed social system," of clique formation, the prison code, and institutionalization. In several prisons they have been employed and are working effectively in administrative positions as well as in the treatment program, either as guidance counselors or as group therapists.[13]

What we have been discussing thus far is the service rendered the prisoner by the conventional professional staff members of the clinic. In recent years a service has been developed known as counseling which envisages an intimate man-to-man relationship or guidance of group discussion. This cuts across group therapy. But there are certain borderline individual or group relationships that differ somewhat from the more formalized types of treatment. Even in the early prisons there were certain individuals who gained a sort of *rapport* with an occasional inmate. Guards, maintenance personnel, the steward, the block man, as well as other non-professional members, often engaged in a kind of counseling. The contribution of such persons to morale cannot be overlooked in appraising even the old-style prison. There are many incidents in prison history in which the best friend a prisoner had was a member of the custodial or maintenance staff.

Counseling men in prison, all of whom have problems, is a vitally important adjunct to the classification clinic. Aside from those skilled in the work, much can be achieved with certain of the non-professional staff through in-service training programs. Obviously all members of the person-

---

[12] Donald Clemmer, *The Prison Community* (Boston: Christopher Press, 1940; reissue New York: Rinehart, 1958).

[13] See Lloyd E. Ohlin (ed.), "Sociology and the Field of Corrections," prepared for the American Sociological Society (New York: Russell Sage Foundation, 1956). See also references to other contributions of the sociologist, pp. 361-362.

nel force are not suited for this work. But in any prison there are some persons who can be of real service in advising or guiding those who are in need. Several prisons already have counseling programs in operation, in charge of trained counselors.

6. The Treatment Staff and Immunity

We have deferred until this point the perplexing question of immunity. By that is meant to what degree the inmate can confide in and trust members of the professional staff, including psychiatrist, psychologist, social case worker, or chaplain.

In discussing the classification clinic we referred to the interviews to which the prisoner is subjected. Records are an important adjunct to effective treatment. One of the perplexing questions in the prison revolves around the typing of these records. Many institutions use inmate typists and stenographers. It is obvious that confidential records cannot be viewed by inmates who work in the diagnostic offices. Just how far inmate help can be used is a question confronting all professional workers.

But the larger issue of immunity bears down relentlessly on the professional staff. Due to the prison code the professional worker is a member of the enemy—the administration. Thus most inmates are suspicious of all the personnel of the clinic. This often applies to the chaplain too although, in a free society, this individual really has immunity before the law.

To what degree immunity is extended any member of the prison staff is unknown. Doubtless many confidences are scrupulously maintained but if information of a serious nature such as an escape plot, or a crime previously committed, is imparted by an inmate to a social worker or a chaplain, the worker is hard put to hold it inviolate. It is related of Thomas Mott Osborne that he permitted the most intimate confidences from his men but he always prefaced his interview by urging them not to place responsibilities on him greater than he could bear.

Professional personnel in an institutional setting have the same problems of treatment that holds for those operating in a private civil capacity. Any effort to bring about a change in another person operates within the structure of the child-parent relationship. This relationship must be skilfully manipulated between therapist and patient. In both the private and institutional setting the skill of the therapist will depend upon his capacity to penetrate this relationship and to assess readily his own role as a parental figure. It is obvious if one hopes to change another he must avoid, or at least control, a role which is imitative of the original parent with whom the patient or prisoner had so many previous frustrating experiences. It is in this connection that every therapist must, in some degree, create a relationship of permissiveness which at times may come nearer to the borderline of complicity. How the latter situation can be avoided is largely a matter of the therapist's skill.

The difficulties of the therapist are formidable in manipulating the relationship with the prison client. It would seem obvious that if the therapist is readily identified with the disciplinary authority his task is even greater.

In the therapeutic setting the success of the psychiatrist or the social case worker depends upon his skill in facilitating release of unconscious material in the prison patient. Some of this material will be more or less repugnant by traditional standards of thinking or of conduct. Much of this will never attain a substance greater than fantasy, but there will be occasions in which the prisoner will impart information based on reality such as past offenses, even of a serious nature, or of statements of intended antisocial behavior. It is obvious that the therapist, although permissive, nevertheless is a representative of the moral order and his conscience is bound to it; therefore he carries a responsibility to maintain it.

In the event a prisoner should confide a plan that will be injurious to others the therapist is confronted by a very real problem. His first task is to determine its implications. The prisoner may be putting the therapist to a critical test or he may be attempting to make the therapist an accessory, or to erect a formidable resistance further to penetrate his problem. The therapist has an obligation to the prisoner to interpret the material and confront him with its consequences in reality. If the prisoner cannot use the interpretation, the therapist has recourse to persuasion to induce the abandonment of the intended antisocial plan. If this fails, it is the therapist's obligation to report the matter to the administrative authorities.

It is clear, however, that in the process of treatment any confidences of this nature would be an indication that a critical point had been reached in the relationship that would either affect a change for the better in the prisoner or call for a complete abandonment of the treatment.

It may be seen from the above discussion that professional immunity represents a knotty problem within an authoritarian setting. There is bound to be privileged information that is on a higher plane than confidential data. It is in this area of privilege that real therapy takes place. If the therapist possesses a high quality of professional integrity and if he has not allowed the prison to institutionalize him, he can meet the test in this dilemma. It is a problem that calls for rare understanding by the administration and a high sense of responsibility in the members of the professional staff.

7. Summary Appraisal of Classification Procedure

Classification began as an intramural process—the movement to divide prisoners within one institution into grades. Distinguishing them by the insignia they wore on their sleeves or by type of uniforms was a crude and perverted form of classification, because it was based on good behavior that could be achieved more easily by the inured convict than by the sensitive and demoralized first offender.

The next stage in classification, developed within the past 25 years, calls for a clinic composed of specialists who know the prisoner's potentialities, background, and chances to succeed, and for some sort of statewide system made up of a number of penal institutions to which various types of prisoners can be transferred for treatment and training.

In today's classification procedure there is real danger of forgetting the prisoner except to ask him innumerable questions about himself, many of which are callously repetitive. It is the inmate who should be helped, not the professional standing of the men comprising the classification board, and professional jealousy has no place in such a program. Real classification means the best possible help that can be extended to the inmate. A group of specialists is important; so is a trained custodial force, a staff of men and women who have some notion of what the program means, who are in sympathy with it, and will manifest a degree of sympathy with it, and will manifest a degree of sympathetic insight. The fact that their services are mainly custodial should not prevent them from being helpful.

Treatment inside the prison means much more than a mere paper program, much more than mechanics. Too often the inmate is called upon to *accept* what is judged *best* for him by the clinic—all too frequently dominated by the warden—and there is little actual participation in the treatment plan by the one who is to be helped, the one for whom thousands of dollars are spent annually in professional personnel.

Here, then, is what classification must develop when dealing with prisoners, and it is just as true in planning for education or the inmate's work program as for parole. The inmate must share in the planning, must assume responsibility for it. The staff of the clinic can help him, but the real decision *must rest with him*. This is not new in social case work or psychiatry, but it is almost foreign to penal institutions. Now that we have classification clinics, we must embrace this philosophy. Otherwise, classification will remain as sterile as it now is in most prisons.

While the classification clinic is here to stay, it is possible that it may be only a step in the general direction of personalized treatment. Out of it may emerge a less formalized and more intimate technique of helping the inmate. The techniques of group therapy and group counseling are examples of this. At present it is questionable whether the clinic's good intentions can be fully appreciated by the prisoner. He walks into a room where he sees a large table, around which are seated the members of the professional staff. He is asked to sit at the head of the table and accept the plan worked out for him by these experts. How much of such a plan he can or will accept is at present unknown. So it would seem that the next step will be to eliminate the dull formality of the clinical procedure and to personalize the process. This might be accomplished by having a member of the staff develop the treatment plan privately with the inmate.

The most trenchant criticism of modern classification clinics, as we find them in prisons today, is that they fail in treatment. There is little criticism regarding diagnosis, but it is debatable whether the prison will ever be able to do very much in the field of treatment. The clinic can point out many physical defects in prisoners, and the facilities available will be able to clear up many of these. But there is little evidence that a program can be developed that has meaning for the inmate except in a few isolated cases. One can admire the shiny filing cabinets in which are kept the complete records of each inmate; one can respect the professional staff for their skill and dedication to long hours of toil. But he can have little enthusiasm for results. Imprisonment nullifies most of the efforts of even the most conscientious members of the classification clinic. Dr. Ralph Brancale, Director of New Jersey's Diagnostic Center at Menlo Park, is especially harsh in his appraisal of the classification process. He writes:

> The average classification process, when boiled down to its essentials, consists of little more than a check list of the inmate's deficiencies and assets, with specific recommendations for rectifying or exploiting each specific finding. The assumption behind this approach seems to be that an atomistically constructed picture of the patient's assets and liabilities, followed by a forced cementing of this into a hodgepodge semblance of personality structure, will automatically provide the basis for a prescription of treatment which is expected to solve all of the offender's problems.
>
> This segmental approach to classification which is so prevalent in correctional institutions should be recognized for what it is: a *superficial, impractical, nonintegrative* approach which ignores the core of the problem of offenders and their personality structure—too often one of obscure psychopathology. Furthermore, it ignores the basic mental processes which are probably responsible for antisocial reaction.[14]

What is probably needed is a reappraisal of the entire diagnostic, classification, and treatment complex. While we stated above that there is little criticism with diagnosis as an objective, there are pertinent questions that must be asked. For instance, do we have adequate diagnostic facilities and personnel? Is it possible to diagnose without considering certain areas of science that have considerable bearing on human behavior, such as those usually referred to as the constitutional, or biological? In several countries, notably Italy and certain Latin American countries (Brazil, Argentina, Chile), we are likely to find anthropologists, endocrinologists, and "body-build" experts on diagnostic clinics. It is doubtful that any

---

[14] Ralph A. Brancale, "Psychiatric and Psychological Services," by permission from: Paul W. Tappan (ed.), *Contemporary Correction* (New York: McGraw-Hill Book Co., Copyright 1951), p. 193; italics added.

can be found in American prisons. Another pertinent question is: can we telescope diagnosis into a six-week quarantine or reception period as is attempted in practically all correctional institutions today? We sometimes refer to a prison as a hospital; yet those sent to prison are there against their will whereas patients in a hospital usually enter voluntarily and are anxious to cooperate in the diagnosis of their ailments.

The rapid tempo of processing the quarantined inmate, due to pressure of intake from the courts, overworked personnel, and inadequate facilities cannot help but make diagnosis little more than perfunctory. Perhaps treatment, of a kind, should begin the first day a person enters a prison. Accurate diagnosis can wait and proceed at a slower tempo, more closely integrated with treatment. There is plenty of time in a prison; this is one way in which it differs from a hospital.

We have placed most of our faith for re-socialization of the convicted offender in professionally trained treatment personnel. Yet correctional institutions are woefully weak in this very treatment personnel which must diagnose, classify, and treat. In a study released in 1957 but which covers the year 1954, we find the following:

Professionally trained treatment personnel in American prisons as of 1954

1. Psychiatric Services

Full time, State Prisons	31
Full time, Federal Prisons	12
Part time, State Prisons	34
Part time, Federal Prisons	5
Consultants from outside State Prisons	42
Consultants from outside Federal Prisons	9
Total Psychiatrists	133

2. Psychological Services

Psychologists in State Prisons	85
Psychologists in Federal Prisons	5
Total Psychologists	90

3. Social Workers

State Prisons	96
Federal Prisons	66
Total Social Workers	162

The following states reported no psychiatrists, psychologists or social workers in their prisons: Arizona, Arkansas, Florida, Georgia, Idaho, Kansas, Mississippi, New Hampshire, North Dakota, Oregon, South Dakota, Wyoming.[15]

---

[15] Abstracted from Warren S. Wille, "Psychiatric Facilities in Prisons and Correctional Institutions," *Amer. J. Psych.,* Vol. 114, No. 6 (December, 1957), pp. 481-487.

The study further indicates that of the 133 psychiatrists employed in American prisons, only 53 were reported to be certified by the American Board of Psychiatry and 27 to be "board eligible." It was not ascertained how many psychologists were academically qualified. While many persons doing treatment work in our prisons are labeled "social workers," there are actually few who are entitled to that appellation. In other words, not many are graduates of an accredited school of social work or members of the National Association of Social Workers.

8. Psychotherapy and Group Counseling in the Re-Socialization Process

We have alluded to group therapy, group counseling and guidance earlier. It is important that a few words be added to describe these approaches to treatment and to make an appraisal of their effectiveness in the correctional program.

Individual psychotherapy at the hands of a psychiatrist or a psychiatric social worker has long been considered a valuable tool in dealing with maladjusted children or adults and, for that matter, the group concept of therapy is not new in several other areas that involve people with problems.

Group therapy had its beginnings many years ago when it was utilized in dealing with sick people—the tuberculous in particular. But it gained widespread use during World War II when the armed forces were confronted with a limited trained personnel and a serious problem of battle-fatigued soldiers. Forms of mass or group treatment had to be adopted, especially in processing certain men for combat duty. Thus group therapy was applied to those who found it difficult to conform to arbitrary discipline, so characteristic of military life. After the war it was but natural that the technique should be accepted in connection with the inmates of correctional institutions. The technique and its variations have become widespread in several states where it is considered by many correctional people as one of the most fruitful innovations to the field since the development of the diagnostic clinic.

When discussing group therapy one is likely to think of a wide variety of techniques ranging all the way from Alcoholics Anonymous, testimonial meetings, and Coué-ism, to the highly analytical varieties employing the dynamics of Freud or Adler. Regardless of philosophy or technique, all types fall within two limits, according to Dr. Giles Thomas of Columbia University's Department of Psychiatry and Medicine. These are: (a) the repressive-inspirational (Christian Science, Coué-ism, Alcoholics Anonymous) and (b) the analytic (orthodox Freudian analysis).[16] In the for-

---

[16] The reader is referred to the cogent article by Dr. Thomas on group therapy in *Psychosomatic Medicine*, Vol. 5 (1943), pp. 166-180.

mer type the patient is urged to control himself and to suppress his worries or frustrating desires with the suggestion that he find a satisfying outlet in his work, community activity, or in religion. The latter type calls for release or catharsis so the individual is free to follow his own social objectives. As Dr. Thomas puts it: "Analytic therapy urges the loosening of repression, the conscious recognition and analysis of unconscious wishes; it aims to free energy bound in needless repression . . . so that the individual will himself find suitable social outlets."[17]

The experimentation with group therapy in the armed forces was found to be highly successful. While it is assumed that the therapist or group leader or integrator must have psychiatric training, this is not necessarily the case. It is essential, however, that the leader be skilled in integration. Whether he guides the verbalization of group members or chooses the non-directive technique, he must have the ability to synthesize. Potential group therapists may be drawn from psychiatry, psychology, sociology, social case work, counseling, or guidance.[18]

This is exactly what is being done in many states where group therapy and group counseling are being developed and expanded. California, for example, is drawing into the project many types of staff personnel including members of the custodial and maintenance forces. Starting some years ago with highly trained behavioral personnel, the Department of Correction took the pioneer step of training a staff of group counselors recruited from all services. There are approximately 500 groups regularly attending sessions in the various institutions in California. This calls for approximately the same number of leaders drawn from the staff. In 1957 about one-half of all the state's prison inmates were enrolled in these sessions. The chart on page 481 indicates the extent of this service in this one state of California.

---

[17] *Ibid.,* p. 167.

[18] Selected bibliography on group therapy, guided group interaction and guidance counseling: Marshall H. Clinard, "The Group Approach to Social Reintegration," *Amer. Soc. Rev.,* Vol. 14, No. 2 (April, 1949), pp. 257-262; Joseph Abrahams and Lloyd W. McCorkle, "Group Psychotherapy of Military Offenders," *Amer. J. Soc.,* Vol. 51, No. 5 (March, 1946), pp. 455-464; for the work of S. R. Slavson in this field see his *An Introduction to Group Therapy* (New York: Commonwealth Fund, 1943), *The Practice of Group Therapy* (New York: International Universities Press, 1947), and "Milieu and Group Therapy," *Yearbook,* N.P.P.A., 1948, pp. 119-130; Samuel B. Hadden, "Group Therapy in Prisons," *Prison World,* Vol. 10, No. 5 (September-October, 1948), pp. 174ff.; James J. Thorpe, "Group Therapy," *Proceedings,* American Correctional Association, 1948, pp. 184-189; F. Lovell Bixby and Lloyd W. McCorkle, "Applying the Principles of Group Therapy to Correctional Institutions," *Federal Probation,* Vol. 14, No. 1 (March, 1950), pp. 36-40; Carl R. Rogers, *Therapy Implications and Theory* (Boston: Houghton Mifflin Co., 1951); Arthur Mann, "Group Therapy—Irradiation," *J. Crim. Law,* Vol. 46, No. 1 (May-June, 1955), pp. 50-66; Morris G. Caldwell, "Group Dynamics in the Prison Community," *J. Crim. Law,* Vol. 46, No. 5 (January-February, 1956), pp. 648-657; and Norman Fenton, *An Introduction to Group Counseling in State Correctional Service* (Sacramento, California: State Department of Corrections, 1957).

Aside from the counseling with inmates California is pioneering in group counseling with some of the wives of inmates. This initially started with individual counseling with certain inmates' wives and then gradually was extended to group conferences with several of the women.

	JAN.	FEB.	MARCH	APRIL	MAY	JUNE	JULY	AUG.	SEPT.	OCT.	NOV.	DEC.
No. Groups	488	500	512	511	523	498	492	494	496	497	502	537
No. Leaders	431	446	425	460	463	469	453	459	466	463	473	509
No. Leaders Overtime	104	112	107	109	108	109	107	109	129	84	108	128
No. Inmates	6960	7114	7547	7535	7717	7477	7659	7634	7739	7800	7909	8441
Camp Groups	33	33	33	35	34	31	27	31	31	33	35	33
Inmates in Camp Groups	436	446	443	482	508	433	408	444	418	456	480	490

Experience with group counseling is beginning to show results that appear gratifying. There is evidence that both inmates and custodial officers who participate with a degree of sustained interest become more relaxed and tolerant toward their tasks and toward each other. Infractions of rules diminish where group counseling operates on a wide scale. In a survey of results of widespread counseling in California, conducted by the Department of Correction, the following samples were indicated:

1. At Folsom prison members participating received only one-half the number of disciplinary infractions as inmates who applied for counseling but were not enrolled and only one-third as many as inmates who had neither requested nor participated in the sessions.
2. At Soledad prison a chart indicated a 68 per cent increase in inmate participation in counseling between 1956 and 1957 and a 26 per cent increase in employee participation. Disciplinary infractions decreased gradually through the period.
3. At Chino institution for men a statistical analysis of infractions between July 1, 1956 and July 1, 1957, showed that there were .086 disciplinary reports during the period per man not in counseling and only .052 disciplinary reports per man in counseling.
4. At Tehachapi unit it was decided by the Department to try 100 per cent group counseling for one year rather than to construct guard towers. During 1955 there were eight escapes and during 1956 the same number. In 1957 there were three escapes. This represents a monetary saving during the past year of almost twice the proposed group counseling budget for the entire Department. It also suggests that group controls may operate to reduce the number of escapes, which are expensive financially and in terms of institutional morale.[19]

[19] Supplied the writers by the Department of Correction, Sacramento, California 1958. A specially-designed textbook has been prepared for classes in group therapy. See Norman Fenton, *What Will Be Your Life?* (Sacramento, Cal.; Department of Corrections, 1955).

29 ——————————

Education and
Religion Behind Bars

*Education for adult prisoners has an aim and a philosophy.
Its philosophy is to consider the prisoner as primarily an adult
in need of education and only secondarily as a criminal in need
of reform. Its aim is to extend to prisoners as individuals every
type of educational opportunity that experience or sound
reasoning shows may be of benefit or of interest to them, in
the hope that they may thereby be fitted to live more competently, satisfyingly and cooperatively as members of society.
(From* The Education of Adult Prisoners, *New York: Osborne
Association, 1931.)*

AUSTIN H. MACCORMICK

1. Early Prison Schools

John Howard advocated educational and religious instruction for prisoners when he envisaged the penitentiary as a rehabilitative
device for criminals. In the Walnut Street Jail in Philadelphia a school
was opened in 1798. The prison board was casting about for some
"kind of leisure time occupation" for the inmates and believed that "the
most beneficial employment [aside from shop work] was to establish
schools of learning for some and improving others in the first Principles
of reading, writing, and arithmetic." Later, books and desks were pro-
cured and thus was opened the first prison school in America.[1] Lectures
were also provided the inmates.

In 1801 New York State elementary education was provided for "meri-
torious convicts" during the winter months by the better educated pris-

---

[1] From the Minutes of the Board of Inspectors, July 27, 1798.

oners.[2] This is probably the first time inmate teachers were ever used in prison schools, a practice that has since been widely adopted in this country's correctional institutions. New York State passed a law in 1847 creating a school for its prisons although much earlier the chaplains assigned to the various prisons in Pennsylvania, Massachusetts, and New York were giving elementary instruction in reading and writing.[3]

Early prison schools could hardly be dignified by the name. They were crude and woefully weak in curriculum and teaching methods. No appropriations were made for books or supplies, reliance being placed on donations from friends and Bible societies. Yet education was, from the beginning, regarded as one of the methods to be used for redeeming prison inmates. At the first conclave of correctional administrators held at Cincinnati in 1870, the following principle was endorsed and incorporated in the Declaration of Principles:

> Education is a vital force in the reformation of fallen men and women. Its tendency is to quicken the intellect, inspire self-respect, excite to higher aims, and afford a healthful substitute for low and vicious amusements. Recreation is considered to be an essential part of education. It has come to be recognized that recreation and education are, therefore, matters of primary importance in prisons, and should be carried to the utmost extent consistent with the other purposes of such institutions.

But even as recent as 1927-28, Austin H. MacCormick found that there were no schools in 13 out of the 60 prisons surveyed and that not one single prison made adequate provision for vocational education. He wrote: "Taking the country as a whole, we are tolerating a tragic failure. . . . Not a single and well-rounded educational program, adequately financed and staffed, was encountered."[4]

Mr. MacCormick, realizing that our system of prison education had to be constructed from the ground up, outlined a comprehensive program that included rudimentary academic, vocational, hygienic, social, and cultural education. The fundamental principles on which the new prison education was to rest were: (1) avoidance of mass education and the adoption of individualized instruction; (2) avoidance of reliance on mere stereotyped programs and routine; (3) recognition that convict education is adult education and not the feeding of juvenile instruction to grownups; (4) a broad and inclusive curriculum designed to meet all needs; and (5) making *interest* rather than *compulsion* the psychological basis of the system.

---

[2] Orlando F. Lewis, *The Development of American Prisons and Prison Customs, 1776-1845* (Albany: 1922).

[3] See W. M. Wallack, G. M. Kendall, and H. L. Briggs, *Education Within Prison Walls* (New York: Columbia University Press, 1939), p. 4.

[4] Austin H. MacCormick, *The Education of Adult Prisoners* (New York: The Osborne Association, 1931), p. 5.

2. The New Day in Prison Education

This pioneer exposition and criticism sounded a real challenge to those interested in the education of adult inmates. The Federal Bureau of Prisons began to take an active interest in prison education in the year 1930. Congressional appropriations made it possible to install trained supervisors in the major federal penitentiaries and reformatories. In addition, trained librarians were employed so that library facilities could be placed on a competent and quasi-professional basis.

The State of New York appointed the Englehardt Commission in 1933 to study the scope and ramifications of education in adult penal institutions of that state. Its recommendations were far-reaching and crystallized the attention of penologists on the new education, an integrated technique in dealing with the individual's specific needs. One recommendation succinctly states:

> The term "education" as used in correctional work should be interpreted very broadly. Education in terms of the three R's alone, or of vocational training organized and administered in the manner of traditional schools, is inadequate in correctional work. Education can and must be administered in terms of individual needs. There must be a complete background of information upon which an educational diagnosis can be made and the administrative set-up for training must permit the use of teaching methods applicable to diagnostic procedure. Education must be directed purposefully toward specific objectives. Teaching must be very largely in terms of guidance. It may be, but is not necessarily, concerned with textbooks, classrooms, and the ordinary appurtenances of the traditional school. The objective is always the attainment of some well defined end, such as changing attitudes, increasing vocational efficiency, elimination of complexes, the development of willingness and skill for cooperative living after release.[5]

It is obvious that the educational needs of the inmate must be correlated with his other needs as diagnosed by the classification clinic. If the man is illiterate, it is worth something both to himself and to society to teach him the rudiments of reading, writing, and arithmetic; but he may be lacking in some more vital prerequisites for a satisfactory adjustment —may need immediate medical attention, be suffering from a number of personality shortcomings, be confused about his rôle in his family, or be worried about his former job. A fact-finding case study of the man will recommend the steps necessary to clear up each inmate's problems.

Taking the New York State experiment in penal education as an example of what may be accomplished, we summarize briefly some of its essential features. The report of the Division of Education, as interpreted by Wallack, Kendall, and Briggs, states: "[We] accept the modern point of view which holds that the purpose of education is to develop the in-

[5] Quoted by W. M. Wallack, G. M. Kendall, and H. L. Briggs, *Education within Prison Walls* (New York: Columbia University Press, 1939), p. 19.

dividual as a total personality. It is for administrative purposes only that we have divided our work into the two categories: general education and vocational education.". . . "Under the heading of general education the Division of Education includes all types of educational activities which do not aim to develop vocational skills" but are intended "to revise attitudes toward social institutions and the individual's relationship to them, to develop interest and skills needed in acceptable social living," [and] "to provide the training basic to all desirable human relationships."[6]

Aims of Social Education in Correctional Institutions

Activities in the social education field are:

1. The so-called academic subjects: the social studies, history, geography, English, mathematics, general science.
2. Mental hygiene: classes in personality development, consultation, guidance.
3. Health and physical education: classes in personal hygiene and health, public health, correction of physical defects, games and play.
4. Recreational activities: sports, games, entertainments, hobbies, clubs, publications, reading, organizations.
5. Arts: music, dramatics, sculpture, painting, sketching.
6. Classes for physically and mentally handicapped.
7. Cultural development: discussion groups, library research and reading, personal relations.[7]

These social educational activities aim:

1. To bring the inmate to adopt goals and attitudes which are in accord with those of society and which will further the improvement of society.
2. To show the inmate the desirability of furthering the interests and standards of worth-while social groups. Social education must bring about a transfer of allegiance from anti-social groups and methods to socially desirable ways of achieving desired goals.
3. To develop points of view which will make apparent to the prisoner the futility of commiting criminal acts, and the advantages of law-abiding post-institutional living.
4. To stimulate and develop desirable interests which will enable the individual to live a worth-while and yet an interesting life. This has reference particularly to leisure-time interests and activities.
5. To stimulate and make possible sustained interest and effort toward self-improvement.
6. To develop skills, understandings, and knowledges which will enable the individual to perform the ordinary duties of every efficient citizen.[8]

This paper program is naturally ambitious. Nevertheless the goals are set up, and many of them may be achieved in such a subtle manner that they may not really seem to be accomplished. The educational process is a day-by-day activity, wearing down much of the resistance of the

---

[6] *Ibid.*, p. 22.
[7] *Ibid.*, p. 23.
[8] *Ibid.*, p. 24.

student and changing life habits slowly but inexorably—provided the teachers are well trained in their skills and objectives. In other words, education is also therapy.

Aims of Vocational Education in Correctional Institutions

These aims of vocational education are not quite so clear or, at least, they are not stated in tangible terms. The trade-learning mania that characterized the reformatory ideal of the past half century has been found expensive and futile. The New York State educational authorities do not emphasize trade-learning. They point out the problems that confront industry:

> It is evident that the philosophy governing a program of vocational education within the prison must be influenced largely by conditions as they actually exist in the occupational world in which the released inmate must eventually find employment.[9]

After pointing out the rapid changes that have taken place in the industrial world, they add:

> The prison inmate upon parole, if his period of imprisonment has been of any great length, faces an unfamiliar world. He must be trained to live successfully in that world . . . must be equipped to earn a living in a highly competitive and rapidly changing world if [he is] to use fair means in securing food, shelter, and clothing. This requires a program of vocational training in the prison which is based upon individual inmate needs as well as upon the existing employment situation. . . . The current employment situation demands workers trained in a wide variety of skills. . . . The worker released from prison should have acquired not only usable skills but that pride in high-grade performance which enables one to "get a kick" out of doing a job well. He must be trained to the point where he gains more satisfaction from the performance of legitimate work than from criminal activities.[10]

Trades, as such, should not be taught. Rather, a wide variety of vocational skills should be offered to meet the needs of a maximum number of individual cases. This may be gauged in terms of personal satisfaction and potential economic reward in the future. The establishment of good work habits is essential, if that is possible with the human material at hand. Motivation also is essential. To develop these attributes or virtues is more difficult in prison than in the free community; difficult both for the educational director, who must attempt to meet this challenge, and for the inmate, who is frequently fairly well disillusioned with life. It is relatively simple for the free man to "decide" that he is going to improve his life by taking a course in some adult education program, but it is seldom that the free man sticks to his training course. We only

---

[9] *Ibid.*, p. 27.
[10] *Ibid.*, pp. 28-29.

hear of those who "stick it out." No system of intramural prison education can completely break down the prisoner's feeling of despair. Some service within the prison must patiently work with the inmate in such a way that he will accept his situation as he finds it.

Another serious problem—one that is usually overlooked by classification clinics and centralized correctional protocol—is that many men in prison start on an educational plan after serious deliberation and then suddenly are "transferred" to some other facility, thus bringing to an abrupt stop the individualized program. This happens hundreds of times in prison and the frustration experienced, not only by the inmate students but the teachers as well, is very real.

Any constructive system of education can be of great service to a well-adjusted individual. So inmates in correctional institutions must first be adjusted, if this is at all possible. There must be someone in the prison set-up with insight into the personality problems of the individual. In the old days there was no one to whom the prison could turn but the chaplain. This need is still often supplied by the chaplain, and clergymen preparing for prison chaplaincy receive extensive training in pastoral counseling in addition to their normal studies. However, the best prisons employ trained personnel so that the bewildered inmate may find the encouragement he needs from a correctional officer who possesses that rare attribute, insight. But there should also be case workers available to direct those who are able to be of real service to the intimate needs of the prisoner. Close cooperation between case worker and educational director will be of inestimable value to the inmate who is trying to adjust himself to the prison routine and prepare himself for release.

Much progress has been made in the field of correctional education since MacCormick's study of 1927, but most prisons are still quite backward. While the employment of prison inmates as teachers is obviously better than nothing, it is recognized by correctional people that if education is to have any meaning, not only to prison authorities, but to the nation as a whole, the same standards expected in our public schools should obtain in our reform schools and adult correctional institutions. Education is, in reality, a national problem. In some states the teachers employed in the prisons are recruited from local boards of education and are paid comparative salaries. This is certainly as it should be.

It is gratifying to note the serious interest that has been demonstrated by the American Correctional Association. Some years ago this organization published two Yearbooks, covering various phases of this most important field, which marked a milestone in the long road that has been charted by correctional reformers.

The first Yearbook "reviewed the history and the progress of correctional education and outlined its accomplishments and possibilities.[11] It

---

[11] *Correctional Education Today*, 1939.

set the standards which every conscientious prison official desires to adopt." The second Yearbook was prepared "primarily for wardens of our prisons, the superintendents of our reformatories, the keepers of our three thousand jails and the staffs of all penal institutions.[12]

3. Prison Libraries

The library of a prison is a sort of stepchild of the program of rehabilitation. Seldom do we find a well-stocked library adapted to the several types of inmates who will use it, and even more infrequently is a trained librarian in charge. It is a lamentable fact that the same situation too frequently occurs in the free libraries of the nation.

A good library in an institution need no longer be defended as a handmaiden of good educational philosophy. Adequate supervision is just as urgently needed as in the prison schools—and is not being provided. The best we can expect or urge at the moment is a reasonably adequate budget and a plan that calls for a wise selection of books to meet the peculiar demands of the inmates.

Early libraries in prisons suffered in the same manner as did the early schools. What books were available were of little interest to the average prisoner. Many titles were religious in nature or filled with moral platitudes. Some inmates, in order to make an impression on the overworked and well-meaning chaplains, chose such books and pretended to read them avidly. Where shelves or book-cases were furnished they were likely to be in dark, dingy corners, the books often uncatalogued and seldom properly repaired when worn out by constant use. Restrictions as to time of reading, number of books borrowed, and so forth, often discouraged many of the more intelligent inmates from taking advantage of the service. Most of the dull and obsolete books received through charity were given away so as to be rid of them.

The prison library in most state prisons is little better than its early predecessor, as compared with improvements in other facilities. It still suffers from lack of funds, worn-out books, unwanted titles, and untrained personnel.

In the federal prisons the libraries are far better. The book selection is good and there are trained librarians in most institutions. Dr. Benjamin Frank, superintendent of vocational education, states that prison inmates read five times as many books as the general public. A survey made by him in the Federal Bureau of Prisons shows that 75 per cent of the inmates make use of the facilities and that the average inmate reads 70 books per year as compared to 10.9 per capita in the general population.[13]

---

[12] *Prison Administration—An Educational Process,* 1940. See also, *Manual, 1956,* Chapter 20, "Education."

[13] From *Federal Probation,* Vol. 14, No. 1 (March, 1950), p. 67.

The library at the Federal Penitentiary, Lewisburg, Pa.

Good library procedures may be found in various progressive prisons throughout the country, but on the whole the story of the prison library is none too encouraging. The library committee of the American Correctional Association, however, is doing yeoman service in raising the standards of this much neglected service to prison inmates. This organization has prepared a manual for libraries to follow in setting up and developing institutional procedures.[14]

4. Letter Writing and Censorship of Mail

In the early days of prisons not many inmates were literate enough to write letters. But with schools and chaplains dedicated to the fundamentals of education, more and more inmates learned to write. And, with compulsory schooling, which began as early as the 1840's, literacy became more common.

In a modern prison with a population of as many as three to four thousand, letter writing is of tremendous importance to both the inmates and the staff. So long as the responsibility of safe custody weighs heavily on warden and staff, some type of censorship of prisoners' mail is necessary. Some limitation on the number of letters sent and received by the inmates is also justified. There is little uniformity in current practices. In a survey conducted in 1952 the following policies were discovered regarding letter writing.

---

[14] *Library Manual for Correctional Institutions* (New York: American Correctional Association, 1950).

Prisoners awaiting letters from home.

Number of letters inmate may write during quarantine period

8	institutions	—no limits placed on correspondence
12	"	—one letter per week
7	"	—two letters per week
10	"	—three letters per week
2	"	—one letter per day
1	"	—a "reasonable" number
1	"	—two letters per month
2	"	—six letters per month

Number of letters inmate is permitted to write after quarantine period

5	institutions	—no limit
11	"	—one letter per week
10	"	—two letters per week
11	"	—three letters per week
1	"	—varies "with grade" of the prisoner
1	"	—a "reasonable" number
3	"	—one letter per day
1	"	—one letter per month
1	"	—two letters per month
1	"	—six per month
1	"	—eight per week
1	"	—six per week
1	"	—ten per week
1	"	—"depends on conduct"[15]

---

[15] "A Limited Survey of Some Prison Practices and Policies," *Prison World,* Vol. 14, No. 3 (May-June, 1952), p. 6.

Upon the prisoner's reception in the prison he is informed of his right to receive and send the permitted limited number of letters and he must designate the names of those persons with whom he wishes to correspond. Usually he may write additional letters to lawyers and to others, such as a former employer, to clear up unfinished business. In many institutions a stamp is furnished by the administration for one letter per week; other postage must be paid by the inmate himself. Some institutions furnish a copy of rejection regulations to be sent to the prisoner's correspondents. A sample is the following that was used at Sing Sing prison some years ago:

This Letter Is Returned to You Because:

1. Special letters must not be submitted Saturday, Sunday or Holidays.
2. You did not sign it properly.
3. You did not fill out stub properly.
4. Permission to write the person addressed has not been filed.
5. It contains criminal or prison news.
6. Letters addressed to General Delivery cannot be mailed.
7. Begging for money or packages is not allowed.
8. You are not permitted to receive the articles requested in your letter.
9. The articles requested can only be sent New from Dealer.
10. Correspondence with newspapers or newspaper employees not permitted.
11. You cannot have a visit with the person named without permission.
12. You did not stick to your subject.
13. Inmate who wrote letter for you did not sign his number.
14. You have no stamps on deposit.[16]

The following are some of the precautions taken:

Letters to persons not approved are rejected; letters approved are examined under an ultra-violet ray device in order to detect possible invisible writings. This is especially important in the thwarting of escape plots and other disturbances.

Each inmate may send one letter each week for which the State pays the postage. These are called "Sunday letters." The sending of additional or "Special" letters depends entirely on the conduct grading of the inmate. Those in Grade A may write a reasonable number of letters provided they pay their own postage. Restrictions are placed on inmates in other conduct grades.

Prisoners with a history of possessing, using or selling narcotics are especially suspect so that their letters are more carefully scrutinized. To quote the warden: "Letters for such inmates are given the utmost attention, such as the removal of stamps from envelopes, and the subjection of the letter and envelope to the telltale eye of the ultra-violet ray machine is made. In extreme cases the material is sent to the institutional laboratory for further analysis. The narcotic problem is a grave one, especially to institutions receiving prisoners from metropolitan areas."

Magazines and newspapers must be sent from the publishing office only.

---

[16] Lewis E. Lawes, "Censor Mail in My Jail?" *Jail Association Journal* (November-December, 1939), pp. 30 f.

Objectionable articles "are occasionally deleted and publications not in harmony with the rehabilitative philosophy of the institution are rejected."[17]

Here we have an efficient and comprehensive system of censorship in a large correctional institution. Commenting on such a system of censorship, the *Survey of Release Procedures* warns:

> The system sometimes defeats its own purpose. Setting a limit on the number of letters permitted is usually thought necessary in order to keep the work of censorship within bounds. However, it often necessitates elaborate record-keeping which is more work than many extra letters. To limit an inmate's correspondence to members of his immediate family, as is often done, conceals from authorities a great deal of valuable information as to other contacts and interests, both good and bad, which the prisoner may disclose in letter-writing. Especially if the inmate is an unmarried man, frequently the most valuable insight into his weaknesses and his needs may be found not in his letters to his family, but in correspondence with both his men and his women friends.[18]

We argue for fewer restrictions on letter-writing. Letter-writing keeps the prisoner in contact with the outside world, helps to hold in check some of the morbidity and hopelessness produced by prison life and isolation, stimulates his more natural and human impulses, and otherwise may make contributions to better mental attitudes and reformation. Letters sent out of the prison have some bearing on the thought processes of the inmate and, if the prison is genuinely interested in therapy, the contents may prove both revealing and helpful. Naturally, the inmates must be informed upon entering the institution that a fair sampling of incoming and outgoing letters will be censored.

5. The Prison Chaplain and Religious Instruction

Next to the warden and his custodial staff, the chaplain and the physician were the first paid prison officials, although legislators were somewhat tardy in appointing the chaplain. For many years ministers in the communities near the prisons volunteered their services. The public was slow to support paid "moral instructors," partly owing to the suspicion that the inmates might be proselytized by zealous sectarian chaplains.

As we mentioned earlier, chaplains were the first school teachers and the first librarians. Perhaps the first chapel ever erected in a correctional institution was built in the Walnut Street Jail in Philadelphia in 1817. It was reported that at the opening of the chapel the Rev. Doctor Stoughton "held prayers and appropriate discourse, during the whole performance of which the Convicts behaved themselves with the greatest propriety and decorum."[19]

---

[17] *Ibid.*

[18] *Attorney General's Survey of Release Procedures,* Vol. 5, "Prisons," p. 104.

[19] From the Minutes of the Board of Inspectors, September 1, 1817. See Negley K. Teeters, *The Cradle of the Penitentiary* (Philadelphia: Pennsylvania Prison Society, 1955), p. 99.

Bishop William White of Christ's Church, Philadelphia preaches a sermon in Walnut Street Jail in 1787 under a simulated "handicap" set up by the unsympathetic jailer. "A cannon, apparently loaded—for a man with lighted linstock stood by its breech— was pointed down the path in the faces of an array of all the motley of the establishment. . . . 'The keeper's battery was a standing joke for the prison for all time." (From "Reminiscenses of Old Walnut Street Prison" by one of its keepers, William Webb, Philadelphia "Sunday Dispatch," October 16, 1859–February 5, 1860.)

Director James V. Bennett of the Federal Bureau of Prisons, in an article dealing with prison chaplains, states that these men were the "first to introduce social case work technique in prisons and the first to recognize the reformative value of individualized treatment of the offender."[20] There can be no doubt that the early chaplains were called on to perform a wide variety of services for the inmates and that many dispatched them courageously and sincerely.

We may smile at the stereotype often presented in cartoons of ministers, especially of those working with hardened and depraved criminals, but the fact remains that many of them have accomplished much good. Among the early chaplains were men possessing dynamic convictions who commanded the prisoners' respect. But some of them were not in the chaplaincy because of real commitment to that special work and were therefore inept and sometimes absurd. It was doubtless from this group that the stereotype was created. Unfortunately the chaplain, in some instances, has permitted himself to become institutionalized to such a degree

---

[20] "The Role of the Modern Prison Chaplain," *Proceedings,* American Correctional Association, 1937, pp. 379-388.

that he has often been very ineffective in dealing on an honest level with the inmates.

In the early days, the chaplain's success was usually measured by the number of conversions he was able to record. The wily inmate became adept in simulating religion for the purpose of gaining concessions or securing an early pardon. Some naive chaplains were all too eager to accept such outward expressions of religion as the genuine article. This made the chaplain appear ridiculous, and lowered the effectiveness of the religious program in the institution. A prisoner writes thus about the prison chaplain:

> My impression of the average chaplain is that he is a broken-down failure, half a man of God and half a politician with a very anemic mind. If religion is of any value at all, it ought to be put to use in changing the attitudes of the prisoners. But with the kind of chaplains with which I am familiar, this process is something like trying to cut glass with a piece of soap.[21]

And this is what a chaplain in one of our large prisons thinks of the prisoners' attitudes toward the chaplains and the church services:

> We, as chaplains, know very well that only about 25 per cent of the prisoners attend our Chapel Service with any real sincerity. We know too, that of those who do attend, religion may or may not be the inducing factor. Many men merely attend to get out of their cells, some go to hear the music, quite a few feel that Chapel attendance may influence the Parole Board in their favor. Others go because relatives and friends urge them to. With many, attendance is nothing more than a superstitious habit with no comprehension of the real significance of religion. Only those interested in religion accept our sermon in the spirit it is given. Thinkers are critical, opposing or disapproving, according to their convictions. The dull are indifferent; and the in-betweens respond, perhaps, when we illustrate our point with a funny story. This sounds discouraging; but until religion becomes not only an ideal, but a practical application in everyday life . . . it will not influence our prisoners.[22]

In our discussion of the prison chaplain, we may fruitfully examine the role of religion in making an analysis of crime. From time to time we hear of commissions, committees, or surveys, generally composed of ministers with perhaps some laymen added to give them a certain public acceptance, whose purpose is to answer the age-old question: "Will religious training keep men and women from committing crimes?"

Such commissions and other worth-while efforts may be warmly welcomed by all who are interested in the understanding and repression of criminality. Any sympathetic and humanitarian effort to assist people, especially young people, to go straight, is highly commendable and should

---

[21] *Prison World,* Vol. 2, No. 4 (July-August, 1940), p. 19.

[22] Francis J. Miller, "The Inmate's Attitude Toward Religion and the Chaplain," *Proceedings,* American Prison Association, 1941, pp. 431-435.

be encouraged. In some cases, traditional religion, with its emphasis on spiritual and moral regeneration, can lead to creative reformation of the prisoner. In others, the traditional coupling of religion and authority precludes any constructive response from the prisoner.

Serious attempts to evaluate the effectiveness of religious work among prison inmates have never been made. Success has been too often assumed. However, many concerned chaplains have recently given much thought to their work. It is debatable just what the chaplain can or should attempt to accomplish in prisons. The question of morality versus religion has never been satisfactorily faced. Bonger, the eminent Dutch criminologist, quotes the findings of European criminologists in support of the conclusion that "morality lies anchored more deeply in the human mind than does religion, and therefore does not rest on it," and that "there exists no proof of the contention that strongly religious tendencies imply any strong moral predisposition"; and adds that "too little is known about the psychology of religion . . . and therefore the question remains an open one."[23]

It is also debatable whether the chaplain should assume the highly specialized duties of others on the prison staff, such as guidance officers, social case workers, the educator, or parole officer.[24] The writers know of at least one penitentiary in which the chaplain for years handled parole work; his assistants counseled the inmates spiritually.

Unless prisons develop trained chaplains, as is being done in the federal system and a few of the states, it might well be recommended that those who are state-supported be eliminated. In their place, outside chaplains, trained by the various denominations, should be introduced. If the denominations wish to gain a state subsidy for this work, it is not unreasonable, but they should assume direct control of the chaplain's activities within the framework of their profession rather than relinquish organizational guidance to the warden.

It should be remembered that the chaplain's professional training is in the field of theology, rather than criminology or social work. His function should be primarily pastoral, unless he is also adequately trained in another discipline. At the very least he can provide opportunities for worship and spiritual guidance; at best, he can offer to the prisoner a kind of acceptance and a possibility for growth beyond the potential of any other member of the prison staff. The prison chaplaincy, while of some promise to an enlightened criminology, is in need of adequate evaluation and increasingly realistic standards of preparation.

---

[23] W. A. Bonger, *An Introduction to Criminology* (London: Methuen & Co., 1936), pp. 127-136.

[24] See A. E. Kannwischer, "The Role of the Protestant Chaplain in Correctional Institutions," *Amer. J. Corr.*, Vol. 19, No. 1 (January-February, 1957), pp. 12f. See also *Manual, 1956,* Chapter 22, "The Religious Program."

30

Self-Government in Prisons

Under [self-government] all privileges are utilized as means of obtaining responsibility for the good conduct of the prison community. (From Prisons and Common Sense, *Philadelphia: Lippincott, 1924, p. 80.)*

THOMAS MOTT OSBORNE

1. The "Trusty" System

The idea of trusting men in prison to do something on their own responsibility is said to have originated in the Wethersfield, Connecticut prison as early as 1835.[1] The warden at that time, Amos Pilsbury, developed a system whereby he picked certain inmates in whom he had confidence to go, unescorted, into the nearby town on errands. This practice developed into a "trusty" system, in which good behavior became an inmate objective. It was copied in nearly all early prisons and is today quite prevalent. Many institutions publicize their "trusty" system and give the impression that it is an "honor" system. While there is honor implicit in the trusty system, it is a far cry from what is more precisely known as an honor system. In fact, a respectable honor system probably exists nowhere in a modern prison.

There is a general notion that trusties are the prisoners who have given the most evidence of reformation. More often than not, they are the worst criminals in the institution. One such case, mentioned by Courtney Riley Cooper, developed a notorious reputation in penal circles. A prisoner sentenced for manslaughter was made a "trusty" because there was a

---

[1] In the interest of historical accuracy, a type of "inferior" self-government among the inmates was set up in the Walnut Street Jail in Philadelphia following 1790. Robert J. Turnbull describes this system in his pamphlet, *A Visit to the Philadelphia Prison; Being an Accurate and Particular Account of the Wise and Humane Administration Adopted in Every Part of the Building,* Philadelphia, 1796. See Negley K. Teeters, *The Cradle of the Penitentiary,* Philadelphia: Pennsylvania Prison Society, 1955, pp. 46, 95.

shortage of paid guards and because he was a crack shot with a rifle. An unwritten law in this farm-institution was that a trusty who wounded or killed a prisoner attempting to escape was entitled to parole. This trusty shot an escaping prisoner and was immediately paroled. Shortly after his release, he was returned to the prison for shooting a woman. Again he was made a trusty and once again was released for killing an escaping fellow inmate. It had been rumored that trusties in this establishment encouraged men to escape so they could shoot them and thus be granted a parole. Cooper claimed this story came from the files of the United States Department of Justice.[2] Such "trusties" have been used as guards for years in several states. The system was used in the Angola, Louisiana farm-prison until the state correctional system was completely overhauled a few years ago.[3]

Another type of trusty system was prevalent in one large state prison many years ago. During the regime of a warden who had considerable political influence, the public was offered a "window dressing" tour of the prison at twenty-five cents per head. The first sight that met their eyes was many prisoners working outside the walls, tending flowers, mowing the lawns, cleaning windows, and performing other small chores. Proudly an attendant would explain that these men were "trusties" and that many of them ran errands uptown and had considerable freedom. He did not explain that they slept outside the prison proper in a dormitory and that they were, in reality, stool-pigeons. In fact, at the time, the warden himself never went inside the walled institution unattended, for fear of his life. This warden was finally removed after he had demoralized his entire inmate population. The citizens of the state in which this situation existed were totally unaware of the perverted effects of this show-window "trusty" system.

Still another method of maintaining a trusty system is to place "on trust" only those inmates whose sentence is within a few months of expiration. Their good behavior is almost guaranteed by their fear of incurring a punishment that will delay their release. The administration of any correctional institution that employs the trusty system is suspect, for it is often merely a spy system. The books written by ex-prisoners tell how far inmates will go to curry favor, despite great risks to their persons.

Many repeaters are already institutionalized when they enter a prison for their second, third or fourth time. They know the ropes and adapt easily to the abnormal routine. To the first offender, this facile adjustment is impossible. He breaks the rules through emotional frustration. The so-called good prisoner is likely to be a wily crook who outwits the officers,

---

[2] Courtney Ryley Cooper, *Here's to Crime* (Boston: Little, Brown, 1937), pp. 429-431.

[3] See Harry Elmer Barnes and Donald Powell Wilson, "A Riot Is an Unnecessary Evil," *Life,* Vol. 33, No. 22 (November 24, 1952), p. 140.

knows how to play the game, and has no scruples about treason toward either his superiors or his fellow-inmates. By various means such schemers gain the friendship of unscrupulous officers and guards and maintain their position through collusion with them in whatever duties and intrigue are involved in this relationship.

The newcomer to the prison discovers a situation that is both bizarre and confusing. He finds that he must align himself with someone to whom he may be loyal. The prison community is a shifting mosaic of cliques and gangs, as Hans Reimer has described it:

> . . . the general population is broken up into several cliques and separate groups . . . each man is "classified" and . . . he acquires status in terms of his reactions to the prison situations. The writer found that men who gave information had considerable difficulty in gaining confidential and satisfactory relationships with other inmates . . . The prison population is largely in control of a small group of men which has two divisions. There are the "politicians," "shots," or whatever they may be called in varying institutions, who hold key positions in the administrative offices of the prisons. They wield a power to distribute special privileges, to make possible the circulation of special foods or other supplies. They, in frequent instances, become "racketeers" and use their positions to force money and services from less powerful inmates. These men are seldom trusted by the top level of the prison hierarchy, are frequently hated by the general population because of the exclusiveness and self-seeking behavior characteristic of them. . . . The other section of this controlling power is held by the so-called "right guys." These men are so called because of the consistency of their behavior in accordance with the criminal or prison code. They are men who can always be trusted, who do not abuse lesser inmates, who are invariably loyal to their class—the convicts. They are not wanton trouble-makers but they are expected to stand up for their rights as convicts, to get what they can from the prison officials, to never permit an opportunity to pass from which they might secure anything from a better job to freedom. . . . These men, because of their outright and loyal behavior, are the real leaders of the prison and impose stringent controls upon the definitions of proper behavior from other convicts.[4]

The results are bound to be evil in any event. If the newcomer aligns himself with one of these inner gangs, based on special favoritism and snitching and squealing on his fellow-prisoners, he may get along fairly well in most prisons. But, instead of being trained to be a decent citizen, he is getting a most effective education in fundamentally antisocial and corrupt conduct. If the inmate attempts to remain a decent and honest man and refuses to co-operate with the schemers of prison politics, he is hopeless, unprotected, and the legitimate prey of anyone who desires to secure special favors by passing on stories about him to the prison authorities. The authorities themselves are all too frequently involved in the intrigue and conspiracy that permeate the institution.

---

[4] Hans Reimer, "Socialization in the Prison Community," *Proceedings,* American Prison Association, 1937, pp. 152-153.

The decent prisoner finds it extremely difficult to pick his friends within a prison that is honeycombed by the trusty system of treachery and snitching. The sincere and resolute prisoner might like to participate honestly in such intramural activities as singing, writing for the prison magazine, acting in dramatic shows, but he must weigh carefully everything he does so that he does not lose caste with his fellow convicts.

The feeling of utter hopelessness and helplessness developed in the mind of such a prisoner promotes mental and personal disintegration, and, in many cases, mental illness. The administrative system of the average prison, then, far from promoting or realizing its objective of rehabilitating the criminal, actually results in a most efficient training in crookedness and deceit or in the gradual breakdown of the body and mind of the inmate.

2. The Introduction of Self-Government into Prisons

No one would call a trusty system "self-government," regardless of any merit that may be attached to it. Self-government in an institution presupposes group responsibility as well as administrative courage. These two attributes are difficult to nourish in a medium like a prison where personality is submerged. But wardens with initiative and vision have, from time to time, experimented with forms of self-government.

Perhaps the real forerunner of inmate self-government may truly be seen in a short-lived but unique experiment inaugurated by an unsung warden of the Michigan penitentiary at Jackson in 1885. The warden, Hiram F. Hatch, actually devised a form of self-government that he called "The Mutual Aid League." His set of principles anticipated by many years the widely publicized "Mutual Welfare League," introduced by Thomas Mott Osborne in the Auburn and Sing Sing prisons.[5]

Yet it is Osborne whose name is inextricably linked with the concept of inmate responsibility in prisons. Realizing the vicious effects of abortive attempts at honor and self-improvement, especially as found in "good time" and the trusty system, Thomas Mott Osborne became convinced that it was really possible to develop responsibility on an honorable scale among prison inmates.

Osborne was a native of Auburn, New York, home of the famous prison. He grew up in this city, became a prominent industrialist, was elected mayor for two terms, and for many years was a trustee of the famous George Junior Republic, a progressive school for delinquent boys, at Freeville, New York. William Reuben George (1866-1936), founder of this Republic, had adopted a form of community organization in which the inmates were *citizens,* with a responsibility for order within the community.

---

[5] For details see Harold M. Helfman, "Antecedents of Thomas Mott Osborne's Mutual Welfare League in Michigan," *J. Crim. Law,* Vol. 40, No. 5 (January-February, 1950), pp. 597-600.

On Mr. George's suggestion, Osborne decided that there were possibilities for the same type of program in adult prisons. Donning the uniform of a prisoner and taking the name of Tom Brown, Osborne entered Auburn prison. Here he mingled with the inmates, questioned them tactfully, shared in their institutional life, and became convinced that he could be of service not only to them but to society as well. After much thought he finally conceived a plan he felt would effect the reforms necessary to rehabilitate the prisoner. He called his plan the "Mutual Welfare League." The Auburn warden and the governor of New York were willing to cooperate with him, and in 1914 his experiment in prison democracy was launched.

Under the League the rules of discipline of the institution were chiefly left to a body of delegates, 50 inmates elected by the prisoners on the basis of the representation of the several work or shop gangs. Infractions of discipline were dealt with by a board of five judges chosen by the delegates, though an appeal might be taken from their decision to the warden. The decisions of the judges were carried out by the regular officers of the prison. No keepers or guards were allowed in the shops, where discipline was wholly in the hands of the prisoners. A system of token money was introduced and a commissary was organized from which inmates could purchase articles of comfort. An employment bureau was maintained by the members of the League. Outdoor recreation, lectures, and entertainments were provided. The privileges of the inmates depended upon the degree of their adherence to the regulations of the system. Released members of the League were expected to make an effort to find employment for their fellow members upon the expiration of their sentences and to sustain the efforts at reformation begun within the institution. In this way it was hoped that an adequate system of social education might be provided that would restore to a normal life the great majority of inmates who had hitherto been released only to be returned for another offense after a brief period of freedom.

The success of the League at Auburn prison became almost a sensation in correctional circles. Correctional people throughout the country were watching it with tremendous interest—some of them hoping it would fail. Osborne was appointed warden at Sing Sing. He took the position against the advice of his friends who felt he might run afoul of politics. But he immediately set to work to build up the same type of inmate responsibility he had developed at Auburn. The guards were taken out of the shops and the discipline there placed in the hands of the elected delegates of the League. The atmosphere of the prison changed almost overnight. There was an inmate court that was responsible for all offenses within the prison. The unruly prisoner was disciplined by suspension from the League which automatically denied him certain privileges, such as "yard-out," attendance at movies, school, and the right to vote at elections.

But the new regime was not all ease and contentment. The Superintendent of Prisons interpreted Osborne's efforts as a form of "coddling." Newspapers criticized the new system of prison management. Finally, charges were lodged against Osborne and he resigned to stand trial. He was later vindicated but the publicity left its mark. He returned to Sing Sing but did not remain long. He resigned in disgust as politicians who were out to discredit him and his work made his efforts distasteful.[6]

There are many criticisms of any form of self-government in a prison. First we might state that it should not be applied to all inmates now imprisoned in our penitentiaries. One weakness of the Osborne scheme, which he contended was not "self-government" but rather training for citizenship, was that he applied it to the nonreformable as well as the reformable. At the time Osborne administered the system, he could scarcely do otherwise, because he had few facilities for classifying and segregating the prisoners. Nonreformable inmates must be ruled decently but firmly in a separate institution. The self-government plan, if it is used at all, should be utilized only with the distinctly reformable types. Even these should not be put into the system of democratic control immediately upon entry into the prison. They should go through a preliminary period of observation, isolation, and supervision. The ability of a prisoner to get along well under self-government should be one of the chief tests of his progress toward reform and of his fitness for release. Any man who cannot do well in the prison democracy should not be set free to live in the more difficult situation outside prison walls.

Secondly, self-government by prison inmates is exposed to the danger that a few will use important offices for personal advantage. This is likewise true outside prisons. It can be coped with in a free society but it is difficult to surmount when found inside prison. Self-government is theoretically sound, but in practice it can rest only upon a satisfactory personal relation between the inmates and the prison authorities.

It is doubtful whether self-government will work in any institution that is as bankrupt in program as most of the large maximum-security prisons are today. In medium- and minimum-security institutions, the better reformatories and prison camps, managed by officers of marked ability, it might conceivably work; but with large populations of all stripes and brands, it will rarely prove successful. It is obvious that there is more to reformation or rehabilitation than the observance of some sort of self-government. Honor systems will work only when those who possess zeal, conviction, energy, and honor are placed in charge of our prisons and are permitted to experiment.

Perhaps the most important experiment in institutional management

---

[6] For a complete story of Osborne's work, see Frank Tannenbaum, *Osborne of Sing Sing* (Chapel Hill: University of North Carolina Press, 1933); Rudolph W. Chamberlain, *There Is No Truce* (New York: Macmillan, 1935).

since the days of Osborne was that inaugurated in 1927 at the Norfolk, Massachusetts cottage prison. It was conceived and directed by the first superintendent, Howard B. Gill. He set up a system of "cooperative self-government." With the injunction of "No Escapes! No Contraband!" the inmates, stimulated by the superintendent's interest in self-expression, elected an Inmate Council to treat with him. The system thus started as a practical means of helping to control escapes and contraband. Later it developed to include expediting the construction program. In its raw beginnings, it was merely an arrangement between the inmates and the superintendent, but very quickly the weakness in this was apparent, and it was expanded to include *all* the staff members, especially those affected by the various institutional activities, and the house officers. Only the guards or watch officers were not included in the system.

Individual disciplinary cases were not handled by the Inmate Council, although it considered *general policies* of discipline. Occasionally the chairman of the Council would act as go-between to help clear up a disciplinary situation such as those involving an escape or the introduction into the institution of contraband. In time, subcommittees, composed of representatives of the inmates and members of the administrative staff, were appointed to take care of the various maintenance jobs and of all disciplinary policies. In short, this form of intramural government was based on a *mutual* sense of responsibility. The joint committee of construction, for example, set about getting the prison wall built speedily and harmoniously for only $90,000, compared to well over a million dollars for the wall of some prisons built in recent times.

As activities within the institution increased, various committees were inaugurated, each dealing with some phase of life close to the inmates and each representing the inmates and the administration. Committees dealt with maintenance, education and library, entertainment, sports, food, prison publication, store, avocational, and the like. *Joint responsibility* was the keynote of the Norfolk plan. One unusual feature of the plan was that the custodial officers, or Watch, were not included but were purposely kept aloof from the inmates. Their sole duty was one of maintaining custody.

The Norfolk plan was not an honor system nor a self-government system. Inmates and staff worked together, equally responsible and equally represented, for the good of all. This plan, as developed by Superintendent Gill, differed from the Osborne system in that staff officers were made an integral part of the institutional government; Osborne alone assumed this rôle at Auburn and Sing Sing. Thus the Norfolk plan was not dependent on an outstanding leader who may pass out of the picture by removal or death, as was the case in the Osborne system. The staff officers at Norfolk had as much responsibility as the inmates.

Superintendent Gill thought of his institution at Norfolk as a "supervised community within a wall," a *community* prison. His philosophy of correctional treatment justified a walled institution, since it guaranteed safe custody and made possible a free community without undue restraints. The *community* contained an administration building, a hospital, a jail, a police station, a school, a church, a community building, shops of various types and sizes, 23 separate living units—each designed for 50 men and two house officers and each having its own dining room, "tinkering" room, individual rooms (not cells), bathrooms, and so forth—and six or seven major recreational fields with play space.

In this Norfolk community, life was as normal as possible. Men did not march to and from work, chapel, or other places; they were expected to be at their assignments as in the outside world. Prison uniforms were all but abandoned in favor of ordinary clothing. Men were allowed to say and do and have whatever was normal so long as they did not violate the two fundamental rules about escape and contraband.

Conclusion

Honor systems in various forms are scattered all over the country, most frequently on farms operated by state correctional programs or in the juvenile institutions where *honor cottages* are often the selling point of the reform program. Around Christmas time honor inmates are permitted to go home for a few days' vacation, notably in Alabama and Mississippi, where the governor grants this privilege to a few of the state's prisoners.

Here and there throughout the country may be found wardens experimenting with forms of self-government, limited in scope, in which inmate councils, elected by the prisoners, are given responsibility along various lines. Programs for "town meetings," types of recreational activity, prisoners' complaints, etc., are discussed by these councils. This is, of course, quite healthy, but it is a far cry from what is generally meant by self-government.[7]

We cannot help being suspicious of any *trusty system*. Self-government schemes will work only if the administration is honest and sincere and definitely recognizes the fallibility of human nature. If there is graft and shady politics in an institutional election, it must be dealt with patiently, for there is the same phenomenon in the free community outside the prison. How can we expect prisoners to eschew gang politics when we cannot rid our own municipal campaigns and elections from this evil?

Another element in human behavior that must be sympathetically dealt with in correctional institutions is *moral lapse*. In and out of prison there

---

[7] See Fred R. Dickson, Norman Fenton and Alma Holzshuh, "The Inmate Advisory Council," *Proceedings, American Correctional Association,* 1955, pp. 142-146.

are persons who in most situations will behave honorably but on some occasions find it difficult to measure up to their own ideals. In the free community we have to be tolerant of these moral lapses. The same tolerance must be afforded inmates in a prison. In children's institutions, many inmates just walk away and do not know why they do it. In adult prisons many rules are broken without any good reason. Culprits themselves frequently cannot understand why they act as they do.

Because so few administrators really understand men, it is only rarely that a system of self-government or an honor system deserves the name. They are in many respects mere travesties of the genuine article. We warn the reader to take magazine articles on such subjects with the traditional grain of salt, and to be critical of speakers who relate what is going on in various prisons or reformatories. It should be remembered that when such glowing tales are being related there are no inmates present to tell their version of the story.

31 ───────────────────────────────

Prison Visiting, Citizen

Participation and Inmate

Contributions to Society

You must give some time to your fellow man. Even if it's a little thing, do something for those who have need of help, something for which you get no pay but the privilege of doing it. For remember, you don't live in a world all your own. Your brothers are here, too. (Quoted in Federal Probation, *Vol. 20, No. 2 (June, 1956), p. 9.)*

ALBERT SCHWEITZER

1. The Prisoner's Family as Visitors

The practice of permitting prisoners to see an occasional friend or relative is an old one. The origin of the custom is obscure. We have records that visitors were permitted to call on immured prisoners in the Old Stone Prison in Philadelphia, erected in 1718. Visitors were later permitted in the Walnut Street Jail, the first penitentiary. In the minutes of the Board of that establishment we find as early as 1808 that "prisoners who conduct themselves properly and are diligent in their work" were eligible for visits from their wives, parents or children "once in three months by orders signed by the two visiting inspectors." All conversation between prisoner and visitor "shall be through both of the grates with a keeper at the entry to hear all that passes in such interviews." The time limit of each visit was set at fifteen minutes.[1]

No doubt visiting was introduced at the outset as a privilege, but it is generally agreed among most correctional people today that it is a prisoner's right. The question, however, has never been settled. In many prisons visit-

---

[1] From the Board's Minutes, March 14, 1808.

ing has merely been tolerated and some wardens have definitely frowned upon it. An early complaint made by a prisoner in the Eastern Penitentiary at Philadelphia, dated September 15, 1845, has the following to say about the few visits afforded the inmates.

> The inspectors say it is their desire to treat every prisoner with humanity and kindness and I believe this is fulfilled on their part, with one exception, and that is the visitation of prisoners. It appears from the latter end of the 25th chapter of Matthew that visiting the prisoner was of an importance inasmuch as for the neglect thereof an everlasting curse is pronounced in verse 31st. I will not, however, censure the inspectors here but I merely observe that if the prisoner's family and friends were more frequently suffered to visit him there would be fewer cases of insanity than now are. I am very confident that this assertion can be substantiated by philosophical reasoning.

The average person usually thinks of prison visitors as those who are closest to the inmate: parents, children, wives, relatives, and others definitely known to the man before his incarceration and closely identified with him. They bring him news from home. They try to bridge the gap between the free community he once knew and his lonely spirit enmeshed in problems they can only know superficially. It has been a tradition in prisons to regulate such visits rigidly, limiting them in number and in length. Closely supervised by custodial officers, they usually take place between finely meshed screens so that contraband articles such as drugs, small saws, or pistols may not be smuggled into the prison. Many prisons search the visitors prior to and immediately following visits. At San Quentin prison in California there is a fantastic fluoroscope machine to which each visitor is subjected. This penetrates right through the individual so that it is impossible to smuggle anything into the inmates.

Those who have witnessed a visiting day in a prison agree that the bedlam due to several visitors talking loudly, sometimes in several dialects and languages, is not a very wholesome picture. To the credit of some wardens, this madhouse type of visitation has been modified by providing desks or tables, or in unusually progressive institutions, small cubicles, where visitors may sit comfortably and talk without screens.

Almost never are facilities available for the children who accompany their mothers. After the initial greeting children are left to shift for themselves. Some provision, such as playground equipment, should be installed and, if possible, volunteer supervisors placed in charge. Some community women's organization could be induced to render this service if only prison officials would take the initiative.

In recent years the Federal Bureau of Prisons has humanized visiting in some of its minimum- and medium-security institutions by installing comfortable lounges and easy chairs in the visiting rooms. But guards are still present.

Plan of "tough" visiting gallery.

The Federal Bureau of Prisons, in its *Handbook on Correctional Institutions, Design, and Construction,* recommends three types of visiting room equipment, each adapted to the type of custody for which the prison is built:

1. The informal room: for minimum-security prisons—davenports, large easy chairs arranged in such a way as to insure ease of supervision.
2. Modified informal room: visiting table with a centered division extending from floor to 4 to 6 inches above table; the sole purpose of the table to prevent the passing of contraband articles such as narcotics, knives, etc.
3. Formal room: (a) counters table height on each side of a metal partition with large panels of stainless steel wire screening inserted between visitors and inmates for vision and sound transmission, or (b) counters table height, on each side of a metal partition with large sheets of tempered plate glass inserted in the partition for vision and with voice-powered telephone handsets for voice transmission.

A study of the mechanics of prison visiting by members of the prisoners' families was made some years ago. Of the 38 institutions that replied to a questionnaire, half reported that visitors were still separated from inmates

Relaxed type of visiting room. Picture posed by personnel and their families.

Another type of prison visiting.

by screens. In most prisons where the screen has been eliminated, inmates and their visitors exchange talk at a table. In some few institutions visitors and inmates sit in a bullet-proof room, separated by steel partitions equipped with amplifiers.

As to length of visits there was a wide disparity—20 minutes to all day. The latter situation occurs in a few institutions where special provision is made for visitors who come from a long distance. Frequency of visits ranges from once a month to twice a week. There is little uniformity in prison visiting. As to the number of visitors at a time there is also a wide range. Four prisons restrict the number of visitors at a time to two; five to three; two to four; and two prisons allow more than four at a time.[2]

It is possible that local conditions make this wide disparity necessary. But it appears more reasonable to believe that where restrictions are so obvious, administrators are indifferent or fearful of adopting a more liberal plan.

Visiting facilities, California Institution for Men, Chino.

Perhaps the most striking innovation in prison visiting in this country is to be found in the California Institution for Men at Chino. Under the initiation of the first superintendent, Kenyon Scudder, inmates are able to greet members of their families in civilian clothes without any restrictions or "shake-downs." Picnic benches are provided and every Saturday afternoon hundreds of women and children visit their husbands and fathers. This institution is minimum-security in its nature, but there is no reason why visiting approximating this humane practice cannot be developed in all but the clear-cut maximum-security prisons.

2 For details see Negley K. Teeters, "Prison Visiting in American Prisons: A Preliminary Note," *Prison Journal,* Vol. 20, No. 1 (January, 1940), pp. 35-37. See also Eugene Zemans and Ruth Shonle Cavan, "Marital Relationships of Prisoners," *J. Crim. Law,* Vol. 49, No. 1 (May-June, 1958), pp. 50-57.

2. Conjugal Visiting

Periodically one reads of sexual or "conjugal" visiting in some of the Latin Countries. This sensible practice is consistent with the mores of several countries but it would be inappropriate in our Anglo-Saxon culture. But the same objective, however, could be realized by granting furloughs periodically to selected inmates so they could visit their families. This innovation would not violate our traditional mores. The granting of furloughs, which is actually not unheard of in American prisons, could be developed through a screening process placed in the hands of the institutional clinic. Safeguards against favoritism would, of course, also have to be established. Often, in the past, in some of the states where Christmas furloughs were granted, some of the fortunate inmates were selected by dubious means.

In Sweden a furlough is the inmates' "right," not a privilege. Risks are naturally involved, but authorities in that country maintain that the advantages of the practice far outweigh its disadvantages. Ireland and Belgium also maintain a system of furloughs to certain prisoners who have "earned" them or who are close to the time of their discharge.

Just how prevalent is conjugal visiting? The practice exists, or has existed in several Latin American countries: Salvador, Mexico, Colombia, Argentina, Brazil, and perhaps others.[3] Since the privilege is sometimes abused it is withdrawn and later, perhaps, reinstated.

In Colombia the inmate leaves the prison under guard, wearing his civilian clothes, and meets his wife at a certified rooming-house or in his own home if he lives in the city where the prison is located. He may remain two hours. The couple has complete privacy. A single man may visit a prostitute. In Brazil small rooms are in the front of the prison, specifically set aside for the purpose of the visit. Prostitutes are banned. In a new institution erected in Mexico City and known as *"Fabrica de Hombres Nueves"* (factory of new men) a special hotel-like building was erected for overnight visits of the men's wives. This is likewise true in the progressive Mexican *Islas Marias* prison colony in the Pacific ocean. Prisoners are permitted to have their families with them in this establishment if they so wish. Of the 600 prisoners incarcerated there, 150 have their families with them.

A policy of permitting the families of prisoners to move to the prison compound has long been in operation in several countries. It was the

---

[3] On Salvador, see *Prison Journal*, Vol. 12, No. 3 (July, 1932), p. 17; on Mexico, Norman S. Hayner, "Mexican Prisons Allow Conjugal Visiting," *Life*, Vol. 11, No. 17 (October 27, 1941); Donald P. Jewell, "Mexico's *Tres Marias* Penal Colony," *J. Crim. Law*, Vol. 48, No. 4 (November-December, 1957), pp. 410-413, on Colombia and Brazil, see Negley K. Teeters, *Penology from Panama to Cape Horn* (Philadelphia: University of Pennsylvania Press, 1946), pp. 33-34, 86-88, 236; on Argentina, *Prison Journal*, Vol. 31, No. 4 (October, 1951), p. 97.

practice, at least during the 1930's, in the U.S.S.R., especially in the Bolshevo prison near Moscow (see Sanford Bates, *Prisons and Beyond*, Macmillan, 1936, pp. 220-221). Families of prisoners came to live at the Nevez prison at Bello Horizonte, Brazil, but the privilege was never considered very popular.

There are many practices in the correctional field that seem to work well in other countries yet would not be popular, or even acceptable, in the United States. It is quite doubtful if members of the families of prisoners would wish to "move in" to a prison compound, even if houses were provided.

Perhaps the most dignified type of conjugal visiting was established in Argentina in 1947. A specially built structure was set aside for the purpose in the National Penitentiary in Buenos Aires. Each inmate who maintains good behavior is entitled to periodic visits from his wife, if she agrees; the visit must be volitional on the part of both. An intricate architectural plan provides scrupulously for privacy and staff members detailed for the maintenance of the system are specially selected. Women staff officers greet, examine, and brief the females while male officers handle the inmates as they emerge from the prison proper. The rooms are attractively furnished, each with a private bath and toilet.

One hears rumors that here and there, in our own country, clandestine conjugal visiting sometimes occurs. In no state is the practice permitted by law, yet reliable correctional authorities report that it has existed for many years in the Parchman, Mississippi, state prison. But if conjugal visiting in this country is ever contemplated it must first be officially sanctioned by state law—which would be extremely difficult—and second be set up in a dignified manner so that proper safeguards to insure privacy could be developed and maintained.

But a system of furloughs home is not out of the realm of possibility. Correctional personnel are united in the thesis that one of the best guarantees for rehabilitation of the offender is a loyal and devoted wife and family. The furlough would be of great assistance in cementing and nurturing family ties and would, of course, go a long way in nullifying the disastrous effects of sexual starvation of prison inmates. If Sweden finds it helpful, it is certainly worth a trial in our own country.

3. The Professional Visitor

The second type of visitor is the professional, including representatives of various community agencies. The personnel of such groups run the gamut from sentimentalists to objective case workers. Staff workers of such organizations as the Salvation Army, prison welfare societies, churches, and missions all feel that they are making a definite contribution to the welfare of the prison inmate. In this group may also be included representatives of Alcoholics Anonymous.

It is not our purpose here to evaluate the work done by such groups, but a wise procedure would be for the administration of the prison to require periodic reports from such visitors. Much emotional harm can be done the individual prisoner by promiscuous unsupervised visitors. Tactless remarks regarding religious or sentimental appeals to "loved ones at home" may develop unhealthy guilt feelings. Tampering with the personality of an individual is dangerous and thus supervision or some type of control is desirable within the prison setting. Those who represent a more professional approach to adjustment may be encouraged; and those whose technique is questionable by any reputable standards should be discouraged if not eliminated. Many institutions are bedeviled by this problem.

4. The Lay Visitor

The third type of visitor is the layman. This practice of visiting prisoners by outsiders who have interest in their plight goes back to antiquity as witness the Biblical verse "I was in prison and ye came unto me." Certainly there is nothing new in lay visiting, since it has been carried on in British jails and prisons for 200 years.

Perhaps the most famous lay visitor was Elizabeth Gurney Fry (1780-1845), known as "Newgate's Angel."[4] Others who have remained obscure, some of whom have pre-dated Miss Fry in lending aid and succor to the prisoner, were Jonas Hanway (1712-1786), James Neild (1744-1814), Thomas Shillitoe (1754-1836) and the humble Sarah Martin (1791-1843). These persons are mentioned since they have been badly neglected by history.[5] None of these humanitarians belonged to any professional group but all had a deep concern for those who languished in prison. The great John Howard, whose indefatigable prison visiting of European prisons we described in an earlier chapter, is the best known of all visitors. He, however, was more concerned with improving prison conditions than in taking up the problems of individual convicts.

But friendly visiting began early in the United States also; at least in the city of Philadelphia when prominent citizens made visits to the Walnut Street jail as early as 1776. They lent material aid rather than counsel and spiritual comfort.[6] After the Revolution the Philadelphia Society for Alleviating the Miseries of Public Prisons (established in 1787) continued to carry on the visiting program started before that great conflict began.

---

[4] Patrick Pringle, *Prisoners' Friend: the Story of Elizabeth Fry* (New York: Roy, 1953).

[5] For an account of the labors of Hanway and Neild, see Edward G. O'Donoghue, *Bridewell Hospital, Palace, Prison, Schools from the Death of Elizabeth to Modern Times,* Vol. 2 (London: John Lane, 1929); of Shillitoe, see Auguste Jorns, *The Quakers as Pioneers in Social Work* (New York: Macmillan, 1931); of Sarah Martin, see John Watson, *Meet the Prisoner* (London: Jonathan Cape, 1939).

[6] See Negley K. Teeters, "The Philadelphia Society for the Relief of Distressed Prisoners," *Prison Journal,* Vol. 24, No. 4 (October, 1944), pp. 452-460.

Lay prison visiting gained a decided momentum when the Eastern Penitentiary of Pennsylvania was opened in Philadelphia in 1829. Established on the principle of separate confinement, one prisoner from the other, a keystone of that system was *prison visiting by laymen,* and the Philadelphia Society developed an elaborate system of visiting.

The Visiting Committee of the society was divided into small groups, one of which visited the inmates of each cell block. The management gave these men access to the cells so that they could have direct contact with the men in private. The visits were, in most cases, relatively short, about 15 minutes; some were longer, embracing as much as an hour. The modern criticism of these visitors would be that their approach was too often sentimental and religious, and that they were often hoodwinked by the more wily convicts. But the visitors made many contributions to the physical and social welfare of the immured inmate that deserve a word of belated and unstinted praise. There is little volunteer work done in our time that can compare favorably with what was done by these visitors, all busy men of the community.

These early visitors to the Eastern Penitentiary served as a check on the warden and overseers. Very early in the history of the institution the visitors occasionally noted that two convicts were placed in the same cell. Since this was a direct violation of the law, the visitors called upon the authorities to explain this irregularity. The visitors maintained a genuine interest in the comfort of the inmates, concerning themselves with the quality of food served, clothing, heating of the cells, sanitation, medical care, and anything bearing directly with the well-being of the prisoners.[7]

5. Citizen Participation

Lay visiting merges into citizen and community participation.[8] This is a field that has been somewhat spotty as we survey the development of prisons. Here and there we see a lively program of activities developed and participated in by prisoners with the aid of interested citizens who give of their time and abilities to make incarceration more bearable. Perhaps the first large-scale citizen participation on record was that developed soon after the opening of the Elmira Reformatory in New York State. Citizens of the community and surrounding neighborhood were enlisted to conduct classes in Brockway's experimental prison.[9]

---

[7] See Negley K. Teeters and John D. Shearer, *The Prison at Philadelphia: Cherry Hill* (New York: Columbia University Press, 1955), pp. 161-169. For more details on lay visiting, especially in England, see Watson, *Meet the Prisoner,* and Gordon Gardiner, *Notes of a Prison Visitor* (New York: Oxford University Press, 1939).

[8] We are aware that another area of citizen participation involves active cooperation with prison reform groups, assisting in progressive legislation and other such activities. We do not propose to discuss this type here.

[9] These early days are excitingly recalled in W. Charles Barber, "The Dream of Z. R. Brockway," *Correction,* New York State Department of Correction (February, 1957), p. 3 f.

During these past 80 years many correctional institutions have set up classes in various subjects from academic courses to arts and crafts, from model airplane classes to sculpturing and portrait painting. College glee clubs and debate teams have performed in prisons; lecturers and outstanding musicians have appeared before inmate assemblies. Naturally some wardens capitalize more on this type of participation than do others. Admittedly there is an element of risk involved since there have been a few instances in which a small core of inmates have taken advantage of some outside group and held a few hostages. But more and more progressive correctional administrators are exploring this area and thus enrich the otherwise drab routine of the modern prison.[10]

6. The Inmates' Participation in Community Services

Back in 1894 an English prison administrator, Major Arthur Griffiths, entitled a two volume work on prisons, *Secrets of the Prison House*. This was an apt title since, traditionally, the prison has been considered both by its administrators and the public as a "secret place." Aside from Brockway at Elmira, it is only recently that a few courageous wardens have attempted to acquaint the public with their day by day problems. They have done this at the risk of being ridiculed as "publicity hungry." Formal reports from prison boards or departments of correction are seldom read by citizens so that few, aside from professional correction people, actually know what goes on behind prison walls. Practically nothing, aside from a riot or escape, ever gets into the newspapers. Because of the secrecy fostered by most wardens, few citizens care so long as the inmates are kept secure.

Those who know prisons and prison inmates are familiar with the many contributions to the well-being of society made by hundreds of self-sacrificing inmates, in most cases anonymously. In the field of medical research alone many long-termers and lifers have time and again exposed themselves to disease and possible death for the good of humanity. Their motives no doubt vary; perhaps no different from the reasons free people make sacrifices for their fellow-men. However, in no case is a prison inmate promised a pardon, commutation, or early parole for volunteering in medical experiments. Austin MacCormick, noted penologist, attributes some prisoners' sacrifices to a "social conscience which many of them do not realize they have"; Warden Joseph Ragen of the Illinois state prison has stated that "men are not entirely lost if they are willing to help other people." Conversely, a pertinent remark made by one correctional official after

---

[10] For details concerning such a program set up in the Oahu prison in Hawaii, see Joe E. Harper, "Community Participation in the Prison Program," *Proceedings, American Correctional Association*, 1956, pp. 33-36; also *Handbook on the Inmate's Relationships with Persons from Outside the Adult Correctional Institution* (New York: American Correctional Association, 1953).

enumerating the contributions of inmates in his prison: "While not meaning to discredit the seemingly apparent magnanimity intrinsic in the nature of such activities, it must, nevertheless, be kept in mind that there are, in this prison and most likely in every prison, any number of inmates who participate in avowedly worthwhile projects out of no higher motive than simple curiosity or of seeking a novel way to spend an otherwise boring afternoon."

In an informal survey, by one of the writers, of "extracurricular" activities undertaken in American prisons, one official very frankly stated: "We regret to inform you that we cannot discover an instance in the past five years that would warrant your consideration." Another official stated: "This is to advise you that there are no known instances of outstanding services rendered by any inmate of our prison system." But in the majority of state prisons it was found that many inmates were actually donating their time and their bodies to medical science, in addition to donating thousands of pints of blood to the Red Cross.

There are three areas in which prison inmates are engaged in making contributions that have great meaning to the free society: (1) medical research, (2) community and public service, and (3) cultural.

Medical Science

Ever since World War II thousands of inmates in hundreds of correctional institutions have contributed vast amounts of blood for Red Cross blood banks and in many instances special blood types have been furnished hospitals for emergencies. Some inmates sell their blood; others refuse to take money. They differ in no way from those in free society. The calls for skin to be grafted on persons needing it desperately are many and such requests are courageously heeded by inmates in many institutions. The prisoners in the Ohio penitentiary, for instance, have donated over two thousand square inches of skin to persons who were badly burned in fires.

Medical experiments that have been conducted since the last war in American prisons have involved, among other objects, a vaccine for tularemia, or "rabbit fever"; cancer, malaria, leukemia, radiation burns, cardio-vascular, sensitivity tests for allergies, irritating deodorants, effectiveness in relieving pain of various drugs; and diets.

Community and Public Service

This represents a wide area and many projects are carried on continuously by men and women in prison. Fire-fighting, aside from being a definite type of work responsibility—notably in California—is not unknown in many prisons. While California's prisoners have contributed more than 200,000 man hours in fighting nearly 300 forest fires since 1953, much of it is actually an integral part of the rehabilitative program of the several camps and prisons. It is a valuable and even dangerous activity.

The men's lives are placed in jeopardy and several have actually lost their lives in fire fighting. Flood control and rescue work is another area of community service needed in many states and participated in by hundreds of inmates. For instance in Marietta, Ohio, prisoners fought floods for many hours and salvaged the belongings of the victims. The same type of work was done by inmates of the Iowa prison at Fort Madison.

We also find toy-making and toy-repairing in many state prisons. Hundreds of inmates spend thousands of hours in this activity. The Salvation Army supervises much of this work in the various prisons. Other projects worth recording are listed here.

Seagoville Federal Institution, Texas: the inmates contribute $10 per month to help "possible juvenile delinquents" adjust to community life through identifying with clubs or other character-building agencies.

Leavenworth Federal Penitentiary: the inmates here have adopted two children, one a two-year-old Korean child, the other a twelve-year-old French boy.

Iona, Michigan Reformatory: the inmates donated over $400 to the Foster Parents' Association of New York City in order that they might "adopt" a child in Greece; this is remarkable considering the meager wages prisoners receive for their work.

Virginia prisons: in the women's prison the inmates have sent many CARE packages overseas at their own expense. One of the institutions has an obsolete but serviceable fire engine for its own protection. The institution is located in a rural area that does not have adequate fire equipment. On several occasions the fire engine, manned by inmates, has been used to fight community fires and has been the means of saving considerable property. One or two missionaries are partially supported in foreign fields by religious groups in Virginia's prisons. In one of the state's institutions a little girl was rescued from quicksand by several inmates. They formed a human chain and were thereby able to extricate the child from certain death. Several men awaiting the death penalty have contributed the corneas of their eyes that others might see; also one man donated his aorta (the large artery of the heart) to be removed after his death.

Maine: aside from conditioning toys for service clubs, Maine prisoners have volunteered for hurricane and disaster work; they also respond generously to charity drives.

Marquette, Michigan: the prison vocational school reconditioned sleds and skis for the children of the Holy Family orphanage at Marquette. An inmate at the prison's camp took a rowboat alone in choppy, icy water of Lake Superior and rescued two fishermen.

McAlester, Oklahoma: occasionally inmates contribute defense funds for condemned men who have exhausted their own funds; money has also been raised for illnesses among families of inmates. Three musical organizations in the state's prison give public concerts for certain organizations.

Money collected is used to buy school uniforms, both musical and athletic. The men receive no money for their efforts.

Iowa State Penitentiary, Ft. Madison: in the institution's excellent paper, *The Presidio,* a column entitled "On the Credit Side," frequently lists prisoner activity in community service. For the past several years an all-inmate musical show is produced to which the public is invited. Proceeds go to the Shriners' Hospital for Crippled Children. It is not unusual for as much as $2,000 to be netted for this charity. About 75 inmates participate, the concert gives six performances and some 6,000 people from the community attend. The inmates also give to flood relief; they collected $250 one year to help a stranded migrant family that was found in the community in a destitute condition. Two "anonymous" inmates have produced handsome hand-tooled leather goods which they turned over to the Salvation Army for sale. They received none of the proceeds themselves.

Wethersfield, Connecticut: The inmates took up a collection of over $200 as a contribution to the Hartford summer camp. It was made as a memorial to a man who had lost his family in a fire. As the *Monthly Record,* the inmate paper remarked: "The present publicity of our donation, a front page story in the Hartford *Courant,* helps to counterbalance the unsavory publicity so often given to the inmates here and to ex-cons in general."

Cultural Contributions

Many great books and essays have been written by persons who "languished" in jails and prisons. One need only mention John Bunyan's *Pilgrims Progress,* written while he was confined in Bedford Gaol in England, or Oscar Wilde's *Ballad of Reading Gaol.*

In our own time Jan Valtin, who wrote a best-seller *Out of the Night,* admitted he improved his time while in San Quentin prison by using the library facilities in his spare time, and of course, Sidney Porter, as O. Henry, mastered the art of the short story while in the Ohio penitentiary at Columbus. Others were Alexander Berkman, author of *Prison Memoirs of an Anarchist,* Kate O'Hare, who wrote *In Prison,* Victor Nelson, author of *Prison Days and Nights,* Edna O'Brien, who wrote *So I Went to Prison,* and more recently, Helen Bryan, author of *Inside,* although most of these contributed to the literature after they were released.

The writings of Tom Runyon, an inmate of the Iowa Penitentiary at Fort Madison, are well known to many people who know prisons. His *In For Life* has been characterized as possessing more insight into the problems of men in prison than any for many a decade. Runyon was serving a life sentence with a Federal detainer for another possible life term. His case caused considerable public attention because of the interest of the detective-story writer, Erle Stanley Gardner and his Court of Last Resort.

In the midst of the efforts of this group and many others who were vitally interested in Runyon's release he died suddenly, in 1957. He found himself as a person while in prison, due to the efforts of an understanding warden, and wrote many short stories for the popular magazines. He was also responsible for elevating the prison's publication, *The Presidio,* to one of high rank throughout the prison world.

Another prison inmate, Caryl Chessman, wrote a book that was on the best-seller list for many months, *Cell 2455, Death Row* (1954). Subsequently his *Trial By Ordeal* (1955), and *The Face of Justice* (1957) were published.[11] All were written while he was fighting for his life against a first degree kidnap death sentence in San Quentin's death house. We need not agree with this psychopathic criminal in his rationalization of his life's misdeeds, to marvel at his brilliance and literary gift.

An Ohio penitentiary inmate, publishing under the pseudonym of Stephen Schmiedl, wrote *Bighouse Banter* (1952) which deals with that not-too-rare commodity, prison humor. The author, who has spent over thirty years in prison and who lost his right arm at the age of ten, claims his literary talent is the result of having been placed in the same cell that was occupied many years ago by O. Henry. When, for some reason, his housing was changed, he seems to have lost the "magic touch." In the same prison at Columbus, Ohio, another inmate has fourteen original popular songs under copyright and at the same time has obtained patents for surgical instruments he has invented. The Ohio officials reported that several other inmates have marketed paintings, short stories, and articles in recent years. The Virginia prison at Richmond also has an inmate, signing himself Paul N. Deadwiley, who published a book in 1955 entitled *Sin Street.*

Two unusual contributions from inmates in the Tennessee prison at Nashville are of interest. Some years ago Frank Grandstaff, a fourth offender in for life due to the state's habitual criminal law, composed a cantata about Big Springs, Texas. At the request of the governor of Texas he was granted a leave from the prison to go to Big Springs to conduct the premiere. In 1950 Governor Gordon Browning granted Grandstaff a pardon, partly due to the interest of many citizens of Texas. The governor, in the pardon, stated: "It is true that he was a repeater. All of his offenses had been relatively small, about the largest thing he was ever guilty of taking was a radio. He has served ten years."

The other note of interest from the Tennessee prison is that two Negro songwriters published "Just Walking in the Rain" which was a top hit tune in 1956. These men are Jonny Bragg and Robert Riley.

An inmate in the Virginia prison received the Dr. Christian Award for a radio script he submitted. A prisoner in the South Carolina prison de-

---

[11] All of these are published by Prentice-Hall, Inc., Englewood Cliffs, New Jersey.

signs exceptionally true models of early side-wheelers that once plied between New York and Albany—the area in which he was reared—and they have been placed in a New York State museum.

These are samples of what many prisoners do in their spare time and, in most prisons, they have plenty of that commodity. There is no doubt that what is recorded here represents only a small part of the worth-while activity of the prison inmates of this vast country. Literary and other cultural talent is found in all walks of life and, of course, this statement applies to those who have violated the law and have been sent to prison. The public seldom gives much thought to this rather important phase of prison life.

7. The Penal Press

Just when the first inmate newspaper appeared in prison is debatable. It is believed that the Elmira, New York Reformatory published the first paper, a four page affair called *The Summary,* on November 22 (Thanksgiving) 1883. Four years later, in August 1887, the first paper to be published in a penitentiary appeared in the Minnesota prison at Stillwater. Known as *The Mirror* and started by a few inmates, its objectives were: "to be a home newspaper; to encourage moral and intellectual improvements among the prisoners; to acquaint the public with the true status of the prisoner; to desseminate penological information, and to aid in dispelling that prejudice which has ever been the bar sinister to a fallen man's self-redemption."[12]

There are at present approximately 200 papers and magazines printed in reformatories and prisons, ranging from crude mimeographed sheets of two or four pages to elaborate newspapers and slick-paper magazines. Notable among the latter is *The Atlantian,* published in the Federal Penitentiary at Atlanta, Georgia, *The Presidio,* at Fort Madison, Iowa, and *The Eastern Echo* at Philadelphia. According to a recent survey of the prison press, these publications may be divided into two groups: (a) publications expressly written for the inmates; and (b) "front" publications, which may appeal to inmates as well as to the outside public—a sort of "public relations" medium.[13]

The prison press affords an opportunity for inmates to release their latent talents along creative lines, such as short stories, jokes, editorials,

---

[12] From an early account of the penal press, prepared by Mrs. Isabel C. Barrows, "Periodicals in Prisons and Reformatories," in C. R. Henderson (ed.), *Correction and Prevention,* Vol. 2 (New York: Russell Sage Foundation, 1910), pp. 236-260.

[13] Walter A. Lunden and Oliver A. Nelson, "Prison Journalism," *Prison World,* Vol. 12, No. 3 (September-October, 1950), pp. 7 ff. For cogent comments on the status of the penal press and its editors, see: "The Penal Press," the *Prison World,* Vol. 13, No. 1 (January-February, 1951), pp. 4 f.; "The Penal Press," *Prison Journal,* Vol. 35, No. 1 (April, 1955).

poetry and dramatics. Sports events are unusually popular when written up in these papers. Articles on prison life and crime show up in from 23 to 50 per cent of the space.[14]

As an analysis of the penal press we can not improve on the following editorial, written by the editors of *The Howard Times,* published in the Rhode Island state prison, and labeled *Struggle of the Penal Press:*

> During the life span of nearly a century, penal publications have increased in quantity, seldom in quality. Today they are branded with the dubious distinction of never having accomplished the purpose for which they are intended—the reform of the penal system. It is still submerged in a morass of tradition imposed upon it by successive administrations appointed by many decadent systems of state and local governments.
>
> Most penal institutions condone the press only as a necessary evil. It exists because the administration which fosters a publication is considered progressive and in line with the modern trend in penal indoctrination.
>
> It is often referred to with pride. It should be encouraged as an actual step toward rehabilitation. In order to accomplish this, it would be necessary to lift the curtain of secrecy and censorship which is maintained.
>
> The most perplexing problem facing the penal press is how it can operate in an effective manner when it is forbidden the privilege of printing the truth. Restrictions enforced, in addition to suppression of truth, quite often are a limit of their circulation by curtailed mailing lists and outright refusal to place the printed issues in competition with other news material.
>
> Usually a limit is set as to the use of the facilities and materials to maintain a regular schedule. Because of these restrictions and strict manuscript censorship more than two hundred publications have been doomed to the role of silent ghosts in a country which prides itself as a champion of a free press.
>
> In some instances, methods are used to circumvent at least a few of these restrictions in order to present the plight of over one hundred and fifty thousand men and women confined in the United States. Usually these attempts are apparent only to those who are schooled in subterfuge.
>
> Occasionally, the penal press will seem to win a minor point in its constant and unequal fight for humane justice. Whether the credit is rightly theirs is debatable. However, it does spur them to greater efforts and gives them hope and encouragement. Every penal publication is constantly trying to rise above the adage that they are the "stepchild of the fourth estate."
>
> Do not doubt the intrinsic worth of penal publications. The material used often comes from the frustrated pen, but it also comes from a contrite heart. If you will look close enough, you will have had a glimpse into the soul of suffering humanity.
>
> The penal press will show its true worth when encouraged, and will fail so long as it is suppressed.[15]

We see from the above that those who edit and publish the papers circulated in prisons find their mission sharply curtailed by the administration. Yet it is surprising how free many of the publications are. Some of the age-old questions of prison administration are freely discussed in some papers.

---

[14] Lunden and Nelson, "Prison Journalism," p. 8.
[15] March 22, 1949, p. 2. Reprinted by permission.

Such pressing problems as the quality of the food served, parole planning, loss of privileges, punishments, the program of activities, and the like, are considered. In some cases editors are permitted to interview visiting penologists, state officials, members of boards of pardons, and even to make surveys by correspondence, and thus print their frank findings where all inmates may read.

The prison press serves as a safety-valve. Aside from mirroring the events of the day, it can, and should, be the medium for the inmates to air their views on subjects closest to their experience such as crime, court trial, sentencing practices, prison program, and release procedures such as parole. The prison administrator who affords the editors of the penal press the widest latitude within the confines of good taste and recognized standards of enlightened journalism need have no fear that there will be riots or overt action from embittered prisoners.

32

Inmate Labor
in the Correctional Program

The old-line industrial prison is dead, and, in view of its glaring faults, its demise was warranted. The modern concept of prisons as institutions for treatment does not contemplate the "busy prison factory" or the self-supporting prison as a goal. Nevertheless, in any well-rounded program directed toward the needs of those confined, some employment projects have their place. (From "The Federal Government and the Prison Labor Problem in the States," Social Service Review, Vol. 24, Nos. 1 and 2, (March and June, 1950), p. 236.)

FRANK FLYNN

1. "Hard Labor" or "Productive Labor"?

While we read on numerous occasions of criminals being sentenced to a term of "hard labor," we know that aside from a few prison camps in the southern states there is no hard labor. Furthermore, there is very little productive labor in any of our prisons.[1]

The public believes, in theory at least, that prisoners should work—and work hard. It is actually ambivalent about the matter; when jobs are plentiful, prisoners should work but during periods of recession, criminals should not "take jobs from law-abiding citizens." While the stereotype of prisoners in striped garb dragging a ball and chain and "making little stones out of big ones" with a sledge is no longer accepted, there are many citizens who demand that prisoners work. It is repugnant to them that prison in-

---

[1] Of the 191,776 average number of prisoners under sentence for the year 1940, only 83,515 were productively employed. See "Prison Labor in the United States," *Monthly Labor Review*, U.S. Dept. of Labor, Bureau of Labor Statistics, September 1941, Table 3, p. 585. These are the latest over-all official statistics available and will be referred to in this chapter.

mates should live in idleness. Yet most of them do. Aside from the few who are productively employed, many are provided with "made work" or are in complete idleness.

It should seem obvious to any intelligent citizen that sustained idleness is extremely detrimental to public policy. Yet legislators during the past hundred years have seemed impervious to demands of penologists and prison administrators to develop a sane and productive labor policy for our prisons. Aside from the deteriorating influence of stagnating idleness on thousands of prison inmates, much of the spasmodic rioting and discontent in our institutions has been due to this situation. Every warden knows that if work is not provided the inmates, he is sitting on a powder keg, apprehensively waiting for the fatal day when the latent and unleashed energy and abject bitterness will rend the prison routine asunder. Many are the stories that could be told of thousands of dollars worth of machinery destroyed by rioting prisoners because they were forced into idleness. More tirades have been hurled at the apathetic public in conferences, after-dinner meetings, in books and pamphlets, describing the chaotic and demoralizing conditions in our prisons due to the problem of idleness, than on any other topic in the correctional field.

The public is no more to blame for this condition than for any other social problem. It does not understand the intricate ramifications of prison labor. Few citizens know that the prisoner is sitting in his cell or in an idle company, listening to the clock tick off the minutes and hours, day in and day out. If he does hear of such a situation, he may merely shrug his shoulders and inconsistently suggest that it serves the criminal right for violating the law. In his lack of orientation he cannot visualize the price *society* must pay for a short-sighted prison labor policy through increased taxes or the release of embittered prisoners.

Many reasons have been given at various times for working prisoners:

1. *As punishment:* it is assumed that if hard and onerous tasks are the lot of the convicted criminal, they will serve as a deterrent to crime.

2. *As discipline in prison:* if all inmates are at appointed tasks, regardless of what they may be, it is simpler to operate the institution and thus discipline is maintained.

3. *To relieve the monotony of a prison term:* it is axiomatic that time passes more quickly if men are busy.

4. *To reduce operating costs through the production of goods that may be sold:* this is a sound economic concept but, as we shall see, there are limitations to this theory.

5. *To assist inmates to aid in the support of their families:* by paying a slight wage for their work the prisoner maintains his self-respect and assumes his responsibility for supporting his dependents.

6. *To give the prisoner a small sum,* through wages however small, so he may purchase notions, candy, cigarettes, etc. from the prison commissary.

7. *To teach prisoners trades:* this is based on the assumption that those coming to prison are unskilled and the least the prison can do is to develop good work habits, manual skills and stimulate an interest in some trade.

8. *As a reformatory device:* work of some sort can be regarded as therapeutic and be of aid in restoring a man to society as a social asset.

As we trace the history of prison labor we shall see that each of these objectives has been used at one time or another, or a combination of two or more. It is because of these reasons that the problem of inmate labor is so complex.

Aside from arduous toil in the galleys, described in our chapter dealing with transportation, and other similar types of sheer drudgery, there was no meaningful labor for prisoners until the rise of the house of correction. This institution was established for the purpose of putting "idle rogues" and other elements of the social debtor class to work at productive labor.[2] With the rejuvenation of the Walnut Street Jail in Philadelphia in 1790, a progressive system of "contract" labor was installed. This system was heralded as the wonder of the age. For the first time in history, convicted felons were employed at dignified productive tasks and paid wages. The system used in this little jail served as a pattern for later penitentiaries.[3]

Amazing as this system was, it was conceived primarily to help pay for the prison although in the thinking of some of the inspectors its reformative value may have been apparent. We must turn to two European prison administrators to see how labor was conceived as a meaningful therapeutic instrument for the prisoner. An early form of vocational training was established during the 1830's by Col. Manuel Montesinos, director of the prison at Valencia, Spain, and Georg Obermaier, superintendent of the Kaiserslautern prison in Bavaria. The former took over an old Augustinian convent and supervised the work of some 1,500 prisoners.

---

[2] See pages 330-331.

[3] It was Robert J. Turnbull in his pamphlet, *A Visit to the Philadelphia Prison* (1796), p. 16, who records: "Some prisoners make from one dollar to a dollar and a half a day," and adds: "There was such a spirit of industry visible on every side and such contentment pervaded the countenances of all, that it was with difficulty I divested myself of the idea that these men surely were not convicts, but accustomed to labour from infancy." La Rochefoucauld, the French moralist, writing the same year in his description of the prison, *On the Prisons of Philadelphia by An European,* states: "Each convict has a book in which he enters his bargain with the outside employer, and in which his earnings are also set down in order. The convict's outgoings, whether an account of his prosecution, his fine, the price of the instruments which he breaks, or injures, of his cloathing, and of his board, are likewise set down in his book which is audited every three months in the presence of the inspectors." (Pp. 15-16.)

He installed forty trades by which the prison was made self-supporting. Obermaier was also a great advocate of trade training and his prison, too, was one of the finest on the continent. These two men developed trade training and productive work when exploitation of prisoners was the prevailing practice.

The situation in England was somewhat different. Overcrowding of jails, houses of correction and later, penitentiaries, forced penal authorities to develop unproductive machines such as the treadwheel and crank. These "engines of damnation" accomplished nothing in most instances, save grueling drudgery for their victims. The treadwheel, as invented about 1818 by the engineer, Sir William Cubbit, and presumably used first in the Suffolk County Gaol at Bury, consisted of 24 steps, fixed lengthwise, like the floats of a paddle-wheel, to a wooden cylinder 16 feet in circumference, the steps being eight inches apart. The wheel made two revolutions per minute and there was a mechanical contrivance by which, at the end of each thirtieth revolution, a bell rang; the twelve men at the wheel then stepped off and were replaced by twelve more. While off the wheel the men could doze or read, but not communicate. Sometimes these treadwheels did accomplish something such as grinding corn or pumping water; more often they did nothing. The effects of this device on the prisoners are described from an early source:

> The weariness of the employment results from two causes. First, the want of a firm footing for the feet—a want painfully experienced in walking through a deep soft snow; and secondly, the strength that is expended to keep the body from sinking with the step—which is equal to that re-

Treadwheel and oakum sheds, England; a familiar sight in early separate prisons.

quired to lift a man's own weight, say 140 pounds. . . . No wonder the prisoners maim themselves or feign sickness (at the rate of 4,000 instances in a twelve-month) to escape such a brutal use of their bodies.[4]

The treadwheel was resorted to in only a few instances in the prisons of this country. There is record that it was used in 1824 at the Simsbury, Connecticut copper-mine prison. Richard Phelps, writing about this bizarre institution, states: "Of all the labor required, the treadmill was dreaded the most, and the most stubborn were put to this employment. In extreme cases, one of the *lady-birds* [females] was put on the wheel among the men as punishment.[5]

Useless labor in early prisons in England—the "crank."

But even more widespread and diabolical, was the "crank," a device admirably adapted to cellular confinement. It was invented by a man named Gibbs at Pentonville, London, prison, probably about 1846. George Ives describes it:

The newest monster might be compared to a churn in appearance, or to a chaff-cutter, or twenty other things, made up of a case and a handle; it

---

[4] *Journal of Prison Discipline and Philanthropy* (January, 1857), p. 40.
[5] Richard Phelps, *A History of Newgate of Connecticut* (Albany: 1860), p. 77. This prison is described in Walter D. Edmonds, *Drums Along the Mohawk* (Boston: Little, Brown, 1936), pp. 99 ff.

was just a metal box raised to a convenient height by a support. It had a handle to be turned and a clock-like face upon one side to count the revolutions made. The requisite amount of resistance was secured by a metal band, which could be tightened with varying force pressing inside, upon the axle; it might perhaps be compared to a hand-brake sometimes seen upon the back wheel of a bicycle. The ordinary resistance was nominally from 4 to 11 lbs.

The usual number of revolutions required was 14,000 a day, being at the rate of 1,800 per hour. . . . At first it had been intended only for vagrants and short-sentence men, but after 1848 it tended more and more to oust labour altogether: a noble model for the admiration of God and the imitation of men.[6]

As time went on, the crank was improved by various ingenious prison administrators. A person could make about 10,000 revolutions a day with ordinary effort, with the resistance varying from five pounds for boys to ten pounds for men. Daily food had to be turned for—1,800 revolutions for breakfast, dinner, 4,500. Often men turned the crank for long hours after dark in their cells to make up for time lost during the daylight hours.

Other onerous tasks, albeit with some productive value, were "picking oakum," pounding hemp, and rasping logwood. "Oakum" is the fiber from old rope, often encrusted with pitch or tar which the worker untangled slowly and piled into a heap. It was used to caulk ships. Logwood was pounded into small pieces for the preparation of dyes. Picking oakum was widely used both in England and America and one may still see prisoners in Europe, working in their solitary cells engaged in this task.

2. The Systems of Prison Labor

Aside from agriculture, there have been six systems by which prisoners have been employed in this country. They are the contract—with its variation, the piece-price system—the lease, the public account, state-use, and public works and ways.

Development of Contract Labor and the Piece-Price System

The earliest form used was the contract and piece-price systems. The contract system involved the letting-out of the labor of the prisoners to an outside contractor who usually furnished the machinery and raw material and supervised the prisoners' work by men in his employ. The prison management had nothing to do with the operation, its responsibility being only to guard the prisoners. This system was particularly vicious, since the inmate workers were at the mercy of the contractor's supervisors who sometimes added to their petty graft by "short-changing" the inmate in his work tally.

---

[6] George Ives, *History of Penal Methods* (London: Stanley Paul, 1914), p. 190.

The piece-price system is merely a variation. In this, the contractor furnished the raw material and received the finished product, paying the prison so much for each unit accepted by him. Exploitation was minimized since the prison hired its own supervisors. Technically, it was the piece-price system that was used in the Walnut Street Jail in Philadelphia between 1790-1800.

The real beginning and expansion and high prosperity of the contract system did not come until the years 1825 to 1840. In the earlier epoch, sales were difficult because there was no well established connection between the prison shops and the customers in the outside community. Then, too, the Industrial Revolution, with the change from handcraft to power machinery, was well underway at these later dates.

But it was the rise of the merchant-capitalist in America, in the period following 1825, that breathed life into prison industry. Here was an intermediary who was glad to furnish the raw material and take the finished product at an agreed-upon rate. The system of contract labor was only one incident in the rise of the merchant-capitalist, of which home labor and the sweat-shop were other phases in his attempt to obtain cheap labor.

The prisons of New York State enthusiastically adopted this new method of working prisoners. As early as 1819, the Auburn prison was authorized by the legislature to provide inmates with "joint labour" by day, with solitary confinement by night. Deficits in operating prisons soon gave way to actual profits. By 1828 the Auburn and Sing Sing prisons were paying for themselves.

Other states, impressed by the financial results obtained by New York's prisons, introduced the contract system in their establishments—but Pennsylvania still frowned upon it. Dedicated to the principle of separate confinement with each prisoner in his own cell, this state could scarcely take advantage of congregate labor and machine production that was imperative if contract labor was to be exploited to its full advantage. This inability of the Pennsylvania System to pay for itself was one of its main weaknesses in a country frankly espousing rugged individualism.

Aside from the partisan criticisms hurled at the contract system by advocates of the Pennsylvania System, there was little opposition to this new philosophy, but it was not organized, and practically no attention was paid to it. Here and there, where trade unions were strong enough to be articulate, resolutions were passed condemning the system as being unfair to free labor.

By 1870 the Industrial Revolution had become deeply rooted in the economic development of the country. This created a much larger industrial proletariat and weakened the influence of the merchant-capitalist. The greatly increased laboring class began to build up its strong national organizations and eventually became a sufficient power to command the attention and solicitude of astute politicians. Their old enemy, the contract prison labor system, was a special target of attack. With the formation of the American Federation of Labor in 1880, labor and its political lobbyists went all-out to destroy all vestiges of the system.

The abolition of the contract system was desirable, but the pendulum swung too far toward state control and the fatal restriction of prison labor, to the detriment of the state and the prison inmate and with no corresponding benefit to free labor. To replace the abuses of "forcing" labor under the contract system the tendency was toward an expensive and demoralizing idleness.

Aside from the vocal objections of free labor and the politician to the contract system, the humanitarian also had a valid criticism. The prison factory had exploited the prisoner; and according to the philosophy of the times, he deserved to be exploited, or at least, he needed strict discipline and hard and onerous manual labor for deterrence and reformation. A few intelligent ex-prisoners exposed the evils of the system. Kate Richards O'Hare, political prisoner in the Missouri State Penitentiary during World War I, pictured the system as one of rank exploitation and brutality, motivated only by the greed for profits at the expense of the helpless prisoners.[7] Another excellent description of the contract system with its task, or stint, is given by Alexander Berkman in his *Prison Memoirs of an Anarchist.*[8]

Although John Howard and other early prison reformers had believed in labor for prisoners, they were strongly opposed to exploitation. Howard believed "that there is a great error in expecting that a prison should be made to maintain itself since with the strictest economy, a considerable annual sum will be found necessary for its proper support." Some of the early administrators saw the dangers and abuses in the contract system after they had seen it in operation for some time. Amos Pilsbury, warden of the Wethersfield, Connecticut prison had praised the prison factory, but he changed his mind and regretted "that he had ever paid into the public treasury the earnings of prisoners instead of using the surplus funds for the prisoners' benefits."[9]

The Lease System

Another variation of the contract system, one even more vicious, was the lease system. Under it the contractors assumed entire control of the

[7] Kate Richards O'Hare, *In Prison* (New York: Knopf, 1923), pp. 100-107.
[8] Alexander Berkman, *Prison Memoirs of an Anarchist* (London: C. W. Daniel Co., 1926), p. 134. Also published in New York: Mother Earth Publishing Association, 1912.
[9] Z. R. Brockway, *Fifty Years of Prison Service* (New York: Charities Publication, 1912), p. 33.

prisoners including their maintenance and discipline, subject, of course, to the regulations fixed by statute. The prisoners were removed from the prisons and employed in such outdoor labor as quarrying, agriculture, mining, bridge and road construction, and in turpentine camps or sugar cane plantations. This form of labor approached peonage, if not slavery, and was characteristic of the "chain gangs" of the South.[10]

As leasing became more widespread some states passed laws turning the prisoners over to the counties, thus avoiding the maintenance of a state prison. The system of leasing to private contractors was finally abolished in every state by 1936.[11]

The Public or State Account System

Under the "public account," or "state account" system, there is no intervention of outside parties. Employment in all respects is directed by the state and the production is sold on the open market for the mutual benefit of taxpayer and prisoner—though the prisoner receives a very small share. This system has the advantage of being wholly acceptable to some of the large pressure groups of the state, usually the farmer or truck-farm populations. For example, binder twine has been manufactured for years in the Minnesota and Wisconsin prisons, so that farmers may buy this much-needed product as cheaply as possible rather than be at the mercy of the private producers in an industry that tends to be noncompetitive in its pricing.

Another advantage of this system is that usually enough productive work can be given the inmates so that the bulk of them may be kept gainfully employed. The objection is that it has doubtful value in trade training since there are no factories to which discharged prisoners may apply for employment.

The State-Use System

The most rational and hopeful prison labor system is the *state-use,* under which articles produced in the prisons are disposed of only in state-supported institutions and bureaus. In 22 states it is mandatory that institutions and bureaus purchase commodities that meet their needs from the prisons if such products are manufactured by the prisons. Pennsylvania is the outstanding state operating an extensive state-use system that has not yet adopted the *compulsory purchase plan.* The success of the state-use system of prison industry depends upon the type of legislation that creates the system. As the needs of state institutions and bureaus are

---

[10] For further details of the lease system and early southern prison camps, see J. C. Powell, *The American Siberia* (Chicago: H. J. Smith & Co., 1891). Powell was superintendent of an early Florida camp and wrote of his experiences.

[11] "Prison Labor in 1936," *Monthly Labor Review,* Vol. 47 (August, 1938), p. 251.

Tire recapping shop, Federal Reformatory, Petersburg, Virginia.

almost without limit, a well-integrated and efficient system of prison industries can be set up where the will to do so is present.

But pressure groups among the ranks of free enterprise, as well as petty graft, nullify most efforts to establish an effective compulsory state-use system. And even in those states that provide for mandatory state-use, eternal vigilance is necessary to see that the state-supported customers actually make their purchases from the state prisons on good faith. On the average, if the correctional institutions could supply one-tenth of the total public institutional purchases in any year, all the able-bodied inmates could be put to work in productive enterprises. Thus a really effective state-use system could easily solve the problem of prison labor and industry.

Public Works and Ways

By the "public works and ways" system of prison labor, prisoners are employed in the construction and repair of public streets, highways, and similar public endeavors. This is the sort of work the average layman usually recommends when he discusses the problem of inmate labor. Many southern counties have their road gangs, and they are still a familiar sight in Virginia, North Carolina, and other states. Sometimes the prisoners wear the traditional stripes, but in some states they have a less loathsome color combination. Many states are proud of the amount of public work accomplished by their prisoners.

California has, for many years, maintained work camps for her prisoners, as annexes to the regular prisons. There is no opposition to this practice from labor unions. The prisoners, working sometimes one hundred miles from the prison, accomplish a valuable function in fire fighting, trail clearing, and parasite control. The federal government also employs many of its prisoners in work camps scattered across the country.

Prisoners have often been used to erect new cell blocks within their prisons—often to do much of the construction work on new establishments.

Farming

Agriculture is a special form rather than a separate system of prison labor. It is generally prevalent in all states in one or more of their correctional institutions. In recent years scientific agriculture, horticulture and animal husbandry have been applied in federal and state systems with excellent results. About 250,000 acres of land are actually under cultivation on prison farms. The value of the food grown and canned is impressive.

Gardening as productive labor for prisoners.

It is probable that agriculture will greatly expand as a phase of prison economy. Prison agriculture has many advantages. There is no legislation restricting it. Through raising their own crops and vegetables, prisons can supply a more ample and better-balanced diet than can be purchased with the funds available in the prison budget. The exercise and outdoor life improve the health of prisoners employed in agriculture. The work outside prison walls aids in rehabilitation and is a first step toward readjustment to community life. With present legislation against prison labor, agriculture must provide the main relief. To give the best advantages, prison farms must produce at least 40 per cent of the food needed by the institution. It follows, too, that agricultural products must be processed and canned by prisoners.

Hobby Shops

The maintenance of hobby or trinket shops, quite prevalent in some prisons, is not a system since it is frankly developed to keep men busy. Despite the fact that thousands of dollars accrue to the prisons or to the men who work in these shops it cannot be considered gainful employment. Some ingenious wardens have developed methods of disposing of the products of hobby shops such as blankets, rugs, trinkets made of wood, clay, bone, leather, and the like although under a strict interpretation of the law this would be considered illegal. Some wardens entice visitors to their institutions, where they are taken through for a nominal admission fee after which they are guided to the "store" where they are encouraged to buy gadgets made by the prisoners.

No one should object to this endeavor but it should be stated that such a system solves nothing vital to prison administration. The men thus engaged keep busy during their spare time—often shops are set up in their cells—and small sums of money eventually trickle down to those who manufacture these products. This is *not* prison labor as interpreted by students of the subject.

3. The Passing of the Prison Factory

The early penal administrators conceived a brilliant method when they evolved the idea of working prisoners. It was so simple. Idleness would be abolished, the prisoner employed and disciplined. Labor would serve as punishment and as an aid to reformation. Prisoners would learn trades or, at least, good work habits, so they would have something to fall back on when released. Best of all, prisons would be nearly or completely self-supporting.

In virtually every institution, workshops were installed and the prison factory came into existence. Every conceivable commodity was manufactured; and some commodities were produced almost exclusively in prisons, such as cheap trousers for men, hollow iron ware, stoves, bolts, and binder twine. The heyday of prison labor was the last half of the nineteenth century and the early years of the twentieth. Then, slowly but inexorably, pressure from without began to throttle the system of prison labor. More and more restrictions were placed upon the prison administrators so that they found it harder and harder to keep their inmates working. If administrators and representatives of labor and management had conferred from time to time on the matter, instead of working at cross purposes, a satisfactory solution might have been forthcoming. During the boom days of prison labor, their success made many prison administrators bumptious and cocksure, so that they were unprepared for the restrictive measures that gradually overthrew the prison-factory.

Restrictive legislation took a wide variety of forms: here the prison was forbidden to produce any commodity that competed with free labor; there diversification, with percentage restrictions, was made mandatory. In one state, certain types of machine work were forbidden; in another, a very short work day was forced upon the prison. Sometimes the prisons were not allowed to set up any industry which was prominent in the state; in many states, prison goods had to be labeled "made by convict labor."

But the most serious blow of all was the passage of three federal statutes, the Hawes-Cooper Act in 1929 (in effect January 19, 1934), the Ashurst-Sumners Act passed in 1935, and the Act of October 14, 1940. The first divested state prison products of their interstate character on their arrival at the destination point; the second prohibited transportation companies from accepting prison-made goods for transportation into any state in violation of the laws of that state and provided for the labeling of all packages containing prison products in interstate commerce. This measure was primarily aimed at the contract and piece-price systems, although the state or public account system also suffered. The large farm machinery companies were especially articulate in squeezing out the prison labor that produced binder twine and farming implements.

By 1940 every state had passed some type of restrictive legislation to prohibit the sale of prison-made goods on the open market. In that same year, the burial of prison labor was completed when President Roosevelt signed the Act of October 14, 1940, excluding almost all products made in state prisons from interstate commerce. Farm machinery parts, farm commodities, goods manufactured for the use of states and political subdivisions, and articles made in federal institutions are exempted by this act.[12]

Such restrictions, dictated by labor interests and manufacturers exerting pressure on legislatures, practically forced the various states to prevent the shipment of prison-made goods across state lines and to make state use the dominant system throughout the country. The following table indicates the work status of prisoners in 1940, the data for which come from the latest *official* figures.

---

[12] For an excellent analysis of the effects of the federal restrictive legislation, see Frank Flynn, "The Federal Government and the Prison Labor Problem in the States," *The Social Service Review*, Vol. 24, Nos. 1 and 2, (March and June, 1950), pp. 19-40; 213-236. For a partial survey of prison labor policies throughout the world, see Ralph England, *Prison Labour*, issued in 1955 by the Department of Economic and Social Affairs, United Nations.

EMPLOYMENT STATUS OF PRISONERS IN STATE PRISONS, 1940

Total Prisoners	173,284	Piece-Price	308
Number Employed	76,775	Maintenance	60,268
In State Use	44,345	Attending School	11,673
Public Works	22,066	Sick or Unavailable	14,127
State Account	10,056	Idle	10,441

Monthly Labor Review, Vol. 53, No. 3 (September, 1941), U.S. Department of Labor, Bureau of Labor Statistics, Table 5, p. 588 (adapted), "Prison Labor in the United States." These are the latest official figures on prison labor. The over-all picture today is probably little different in proportion to the number in state correctional institutions.

The next table shows the shift of prison labor from the lease and contract systems in operation during the early days of the industrial prison to the more rational and less exploiting systems functioning today.

PER CENT OF PRISONERS EMPLOYED AT PRODUCTIVE LABOR
UNDER DIFFERENT SYSTEMS IN SPECIFIED YEARS

System	1885	1895	1905	1914	1923	1932	1940
Lease	26	19	9	4	0	0	0
Contract	40	34	36	26	12	5	0
Piece-price	8	14	8	6	7	11	**
State Account	0	0	21	31	26	19	12
State-Use	26	33	18	22	36	42	59
Public Works and Ways	0	0	8	11	19	23	29
Total Percentage	100	100	100	100	100	100	100
Per cent of all prisoners under sentence, engaged in productive labor	75	72	65	*	61	52	44

* Not reported.   ** Less than 1 per cent.

Monthly Labor Review (September, 1941), Table 1, p. 582.

This shift to the state-use system had to be universal: by 1929 it was already functioning in the federal prisons and in widely separated states. The state-use system involving the manufacture of automobile tags grew out of the phenomenal development of the automobile in the late 1920's. Here was an opportunity to put hundreds of men to work without opposition from the labor unions. But immediately the question was asked: "What reformative value is there inherent in making license tags? What trade-training does it supply if the prisons have a state-wide monopoly?" This system also induced some administrators to "overassign" inmates to this type of work.[13]

World War II gave a marked impetus to prison industry. Due to the efforts of the Director of the Federal Bureau of Prisons, James V. Bennett,

---

[13] For an analysis of the practice of using inmates to make tags, see Louis N. Robinson, *Should Prisoners Work?* (Philadelphia: Winston, 1931), p. 20.

and the Prison Industries Section of the War Production Board, prisoners in both federal and state prisons were put to work manufacturing war goods. Practically every prisoner able to work was mobilized for production. The total output for such prison-made goods amounted to $138 million in addition to some $75 millions of farm products. The percentage of farm production was twice that for the country as a whole. Special progress was made in the processing and preservation of foods.

Inmates cooperated patriotically with the war production program. Prison morale was high. It was noted, perhaps for the first time, that the nation's prisons represent a vast pool of manpower as well as technical skill and equipment that can be called into production on a moment's notice. Yet after the war, orders fell off and prison industry immediately lapsed into a state not far different from that of pre-war days. But it did demonstrate the fact that if there was the will to utilize manpower efficiently and honestly, the large number of prison inmates who are now languishing in idleness or dawdling over "made work" tasks could make a different contribution to the productivity of the nation.[14]

4. The Situation Today

Today in almost any penitentiary the investigator will find the work situation as follows:

1. A certain large percentage of men are completely idle. Some play checkers or dominoes, a few actually play baseball in an organized game, and some read. These idle men do the same thing day in and day out for months.

2. A large percentage (greatly exceeding the optimum) is employed in maintenance work. But maintenance in almost every prison is executed by two to ten times the necessary force. Ten men sweep out a cell block that could be efficiently accomplished by two. Next to complete idleness in prison, maintenance work is the most demoralizing job imaginable.

3. A certain small percentage is in industrial pursuits, actually producing commodities. But this work rarely involves a full working day—three hours in some prisons, four in some, rarely six, and almost never eight. In some prisons distasteful factory work (jute mill) is reserved for the hard-boiled prisoner and is disciplinary rather than reformatory.

4. A very small percentage is employed at clerical work, in the school as teachers, in the library, in the warden's office. This group is probably better assimilated than any other. This is necessary work, but it cannot be considered part of the prison labor policy.

5. A number of inmates, small in some prisons and larger in others, work in trinket shops or cells, their products being disposed of through a variety of means. San Quentin prison in California is conspicuous for this type of hobby work.

This quick bird's-eye picture of what actually exists today in our penitentiaries brings us to the summary statement that prison labor is a disappointment. The summary table for three decades shows: *first,* that

---

[14] For a complete summary of state prison activity during World War II, see H. E. Barnes, *Prisons In Wartime* (Washington, D.C.: War Production Board, 1944). For a cogent and provocative article dealing with the "isolated view" held by society and by some prison administrators of prison labor, see Manuel Lopez-Rey, "Some Considerations on the Character and Organization of Prison Labour," *J. Crim. Law,* Vol. 49, No. 1 (May-June, 1958), pp. 10-28.

read a test verse in Psalms 51, beginning "Have mercy upon me."[2] Benefit of clergy may be classed as a jurisdictional adjustment or an early form of the suspended sentence.

There is much confusion concerning the suspended sentence and probation. The terms are often used interchangeably. A suspended sentence is not probation. As we shall see, probation must carry with it some degree of supervision. So far as the suspended sentence is concerned, judges are restricted by statutes in invoking it. In some states, the imposition of the sentence is suspended; in others, the execution of the sentence that has been imposed is suspended. In still other states, either the imposition or the execution of the sentence may be suspended. While it is a moot question as to the desirability of one form over the other, it is in general agreement that of the two, the suspension of the imposition of the sentence is the more desirable. This is due to the fact that in this type there is less stigma attached.

It is beginning to be realized that the suspended sentence is a vestige of the era of a retributive jurisprudence and should either be abolished or at least re-interpreted in the light of the newer philosophy of probation. When certain jurists began to place restrictions on the quasi-freedom of the recipients of the suspended sentence, especially some type of supervision, the rudiments of probation began to emerge.

2. Early Probation

Before discussing early probation, let us first define just what it means:

> As applied to modern courts, probation seeks to accomplish the rehabilitation of persons convicted of crime by returning them to society during a period of supervision, rather than sending them into the unnatural and too often socially unhealthful atmosphere of prisons and reformatories. . . . [Successful probation entails] an adequate investigation into the facts of the defendant's environment, character and previous record; a wise selection by the court of the offenders capable of benefitting by the treatment, and a zealous but sympathetic prosecution of his duties by the supervisory officer.[3]

The term "probation" comes from the Latin, *"probare,"* meaning to test or to prove. It is said to have been used first by a Boston cobbler, John Augustus, who introduced the practice of friendly supervision in the city courts as early as 1841. He spent much time visiting the courts. Becoming interested in the common drunks, in jail because they could not pay their fines, he came forward and paid the fines himself. By "1858 he had bailed

---

[2] Frank W. Grinnel, "The Common Law History of Probation," *J. Crim. Law,* Vol 32, No. 1 (May-June, 1941), pp. 15-34. See also Jerome Hall, *Theft, Law, and Society,* 2nd. ed. (Indianapolis: Bobbs-Merrill, 1952); also George Dalzell, *Benefit of Clergy in America* (Winston-Salem, N.C.: John F. Blair, 1955).

[3] *Survey of Release Procedures,* Vol. 2, "Probation," pp. 1-2.

out 1,152 men and 794 women and girls. . . . In addition to those bailed he had 'helped' over three thousand females who, being neglected by the world, had no sympathy or protection but what he volunteered to furnish."[4]

This humble man, with his strange new ideas, was constantly harassed by caustic criticism. To the charges of selfishness, insincerity, and greed, he answered: "While it saves the county and state hundreds, and I may say thousands of dollars, it drains my pockets instead of enriching me. To attempt to make money by bailing poor people would prove an impossibility. . . . The first two years of my labor (1841-1842) I received nothing from anyone except what I earned by my daily labor."[5]

Dr. Sheldon Glueck, in the "Introduction" to the *Journal* of Augustus, briefly describes how he worked:

> His method was to bail the offender after conviction, to utilize this favor as an entering wedge to the convict's confidence and friendship, and through such evidence of friendliness as helping the offender to obtain a job and aiding his family in various ways, to drive the wedge home. When the defendant was later brought into court for sentence, Augustus would report on his progress toward reformation, and the judge would usually fine the convict one cent and costs, instead of committing him to an institution.[6]

The first probation law, passed in Massachusetts in 1878, enabled the city of Boston to appoint Edward H. Savage, a former chief of police, for this new work. In 1891 a second law was passed in Massachusetts requiring the criminal courts to appoint officers for the extension of probation. By 1900, however, only the following five states recognized this new technique legally: Massachusetts, 1878; Missouri, 1897; Rhode Island, 1899; with New Jersey and Vermont following in 1900.

But the creation of the Juvenile Court, beginning in Chicago in 1899, which re-emphasized the importance of the suspended sentence with adequate supervision, gave great impetus to probation. The problem today is to see that probation is extended to cover more cases and types of delinquency, to impress upon judges its importance, and to raise standards of probation services.

Early probation methods were rather crude, but even the reports of those early officers showed gratifying progress. The personal work has been continuously improved by contributions from social case work, psychology, and psychiatry. There is no excuse for employing untrained personnel, yet many of the officers in various jurisdictions of the country are both untrained and politically appointed. The educational work of the National Probation and Parole Association has set up and improved standards so

---

[4] *John Augustus, First Probation Officer*, N.P.P.A., 1939, p. vi.
[5] *Ibid.*, p. vii.
[6] *Ibid.*, p. xvi.

that now a vast amount of creditable work is being accomplished at national, state, and local levels. It has been important to impress upon judges the necessity of extending the use of the probation technique. Political pressure on judges has brought about many serious miscarriages of justice, for dangerous criminals have sometimes been placed on probation. Slowly, judges are beginning to accept the grave responsibility that this technique demands of them, and the future should see its more hopeful extension.

Despite the steady development of probation, it is not a panacea, but rather a method to be used only when a wise and disciplined therapy can be developed. Charles L. Chute, former Secretary of the National Probation and Parole Association, makes this statement:

> A really statewide probation system, and as yet not a single state has a complete one, would mean that every court dealing with offenders would have both men and women trained, experienced and properly supervised to investigate and report to the judge before sentence, on the history and environment of every single offender. It would mean trying out, under friendly but firm supervision, adapted to the individual's needs, every individual, young and old, whose attitude, past history and general characteristics appear upon investigation to offer at least a fair chance of improvement and the avoiding of further crime.[7]

While probation costs money, no one can effectively challenge the fact that imprisonment costs much more per capita than the most thoroughly-supervised probation. Figures to prove this vary in different parts of the country, but an approximation would show that it costs well over a thousand dollars per year for each person in institutions and less than $100 per year for each probationer. We note that a federal probationer costs 29.6 cents per day or $98.26 per year; while in a penitentiary the costs per day are $3.41 or $1,243.19 per year. The average earnings of a federal probationer amount to $2,649.25, which is of course directly used to support his family, thus generally saving society the further expense of relief, which is often needed by the families of men in prisons.[8]

Despite the success of probation thus demonstrated, all too many judges still send men to prison on the slightest provocation. Many cases have come to the attention of the writers where older boys have committed offenses that have irritated the presiding judge to such an extent that he forgot his dignity, gave them a vitriolic tongue-lashing, and sent them to prison "to think it over."

Probation, then, is an enlightened attempt to cope with crime in some of its aspects. Probation is often misunderstood and underrated by the public, partly because it lacks the drama of sending a man to a penal institution.

---

[7] Charles L. Chute, "Probation versus Jail," *Jail Association Journal* (January-February, 1940), p. 5.

[8] These data are from the *Annual Report* of the Administrative Office of U.S. Courts for the fiscal year 1954, p. 74.

News of probation and pictures of those placed on probation or of probation officers seldom get into the newspapers; no lurid headlines are devoted to them. The practice, where it is seriously administered, is characterized by dignity and reserve. Legally, many melodramatic crimes fall outside the sphere of probation, thus only the more or less humdrum cases receive this special kind of treatment.[9]

3. The Place of Probation in the Correctional Program

Probation means that it is to the community, rather than to the prison, that the convicted criminal must be assigned for treatment. But he is to have help: ideally he is assisted in this adjustment by a probation officer well trained in human behavior problems, in no sense a policeman or guard, but a helpful, well-integrated personality, thoroughly acquainted not only with the case but with the facilities the community has to offer in the struggle to recapture self-respect. Probation work is no mean job and calls for resourcefulness, patience, and tact.

Briefly stated, here are the advantages of probation.

A. FROM THE STANDPOINT OF THE PERSON ON PROBATION:

1. *It affords him another chance.* It must be emphasized, however, that such a sentimental reason cannot be advanced for probation since, in the last analysis, society's stake in the case must be given precedence over the well-being of the delinquent. But it is important to relieve the convicted person of the mental fear of being committed to prison at the outset of planning his probation period. Certainly, having another chance must be mentally healthful to the client.

2. *It makes possible a continuation of those life habits that meet the approval of society.* Such habits include his work, meeting family obligations, and participating in recreational activities and any other pursuits that have meaning to him as a person in the community. These are valuable to the client and also to society. A prison sentence interrupts these most important values and makes it doubly difficult for them to be gathered up after the sentence has been served. Assuming that family loyalty exists, probation is sure to have a salutary effect on all members of this group.

3. *It averts the stigma of a prison sentence.* The shame connected with a prison sentence is quite real, and frequently the innocent members of the family must bear it in an even greater degree than the culprit. In addition, the family can assist in the probation plan by providing sympathy, encouragement, material aid, and, if necessary, certain well-chosen restraints. Discipline is necessary, and such treatment can logically come from mem-

---

[9] See Henry P. Chandler, "Probation: What Can It Do and What It Takes," *Federal Probation,* Vol. 12, No. 1 (March, 1948), pp. 11-16.

bers of the family if they are taken into the confidence of the probation officer.

B. FROM THE STANDPOINT OF THE COMMUNITY:

1. *It can be assumed that the community has a definite interest in well-adjusted individuals who are carrying on a constructive life plan.* This includes participation in an economically useful job that makes the contribution to community life an asset instead of a liability. Participation in family life also is an asset to the community. In the last analysis, the well-being of a community depends on how many of its constituents are leading a reasonably well-integrated life. This must be considered in making a choice between probation and prison.

2. *Financially, probation is much less expensive than institutional treatment.* If the court places only five or six men on probation, instead of incarcerating them within a penal institution, the savings thus effected will compensate a well-trained supervising officer. As we mentioned above, probation at its best costs only a fraction of institutional treatment. Yet it is more difficult to get appropriations for effective probation work than for prison walls or custodial officers.

C. FROM THE STANDPOINT OF THE PROBATION OFFICER:

1. *The supervisor in charge can resort to the use of all the community facilities for rehabilitation.* This would include public assistance, governmental employment services, the facilities of case work and group work agencies, settlement houses, and hobby classes. There is almost no limit to the sources of assistance. In this way, probation excels the prison at its best.

4. The Administration of Probation Service

Through the years students have been giving much thought to probation administration. The effectiveness of probation depends largely on its administration. By what agency, state or local, shall such service be directed? The practice emerged in the local judicial unit, the municipal and county court. The local judge had complete control over the service; he alone could decide the extent to which probation methods should be used. It has been argued that the judge knows the prospective probationer, his limitations and potentialities much better than persons working through the larger unit, the state.

On the other hand, some critics see serious danger of political pressure being exerted on local judges, especially where they are elected. Where judges owe their places on the ballot to the party organization and are obligated to many of the ward leaders, an offender who can muster political influence can put the judge in an embarrassing position. In such a situation the probation service becomes a cog in a political machine.

In practice there is no well-defined line of demarcation between the local and state points of view but rather a mixture of both. There are four possibilities inherent in this administrative problem:

> 1. Local administration in which the state provides by law for the use of probation by the criminal courts but has nothing to do with its administration. This is exclusively local, limited to the courts and local governmental units. . . .
>
> 2. Local administration with state advice brings to the service regularly constituted state bureaus created for the function only. Such advisory service may consult periodically with local probation officers without encroaching on the prerogatives of the local department. By such a method, local officers may learn of new techniques being used in various other sections of the state.
>
> 3. Local administration with state supervision is definitely a recognition that probation service, however local, is a direct concern of the state. . . . Thus while probation remains a local matter, general administration of it is a state function.
>
> 4. Direct state control of probation is the other extreme. In this situation the state recognizes local probation only by permitting the local judge to commit the delinquent to the state probation service. Officers are employed by the state, techniques are developed and used by state sanction. . . . The state is divided into separate units with branch offices, so that the staff can conveniently cover a specified area, although their home office is the state capital. Such a system is usually combined with the state parole service, this service being recognized definitely as a state function.[10]

It is unnecessary to debate here the merits of these four systems of probation administration. Experts are not in complete agreement on the best method. The system employed advantageously in one state may not be the best for another. Smaller states may well use the state control method. Larger states may find this system quite unsatisfactory.

It is not necessarily true that advocates of local control are interested in this form only because of the possibilities it may present for political graft. Many sincere judges and students of probation honestly feel that probation is so closely integrated with community life that state control would impair its effectiveness.

5. Standards for Probation Service

The presence or absence of a well-trained and efficient staff of probation officers is the most important item in the probation picture. If the ultimate objective of probation work is, as one writer puts it, "the realignment of the client to his proper place in society's pattern,"[11] the potential proba-

---

[10] Ralph H. Ferris, *Yearbook,* N.P.P.A., 1939, p. 218 f.

[11] Edward J. Crowley, "Aiding the Adult Probationer," *Yearbook,* N.P.P.A., 1936, p. 233. For a work on probation see David Dressler, *Probation and Parole* (New York: Columbia University Press, 1951).

tion officer must understand his responsibilities. No longer can he be merely a kindly person who possesses a sympathetic understanding for the downtrodden and who has an insatiable urge to be helpful. Many judges have a childlike faith in releasing offenders, especially those of tender years, to volunteer agencies or individuals. One judge has great faith in members of the American Legion; another banks on people with "good intentions," or executives of boys' clubs or settlement houses.

There is no doubt that there are many such individuals and agencies in most communities who can be of distinctive service in this complicated field of probation, but most students will agree that direct disposition of cases in their care is a makeshift. There is no satisfactory substitute for competently trained probation officers endowed with authority to enforce a thoroughgoing program of probation supervision.

In 1945 a committee of the Professional Council of the National Probation and Parole Association formulated "Standards for Selection of Probation and Parole Officers." The Standards suggest as minimum educational requirements: a bachelor's degree, preferably with courses in the social sciences; one year of paid full-time experience under competent supervision in an approved social agency or related field, such as teaching, personnel work in industry, case work in an institution or correctional agency. As to personal qualifications, a probation officer must possess a good character and a balanced personality in addition to "good health, physical endurance, intellectual maturity, emotional stability, integrity, tact, dependability, adaptability, resourcefulness, sincerity, humor, ability to work with others, tolerance, patience, objectivity, capacity to win confidence, respect for human personality, and genuine affection for people."[12]

While some of these requirements are, at first glance, widely separated from probation work, it is important to recognize that when a person is placed on probation, his every need must be investigated by the probation worker. All the resources of the community must be marshaled so that a program of adjustment may be effected. The more familiar the worker is with the law, the easier it will be for him to make a reasonable contribution to the long-range adjustment of his client. During the past few years there has been an interesting controversy waging in professional circles as to whether the probation worker should be a professionally trained case worker. The important prerequisite is that he be nonpolitical and effectively trained for his exacting duties.

It is obvious that good probation work cannot be accomplished even by the best trained workers if case loads are too high. Many clients need more personal guidance and attention than do others, thus an arbitrary number of cases cannot be laid down. In some instances 50 cases may be too many

---

[12] Published by the National Probation and Parole Association, New York (April, 1945). See also *Manual, 1956*, Chapter 4, "Adult Probation."

for an officer; in others 100 cases may not be too many. In general, a satisfactory case load would be about 75. Yet many probation officers are carrying 200 or 300 cases at a time. This is unquestionably too great a responsibility if probation is to have any meaning for the client and society.[13]

6. The Mechanics of Probation

Statutory limitations are, to a degree, placed on most courts in the selection of cases for probation. The offenses excepted are those traditionally repugnant to society: (1) crimes of violence; (2) crimes against morals; (3) crimes involving the use of deadly weapons; (4) mercenary crimes; (5) crimes against the government, and (6) crimes carrying a certain penalty.

In the last analysis, however, the question of whether or not probation shall be granted in a particular instance usually lies with the court. A typical provision is that of California: "if it [the court] shall determine that there are circumstances in mitigation of punishment prescribed by law, or that the ends of justice would be subserved by granting probation to the defendant, the court shall have power in its discretion to place the defendant on probation."

Regardless of how much discretion is extended the court by the legislature, it is certainly incumbent upon the judge to ascertain, so far as possible, that the potential probationer is reformable. This must be arrived at by careful investigation by the probation officer by means of the pre-sentence investigation. Such an investigation places a grave responsibility on the officer and calls for objectivity as well as professional training. Another restriction placed on the court has to do with the question of guilt. In most states probation cannot be considered until after conviction for a specific offense.

In a few states, on the contrary, probation may be granted prior to conviction. Since these cases are so few, it may be of interest to enumerate them. In Massachusetts, the Superior Court (not district or municipal courts) may place on probation anyone appearing before it "charged with crime." Rhode Island courts possess some power provisionally to release defendants on probation before sentence. In Kentucky, probation is extended to any person charged with crime, excepting certain enumerated crimes. In one county in Maine, one accused of a misdemeanor may receive probation without arraignment in certain instances.

Probation prior to conviction is to the distinct advantage of the accused

---

[13] For an analysis of probation methods in California, see "Probation in California," prepared for the Special Study Commission on Correctional Facilities and Services, Sacramento (December, 1957). See also Charles L. Newman, *Sourcebook on Probation, Parole, and Pardone* (Springfield, Ill.: C. C. Thomas, 1958).

person because he need not be stigmatized by a conviction nor lose any of his civil rights, which might be forfeited by conviction. But under such a practice society has everything to lose, unless, of course, the probationer comes through his probation period without blemish. In such a case he becomes a definite asset to society and, in addition, the cost of an expensive trial has been avoided. But if the probation is unsuccessful, the state is at a decided disadvantage. As the *Survey of Release Procedures* states: "in order to mete out a just punishment to the offender, there would have to be at this point a formal declaration of guilt, and if the accused should plead not guilty, the State would have some difficulty in securing a jury conviction after a long lapse of time has scattered the witnesses and caused their memory of events to grow dim."

So long as society views certain acts as reprehensible despite extenuating circumstances or motives, looks for revenge and punishment, and insists on emphasizing acts rather than carefully considering the individual and his potentialities for rehabilitation, probation philosophy will not be extended very rapidly.

So far as the length of the probationary period is concerned, it is well agreed among trained officials that the period should be from 18 months to two years. If the prisoner is eligible for probation it is assumed that he possesses the capacity to adjust to community life in a reasonably short period of time; otherwise he does not represent a good risk. If long periods of supervision seem necessary such a practice merely piles up a larger case load for the officers. Certainly a two year period under adequate supervision should serve the purposes for which the philosophy of probation was introduced.

It is difficult to ascertain the number or percentage of persons placed on probation. An official of the National Probation and Parole Association, in a letter to one of the authors, said: "There simply are no statistics on this point which are at all satisfactory." Courts and state departments do not yet produce uniform statistics that can be compiled adequately, although there are some excellent individual probation department reports.

Interesting figures on the use of probation by the federal courts come from the Federal Bureau of Prisons. Of the 26,624 federal offenders sentenced during the year ending June 30, 1957, 11,148 or 41.9 per cent were placed on probation.[14] The number has been increasing yearly, which shows a healthy sign of probation progress by the various federal judges. There are, however, some marked differences relative to its use by the jurists, as may be seen from the following table for the year 1957.

New Hampshire 57.1
Mississippi (southern district) 62.3
Mississippi (northern district) 27.7

---

[14] *Federal Prisons, 1957,* Table 32, p. 94.

Michigan (eastern district) 37.8
Michigan (western district) 46.2
Texas (northern district) 33.7
Texas (western district) 15.3

If progressive correctional standards are to be developed, there must be a vastly increased emphasis on probation. Its exponents must interpret the philosophy underlying probation more clearly and institute a definite campaign of education that will break down the prejudice against it and explain its wider objectives. Probation, like parole, seems to the average layman a sop thrown to the criminal and a slap at society. Even the average judge looks upon it as a form of leniency. Intelligent students of penal treatment realize the deep-seated prejudice of newspaper editors, many lawyers, and judges against probation and parole. One reason is that often probation does not marshal the best thought and techniques to its fulfillment. In many jurisdictions, one probation officer (often untrained) handles all the cases of two or three counties. In some of our metropolitan cities there are overworked probation officers with too little preliminary training. Each is carrying a case load of hundreds of probationers. The checks on these cases are most elementary, such as requiring a postal card in the mail box on the first of each month, advising the officer whether the probationer is working or has been in any serious trouble. Many absurd stories from such central offices tell of the devices used both by probation officers and their clients to "save the face" of probation. Evasions of the law, slipshod methods of checking up, and poor supervision are all too frequent. It is not strange, then, that the newspapers in such a section criticize probation. Many editors do not know what real probation means, for they have never seen it in action.

There are many studies on the value or the adequacy of probation that lead us to the conviction that the bulk of the cases placed on probation in jurisdictions with adequate and well-trained personnel make a satisfactory adjustment. "Success" and "failure," when applied to the probationer's adjustment in the community, are not adequate terms. What is considered a failure in one jurisdiction might be turned into a success in another, depending largely on the preparation of the community to receive the probationer and the amount of insight of the supervising officer.

Without adequate supervision, however, the only way we can know definitely how many probationers are failing is by their rearrest for some other offense. Sending postcards to the probation officer serves no purpose in evaluating probation success. There will always be some failures in the best of systems, and these present a challenge that cannot be ignored. But the many failures coming out of penal institutions are much more impressive than the probation failures. And since probation is much cheaper than incarceration, we can well afford to spend more money on the individuals who call for special techniques in the field of therapy.

7. The Attitudes of Judges toward Probation

The attitudes of judges toward probation is highly important, for the co-operation of the bench must be enlisted in the crusade for extending probation far beyond the confines in which it is employed today. As the *Survey of Release Procedures* points out:

> The attitude of judges in regard to release procedures is of primary importance in the administration of parole and probation laws. A judge who is not in sympathy with probation as a method of treating those convicted of crime, or a judge who is in sympathy with the principles of probation, but hesitant to apply them in his jurisdiction, because of inadequate facilities, or for any other reason, can render a probation law practically inoperative. This is true because in most jurisdictions, the judge, and the judge alone, has the discretionary powers of granting or denying probation.[15]

One of the tasks assumed by the Attorney General's study was to sample the attitudes held by the judges of the country toward probation. Trained interviewers discussed this matter with a large number of judges and personally observed them in court sittings. The opinions of the prosecuting attorney, parole and probation officers, and law-enforcement officers were solicited in order to acquire as accurate an idea as possible regarding the real attitude of each judge. Schedules or questionnaires were also prepared for use of the field workers and each item was discussed with each judge. Two hundred and seventy judges scattered throughout the country were interviewed. The following statement from the study is an important preliminary to the statements of attitudes:

> One of the most serious limitations of such data is that what judges say in an interview or in answer to a questionnaire is not necessarily a reliable index of their actual attitudes, of their beliefs, or of their practices. The psychological processes of rationalization, the existence of unrecognized prejudices and personal bias, the influence of political considerations, sensitiveness to any encroachment upon existing power, and a lack of understanding as to modern criminological methods are some of the intangible factors that color what many judges say concerning their attitudes toward parole and probation as well as what they do.[16]

The judges interviewed varied as to education, political affiliation, religion, and social vision. Many did not understand the real implications of probation, but still regarded it as *leniency,* particularly suitable for the first offender. This notion is most difficult to erase from the minds of the poorly informed. Certainly, the modern concept of probation embraces the much wider scope of community safety or social health. Not until this broader view is accepted will probation meet with success. The study alluded to

---

[15] Survey of Release Procedures, Vol. 2, "Probation," p. 411.
[16] *Ibid.,* p. 412.

reports that the judges interviewed recognized that the following factors influenced them in sentencing:

Leniency Factors	Number of Judges
Immaturity of defendants	106
Defendant responsible for dependents	71
No record of prior criminality	68
Defendant a woman	31
Defendant makes "a good impression"	27
Crime against property	27
Crime committed without deliberation or premeditation	22
Underpriveleged childhood	16
Good background	9
Old age of defendant	8

Although the majority of the judges took youth or immaturity into consideration when sentencing, many said "that they had begun to doubt the advisability of the practice, pointing out that the criminal statistics generally show the most vicious criminals to be very young men, many of them in their teens and twenties."[17]

Now let us look at the various factors that make for severity, as expressed by the same judges:

Severity Factors	Number of Judges
Record of prior criminality	75
Crimes against person	53
Defendant makes unfavorable impression	39
Sex crimes	38
Deliberate or premeditated offense	29
Mature defendant	25
Crimes against property involving violence	13[18]

Here is the reaction of the *Survey of Release Procedures* regarding this table of severity factors:

> It is not surprising that recidivism heads the list . . . but it is rather surprising that only 75 judges admitted giving more severe sentences when the offender is a known recidivist. Of course, the highly subjective nature of the material on which these tables [the two listed above] are based make them of little value except for showing the intuitive impressions of judges to be vital factors in determining the sentences which they impose upon persons convicted of crime.[19]

Although this study of attitudes was made 20 years ago, it would be safe to assume that such attitudes have changed very little. No doubt some jurists have modified their earlier concepts of the meaning of probation, but if we were to survey the nation we would find many judges and others

---

[17] Table and quotation *ibid.*, p. 419.
[18] *Ibid.*, p. 424.
[19] *Ibid.*, p. 425.

of the legal profession still skeptical of the efficacy of sound probation philosophy.

8. Concluding Remarks

With the gradual extension of the probation philosophy it is evident that the offense will be minimized and the individual considered more carefully. There is a healthy trend in this direction even today with the introduction here and there of the pre-sentence clinic. We have discussed this in Chapter 16 (pages 264–265).[20] Here let us rate it, with the Youth Correction Authority, as one of the most significant contributions to modern penology. The Youth Correction Authority emphasizes the individual youth rather than the offense he commits. In making an analysis of the background of the young offender, it is imperative that the judge call upon experts to help him in sentencing the boy for his own good as well as for the good of society.

One of the recommendations of the pre-sentence clinic that will loom large in its deliberations in the future will be probation. The judge who now finds himself out of sympathy with the philosophy of probation, or the one who still thinks of it as an opportunity to give the first offender "another chance," will, with the assistance of the pre-sentence clinic, be impressed with its possibilities.

Regardless of the reactionary attitudes of some judges and other critics of probation, we are convinced that it is the only completely promising reformative technique. We are frankly pessimistic of all the ambitious and highly publicized experiments carried on throughout the country *within* the prison. The stories coming from such places make good copy, but many of the men emerging fail to make good citizens.

We must employ adequately trained probation personnel, see that they are well paid, and require them to read widely in the various fields where students of human behavior are pioneering with new techniques and methods. Another practice that should be encouraged is getting together in small groups and discussing the many problems incidental to probation. The alert probation officer will not retire to his own office and shut out new ideas. Staff conferences also are very important. These provide professional and intellectual stimulation, which is productive to both the probation officer and the probationer as well.

---

[20] See Edmond Fitzgerald, "The Pre-Sentence Investigation," *N.P.P.A. Journal* Vol. 2, No. 4 (October, 1956), 320-336).

35

Conditional
Release and Parole

I believe there are many [prisoners] who might be so trained as to be left upon their parole *during the last period of their imprisonment with safety. From Philip Klein,* Prison Methods in New York State (*New York: Columbia University Press, 1920*). *Dr. Howe, husband of Julia Ward Howe, was the first person to use "parole" in this sense.*

DR. SAMUEL G. HOWE

1. The Nature of Parole

Parole is a form of conditional release granted after a prisoner has served a portion of his sentence in a correctional institution. It presupposes careful selection, adequate preparation for release, and some type of supervision in the community for a sustained period of time. It is not to be confused with the outright freedom which comes after the full expiration of a sentence.

Parole is also confused with probation and pardon. Parole follows a sentence imposed by a judge and the serving of a portion of it in confinement; probation is an alternative to a prison term. A pardon is an act of grace, usually at the hands of a governor or a state board of pardons, which results in the remission of all or part of the legal consequences of a sentence. It may be either absolute or conditional. In its generic sense, pardon includes all forms of clemency.

Parole is not a new technique. It had its counterpart in England in what was known as "ticket of leave," which originated in a plan worked out by Captain Alexander Maconochie on Norfolk Island in 1840.[1] After serving a part of the sentence under strict surveillance, the prisoner was granted

---
[1] See Chapter 26 on Maconochie's experiments in corrections.

a ticket of leave which he had earned through his good behavior and work. This permitted him to enjoy a kind of conditional freedom under supervision. Maconochie is indeed the "father of parole."

After the penitentiary was adopted, those convicted of crimes and sentenced to prison were expected to serve out the entire fixed term of years. The penal codes stipulated the penalties and the judges pronounced their sentences. The only possibility of mitigating this fixed term was by executive clemency. Ten years in prison meant ten years. At the expiration of the allotted time the prisoner was released without any supervision except by some private interested citizen or by a prisoner aid society. In the main, the discharged prisoner was cast adrift in society, a potential menace.

Long prison sentences developed bitterness and hopelessness particularly among those who had no influential friends to intercede for them. The shrewd criminal, the recidivist, the professional, often succeeded in purchasing this release through the governor's pardon. But the poor and friendless languished behind bars until their sentences were fully spent.

Before taking up the discussion of parole we might list the methods by which a convicted criminal may be released from prison, once having entered. First, he may receive an outright pardon through executive clemency. Second, if he is a member of a group of convicted persons, he with the others may be granted amnesty. This form of "general pardon" is rarely used in the United States except for conscientious objectors or other political prisoners.

A third token of grace is a reprieve, which applies only to persons about to be executed. Again, this is usually at the hands of a governor or whoever by law may share with him this responsibility.

Commutation of sentence and conditional release from prison, both discussed below, may also be tendered the prisoner; in the former case, to mitigate the sentence imposed by a judge and granted usually by a board selected for that purpose; and in the latter case, to shorten the fixed sentence by "good behavior." All or most of these procedures are in operation in most of the states but the conditions surrounding them vary widely.

The first parole law was enacted as early as 1837, in Massachusetts. But up to 1880 only three states provided for this new device. In the decade between 1880 and 1889, nine more states provided for parole. The last state to pass such legislation was Mississippi, in 1944. Today all states make legal provision for the paroling of inmates from prisons.

Today it is axiomatic that every person sent to prison should be released some time prior to the expiration of the court's sentence, but with the qualification that he be supervised from the moment of release until he is either ready to conduct himself as a respectable citizen or until the final expiration of his sentence. Such a policy is known as parole.

Unfortunately parole is often inaccurately associated with leniency. There have been all too many cases in which there has been leniency in granting parole, but from a professional point of view sentiment should have no place in its dispensation. Parole, wisely administered, guarantees to the community that the released prisoner is a potential asset because of his institutional experience. Parole is one of the bulwarks of a progressive penology. Until we find a workable substitute for imprisonment it is imperative that the parole technique be improved and extended. It is necessary that we discuss it carefully and show its effectiveness in present-day penology.

2. Commutation

Fixed, or "time" sentences shattered any hope the criminal had of early release and embittered him against the forces of control. In many cases this bitterness was justifiable, especially where justice went awry and innocent men were given fantastically long sentences by self-righteous and emotional judges. Our early penitentiaries were filled with disappointed and dangerous men awaiting the day of their release to wreak vengeance on society. To make matters worse, criminals with influence sometimes received a pardon from a weak but amiable governor. State executives were harassed all through their terms of office by pressure from politicians, ministers, well-meaning friends of prisoners, and influential citizens "back home." Not infrequently jurors, after convicting a man for a crime, would immediately join in a petition to the governor to pardon him. The pardoning power of the governor became a menace. Sometimes governors found it financially advisable to release large numbers of prisoners to cut the mounting costs of institutional incarceration. It is obvious that many criminals were released who were a danger to society.

These abuses, resulting from sentencing criminals to definite and fixed terms, made it necessary to modify the law in some manner. In 1817 New York state passed the "good time" law, empowering the inspectors of the prison to reduce by one fourth the sentence of any prisoner sentenced to imprisonment for not less than five years, upon certificate of the principal keeper and other satisfactory evidence, that such prisoner had behaved well, and had acquired in the whole, the net sum of 15 dollars or more per annum. With the introduction of the good time system by New York, other states passed similar laws. By 1916 every state and the District of Columbia had passed such laws.[2] The amount of good time and the methods by which it is computed varied from state to state.

There were at least two early objections by prison officials to good time laws. The first was one of bookkeeping, which was certainly difficult. But the more serious objection was that good behavior was easily simulated by

---

[2] For a recent analysis of these laws and objections thereto, see G. I. Giardini, "Good Time—Placebo of Correction," *Amer. J. Corr.* Vol. 20, No. 2 (March-April, 1958), pp. 3 ff.

the more wily criminals, and that although such laws might assist the guards in keeping order, many deserving prisoners would be denied commutation because of their difficulty in adjusting to the routine of the institution. It is axiomatic that the repeater adjusts easily to prison life and is more likely to be credited with good behavior. Although this good-time practice is a form of commutation, it should not be confused with the shortening of a sentence by an official body set up for that purpose. A sentence of 25 years may, for certain reasons, be commuted by the body having such authority to ten years; life imprisonment may be commuted to 20 years or to some other such reduced sentence. Good-time means that the sentence shortened, not by an authority outside the prison, but according to a definite schedule stipulated by statute. It is a form of *conditional release*. In many institutions still functioning under good-time laws (the federal system, for example), they are often held over the inmate as a threat and assist the management in maintaining some degree of obedience to the rules. In the last analysis, they are merely another form of institutional repression.

3. Origins of Parole and Indeterminate Sentence

But good-time laws were only the beginning of the modification of definite sentencing. Even before these laws had become common a bolder concept had been suggested, the indeterminate (or indefinite) sentence. In 1832 Richard Whately, (Anglican) archbishop of Dublin, advocated the indefinite sentence in these prophetic words:

> It seems to be perfectly reasonable that those whose misconduct compels us to send [them] to a house of correction should not again be let loose on society until they shall have made some indication of amended character. Instead of being sentenced, therefore, to confinement for a certain fixed period, they should be sentenced to earn, at a certain specified employment, such a sum of money as may be judged sufficient to preserve them, on their release, from the pressure of immediate distress; and orderly, decent, and submissive behavior should during the time of their being thus employed be enforced, under the penalty of a prolongation of their confinement.[3]

The first legislative action in America on the indeterminate sentence was taken by Michigan in 1867. This was due to the efforts of Zebulon R. Brockway, for 11 years superintendent of the Detroit House of Correction. The Detroit institution detained many prostitutes and it was because of this type of offender that the law was passed. It provided that prostitutes be sentenced to a maximum of three years and that the board of managers of the institution be empowered to release before the end of the term. Brockway, in 1868, presented a bill to the Michigan legislature for the

---

[3] As quoted in *Survey of Release Procedures.* Vol. 4, p. 16.

indeterminate sentence without any maximum or minimum, but it was defeated.

At the first meeting of the National Prison Association in 1870 at Cincinnati, papers were presented by Sir Walter Crofton of Ireland and by Brockway on "The Ideal of a True Prison Reform System," focusing the attention of American penologists on the indeterminate sentence and the possibilities of parole.

The ideas thus expounded were adopted in 1876 in the new institution at Elmira, New York, where the reformatory principles were worked out in practice. Parole was actually realized in this new setting, as prisoners, upon release, were continued under the jurisdiction of the prison for an additional six months.

Though it is generally held that parole and the indeterminate sentence are a fundamental unity in principle and successful practice, their progress and acceptance have been more or less uneven. The parole system has made more rapid strides. The indeterminate sentence was not generally adopted until after the turn of the century. By 1900 such provisions were found in five states only; by 1915, 31 states had adopted such laws; at present all but the federal system and nine states have some form of the indeterminate sentence.

There are many arguments against the strict indeterminate sentence, which calls for a person to be sent to prison without any stipulated sentence and with hope of release based exclusively on the prisoner's readiness for freedom in society. Many of these arguments are based merely on culture lag or timidity. There is a real question, however, of the constitutionality of such a provision unless some definite limit is set even though that limit be for life. Some opposition is based on fear of usurpation of certain prerogatives heretofore vested in the courts. Many judges are unwilling to relinquish their sentencing powers, although some jurists agree technically they are not trained to mete out sentences wisely.

When the first legislation creating this type of sentence was passed, no restrictions were placed on the judges' sentencing power, although definite limits were placed on the maximum number of years a convicted criminal was obliged to serve. This situation invited some judges to fix the minimum at any point within a day of the maximum. A judge could sentence a man under the law to "nineteen years, twenty-nine days to twenty years." However, many states have corrected this unwholesome feature by making it mandatory for the judge to place the minimum at a point no more than one-half of the maximum; thus, five to ten years, ten to twenty years, although, of course, the minimum may be as low as one year.[4]

---

[4] For further information regarding the shortcomings of sentencing in this country, see Sol Rubin "Long Prison Terms and the Form of Sentence," N.P.P.A. *Journal,* Vol. 2, No. 4 (October 1956), pp. 337-351.

4. Early Parole in America

While parole and the indeterminate sentence are generally thought inseparable, this is not necessarily true. Where a quasi-indeterminate sentence is in operation, parole practice is usually integrated with the minimum-maximum concept. But parole has been extended to county jails and county penitentiaries, where the limited indeterminate sentence has not been inaugurated.

Although it is strictly true that there is no indeterminate sentence in the federal system, it is interesting to note that parole may be granted—that is, the prisoner becomes eligible for parole—"upon expiration of one-third of his sentence," providing such sentence exceed one year, less good-time earned.

Because the prisoner is usually eligible for parole after serving the minimum term, parole is often looked upon as a right rather than a privilege. The prisoner feels that after the expiration of his minimum, he is entitled to parole and all the pressure he can exert is brought into play to effect a parole. This is understandable because paroles have been forthcoming in many states immediately after the minimum has been reached.

Certainly with prison guards, tipstaves, judges, prosecuting attorneys, lawyers, politicians, and influential friends passing out erroneous concepts of parole, it is not surprising that convicted criminals and inmates of correctional institutions, as well as the public, come to accept this as fact. The friends of parole are confronted by these handicaps together with many more. Newspaper editorials deprecate parole, especially when some criminal on parole is arrested for committing another offense. In the majority of such cases, the conditions under which the discharged prisoner thus arrested was released from prison are likely to have been very lax. In fact, parole of this sort can scarcely be labeled as such.

Practices are not uncommon where release of this kind cannot, with any degree of accuracy, be called parole: First, those in which parole is granted merely on the basis of political influence; second, where supervision is farcical because parole officers are untrained or overworked by large case loads; third, where funds are inadequate for effective supervision; fourth, where the institution employs slipshod methods in selecting parolees; fifth, where outside sponsors, if required, are not carefully selected by the parole board.

One state spends $25 per year on administering each parolee—which means that the latter may send an occasional postal card to the central authorities advising them that he is behaving himself and working faithfully at some job. In another jurisdiction overworked parole officers have a case load of over 500 ex-prisoners.

In fact, the parole ideal has not been attained in most states. Effective and efficient parole costs no more than one-third that of institutional treatment, but the cry of economy is always raised when attempts are made to inaugurate a system of parole worthy of the name. So long as such conditions exist, parole will be hamstrung, and the public will question it. But despite these shortcomings in parole administration, there are results that are gratifying to the most carping critic. Today parole advocates can truthfully insist that there is nothing wrong with parole *per se,* but only with its administration. Parolees returned to prison may well be evidence of excellent supervision by the parole officer, for shoddy parole work will not detect violations of parole except in an overt criminal act.

5. The Mechanics of Parole

Good parole practice is composed of three fundamental processes: preparation, selection, and supervision. It is easy to write these words down on paper, but it is quite difficult to describe them adequately. Parole application is, in many states, automatic, but students of parole insist that preparation should begin the moment the prisoner enters the institution. Conscientious preparation is essential to a well planned system of parole. Each prospective parolee is entitled to a thorough and honest parole plan. In institutions where trained personnel are employed for this specific purpose, parole ceases to be a hit-and-miss system and gradually develops into something of logic and value.

Where there is a professionally trained staff of social workers, psychiatrists, psychologists, and others concerned with the task of understanding the prisoner, preparation for parole usually begins to function immediately. Initial examinations and interviews are made; inmate replies are systematically checked through community agencies; reports from physician, director of prison industry, educational director, and other staff members are scrutinized and recorded in the individual inmate's file—all this and more make up this preparation.

The American Correctional Association has compiled a handbook dealing with pre-release preparation for parole consideration.[5]

It is obvious that without summaries, carefully compiled, a parole board can do nothing intelligently. But, however excellent the classification clinic or however conscientious the summary compilation, the work of parole preparation will be largely nullified if there is no competent parole board to act on the recommendations and to carry out good parole supervision. While several states maintain excellent parole boards, all too many of them exercise a glaring apathy in appointing capable, conscientious members. Most of them are political in their composition. Many parole boards are

---

[5] *Handbook on Pre-Release Preparation in Correctional Institutions* (New York: American Correctional Association, 1950). See also *Manual, 1956,* Chapter 9, "Parole."

composed of unpaid members and some are paid for part-time service. Good procedure calls for full-time, well-paid boards composed of outstanding persons, highly motivated in the best interests of both society and the prisoner.

In the states where parole has meaning, it is extremely difficult for prisoners "to make parole." On occasions, in some states, over 70 per cent of the applications are denied upon first petition. Even after a second and often a third appearance many prisoners are denied parole. In the federal system two-thirds of the applicants were denied parole during the year ending June 30, 1957. The Director of the Federal Bureau of Prisons states that of 10,306 cases reviewed by the parole board, 3,475 (33.7 per cent) were granted, and 6,813 (66.3 per cent) were denied.[6]

6. Parole Selection

Who should be selected for parole? The answer is: anyone who is expected to return eventually to society. More than 95 per cent of our convicted criminals do return, sooner or later. Some die in prison of natural causes. A few make good their escapes. It is axiomatic that every person sent to prison should be released some time prior to the expiration of the maximum sentence. Otherwise a discharged prisoner, on the completion of his sentence, leaves the institution without any supervision whatsoever. Moreover, if he has been denied parole on several applications he is either broken in spirit or deeply embittered against society—in either case, is a social liability.

If nearly all prisoners are to be released through the parole procedure, it is obvious that they should be adequately selected and efficiently supervised. The question is not *who* will be paroled but *when* the applicant be granted this coveted status.

Selection for parole is generally limited by the type of indefinite sentence practiced in specific jurisdictions. No one can apply for parole unless he has spent the time behind prison bars required by law. Thus the limited indefinite sentence complicates the problem of parole selection considerably. The general practice is to permit the inmate to apply for parole upon the expiration of his minimum sentence or, in states permitting "good time," at the minimum less the good time. A few states recognize parole eligibility at any time after commitment. In such cases this important procedure is in the hands of the parole or other control board.

These legal requirements as to when parole may be granted in no way simplify the decision about parole. Many a prisoner is ready for parole long before the minimum, or one year before the minimum, or three years before the maximum. Much harm may be done the individual and the community because he cannot be paroled before the legal requirements permit

---

[6] *Federal Prisons, 1957,* Table 29, p. 90.

it; and a complete overhauling of the laws on parole release is sorely needed. It is clear that a scientific selection for prospective parole depends on the ultimate realization of the unlimited indeterminate sentence.

Assuming that statutory limitations are no hinderance to the paroling technique, what factors should be considered in preparation for parole? Besides preparation of the prospective parolee, preparation of the community is an important factor, and includes such essentials as location of a sponsor and of a job or occupation that holds interest for the client, a family situation that is helpful and (if possible) sympathetic, a program that attempts to develop all the potential assets of the individual, and a small but helpful group of friends who are approved by the paroling authorities. To establish these prerequisites is no easy task, but all are essential for a successful adjustment. The ultimate objective of parole, then, is to establish as nearly as possible a normal adjustment of the individual.

The potential parolee should also be briefed, while in the institution, as to the meaning of parole. His responsibilities should be carefully outlined. This is done in some states through discussions or guidance programs, or by means of mimeograph material.

It is generally recognized that many parole violators come to grief merely because they were not adequately prepared to face the free and normal society. Dr. Norman Fenton, formerly chief of the Classification Bureau of the Department of Corrections for California, has pointed out the importance of psychologically preparing the man for release. The prison has institutionalized him; thus he must be de-institutionalized. Furthermore the potential parolee must be prepared to accept direction outside the prison from the hands of an official in the community. This transition is sometimes hard to make.[7]

7. Importance of Transition Pre-Release Regime

Aside from pre-release preparations there is real need for the prospective parolee to spend a preliminary period of semi-normality under institutional restraint, to serve as a transition period prior to actual departure from the prison. The *Handbook* has the following to say on this point:

> A perplexing situation in the treatment of prison inmates occurs when men are under custodial supervision in the institution on one day and on the following day leave prison to comparative freedom in the community. A somewhat similar situation would prevail were disturbed patients of a mental hospital released directly to the community because of the expiration of a period of commitment. . . . Prison management probably more than any other aspect of public administration has been curtailed by legal restrictions or public opinion in the freedom of experimentation in methods

---

[7] "The Psychological Preparation of Inmates for Release," *Prison World* (November-December, 1949), p. 9. See also, *Handbook* on Pre-Release Preparation in Correctional Institutions, pp. 43-50.

of treating their charges. In the coming years efforts should be made by correctional workers to try out new devices and procedures for the rehabilitation of inmates. One fruitful area for research lies in the discovery of practical measures for helping inmates to make the transition to family and community life after release from prison.[8]

Some of the devices used in various prisons may be listed:

1. Honor blocks
2. Minimum custodial facilities
3. The separation center (such as at Chino, California and the Federal Penitentiary at Lewisburg, Pennsylvania)
4. Honor camps (road and forestry camps)
5. The day parole or furlough (to work during the day at home, on ranches, in stores or factories in the vicinity of the institution, returning at night)
6. Community dormitories for releases
7. Camps for released prisoners[9]

American corrections is just reaching the threshold concerning this vitally important problem. Some other countries have been engaged in this type of pre-release transition for years. In Colombia, for instance, a "preparatory release" period precedes his "conditional release" period. This may be granted two or three years beforehand. The candidate is given the opportunity of securing work outside the prison, returning to the prison at night. In Argentina an area of the prison in Buenos Aires is set aside for pre-parolees. It is attractively furnished, homelike in every respect. At meals, the men sit at tables for four, and have a lounging and reading room furnished with comfortable chairs with current magazines available. They have individual rooms, not unlike modern hotel rooms. The candidates may go out into the community to look for work and are also aided by trained vocational workers who assist in adjusting each man's situation.[10]

8. Supervision of Parolees

The crux of successful parole is supervision. Without intelligent, trained supervisors the entire system breaks down. All prisoners should be released with some restraints if parole treatment is to mean anything, so it is important to have enough supervisors who understand the responsibility of supervision.

Supervision does not mean espionage; the parole officer must not be a policeman. He should be a friend in need, an adviser who thoroughly understands the individual's peculiar problems arising from his term in prison. Many of these problems would not get a moment's thought from

---

[8] *Ibid.,* pp. 52, 61.

[9] *Ibid.,* pp. 57-61.

[10] For Michigan's program in pre-release see Gus Harrison, "New Concepts in Release Procedures," *Proceedings,* American Correctional Association, 1953, pp. 245-249.

the free citizen. The released prisoner must make hundreds of decisions each day that he did not make before release. Obtaining clothing, food, shelter, recreation, and catching up with the loose ends of the humdrum life of the average citizen are serious, complicated tasks.

Aside from these day-by-day ordinary activities, the parolee must bear constantly in mind the rules imposed upon him by the parole board, which have been carefully explained—in the ideal situation—by his supervisor. Many of these are irksome, but he must conform. Regardless of the sympathy elicited by the parole officer, his success depends largely on his own initiative and ability. A hostile community, a suspicious neighborhood, an almost strange family and an even more strange group of erstwhile friends and acquaintances—all must be convinced that he is going straight. The successful parolee deserves credit, for he has a superhuman task.

According to the *Survey of Release Procedures,* conditions "regularly imposed in one or more jurisdictions" require that the parolee must:

> Submit an explanation for idleness.
> Abstain from the use of narcotics.
> Not carry weapons.
> Not visit or correspond with inmates of penal institutions.
> Be home by a stated hour in the evening (usually 10 o'clock).
> Promptly report the loss of employment.
> Support dependents.
> Submit to medical treatment if ordered to do so.
> Not apply for a license to hunt.
> If improperly treated by an employer, report the fact.
> Not visit places where liquor is sold or drunk.
> Avoid gambling places or pool halls.
> Attend church regularly.
> Not attend public dances.
> Save a certain percentage of his earnings.
> Waive extradition if apprehended outside the state.
> Not borrow money or valuables, or purchase on the installment plan without permission.
> Where possible, pay board and room in advance, and mail in receipts showing such payments.[11]

It is no doubt possible to justify these rules in many cases, but some of them make a day-by-day adjustment in a normal society extremely difficult. The reader might well ask himself if he could possibly lead a normal life under such conditions. Next he should ponder the question how far such repressions contribute to a satisfactory adjustment in the normal life pattern. With adequately trained parole personnel, reliance upon these rules will rarely be necessary. Scolding by the supervisor should be as infrequent as possible, and shadowing should never be necessary.

---

[11] *Survey of Release Procedure,* Vol. 4, p. 212. For a state by state survey of parole rules, see Nat R. Arluke, "A Summary of Parole Rules," *N.P.P.A. Journal,* Vol. 2, No. 1 (January, 1956), pp. 6-13.

It is upon this flexibility of administering the rules of parolee behavior that success or failure of parolees usually depend. Many heartbreaking stories are told of inept parole officers, particularly of their refusal to extend concessions in situations where such actions would insure success instead of failure. A poorly trained parole officer or one who is harassed by too many cases may do the service a great deal of harm. It is needless to point out that parole officers should be carefully selected, have civil service rating with adequate tenure, be well paid, and not be burdened with too heavy a case load. Estimates differ as to the number of cases that can be effectively supervised by a parole officer. Some authorities say 50, others 75, and still others, 100. But there are limits, and it is all too true that in most states officers have case loads much too heavy.

The report of the Committee on Standards and Procedures in Parole Supervision of the National Parole Conference states: "The geographical area covered, available transportation facilities, the amount of time required for pre-parole investigations, and the resources provided by the community to supplement the parole officer's efforts directly affect the number of cases he can supervise effectively. It must be assumed that no officer should have more than seventy-five cases, even under favorable urban conditions." The ideal number, however, is 50 cases.

But the number of parole officers and their case load is not the whole picture. A large staff will not take care of the parole problem unless the members are personally fitted for such work and professionally trained. The ideal qualifications include not only a degree from a college or university with training in sociology and psychology, but graduate work either in these fields or from an accredited school of social work. The tendency at this time is to expect candidates to have had some work in a social agency. There have been considerable differences of opinion regarding the training for correctional work. Some feel that conventional courses in criminology and probation work in college are sufficient to enter the field; others take the position that helping people with their problems is similar in any field, and thus a graduate course in a school of social work furnishes the most adequate preparation.[12]

9. How Successful Is Parole?

There have been few recent over-all studies of success of those granted parole. There is a general feeling that only about 20 to 30 per cent fail to live up to their parole responsibilities during the period of restraint. It is

---

[12] See the following articles on this subject: Walter C. Reckless, "Training Probation and Parole Personnel," *Focus,* Vol. 27, No. 2 (March, 1948), pp. 44-48; "Training Reconsidered" (comments on Dr. Reckless' article) *ibid.,* No. 6 (November, 1948), pp. 180-182; and Reckless, "The Controversy about Training," *ibid.,* Vol. 28, No. 1 (January, 1949), pp. 23-25.

quite true that many criminals carry through this period but, upon release, enter criminal activity once again. The newspapers and many of those who have little use for parole go to considerable length to refer to an arrested criminal as an ex-parolee or as a person "out on parole." Most frequently the suspect has been off parole for some time—often for years.

In making an evaluation of success or failure, we are here concerned whether the parolee met his obligations during the parole period, which often covers considerable time. Parole, as a philosophy or as a technique, cannot be held accountable for relapses of criminals after they have been released from the jurisdiction of the parole board. It would, nevertheless, be gratifying if there were a carry-over of law-abiding behavior for the remainder of the ex-prisoner's life, but this is obviously too much to expect from all who have served time.

In appraising the lapses of men on parole it should be pointed out that in jurisdictions staffed by trained men, in which standards are high, more is expected of the parolee. Thus if he fails to measure up he is returned to the institution on a technical violation. Conversely, if the parole system is lax, the parolee may behave much as he wishes without fear of return. Parole studies must be analyzed in terms of the standards of the paroling authority involved. Students are familiar with the status of parole throughout the country and are able to gauge the validity of figures submitted concerning success or failure of parolees.

A parolee is not indicted merely because he violates rules laid down for his guidance. A good parole officer will work along with a parolee who is sincerely trying to make an adjustment. But if he finds it impossible to comply or willfully refuses to abide by the rules he may be labeled a failure and returned to prison. And, of course, if he commits another offense against the law, he is quite likely to be returned.

Now as to some studies depicting success or failure. Professor Robert H. Gault found in 1915 that 84 per cent of 38,593 parolees from 16 states were successful.[13] W. W. Clark in 1921 found that 72 per cent of those paroled from 19 institutions were successful.[14] These two early studies are restricted only to serious violations for which paroles were subsequently revoked. In the Glueck study of 510 inmates paroled from the Concord Reformatory in Massachusetts, 55.3 per cent were officially known to have committed serious violations of parole.[15] Studies by Professor E. W. Burgess in Illinois, and Frederick A. Moran in New York indicate that

---

[13] Robert H. Gault, "The Parole System as a Means of Protection," *J. Crim. Law,* Vol. 5, No. 6 (March, 1915), p. 801.

[14] W. W. Clark, "Success Records of Prisoners and Delinquents," *J. of Delinquency,* July 1921, pp. 443-452.

[15] Sheldon and Eleanor Glueck, *Five Hundred Criminal Careers* (New York: Knopf, 1930), p. 169.

about 40 per cent of those paroled violated their obligations and were returned to prison.[16]

Austin MacCormick presents some figures from New York state where parole is conducted on a reasonably high level. The pre-war reports showed that about 65 per cent of those paroled five years before had maintained clear records, that about half of the remainder had been returned to institutions for technical violations of their parole conditions, and that the remainder, about 17 per cent, had been convicted of new offenses, of which only half were felonies. During the non-typical war years the percentage of success rose even higher.[17] A study made of paroled men from St. Cloud, Minnesota, reformatory showed 57.9 per cent failed to keep out of trouble.[18]

From the above studies we are able to give an approximation of the success of parole. An open and shut answer, however, is difficult to obtain. Certainly the glowing reports of 85 to 90 per cent success are at least suspect. The diversity of sentencing laws, the differences in time that prisoners must serve before becoming eligible for parole, the wide variations in periods of parole supervision, the efficacy of that supervision, make it almost impossible to judge the success or failure of parole. But even the most carping critic would agree that it is far superior to the practice of releasing men from prison without supervision.

Parole is definitely here to stay. The pin pricks of criticism should have little effect in nullifying the system; rather they should tend to call for parole advocates, progressive judges, and legislators to create scientific legislation to ensure effective parole. So long as we send men and women to prison we must expect them to return to society, and they must be returned with some degree of supervision and not discharged without restraint. If they leave the institution without supervision, at the end of their maximum sentence, we would once more be shackled with the deadly fixed time system.

California has been experimenting with advancing the parole date. Prospective parolees are released three months before they are technically eligible. They are given more intensive supervision in the community by parole officers whose case loads are reduced to fifteen. Later the case load was gradually increased. The study concludes by stating that it is

---

16 See "The Prison System in Illinois," *Illinois Inquiry Commission Report* (Springfield: 1937), p. 617; Frederick A. Moran, "Parole—An Effective and Law-Enforcing Agency," *Proceedings,* American Prison Association, 1937, p. 308.

17 Austin H. MacCormick, "Correctional Treatment of the Criminal," in Boulder, Colorado, Crime Conference, 1949. See proceedings, entitled *Crimes of Violence* (Boulder: 1950), p. 69.

18 Stanley B. Zuckerman, Alfred J. Barron, and Horace B. Whittier, "A Follow Up Study of Minnesota State Reformatory Inmates," *J. Crim. Law,* Vol. 43, No. 5 (January-February, 1953), pp. 622-636.

feasible to release men from prison in advance of their usual time without hazard to the public.[19]

10. Parole Prediction

Predicting the probability of a successful adjustment on parole is, in reality, what every parole board or classification clinic does every time it acts on a parole. This prediction may be arrived at in a variety of ways: the members of the board can guess or give their hunches on a minimum of information submitted to them; they can look the candidate over and make a prediction based on intuition; or they can examine various types of more or less scientific data collected by trained observers, and thus arrive at a decision. Even then, such a decision rests on a majority rule, or in some jurisdictions on unanimous verdict. Scientific parole prediction is, therefore, an attempt to inject some type of valid tests into the selection of prospective parolees.

As early as 1923 suggestions were made that parolability could be predicted. Since that time a core of literature has been developed in this area. The work of E. W. Burgess, Clark Tibbitts, and Ferris Laune is pioneer in prediction. Tibbitts, in 1931, developed 21 factors to be considered by the examining body in weighing parole of a prisoner. These are:

1. Nature of the offense.
2. Number of associations in committing the offense for which convicted.
3. Nationality.
4. Parental status—including broken home.
5. Marital status of inmate.
6. Type of offender.
7. Social type (viz., gangster, socially inadequate, ne'er-do-well, etc.).
8. County from which committed.
9. Size of community.
10. Type of neighborhood.
11. Resident or transient in community arrested.
12. Statement of trial judge and prosecuting attorney with reference to recommendations for or against leniency.
13. Whether or not commitment was on the acceptance of a lesser plea.
14. Nature and length of sentence.
15. Months of sentence actually served before parole.
16. Previous criminal record.
17. Previous work record.
18. Punishment record at the institution.
19. Age at time of parole.
20. Mental age.
21. Personality type and psychiatric prognosis.[20]

---

[19] For details see Ernest Reimer and Martin Warner, "Special Intensive Parole Unit," *N.P.P.A. Journal,* Vol. 3, No. 3 (July, 1957), pp. 222-229.

[20] Clark Tibbitts, "Success and Failure on Parole Can Be Predicted," *J. Crim. Law,* Vol. 22, No. 1 (May, 1931). For an account of the prediction work done in Illinois over a period of years, see Lloyd E. Ohlin, *Selection for Parole* (New York: Russell Sage Foundation, 1951).

On the basis of this table of factors, it is concluded that a prisoner who measures up unfavorably on 15 counts is definitely a poor parole risk. Conversely, a prisoner who possesses less than five unfavorable counts will probably make a fairly adequate adjustment on parole. Three slight criticisms have been made of the system. First, each factor is weighted equally; second, some of the classifications are somewhat ambiguous. For instance, what is meant by *socially inadequate,* or *ne'er-do-well?* But on the whole some such system, taking into consideration a group of categories based on the day-by-day activities of the inmate, together with his previous record, family life, and previous work record, might be of great help in predicting parolability. Certainly, despite criticism it is an improvement over the "hit-and-miss" system still used in most states.

Other studies have since been made by the Gluecks of Massachusetts, and Professor George Vold of Minnesota. These students of parole prediction have developed an elaborate technique that shows promising results and that may or may not be eventually accepted by parole boards. At present, the odds are not in favor of acceptance, partly owing to their confusing categories and even more involved statistical language, and partly to cultural lag that is ever present in the field of penology. At present, however, parole prediction schemata are being used to a limited extent in Illinois and Wisconsin.[21]

So long as we have no clear cut indefinite sentence laws, men will be released from prison regardless of prediction charts. The logical question then is, shall they be released some time prior to the completion of their maximum sentence, under rigid supervision and regardless of the results of prediction charts, or shall they be kept in prison and finally released at the *expiration* of their maximum sentence *without supervision?* No doubt some sort of prediction tables may be of value in attempting to ascertain *when* a man is ripe for parole, but there is some danger of losing sight of the man as a living personality, when we spend our time working out tables made up of qualities we believe he does or does not possess at a certain time during his imprisonment. Tables are not justified when they encourage us to follow them blindly up to the very end of a maximum sentence, without releasing the man for a period of supervision.

---

[21] For a review of the various studies of parole prediction, see: Robert M. Allen, "A Review of Parole Prediction Literature," *J. Crim. Law,* Vol. 32, No. 5 (January-February, 1942), pp. 548-554; Karl F. Schuessler, "Parole Prediction: Its History and Status," *ibid.,* Vol. 45, No. 4 (November-December, 1954), pp. 425-431. For a critique of the studies on parole prediction, see: Lloyd E. Ohlin and Dudley Duncan, "The Efficiency of Prediction in Criminology," *Amer. J. Soc.,* Vol. 54, No. 5 (March, 1949), pp. 441-451; Edwin H. Sutherland and Donald R. Cressey, *Principles of Criminology,* 5th ed. (Philadelphia: Lippincott, 1955), pp. 581-586.

11. Conclusions Concerning Parole

We have stressed in this chapter the contention that parole is not leniency or clemency. It is simply good insurance for society. This, we have pointed out, is based on the fact that the longer a man is in prison, the more difficult it will be for him to adjust to the freedom he must eventually meet when unconditionally released from his court sentence. Parole merely transfers the convict from the restraints of prison to the community where he is confronted with another restraint, that of supervision.

We have contended throughout this book that today's prison is the poorest place we could possibly provide for reformation or rehabilitation. Since we have it with us, the quicker we can get a man released from it, the better. Adequate safeguards, however, must be set up through a rational parole plan. We again emphasize the point that few parole systems envisage thorough-going parole planning.

The one thing that prison authorities can contribute most effectively to the improvement of the parole system is alertness to the exact psychological moment when a man should be admitted to parole. When he exhibits those characteristics and behavior patterns that indicate his sincere interest in going straight, and when his community assets are adequately mobilized for his reception, he should at once have parole privileges. To keep him longer breeds bitterness and may destroy all real desire to reform. It may be seen from this recommendation that the indeterminate sentence must be expanded and made available for the release of the inmate when he is psychologically ready.

A progressive parole board will certainly scrutinize carefully the man's institutional record and give serious consideration to the recommendations of the prison officials—men who know the inmate best. Unfortunately, all too many paroling authorities give little thought to this record but weigh rather the previous criminal record and type of crime committed. Certain types of conventional offenses, fitting into the parole board's prejudices, unfortunately carry more weight than the actual potentiality for satisfactory community adjustment.

The advantage of setting up objective criteria for parole eligibility is that they may be examined by interested parties and minimize pure caprice or prejudice in granting parole. The prison, then, would function as a *proving-ground* for the prospective parolee. The old-style prison does not serve such a function, but in many of the states and in the federal system, the prison authorities should be able to endorse or reject applicants. When classification clinics are properly manned and installed in prisons, they should be given authority to grant parole. The parole boards should have complete authority over the supervision of parolees and power to return them to the institution for parole violation. Only the classification clinic can have the long personal contact and full knowledge needed to determine when an inmate is fit to be paroled.

An all-out indeterminate sentence law would further permit the courts to send those who are reasonably hopeless to an institution for life, or until the time when, by the aging process (maturation), their release would do no harm to society. But, as we have insisted elsewhere, such places of permanent segregation should not be places of punishment, but of safekeeping.

36

A New Look At Corrections—Prospectus

When thee builds a prison, thee had better build with the thought ever in thy mind that thee and thy children may occupy the cells. (Quoted in The Presidio, *Iowa State Penitentiary, Fort Madison (December, 1957), p. 7.)*

ELIZABETH FRY

1. The Dubious Value of Imprisonment

We have taken the position throughout that prisons as we know them in our culture have failed in rehabilitation and, in fact, have been the instruments in hardening many of their victims in antisocial attitudes. We are not prepared to abolish them all at this time, though we are convinced that the swing will eventually be in that direction. Imprisonment as we understand it is less than 200 years old. It was introduced with enthusiasm as a substitute for the physical punishments of the Middle Ages, ushered in through the philosophy of expiation and penitence. We demonstrated that the cruelty of earlier practice and philosophy was carried over to the prison and, in more recent times, supplemented or changed to a degree by psychological punishment. The prison emerged as a place of contamination—exactly what the advocates of the penitentiary tried to avoid—where accidental offenders or novices in crime are debased through association with chronic criminals. Even the best intentioned personnel has not been able to dissipate this serious evil of imprisonment.

It is a fact little known to moderns that it was a Roman jurist, Ulpian, living during the reign of Emperor Caracalla (A.D. 211-217), who protested against prisons as a place for punishment: "Carcer ad continendos homines, non ad puniendos haberi debet—Prisons ought to be used for detention only, not for punishment."[1] Were this wise jurist alive today he would be shocked at the wide use of imprisonment throughout the world for the purpose of punishing criminals. In 1948 Professor Max Grünhut of Oxford University, after surveying the history of imprisonment as an attempt to deal with the offender, stated: "After more than 150 years of prison reform the outstanding feature of the movement is its skepticism concerning imprisonment altogether, and its search for new and more adequate methods of treatment outside prison walls."[2]

The belief that the pentitentiary is a failure or the suggestion that it be abolished are not new. Dozens, or probably hundreds, of persons in official and unofficial positions have denounced the institution as a failure in reformation. Professor Frank Tannenbaum, for many years a close student of prisons, wrote in his *Wall Shadows* in 1922: "We must destroy the prison, root and branch. When I speak of the prison, I mean the mechanical structure, the instrument, the technique, the method which the prison involves."[3] Professor John L. Gillin, a reliable student of corrections, writes as follows:

> What monuments of stupidity are these institutions we have built— stupidity not so much of the inmates as of free citizens. What a mockery of science are our prison discipline, our massing of social iniquity in prisons, the good and bad together in one stupendous *pot-pourri*. How silly of us to think that we can prepare men for social life by reversing the ordinary process of socialization—silence for the only animal with speech; repressive regimentation of men who are in prison because they need to learn how to exercise their activities in constructive ways; outward conformity to rules which repress all efforts at constructive expression; work without the operation of economic motives; motivation by fear of punishment rather than by hope of reward or appeal to their higher motives; cringing rather than growth in manliness; rewards secured by betrayal of a fellow rather than the development of a large loyalty.[4]

All of the prisoners or ex-prisoners who have written on their prison experiences have seen the sophistry inherent in the philosophy of prison reformation. Articles in some of the prison press, where censorship is not too severe, complain of the stultifying sham of imprisonment where initiative is regimented, if not completely stifled. They point out the abnormality of the social life they are forced to lead, all of the innumerable rules they must obey, of the deference they must pay the custodial officers, many of whom are as institutionalized as the inmates. The routine of the prison fits into the concept of punishment or restricted liberty, but not one of reformation or rehabilitation.

---

[1] Max Grünhut, *Penal Reform* (New York: Oxford University Press, 1948), p. 11.
[2] *Ibid.*, p. 449.
[3] Frank Tannenbaum, *Wall Shadows* (New York: Putnam's, 1922), p. 141.
[4] John L. Gillin, *Taming the Criminal* (New York: Macmillan, 1931), pp. 295-296.

We know that those who were responsible for the first penitentiary—members of the Pennsylvania Prison Society—were discouraged regarding the new form of penal treatment they had so enthusiastically endorsed in 1790 in Philadelphia. In 1820 a report was made by one of the Society's committees that dealt with conditions in the Walnut Street Jail. It stated that no reformation had been effected in the institution owing to four reasons that, even today, are often blamed for lack of reformation: the unfitness of the building; the poor classification or inability to classify properly; the crowded condition of the prison; and the want of employment.[5]

Later we find other complaints. Franklin B. Sanborn, secretary of the Massachusetts State Board of Charities, writing in 1866, asked the rhetorical question: "Do our prisons work reformation of the criminal? Go to our prisons and inquire of the officers, hear the stories of the convicts, watch the workings of the system and you will see that instead of reforming they harden the criminal."[6] In 1867, after almost every state had spent millions of dollars on Bastille prisons, a survey was made by two students of corrections, Theodore Dwight and Enoch C. Wines, and their conclusion was: "There is not a state prison in America in which the reformation of the convict is the supreme object of the discipline."

Down to the present day we find despairing criticism of the prison. Edward R. Cass, secretary of the American Correctional Association, writes:

> Two-thirds of the 175,000 offenders in federal and state prisons today have gone *unchanged* through not only one but often two, three or more incarcerations. Or, to put it another way, there is an habitual, roving criminal group in the United States, estimated recently by the Federal Bureau of Investigation to number 3,000,000, *which our penal system is doing nothing to reduce.* Moreover, these same prisons, which now are failing in the job of reformation, will ultimately receive 60 per cent of the estimated 1,000,000 juvenile delinquents now in the country—receive them and discharge them, in most instances, as greater menaces to society than when they entered.[7]

There is some evidence that society's conscience is a little frayed concerning imprisonment. There are today three attitudes concerning the disposition of criminals. These attitudes parallel three stages in the development of correctional philosophy since the emergence of the prison. The first era reflected the attitude that criminals should be punished for their acts—punished in such a way that reformation would result. Many old-time prison men still cherish this notion, doggedly maintaining that those who violate the law should be taught a lesson through punishment.

---

[5] From the report of the Visiting Committee of the Society; see Negley K. Teeters, *They Were In Prison* (Philadelphia: Winston, 1937), pp. 80-81. This is probably the first realization that the prison was failing to achieve its purpose.

[6] Quoted in the 37th Annual Report of the Board of Inspectors of the Eastern Penitentiary of Pennsylvania, 1866, p. 23.

[7] "Prisons Breed Crime," *The Nation*, Vol. 184, No. 19 (May 11, 1957), pp. 410 ff.

The second era witnessed the acceptance of the thesis that a "renovated" prison is the answer to the crime problem. A great many progressive prison wardens, jurists, and other close students of the problem accept this view. They have confidence in streamlining the prison in such a way that it will really be a reformative instrument.

It is this approach that makes it difficult for any advances to be made in corrections. It is this approach that insists on running a clean and efficient prison so as to pass the inspection of a governor, progressive members of the legislature, and some superficially informed reformers, but that fails to eliminate the objections inherent in imprisonment.

But here we must point up the weaknesses even further. In this new dispensation, the machinery of reform is set up inside the institution, which is paradoxical, wasteful, and woefully inadequate. It is paradoxical in that we find reformatory machinery in a contaminated atmosphere still dominated by the locking psychosis. It is wasteful to attempt a treatment program that is nullified by the custody complex. It is inadequate because the treatment personnel cannot possibly handle more than a very few inmates.

The third era has developed or emerged from the disillusionment of the second, which we might term the "humane" period. The thesis held by a fairly large number of prison administrators and penologists in this current era is that true reformation cannot flourish in a punitive setting such as the prison represents. Some administrators holding this view have tried to resolve the many paradoxes of prison life, and have come to the conclusion that the prison cannot reform.

Some critics of the prison suggest farm colonies or "prison hospitals" but such ideas still have incarceration or imprisonment in mind. Our mountain forestry camps in California and in a few other states, worthy as they are, still mean incarceration. No matter how benign the establishment, the weakest link in the chain of imprisonment is the congregation of persons held together under duress in an abnormal atmosphere. Modern corrections, as envisaged in its best phases, employs good classification procedure, a meaningful program for each inmate, intelligent staff, good food, and sleeping quarters. But despite these excellent features of imprisonment, we fail to find reformation. Conversely, we find contamination, bitterness, and the nullification of all decent drives of the personality.

2. The Role of Punishment in the Corrective Process

It must be accepted by society that if we are to develop a scientific philosophy of correctional treatment compatible with the advances made by social and medical sciences during the past quarter century, the concept of punishment must be dissipated, or at least modified.[8]

---

[8] The reader is referred to a provocative book dealing with this subject: Giles Playfair and Derrick Sington, *The Offenders: The Case against Legal Vengeance* (New York: Simon & Schuster, 1957).

If a criminal does what he must do in the light of his background and his hereditary equipment, it is obviously both futile and unjust to punish him as if he had had a real option to go straight and had deliberately chosen to do otherwise. It would be as foolish to punish him for having contracted tuberculosis. This consideration entirely destroys whatever logic there was in social revenge as a basis for punishment.

We realize that society is sadistic and that the uninformed public howls for vengeance, especially when violent and sordid crimes are committed, but we are convinced that the social pulse can be made less jumpy if we have the will. Two illustrations demonstrate the progress discernible in the social thinking of the general public. The first is the general feeling of the persons interviewed on the streets in Chicago in 1946 regarding the disposition of William Heirens, who had been apprehended for several revolting cases of assault and murder.[9] The majority of those interviewed held that he should be sent to a mental hospital rather than to a prison. (He was sent to prison regardless). The second case is that of Howard Unruh of Camden, New Jersey, who ran berserk in August 1949 and killed 13 innocent persons—an all-time high in berserk criminal killing. The average opinion in that vicinity was that he should be segregated in a mental hospital rather than be tried and sent to prison. New Jersey makes just such provision for cases of this type. This deranged ex-soldier was segregated in the state mental hospital. Not many years ago society could not have accepted such a solution for problems of this sort. In fact this solution was not even considered in Nebraska for the disposition of Charles Starkweather, who in 1958 was tried for the wanton killing of 11 innocent victims of his pathological behavior and was sentenced to death.

However, it seems not too much to expect the public to accept gradually the thesis that the criminal of any type is socially ill and needs diagnosis and treatment applicable to his needs. The reader must not assume that we are suggesting that a criminal go "scot free" as an alternative to imprisonment. It is not a case of punishment versus complete freedom. Treatment, involving a re-direction of personal attitudes and habits is necessary.

We are only beginning to learn that we cannot deal with crime in the abstract. A crime becomes such only when a human being comes into the picture and commits an act that society calls a crime. We have already noted elsewhere that what constitutes a crime depends on the penal code —a question of time and geography. Hence, we see that what we must deal with is a human being whom society calls a criminal. One hundred types of men may commit the same crime but the circumstances surrounding the totality of each act are unique and must be studied as such.

---

[9] See our discussion on this case, p. 102.

Thus it is absurd to "make the punishment fit the crime." A sentence that would be just for one criminal might be absurdly light or severe for others.

After all, what we are aiming at is protection of society from criminals and preventing crimes from being committed. Hence, we should approach the problem rationally, and not feverishly demand a more frantic application of the same methods that have been tried and found wanting for so many decades.

If we think of treatment or even meaningful discipline as punishment, then we cannot expect to be rid of punishment. Nor should we wish to. A person suffering from a contagious disease must expect to be quarantined; this is not punishment. A convicted criminal must be held socially accountable, but what treatment is outlined for him must be acceptable by him as in society's interest. Punishment, as we have understood it in the past, has failed in its task. We must explore every avenue in the realm of treatment in order to re-direct the criminal's thinking. If this cannot be accomplished, segregation from society can be the only answer.

Its elimination cannot be accomplished overnight. We shall attempt to be specific in our recommendations and proposals, although it would take the wisdom of a Solomon to prescribe perfectly for the socially maladjusted. Yet we must make a beginning, and that beginning must be premised on the conviction that the prison, as we have known it for 175 years, must be outlawed.

This vital theme of the future elimination of the prison and the substitution of non-institutional treatment for most criminals calls for methods that are rational and scientific. Our fear of criminals and our determination to lock them up is a main cause of opposition to the indeterminate sentence, suspended sentence, probation, parole, conditional release under close supervision, and the employment of ex-prisoners.

We stated above that there are not enough trained personnel to treat adequately the men sent to prison for long terms. But a prison that is merely a *diagnostic depot* could be reasonably well staffed. The work of the staff would be to analyze the prisoner's needs, prognosis, and course of treatment. The beginnings of his treatment might be started in the prison, but the quicker he can be placed in the community, the better for him and for society. Thus, we would have the indeterminate sentence modified by these interesting features. An inmate would be prepared for the earliest release possible, in terms only of his ability to respond to the type of treatment recommended. He would then be fitted into one of a variety of plans. Plan "A" might call for him to work inside the prison and return home in the evenings; plan "B" might embrace work outside the prison with a return each night to the institution; plan "C" might mean the man would work and sleep outside with an occasional return to the prison for conferences regarding his adjustment. Other plans might be made for different types of cases presented to the staff at this diagnostic depot.

Just such a plan is working in the van der Hoeven clinic at Utrecht, Holland under the direction of Dr. P. A. H. Baan. This establishment deliberately receives the most chronic offenders in the country's prisons and develops a therapeutic plan for each person. Many of the inmates go out to work each day in the free community and return at night to the clinic. This institution might be called what Howard B. Gill refers to as a "therapeutic community." The Utrecht clinic is psychiatrically oriented and its program has been accepted by the community.

Many prison administrators have recognized for some time that a man can be ready for release (parole) one, two, or more years before he has served his minimum sentence and is legally eligible, and that when the time finally rolls around, he has become so institutionalized that recommendations from the prison staff are rejected or reluctantly given. This is a serious problem in prisons today.

The partial indeterminate sentence embodied in the minimum-maximum laws found in many states today might well stand under this new dispensation, although a stricter and quasi-limited law, such as is found in California, is more desirable. Our minima today are altogether too high. And perhaps the strict indeterminate sentence would be best although there is honest doubt among progressive thinkers as to its ultimate effectiveness. Then the released prisoner, working outside the prison and reporting periodically to the staff, would continue thus until the date of parole eligibility. He would then receive parole through an Adult Authority and would be under the jurisdiction of the paroling agency until he had served his maximum, just as at present.

The advantage of this system is that the prison (diagnostic depot) would have control over the man until he was finally paroled. He would have problems that the prison staff, with their expert knowledge, could consider and might solve. No one would be released on this conditional form until a thorough plan had been worked out. If the man had a normal home life, a trade, and a good work record, a job would be found for him. If there were no job available, but his home life was adequate, he could work in the prison, or take a vocational course, and live at home. One may see many possibilties in a system that would make the New Prison a real diagnostic center with a fair modicum of treatment.

We are none too optimistic about any speedy introduction of so rational a program. It could be made to work if society were able to accept the diagnostic depot idea and if the lawmakers and judges would entertain such a procedure—but society, lawmakers, and judges almost certainly would not agree to such a plan.

The Adult Authority of California, created by legislative act in 1944, is the closest we have come to non-institutional treatment, with many convicted prisoners working outside institutions in the mountain and forestry camps.[10] The Adult Authority Board, composed of six persons appointed by the governor, is empowered to fix the terms of sentence (within the limitations of the state's indeterminate sentence law) of every adult sent to prison by a court. It is also empowered to grant parole. Integrated with classification clinics operating in each adult prison, the Authority operates in four distinct areas: (1) reception, diagnosis, and orientation; (2) treatment and classification, reclassification, and preparation for release on parole; (3) fixing of the terms of imprisonment and selection of the inmate for final parole; and (4) community supervision in post-institutional treatment.

In the interest of progressive corrections, the fixing of the sentence by the Authority rather than by the judge, with periodic review, is of inherent value. A serious criticism is that the inmate is still imprisoned in a grim institution such as San Quentin and Folsom, large maximum-security establishments, and that he is controlled by a bureaucratic board of six men who have power over him far exceeding that found in any other state.

Aside from the California Adult Authority, which constitutes a break with past concepts, at least to a degree, we find a new progressive philosophy of corrections permeating the armed forces since World War II. In this highly authoritarian, dictatorial culture-complex it has been rigidly traditional for the most repressive penalties to be meted out to those who violated discipline or committed crimes. During the American Revolution soldiers were flogged for wearing a hat uncocked, possessing a dirty gun, or for swearing. One quaint punitive device was known as the "whirligig," a sort of rapidly revolving cage in which the offending soldier was placed. The nausea and agony caused by the whirling motion was recorded as "unspeakable." An old army book records that:

> Raw recruits sometimes cried out or dropped down in ranks from fright at the first sight of an army flogging, but they soon grew scarcely to heed the ever-frequent and brutalizing sight. These floggings were never of any value as a restraint or warning in the Army; the whipped and flayed soldiers were ruined in temper and character just as they were often ruined in health.[11]

The Army, Navy and Air Force have all set up re-training programs, the purpose being to salvage as many of their men as possible. It makes good sense since enlisted men, as well as drafted men cost the services considerable funds for their training. The old style "guardhouse" is slowly disappearing and in its place classification boards, treatment services, and review boards are carefully considering all cases of discipline and antisocial

---

[10] It should be noted in passing, however, that California's work with adult forestry camps long antedates the Adult Authority structure.

[11] Quoted by Clifford V. Oje, "The Air Force Corrections and Retraining Program," *Federal Probation,* Vol. 19, No. 3 (September, 1955), pp. 31 f.

activity. Some 600 service officers have received correctional training at George Washington University under the veteran penologist, Howard B. Gill, now at the American University in Washington, D.C.[12]

3. Substitutes for Imprisonment

Aside from probation, which we have discussed in detail elsewhere and which we believe is the greatest hope of the future, let us explore other possibilities that may serve as substitutes for a prison sentence. Perhaps we might state that there reside within the community many resources that have therapeutic value. There are agencies and services where professionally trained personnel may be utilized. Health clinics, hospitals, social case working agencies, and psychiatric institutes and clinics are among such services. On a different level are settlement houses, hobby classes, and clubs, often staffed by experts or quasi-experts in respective fields. On still another level are missions, Salvation Army facilities, and lodges or retreats which manage to be of service to certain types of social problem cases.

Our thesis throughout this book is that we need trained personnel if we are to deal effectively with those who are socially pathological. Yet we must admit that non-professional personnel can be of great service in working with certain individuals and thus we must learn how to work effectively with all resources within a community.

Certainly one type of prison substitute is psychiatric or casework service outside the prison. Organizations of community therapists could enter into a working agreement with the courts, possibly under the auspices of a public health service or through hospitals, whereby a convicted man or woman would be treated. It is important, however, to stress the voluntary nature of such a relationship. No man should be compelled to take such treatment.[13] Since so little of this type of work has thus far been done, it is merely suggested here as an area that should be persistently explored.

Other substitutes are: indemnity for the victim; fines based on severity of offense or capacity to pay; a training program over a period of time; abrogation of certain rights of citizenship. It must be emphasized that supervision by trained and understanding personnel is presupposed in any of these techniques.

Very little research has been conducted in this country in any of these

---

[12] For a description of treatment programs in the armed forces the reader is referred to Ruth S. Cavan, *Criminology,* 2nd edition (New York: Crowell, 1955), Chapter 22, "Treatment of Offenders in the Armed Forces."

[13] Pioneer work in this field has been done by Dr. Melitta Schmideberg and her associates in New York City. These psychiatrists have developed what is known as "The Association for Psychiatric Treatment of Offenders" (APTO) which is called upon by various official agencies to render treatment to those needing such services. See Jack Sokol, "A Pioneer Approach in the Treatment of Offenders," *J. Crim. Law,* Vol. 45, No. 3 (September-October, 1954), pp. 279-290.

methods. Few, if any, attempts have been made to utilize them except in very minor offenses. But merely because we have not used them does not mean they cannot work. All of them have been tried in other countries and from what evidence we can adduce they seem to work. Some of them have their origin in antiquity, where they were fairly adequate. There can be no doubt that all of them have some merit. If we have the will to experiment boldly we might usher in a new era of correctional treatment.

Restitution or indemnity for an offense against a person would, in many respects, be more sensible than present day practice. As it is today, when a crime is committed it is, in reality, an offense against the state rather than against a person. Thus it is the state that must avenge the wrong. Imprisonment is the easiest answer. If a man is robbed of a sum of money there is little chance that he will recover it. He is merely asked to appear in court against the thief and thus assist the state in convicting and sending him to prison. Many persons balk at this for fear of retaliation by the criminal. A system of restitution would make it possible for the victim to be reimbursed for his loss. The same logic could be applied in cases of physical damage. The court, or jury, could stipulate the amount of damage to be collected. Restitution could be made in installments.

Restitution may sound fantastic to present day readers as a substitute for imprisonment. Yet it is well known that during Anglo-Saxon days it was possible for the offender to forestall prosecution by the payment of what was known as *bot* or *wergild*. Later, however, the idea of restitution of stolen property, for instance, passed into the civil law, where today it is administered under the law of torts. Here is a principle in which, for the social good, civil and criminal law could well develop a more effective co-operation.

One rarely sees the matter of restitution mentioned in any of the modern literature. But it was considered important enough to elicit discussion and lively debate at two world Penal and Penitentiary Congresses, first at Paris in 1895 and second at Brussels in 1900. At the latter conference thirteen reports, covering 147 printed pages, were submitted by some of the world's expert jurors and professors of law, including Judge Simeon Baldwin of Yale University.

Merely because we have not carried through the possibilities inherent in this substitute for imprisonment—aside from a few cases of installment payments of fines in lieu of a jail sentence—it does not follow that the idea is not sound or expedient. That there are difficulties involved will not be denied. The amount of restitution, difficult though it may be to compute, is no more complicated than evaluating the amount of time a man must serve in prison for an offense. The problem of collecting installments over a period of time is not insurmountable. We are able to collect taxes from these same persons; why not their indemnities?

Impinging on this possibility of restitution we find another scheme that has merit but that thus far has not been well received. That is the practice of invoking fines upon a person convicted, according to his ability to pay. It is one of the glaring injustices of our system of jurisprudence that money fines are pronounced by the judge without his knowledge of the hardship imposed on the culprit. Thus a $25 fine may be vastly more formidable for one person to assume than another. Imposition of fines as we understand it today represents one of the most glaring of our discriminations.

Invoking a fine, based on the individual's ability to pay, has considerable merit. The wealthy person, convicted of an offense, may be fined ten times more than a laborer for the same offense. This philosophy of fining could work by establishing units of fines rather than amounts. For example, if two persons have committed the same offense and each has been fined 30 units, the judge may interpret a unit in one case as five dollars and in the other, one dollar. Thus the first person would pay a fine of $150; the second, $30. Education, training, and accumulation of wealth, should hold a person more liable for his acts than others not so fortunate.

Again there is nothing new nor radical in this suggestion. Montesquieu advocated it by stating "A gradation should be established between different penalties corresponding to the resources of the offender." Jeremy Bentham held the same idea. Emmanuel du Mouceau, procurator of the Republic of France, stated at Brussels that if a workman is fined the equivalent of four days' labor, it is only fair that a wealthy person, guilty of the same offense, should be obliged to pay four days' income. He further added that considerable funds could be built up in this manner so as to draw from it to pay the fines of insolvent offenders.

The question of substitutes for imprisonment is thus a lively one in twentieth century correctional debate. To the other international suggestions listed above we should add the following, which are being, or have been, tried in various South American countries. In Colombia when it is determined that the offender is not guilty of a serious crime, he may be prohibited from residing in a determined place; given a conspicuous or special publication of his sentence; prohibition of public rights or functions; suspension of the arts or of a profession; loss of all pensions or wages of retirement of a civil character; required to post guarantee of good conduct; loss or suspension of the rights of citizenship; mandatory labor in certain factories or on public works; prohibition from appearing in certain public places. Ecuador has some of the above penalties, and in addition the court may confiscate a portion or all of an offender's property.

4. Evaluation of Victim's Responsibility

Aside from children under age, the obviously insane criminal and, to a degree, the mentally retarded, all persons committing crimes have, through the years, been held completely responsible. There have been certain extenuating circumstances, of course, but by and large, the thesis has been and still is that the perpetrator of the crime is to be condemned without any thought given to the possible contribution of the victim. Recently there has been some thought focused on the degree of responsibility of the victim in any type of crime, whether it be a murder, burglary, arson or other offense. Some persons who have committed a murder are acquitted by a jury because it was skilfully brought out in court by the defense attorney that the victim had contributed to his death. In many confidence games, robberies, and other offenses against property it is obvious that the victims "asked for it." In short, in many instances the victim sets the stage for the crime.

This thesis is explored quite cogently by Professor Hans von Hentig in his book, *The Criminal and His Victim*.[14] He shows that there is a wide range of interactions between criminals and their victims, so that in a very large percentage of criminal acts we must look for the "subject-object relationship." As he says: "In a sense the victim shapes and moulds the criminal."[15] This can easily be ascertained by objective investigators of a criminal act. In a rape case, for instance, just how far or to what degree did the victim set the stage for her assailant? To what degree do victims of robbery prepare the scene for the thieves? In the notorious jewel theft of the late Agha Khan in France a few years ago, amounting to some $750,000, just how responsible was this colorful spiritual leader for encouraging thieves to relieve him of his baubles? Just how far did the proprietors of the Brinks Corporation of Boston, by their laxity, contribute to the million and a half dollar robbery in January, 1950, reputed to be the largest single theft in American history by conventional criminals? Perhaps the most fantastic robbery of recent years, although not nearly so large as that of the Brinks Corporation was planned by three "bungling" amateurs for a "nice, busy little bank that ought to be robbed" in Port Chester, New York. Apprehending a female bookkeeper who had the key and whose custom it was to open the bank each morning, the trio cruised about town most of the night and forced her to open the door a few hours before the rest of the staff came to work. Two of the men became restive and left the third in the lurch. This lone, determined, "race-track tout"

---

14 Hans von Hentig, *The Criminal and His Victim* (New Haven: Yale University Press, 1949), see especially Part 4.

15 *Ibid.*, p. 384. On the question of "victim-precipitation" in homicide cases, see Marvin Wolfgang, *Patterns In Criminal Homicide* (Philadelphia: University of Pennsylvania Press, 1958), pp. 255-261.

went through with it alone. Almost in sight of police, commuters, and many other townspeople the lone bandit drove off with $180,000, and was later joined by his erstwhile partners. The story of this amazing bank robbery is almost a classic since it demonstrates how easy it seems to be to commit crimes by the perpetrator doing what is least expected.[16]

Americans are notoriously careless about their personal belongings. Despite warnings from the police concerning the dangers of pickpockets, car thieves, burglars, and other types of criminals, we pay little attention to even the rudiments of caution. We purchase first-rate locks for the front doors and tolerate ordinary catches or common locks on the kitchen doors that can be opened by a skeleton key. Windows are left unlocked, property is left unguarded, especially in our cars. People crossing the country in their cars, loaded with expensive clothing, furs and even jewelry, park most anywhere for a meal or for an overnight stand and too often find their valuables gone when they return. Regarding stolen automobiles, the late August Vollmer, former chief of police of Berkeley, California once wrote:

> Of the 7,920 automobiles stolen in 1933 [any year will serve the purpose] in one city, 5,180 of the persons reporting the losses admitted they had not locked their cars when they left them. No intelligent person would put from $1,000 to $5,000 in good money in the street and expect to find it there an hour or so later, yet that is exactly what a large number of people do when they leave an automobile in the street without locking it. Even more; not only are they leaving money at the curb, but they are also putting four wheels under it to make it easier for the thief to take it.[17]

Many crimes of violence seem to be "asked for." In how many cases of "forced" rape, as well as statutory rape, did the victims encourage or lead on the "attackers"? How many persons have been killed merely because they goaded the murderer? We could go on indefinitely with suggestions to appraise, or at least attempt to appraise, the degree of responsibility of the victim in each case.

It is this thought of responsibility or liability of the victim that should give us some pause in discussing not only substitutes for imprisonment, but old-fashioned concepts of punishment as well. We emphasize throughout this book that the philosophy supporting the idea of punishment has been undermined both by modern knowledge of human behavior and by the actual fruits of punishment. Modern psychology, sociology, and psychiatry have clearly shown that we can no longer accept the doctrine of free will since our attitudes and actions are profoundly affected by our heredity, life experiences, and living conditions.

It must not be implied that we propose to punish the victims of crimes

---

[16] For details, see Evan McLeod Wylie's story in *Life,* Vol. 42, No. 22 (June 3, 1957).

[17] August Vollmer and Alfred E. Parker, *The Police and Modern Society,* (Berkeley, California: University of California Press, 1936), p. 65.

either. Nor do we suggest any degree of "blame" merely because an attractive female uses her charms to entice —wittingly or unwittingly—a male into seducing her and thus spoiling her reputation or even bringing about her death. Reviewing the cases we mentioned earlier we wish to state that we do not "blame" the Agha Khan for possessing his tempting jewelry and displaying it before an envious public; nor can we "blame" the proprietors of the Brinks Corporation of Boston if they were lax in safeguarding the money that was so easily lifted by the bandits. Victims of crimes are no more responsible than are those who take advantage of them. We may go even further. Society itself is not to be "blamed" for its philosophy of punishment or of falling back on imprisonment as a security device. As we learn more about human motivation the less we think in terms of blame and more in terms of understanding. In every instance of crime all factors must be appraised, and in the light of these factors we may dispose of the case. Since we take the position that imprisonment has failed so signally to rehabilitate most of those upon whom it has been imposed, we contend that other forms of treatment, especially those that may be utilized outside the prison, must be tried.

5. Scientific Disposition of Criminals

One of the cardinal principles of modern criminal jurisprudence is that no one can be convicted unless it is proved that there accompanied the overt act a *willful intent* to commit such an act. It is further assumed that only those *morally free* to commit an act are liable, so that those who are not responsible—like young children and the insane—should be disposed of otherwise than through the traditional court process. The criminal law requires that before a person can be convicted of a crime, it must be established that he had a willful or evil or criminal intent to commit the act. This philosophy has had a long evolutionary development that need not be expounded here. The fact that it persists is interesting in the light of modern biological science, psychiatry, and medical knowledge.

When a crime is committed, two questions present themselves: "Who committed it?" and "Why did he commit it?" It is the object of criminal law and modern jurisprudence to find satisfactory answers to these two questions. It is no longer necessary to divide criminals into the once popular and convenient categories of accidental, professional, passionate, insane, and so forth. Each individual is a peculiar entity unto himself and must be studied as such. Roughly speaking, for rough it is, we might classify individuals into categories, but, in the last analysis, each individual who commits a crime is in a class by himself.

It is therefore imperative that machinery be set up to make a thorough study of each individual indicted for the commission of a crime. Innocence or guilt of the overt act should be ascertained by the most advanced methods, without the mass of archaic hocus-pocus found today in our court trials. Reform in this field is essential.

Regardless of the methods to be used in ascertaining guilt or innocence the next step is to attempt to understand the "why" of the act. Here we must turn to those sciences that have for their content the analysis of human behavior. Pre-sentence court clinics, clinics of "experts" or "dispositional tribunals," are the only scientific method that can command the respect of intelligent people for the diagnosis and proposed treatment of those individuals who have been found guilty of the commission of a serious offense.

Individualized treatment is the next step. At present, we give lip service to this type of crime therapy. A first offender, who has had an unblemished record as a peace-loving citizen, a good neighbor and a reasonably good father and husband, suddenly commits an atrocious murder. Immediately the neighborhood, the community in which he lives, becomes incensed against him and demands the electric chair. The "worst is too good for him." Yet, after the excitement of the crime has died down and more interesting events detract attention from the misfortune of their neighbor, he could be released on parole without much criticism. Obviously, the only reason his neighbors want him incarcerated is for revenge—that old, universal, sadistic urge.

Probably the reason why the new penology is so popular is because it meets this sadistic urge of the public. It punishes the culprit; "reforms" him by surrounding him with psychologists, psychiatrists, a good hard job, and all the trappings that make the New Prison what it is; and then sends him back to the community chastened and contrite. But that is not what penologists want, since they know full well that the man has been subjected to the most demoralizing and unproductive experience that society can inflict upon him. His experience in the New Prison has stigmatized him, broken him, degraded him, and thrown him back on the community a wasted wreck.

So even the New Prison must go. Each case, regardless of the attitude of the community, must be treated on its merits. After a thorough investigation is made, the plan of treatment must be decided on and immediately put into effect. For the time being, some of these individuals will be sent to prison and probably kept there for life, regardless of the offense. But under no circumstances should they be mistreated or punished in the traditional sense. It can well be argued that those who must be put away should be sent to a hospital-colony where they can lead a life that is as normal as possible.

When we think of a felon, we usually picture a serious criminal, a hardened individual, habituated to committing serious crimes. But many felons are first offenders. Some have committed murder. We lump them together on the basis of the *type* of crime they committed. Instead, we must know them in terms of their backgrounds and potentialities.

We have shown that the bulk of our crimes—about 97 per cent—are either misdemeanors or vices. It is difficult to know how many felonies are committed annually in this country. The best information on this comes from the Federal Bureau of Investigation in its semi-annual *Uniform Crime Reports,* according to which the *estimated* number of serious crimes committed in the year 1957 was 2,796,400. Thousands of these crimes were committed by persons who will never be apprehended by the police; other thousands by persons who will be permitted to plead to less serious crimes; many others will be acquitted by juries or will be placed on probation.

One method by which we might know how many offenders are finally sent to prison for these serious offenses is to examine the figures given out concerning the movements of prison population. Our latest figures for these, however, are for the year 1956. In that year we note that 80,408 sentenced prisoners were received by the prisons from the courts and 85,356 were discharged through parole, conditional release, or through expiration of sentence.[18] As a fair estimate, then, we might put the number of serious offenders each at from 80,000 to 85,000. It is this group which concerns us in this discussion. Out of this number, the thousands of mental defectives and other thousands who are mentally ill do not belong in penal institutions. Others are hardened offenders, a potential menace to society, and need to be segregated for *life* regardless of the offense—with the exception that some of them as they grow older will become more docile and less harmful. So the final release of any of this group of dangerous criminals should depend entirely on careful day-by-day scrutiny of their limitations and potentialities. This can be done by the expert personnel that will be attached to institutions set aside to keep such persons in custody. As time comes for their release, they should be placed in the community, conditionally and under careful supervision. If they fail to measure up, they should be sent back to the institution.

The seriousness of an offense against society depends largely on the elements of time and place. The extent of the social harm done usually determines the attention that crimes receive from the law-enforcing or control group. True, there are various infractions that have always been considered more or less harmful. Garofalo, the Italian criminologist, groups these under the headings of crimes against *probity* and *pity.* Those offenses that call forth social apprehension are likely to be considered most serious. For example, in the ranching and cattle-raising days of the West, horse-stealing was, no doubt, a much more serious offense than murder, which today is generally regarded by the rank-and-file as the most heinous crime. Today horse-stealing is inconvenient to the man owning a horse, but a

---

[18] *National Prisoner Statistics,* Federal Bureau of Prisons, Washington, D.C. No. 19, July, 1958, Table 2.

community does not become highly enraged over such a theft. Tire thefts at the moment create more resentment.

Though murder is traditionally considered the most serious of felonies, many big-time crimes committed against society are much worse in their effect on the average human being than occasional murders. So far as society is affected, murder is not as serious as food adulteration, the whole-sale embezzlement of pre-depression days, employers' evasion of laws to protect the worker from industrial hazards and occupational diseases, the various rackets in the consumer goods field, and a number of other serious offenses against *society as a whole*. The effects of the overt act of a mur-derer can never begin to measure up to the sorrow, misery, and tragedy experienced by thousands of poor persons who lose their life savings through an unscrupulous person who seemed to be a "pillar of the community." A decidedly new pattern of thought regarding various antisocial acts must develop, and the old notions, nurtured by press, radio, and screen, about murder and the "quaint" crimes of petty robbery, burglary and pocket-picking—those aimed at one person or one family—must be overhauled or changed.

In the meantime, the disposition of the case should be determined on the basis of the investigation by experts of the background of the accused, his mental and biological potentialities, and the possibility of his making restitution *to his victim's family*. The plan adopted should involve his amenability to a program of helpful supervision by someone well grounded in the techniques of dealing with human behavior problems. Many prob-lems will emerge. If the offense is such that his social group will have none of him, arrangements must be made for some form of day-by-day adapta-tion to his life outside of an institution. Many persons are placed on pro-bation today, and if the probation staff is capable and adequate, it works reasonably well. The probation technique should be extended.

We do believe that those who violate the laws should not be excused nor permitted to enjoy the privileges of the free man. We must insist, how-ever, that each one must be treated individually. For some, advice is all that this treatment or discipline should be; for others supervision, kind and understanding but strict, may be necessary; for still others, some form of monetary restitution will be recommended by the diagnostic clinic. This restitution might be exacted for years or even for life, a few dollars at a time. Each case will call for some special form of treatment.

Lewis E. Lawes, former warden of Sing Sing, wrote a revealing book in which he substantiates what all penologists and penal administrators know, that the average murderer is an ordinary citizen without any previ-ous notion of committing a crime until confronted with a situation that seemed too much for him to solve in any rational manner.[19] As one reads

---

[19] Lewis E. Lawes, *Meet the Murderer* (New York: Harper, 1949). See also, W. Lindesay Neustetter, *The Mind of the Murderer* (New York: Philosophical Library, 1947).

the various cases presented by Mr. Lawes, one asks, "Why were these persons' lives snuffed out, when so many other more dangerous criminals are permitted to walk the streets free men?"

In the past, intelligent and well-informed persons have begun to bring about the introduction of psychiatry into the juvenile courts and the institutions for the treatment of delinquents, as the most promising avenue of immediate application of science to the repression of crime and reformation of the criminals. Now we must pursue a parallel campaign for the gradual but certain projection of psychiatry into the procedure as regards the adult criminal. The most essential thing here is to educate the public as to the sociomedical nature of the problem of crime. Today we are willing to allow medical men to care for the mental defectives and the insane, but this is the result of a century or more of education emphasizing the medical nature of these problems and disorders. As soon as the public comes to regard crime as a sociomedical problem, we shall have little trouble getting consent to hand over the treatment of crime to psychiatrists and others expert in human behavior. Perhaps a limited form of psychiatric census, achieved by linking up the public schools with child-guidance clinics, might offer some solution to this problem.

In summarizing, however, we may well reiterate the fundamental truth that we shall not make any substantial progress merely by advocating better correctional institutions. There must be a complete escape from *all* institutionalization, save for the few who need permanent segregation or a short period of institutional experience before conditional release under supervision. The rational treatment of criminals demands mainly non-institutional programs—development of constructive public attitudes to replace inordinate fear and contempt and to allow comprehensive and inclusive use of probation, conditional release, and parole. The first great revolution in handling criminals was the replacement of corporal punishment by institutionalization; we are overdue for the second, which will see the replacement of prisons by a flexible program of reformative treatment based on reason and science.

37

The Challenge of Delinqency and Crime: Prevention

It is getting to be commonplace to say that mankind knows more of the stars and planets and the atoms and physical forces than he knows about himself. Yet, it is sadly true. Some years after the Government learns to give five billion dollars to the social sciences for every billion it gives to the physical sciences, America's ingenious young researchers may solve the problem of the awfulness of crime and imprisonment. (*From* Federal Probation, *Vol. 20, No. 1* (*March, 1956*), *p. 24.*)

DONALD CLEMMER

1. The Responsibilities of the School

We have traversed a long road. The eternal challenge of the aberrant behavior of man in society has not been minimized; nor have we held out many hopeful signs for early solutions. Crime is the backwash of our culture, as Tannenbaum states it:

> Crime is the web and woof of society. It is not an accident—not just an accident. The amount, the character, and the kind of crime are socially conditioned. The good people who set out to remake the criminal, to better the police force, to expedite criminal justice, to reform the prisons, begin at the wrong end—and too late. The story starts earlier; it starts within the milieu in which the criminals grow up.[1]

So long as we have people we shall have delinquents and criminals. In no other social problem do we experience so much frustration in assaying causes and prescribing preventive measures and treatment. Yet the student dare not abandon all hope. So long as we can continue to recruit men and women who are willing to accept the challenge of studying human behavior, the fight will go on. Perhaps the analogy of medical science striving

[1] "Introduction," *New Horizons in Criminology,* 1st edition (Englewood Cliffs, N.J.: Prentice-Hall, 1943), p. vi.

for a cure of cancer is not inappropriate. Millions of dollars are poured into research and hundreds, perhaps thousands, of men and women work tirelessly and hopefully in the search for the elusive causes and cure of this dread malady. Those in the behavioral sciences can do no less in seeking the etiology of those types of human behavior that are labeled delinquent and criminal.

The most ardent and hopeful laborer in the field of crime prevention does not envisage the elimination of criminality and delinquency. It seems that only newspaper editors, and those who write letters to the editors, are impatient. They want action now and are at a loss to know why crime continues to soar. In our earlier chapters we pointed out that the causes of crime are not only legion but elusive. We might also add that attempts at crime prevention are legion as well. To say that such effort is many-sided does not shed much light on the over-all problem. Yet it *is* definitely many-sided. There is no "open sesame" to the riddle of why people resort to crime. There is, likewise, no easy answer to the prevention of the behavior society does not like.

Although there are many modes of attack, they might well be narrowed down to two; *first,* realizing the necessity of removing the social and economic forces that induce attitudes leading to delinquency and crime, and *second,* focusing attention on the individual who shows potentialities for antisocial behavior either because of biological and psychological handicaps or lack of social or economic opportunities for attaining a desirable integration.

While it is largely from delinquent youth that most of our adult criminals emerge, there are thousands of adults, both male and female, who transgress the law without a delinquent record; in fact, a fair percentage shows a history of law-abiding conduct and even moral rectitude. It is all very well to focus much attention on programs to cope with delinquency but these alone, however effective, will not prove a solution to the innumerable criminal acts of adults who showed no delinquency in their youth.

In all the books written on delinquency or crime control one finds practically nothing that might be helpful in preventing latent criminality. We have discussed the economic crimes of the upper socio-economic groups, of extortionists, of embezzlers, of "triangle" murderers, of criminals of passion who kill compulsively. Most of these individuals are adults and few indicate a history of delinquency; in fact, many reveal an apparently happy childhood. Thus we do not penetrate the roots of adult crime merely by describing or endorsing community programs for delinquency control. Yet these adults, by their offenses, seem to reveal, either in biological makeup or conditioning, a failure to develop into secure, well-integrated citizens. If the basic personality is laid down at a very early age, as many authorities on child life suggest, certain incipient frustrations and feelings of insecurity may develop unnoticed by the child's parents, teachers and friends.

Later in life, pressed in by the competitive struggle of life, or by accumulated frustrations of various sorts, the adult may develop into a neurotic, a criminal, or a psychopathic personality.

Not all insecure or frustrated non-delinquents will develop into criminals. Yet some undoubtedly do. There seems no way of insuring a race of socially mature, responsible people unless we can detect the mechanisms of insecurity and unhappiness at a very early age and attempt some type of treatment. Our child guidance clinics are clearing up some of these symptoms in young children but only a minute number of those obviously needing such help are receiving it.

How early can we give the troubled child professional help? For the first few years of his life he is with his parents who may be blissfully unaware that symptoms of frustration, such as aggressiveness, withdrawal, or other unhealthy behavior patterns may hold implications of trouble ahead. At six the child enters school, where he may coast along for several years without help. In fact he may continue through adolescence and early adulthood without finding an avenue of wholesome interaction to assist him in facing up to the problems of life. He may or may not become criminal, but at best he may be a disagreeable or neurotic person, causing unhappiness among his acquaintances and to himself. Others, similarly neglected, may have emotional breakdowns, and still others may resolve their difficulties through crime.

The conflicts with society of innumerable children do not become conspicuous enough until they reach adolescence. It may then be too late to be of much help in applying counseling or guidance aids. Authorities and parents can then only hope that they will "snap out of it" before overt damage is done. Yet from the standpoint of their personality much of the damage has already been done.

Perhaps the answer is that children should be identified with the public school much earlier than they now are—as early as three or four, in nursery schools where, through the guidance of teachers trained in social group work techniques with a psychiatric social worker available, danger signals may be detected and elementary treatment begun. In the social setting of the child's peers, he may become socialized. Through skilful guidance by the teacher each child can adjust to the group and learn to take disappointment and frustration in his stride. In short, this area has been all but ignored by the public school. It presents a challenge to society since tangible help can be given the apathetic or bewildered parent in his knotty task of child-rearing.

There is nothing startling in this proposal. For many years the private nursery schools and kindergartens have been engaged in just such programs. The child who fails to make a healthy adjustment to the group is helped by the teacher who is especially trained in early child life patterns. Through consultations with parents certain domestic adjustments are indicated and, if the parent will and can cooperate, the child's difficulties may be overcome. If professional help at a guidance clinic is indicated, this may be recommended. Certainly there is more hope of adjustment at this early age than in waiting for the child to "outgrow" his problem or to defeat it during adolescence.

The National Conference on Prevention and Control of Delinquency questions whether "teachers . . . are alert to telltale signs of potential delinquent behavior." This is probably a fair question to ask, yet we cannot expect the overworked teacher to be a master craftsman in all phases of child development. But here we may list the signs of incipient trouble as compiled by the Conference:

> Is the child unfriendly or seclusive?
> Is he failing in his school work?
> Does he play truant?
> Is he in good physical condition?
> Does he show many fears?
> Is he rejected or unwanted at home, on the playground, or in the school?
> Does he run with a gang?
> Does he have any contact with a supervised recreational program in the community?
> Does he have a church or Sunday School affiliation?
> Does he plan to leave school?
> Does he live in a high delinquency neighborhood?
> Does the child show marks of poverty?
> Does he live in a crowded and unattractive home?
> Does he have academic limitations or special disabilities that interfere with his learning?
> Does he come from a broken or deserted home?
> Does his mother work outside the home?[2]

The *Report* continues by stating that a poorly trained teacher will not detect many of these danger signals. It calls for teachers possessing the necessary insight to identify such limitations to a well-rounded child life. Every effort should be made by school boards to see that properly trained teachers are employed and that a system of counseling be inaugurated to cope with children thus handicapped.[3]

An important adjunct of the modern school in dealing with behavior problems is the counseling service. In many schools this is handled by the visiting teacher, although more frequently it is referred to as guidance or counseling service.

While the services performed by the counselor vary in different school systems there are various phases of child welfare that may conceivably fall

---

2 National Conference on Prevention and Control of Juvenile Delinquency, 1946, "School and Teacher Responsibility," *Report No. 15* (Washington, D.C.: U.S. Government Printing Office, 1947), p. 13.

3 See *Forty-Seventh Yearbook,* Part I, National Society for the Study of Education (Chicago: University of Chicago Press, 1948), which deals with the problem of detecting incipient delinquency in the schoolroom.

within the over-all functional philosophy. These are: (1) acting as attendance officer; (2) working out problems causing nonattendance; (3) adjustment of behavior problems; (4) home-school relationships; (5) referral of problems to child guidance clinics and outside social agencies; (6) direct treatment of children's difficulties. In a report made some years ago covering some 250 cities it was found that the latter four services were the most frequent.[4]

The counselor should be so trained that she speaks the language of the teacher. She should understand child psychology and have no little knowledge of social work and the findings of psychiatry. In general, the training for such positions should be along the lines of psychiatric social work but with more than passing acquaintance with the art of school teaching. Assuming that counselors are well trained, it follows that school boards should not burden them with routine administrative duties that can be handled equally well by clerks. This tendency exists in too many school systems today.

It is not to be inferred that in every school where counseling exists everything is on a high plane. The counseling atmosphere is prepared and nurtured primarily by the principal of the school. If he or she has little insight into counseling philosophy even a good worker can be of little service to the problem child. It is not enough for a city to boast that it employs a hundred counselors or a counselor in every school. Counseling services must be professional. The counselor cannot be asked to do "piddling" administrative duties. Nor can a counselor, on the other hand, permit herself to become institutionalized or routinized by her job. She must accept her responsibility by viewing each child as a personality with potentialities for good rather than as a little scourge to society. She must possess insight and a capacity to work with the agencies in the community with no feeling of authoritarianism that, unfortunately, begins to grip all too many people who deal with children in trouble.

In suggesting that children leave home for school prior to the age of six we appreciate the wave of resistance that deplores such an "unconventional" recommendation as well as opposition from taxpayers who, even now, protest against high school costs. The increased costs for universal nursery schools with trained group workers and psychiatric social workers militate against such a program's being widely adopted. But when we realize that current crime-prevention programs are fighting a losing battle, we realize that a re-orientation and re-appraisal of the problem must eventually call for more drastic and expensive programs. Otherwise we shall continue to lose control of modern delinquency.

---

[4] "The Place of Visiting Teacher Services in the School Program," *Federal Security Bulletin*, 1945, p. 25. See also, Rachel D. Cox, "The School Counselor's Contribution to the Prevention of Delinquency," *Federal Probation*, Vol. 14, No. 1 (March, 1950), pp. 23-28.

Enriched Curricula

The school must share with the family the responsibility of supplying our youth with the tools necessary to meet life. Besides, it has the task of equipping the child with knowledge to assist him in the economic struggle he must inevitably face.

It is traditional that most public schools do a fair job with the pupil of average intelligence, a poor job with superior children, and almost no job at all with the dull group. Subjecting a group of retarded children to a curriculum far above their ability to understand is cruel in its effects on the group, and it is financially wasteful. There should be special classes teaching special skills adapted to the potentialities of such children. Truancy and incipient delinquency in the schoolroom challenge the school to supply a regime where such behavior is not called forth.

Thousands of boys and girls drop out of school because there is no opportunity to learn anything that their limited mental ability can grasp. Relatively few teachers know how to cope with this group. As teachers are generally overworked, that puts a premium on meeting the situation on a low level of adjustment—stern disciplinary measures in the schoolroom or a command to go to the principal's office. The principal rarely has the necessary insight to deal effectively with the problem, so either the situation continues in a "muddling through" manner until the boy drops out of school, or he is badgered into submission.

Vocational guidance should be an integral part of every public school—especially every secondary school—and for all types of pupils. In addition, there should be provided an honest and well-equipped vocational program to meet the needs of the average pupil at the time he finishes school. Too many children who drop out of school stumble into jobs for which they are not vocationally prepared and toward which they develop indifference or even hatred.

Boys and girls leave school to look for work—any work that will relieve them from the utter boredom that permeates the average school curriculum without vocational guidance. Children who contemplate leaving such schools give little thought to the long run of occupational adjustment. A serious problem confronting educators and civic leaders at present is that the group from which most delinquents emerge—as judged by police records, correctional institutions, and crime prevention bureaus—cannot take advantage of skilled jobs or of training for them, because they have neither the special aptitudes nor the intelligence necessary for their successful pursuit.

We cannot expect the school to assume the responsibility of training the boy or girl specifically to go out and land a job. Courses in typing, shorthand, commercial work, and similar subjects do prepare for a specific field of work. But there is a limit to such skills. Manual training for boys, with a knowledge of the use of tools and the properties of metals, woods, and plastics, and home economics for girls are valuable, but they do not prepare for specific jobs. Schools should continue these subjects and enrich their curricula along the lines suggested above.

The schools have lagged behind in offering adequate and compulsory civic education as well as carefully thought out courses in sex education. Attitudes concerning sex are notoriously distorted by school children as well as by most adults. Traditional education has not been forthright enough in its denunciation of crime, except for such offenses as theft, robbery, and murder. On the contrary, it has eulogized the present type of bourgeois society and behavior, especially the leisure-class psychology characterized by the get-rich-quick attitudes of modern life. By not pointing out the immoral and criminal practices of corrupt politicians, and "white-collar" criminals in big business, and greedy labor bosses, the schools perpetuate the idea that "what we have, not how we got it" counts for success.

It is not too late, perhaps, for those who shape the destinies of the primary and secondary schools to develop new objectives that will face up to recent social, political, and economic trends, the impact of which is reshaping the lives of all the people of the earth. The nationwide apprehension of educators and parents regarding the lag concerning the teaching of science to our children in the schools is a serious problem which only indirectly touches on our discussion here, yet this very lag is one more indictment of the educational philosophy that has come under serious attack from many quarters.

The Senate subcommittee on delinquency in its recommendations regarding education calls for more professionally trained personnel in our schools but has certainly not come to grips with the larger implications of delinquency. In the opinion of the writers its recommendations are couched only in general terms. They are:

1. Providing an adequate number of well-trained, qualified teachers and providing that teachers receive training in mental health.
2. Insuring that there are professionally qualified visiting teachers and school psychologists in adequate numbers.
3. Making adequate provision for remedial teaching and special classes and services for deviating children.
4. Making substantial provision for research and experimentation programs.
5. Making sure that vocational and educational guidance services as well as help with other programs are available to every child.[5]

While nothing is to be gained by placing "blame" on the schools for the high tide of delinquency, it does seem that much more money and more serious thought must be allocated to education, especially in developing and revamping the curricula, if delinquency is to be controlled. Former Senator Robert C. Hendrickson, one-time chairman of the Senate Subcommittee on Juvenile Delinquency, states emphatically that "this nation's first

_____

[5] "Juvenile Delinquency," Report of the Committee on the Judiciary, U.S. Senate 85th Congress, 1st Session (Washington, D.C.: U.S. Government Printing Office, March, 1957), p. 242.

Social group work attempts to supply the various community agencies that are dealing with leisure time pursuits, club activity, hobbies, and so forth, with trained leadership. It is not a new approach but rather a new method. It is the conviction of those who have developed this new technique, that such activities furnished by community agencies are merely the means of furthering other socially desirable ends. Group work is an educational process emphasizing the personality development and social adjustment of the participants in various programs. But when it is evident that a specific individual who is a member of certain group work activities—club, ball team, sewing class—has "limited social capacity" or unusual emotional needs, the social group worker, trained in his craft, will recognize in such a case the need of sharing the responsibility with a psychiatrist, a psychologist, or social case worker. It is here that the social group worker makes one of his greatest contributions to society. In dealing with maladjusted persons, he singles them out of the crowd and turns them over to the proper specialists. Group work also can, no doubt, divert certain predelinquent tendencies into socially approved avenues by skillfully directing the wishes and desires of certain individuals into activities where certain satisfactions accrue or where status of the individual is heightened.

Social group work is not essentially a preventive to delinquent behavior and should not be so considered. In a study of the type of clientele served by group work agencies it was revealed that this type of activity is "not in general identified closely with the underprivileged and insecure elements in our population, nor with the age groups among which delinquency is most prevalent."[8]

The question is, then, whether some agencies by the nature of their programs, attitudes, and methods, do screen out the boys and girls who are handicapped physically, mentally, economically, or racially; or who are emotionally maladjusted; or who have an unfortunate and unhappy family background. Yet there is evidence that more purposeful work with social groups at play (recreational agencies), in club work (settlement houses), at summer camps (Y.M.C.A. and Y.W.C.A.), and in counseling (Big Brother and Big Sister), lead to the conclusion that many persons are diverted from frustration or personal maladjustment into more meaningful and socially approved ways of living.

In other words, group work agencies can be expected to take responsibility for both the prevention and treatment of delinquency, but usually only as a part of their present service to individuals and to the community. To the extent that the guidance service is provided and the group work process is effectively utilized, the potentialities for prevention and treatment of delinquency are enhanced.

Group work enters the field of "character building" with a much more professional attitude than most of the traditional organizations. Trained

---

[8] Ellery F. Reed, "How Effective are Group Work Agencies in Preventing Delinquency?" *Focus*, Vol. 28, No. 6 (November, 1949), pp. 170 ff.

personnel is, of course, the most conspicuous feature of social group work. Standards have been increased in the recent past and many schools of social work offer a course of training in the field on the graduate level.[9]

4. Community Action

It has often been stated that the strongest link in the chain of crime prevention is community responsibility. Aside from tax supported services and privately supported agencies working in this important area, the voluntary citizen service, alert to community needs of young people, has long been operative in various parts of the country. Perhaps the best known of these is the community council, or co-ordinating council. The state of California has, perhaps, perfected the co-ordinating council to its highest level.

Fully three hundred councils are functioning in California alone. The Youth Authority of that state has given impetus to the councils since its inception in 1941. But other states also have many in operation. Many of the larger cities have such councils working along neighborhood lines. Their activities are legion and varied. Developing leisure-time pursuits for the youth of the neighborhood or city, regulating questionable media generally recognized as detrimental to children or adolescents, calling for law-enforcement from police, making studies of areas of high delinquency rates, sponsoring worthwhile celebrations of holidays—all are legitimate activities of the councils. Councils can also work actively with youth groups in their attacks on delinquency.

The Chicago Area Project is another neighborhood enterprise that attempts to understand the forces that make for crime and delinquency and to build up adequate buffers against deterioration and demoralization of the area. As early as World War I the stockyard district of Chicago was organized along such lines in what was called "Back of the Yards' Neighborhood Council." More recently, under the guidance of sociologists led by the late Clifford R. Shaw, the Chicago Area Project was developed. Its aims and program are described as follows:

> The Chicago Area Project is a program which seeks to discover by actual demonstration and measurement a procedure for the treatment of delinquents and the prevention of delinquency in those neighborhoods of the city which have for more than thirty-five years produced a disproportionately large number of delinquent boys and girls in the Cook County Juvenile Court. The program is in operation in three such neighborhoods. Its primary feature is the emphasis which it places upon the participation of the residents of the neighborhood in planning and operating the program. The Project seeks to determine to what extent constructive changes can be effected in the social environment of the neighborhood with consequent re-

---

[9] For a good standard book dealing with group work, see: Margaret E. Bennett, *Guidance in Groups* (New York: McGraw-Hill, 1955); also *Reaching the Group: An Analysis of Group Work Methods Used with Teen-Agers* (New York City Youth Board, 1956).

THE CHALLENGE OF DELINQUENCY AND CRIME: PREVENTION   613

duction in the volume of delinquency, by providing to the residents facilities and professional guidance for the development of their own program for the welfare of their children.[10]

The Chicago Area Project and others of its kind are interesting attempts to stimulate the residents of a neighborhood to have faith in their own potentialities for effective citizenship. In a sense, it is the capitalization of the latent resources of leadership in the area with the objective of making the neighborhood a place in which children can be reared decently. No one would call such projects "Lady Bountiful" types of charity, although the propulsion is obviously from the outside and based on arbitrary, although carefully arrived at, assumptions. The activities of the area projects are traditional, such as sports, club work, hobbies, music, dramatics, and civic undertakings.

5. Religious and Moral Training as Preventives

Discussing this subject conversely we find that the lack of religious and moral training has been repeated, *ad nauseam,* as the cause of delinquency and criminal behavior. We find this thesis expounded by jurists, college presidents, law-enforcement officials, and leaders in public life, as well as by religionists and inspirational speakers and writers; it is generally accepted without criticism or question. In our culture one is socially compelled to be "on the side of the angels" so the church with its religious and moral training is placed at the top of the list as one of the insulators against crime and delinquency. It is imperative that we appraise both religion and morality as objectively as possible—a very difficult task—in any discussion of community influences affecting antisocial behavior.

Professor Caldwell makes the following statement which might resolve the sharp differences of opinion regarding the role of religion:

> Statistical investigations have attempted to reveal the effects of religion on crime and delinquency, but they have not succeeded in disentangling the factor of religion from other closely related factors, such as the home, the neighborhood, nativity, education, economic status, and race. Nor do these investigations penetrate beyond such externals as church membership, attendance, and avowals of faith and so reveal nothing regarding the actual influence of religion in the lives of those studied.[11]

One of the difficulties in appraising religion as a preventive to crime is the confusion in defining it. We find the following clustered about religion at various times or with various groups: belief in the hereafter; identification with the supernatural; a code of ethics; one's ideals; church member-

---

[10] E. W. Burgess, Joseph D. Lohman, and Clifford R. Shaw, "The Chicago Area Project," *Yearbook,* N.P.P.A., 1937, pp. 8-28.

[11] Robert G. Caldwell, *Criminology* (New York: Ronald, 1956). p. 246. Copyright 1956 The Ronald Press Company. See also statement on this subject by Herbert A. Bloch and Frank T. Flynn, *Delinquency: The Juvenile Offender in America Today* (New York: Random House, 1956), p. 228.

ship or attendance; the Beatitudes; the Golden Rule; leading a devout life; and one's moral code. What the social group desires from its members is human decency, a reasonable degree of ethical conduct or decorum, and a profession of respect for law. The group leaves much of the rest to the individual as his personal business. Society is really unconcerned whether the individual is a church member or not, whether he goes to church frequently, infrequently, or not at all, or what his fate is after death. Agnostics or atheists have no trouble in society if they are willing to live and let live, although some may experience embarrassment in connection with their jobs or professions.

Studies by two distinguished European criminologists, Gustav Aschaffenburg of Germany and W. H. Bonger of Holland, show that Roman Catholics tend to be more criminal than do Protestants, and Jews the least criminal of the three groups.[12] Such generalizations must be accepted guardedly since other factors are always present in attempting to appraise criminality of large groups of people. In many countries where economic conditions are poor the dominant religion is Roman Catholic. Hence, since most criminals who have been caught and convicted come from low economic levels, Roman Catholics tend to show unfavorably in the conventional statistics. It has been argued that the low crime rate among Jews is due to the close family and community control exercised over this group.[13]

Professor von Hentig has collected some interesting data on the religious affiliation among inmates of juvenile reform schools and prisons.[14] His figures show that among delinquents and reformatory inmates, Catholics are 2½ times more numerous than Protestants. There are relatively few studies dealing with religion and crime, especially among adults. Those that are available are quite old, yet should be mentioned as long as newer ones are not available.

A study made some thirty years ago by C. V. Dunn shows that Protestants are more numerous than Roman Catholics in the 27 adult prisons and the 19 reform schools covered. This is contrary to von Hentig's data but there is nothing to indicate the real numbers of both groups in terms of total memberships in the general population.[15]

There are surprisingly few non-believers in prisons. Again, from an old study, Franklin Steiner found that of 85,000 prison inmates, 68,000 or 80 per cent decisively expressed their preference as Christians. There were at

---

[12] For sources, see J. L. Gillin, *Criminology & Penology,* 3rd edition (New York: Appleton, 1945), p. 207.

[13] This is borne out in a study by N. Goldberg of the crime rate of Los Angeles Jews, which is much lower than the expectancy rate. The study covers the years 1933 to 1947. See N. Goldberg, "Jews in the Police Records of Los Angeles, 1933-1947," *Yivo Annual of Jewish Social Science,* Vol. 5, 1950, pp. 266-291.

[14] Hans von Hentig, *The Criminal and His Victim* (New Haven: Yale University Press, 1948), Chapter 10, "Religiosity and Crime."

[15] C. V. Dunn, "The Church and Crime in the United States," *The Annals,* Vol. 117 (May, 1926), pp. 200-228.

the time (1924) 5,389 of the Jewish faith. There were only 8,000 who were not affiliated with some faith. The proportion of avowed atheists was microscopic, some 150.[16] The study reveals that the proportion of religious affiliates is at least 50 per cent higher among prisoners than among the general population.

A large number of people take their church membership quite casually; this applies to criminals and non-criminals alike. Questionnaires of college students which call for "church preference" actually indicate merely the sect or church in which they grew up. In any survey dealing with religious affiliation there are likely to be many false statements or casual answers. This is especially true of inmates in our prisons or reform schools. Inmates are likely to fake their religious connections in the hope of making a favorable impression on the authorities.

But it is abundantly clear that a majority of our criminals—certainly our prisoners—are brought up in orthodox religious surroundings. Conversely, it is well known that the majority of scientists and academicians have at least discarded religious orthodoxy, and certainly, as a class, are distinguished not only for their achievements but for their law-abiding and moral behavior.[17]

So far as actual church membership is concerned it has been gradually increasing through the years, even in proportion to population. But a high percentage of church membership in the total population has no apparent influence in reducing criminality in the community. There are many isolated instances of apparently devout persons, Sunday School teachers, and even a few ministers of the gospel who become enmeshed in crime. This, in itself, is no indictment of religion or the church. Conversely we have no sure way of knowing how many people are restrained from criminal activity by church membership or regular attendance.

Regarding religious training through attendance at Sunday School, we find from several studies that it seems to make little difference whether children go to Sunday School or not, so far as delinquency is concerned. A study made by Mursall of boys in the Ohio reform school indicated no significant relation between religious training and delinquent or non-delinquent behavior.[18] Another by Hightower showed definitely that the tendency to lie and cheat among 3,000 children tested, was in direct proportion—not in inverse ratio—to their knowledge of the Bible and scriptural precepts.[19] He concluded that "mere knowledge of the Bible is not sufficient to insure proper character attitudes." Kvaraceus made a

---

[16] Franklin Steiner, *Religion and Roguery* (New York: The Truth Seeker, 1924).

[17] In a study of the religious beliefs of American men of letters, J. H. Leuba, in 1916 published his findings as *The Belief in God and Immortality* (Boston: Sherman, French & Company).

[18] George Rex Mursall, *A Study of Religious Training as a Psychological Factor in Criminality,* Ph.D. dissertation, The Ohio State University, 1940.

[19] P. R. Hightower, *Biblical Information in Relation to Character and Conduct* (Iowa City: University of Iowa Press, 1930).

study of 761 delinquent children in Passaic, New Jersey; he found nearly all of them affiliated with some church, two-thirds of them Catholic—either Roman or Eastern Orthodox. Not only were they members but 54.2 per cent of them attended church regularly and an additional 20.4 per cent attended occasionally. Only 8.8 per cent admitted non-membership and only 25.4 per cent of the total stated that they rarely visited any church.[20] Middleton and Fay studied 83 delinquent and 101 non-delinquent girls in the eighth to the tenth grade levels; the delinquents (as determined by three Thurstone Scales measuring social attitudes) manifested more favorable attitudes toward Sunday observance and the Bible.[21] Wattenberg found in Detroit that 69 per cent of 2,137 delinquent boys attended church regularly or on occasions; of the repeaters 65 per cent attended regularly or on occasions, as compared with 71 per cent of the non-recidivists.[22] The Gluecks found that 39 per cent of the delinquents they studied attended church regularly, 54 per cent attended occasionally, and 7 per cent never attended. In their control group among the non-delinquent 67 per cent attended regularly, 29 per cent attended occasionally, and 4 per cent never attended.[23]

There are many interpretations that could be drawn from the preceding studies. Sunday Schools are maintained by the most orthodox churches as well as the most liberal. The programs vary widely. In general, Biblical history is taught. In many, considerable time may be devoted in preparing the child for after-life. In most, perhaps in all, morality is also inculcated. Responsibility for citizenship may be emphasized. In many instances what the child receives may be a reinforcement of precepts inculcated in the home, whereas in others the program may be somewhat foreign to the kind of conversation and training the child hears and receives in his family life. So far as society is concerned it is hoped that the child may develop into a good citizen, responsible to the laws of his city, state, and nation and to an ethical and moral code that conforms to integrity and trust. The child of an atheist or agnostic may develop into an honest man, law-abiding and incorruptible. The question of religious instruction, then, is a knotty problem to solve so far as the prevention of delinquency and crime are concerned.

The actual influence of churches and religious education on morality itself must be considered. This topic is dealt with in an old but defini-

[20] W. C. Kvaraceus, *Juvenile Delinquency and the School* (New York: World, 1945), pp. 101-103.

[21] Warren C. Middleton and Paul J. Fay, "Attitudes of Delinquent and Non-Delinquent Girls toward Sunday Observance, the Bible, and War," *Journal of Educational Psychology*, Vol. 32 (1941), pp. 555-558.

[22] William W. Wattenberg, "Church Attendance and Juvenile Misconduct," *Sociology and Social Research*, Vol. 34, No. 3 (January, 1950), pp. 195-202.

[23] Sheldon and Eleanor Glueck, *Unraveling Juvenile Delinquency* (New York: The Commonwealth Fund, 1950), p. 166.

tive study by Hartshorne and May, *Studies in the Nature of Character*.[24] These authors attempted to make a scientific study of moral and religious instruction as an element in character education.

Are Sunday School children more honest than other children? The authors came to the conclusion that there was no great difference between the two groups. To quote: "Apparently, then, the tendency to deceive is about as prevalent among those enrolled in Sunday School as it is among those who are not, in one community, and in another, those enrolled are less deceptive than those not enrolled."[25]

One gathers from this study that we have not yet devised any mechanical character-building program to replace the day-by-day relationship between parents and children. Home training is still of vital significance. A few hours in school each day and an hour or two each week in Sunday School, with an evening once a week in some character-building organization, cannot offset the influence of a home where the moral tone is on a low level. Where home influences are good, the child will profit little in morality by membership in these organizations. Obviously he may acquire other culture traits, such as nature lore in the Boy Scouts or Biblical culture in Sunday School, but morality and honesty come in a much slower tempo than these various organizations and groups can function. Rewards for honest conduct, morality, daily good turns, and other character traits frequently become so socially compelled that their pursuit often develops behavior in children at the opposite extreme. In other words, a child who must be rewarded for being honest and truthful is under such emotional strain that he tends to be such only by stealth or fraud.

It is interesting and encouraging that the study made by Professors Hartshorne and May was sponsored by the Institute of Social and Religious Research. Both young people and adults are likely to keep their religion and moral code quite separate. Do we want religious orthodoxy or moral integrity? This question must be faced by religious leaders and answered with conviction before religion can cope in any appreciable way with crime. As Eduard C. Lindeman wrote: "I reject the notion that an increase in the number of persons receiving the type of religious education now prevalent will automatically result in the diminution of crime."[26]

Nonetheless it may be assumed that the church, regardless of its point of view regarding theology, dogma, supernaturalism, Sunday observance, or the baptismal rite, is deeply concerned with the well-being of the community and its constituents. With this premise established, the church, like the home and the school, can be and is enlisted in the prevention of crime

---

24 Hugh Hartshorne and Mark A. May, *Studies in the Nature of Character*, 3 vols. (New York: Macmillan, 1929-30). See especially the volume on "Studies in Deceit."

25 *Ibid.*, Vol. 1, p. 359. By permission of The Macmillan Company, publishers.

26 E. C. Lindeman, "Underlying Causes of Crime," *Yearbook*, N.P.P.A., 1941, p. 111.

and delinquency. Regardless of its denomination, it stands for law and order and is more or less articulate in endorsing social amelioration. Ministers are frequently on various boards concerned with good housing, slum clearance, fair employment, non-discrimination, wholesome recreation, and crime prevention.

Many churches supplement municipal recreation programs by opening their facilities to young people's leisure-time activities which often include dancing, forums, clubs, and other interests of adolescents.

There are, in addition, other areas in which churches can collaborate on the community level in combating conditions that undermine family life and are dangerous to the morals of children and adults alike. Often they take the lead in securing proper law-enforcement, especially regarding gambling, prostitution, the sale of obscene literature, and the cruder types of comic books.

Since the church is, in essence, the community in action, it has come to realize the impelling need to gear the application of its religious tenets to the social exigencies of everyday life. The challenge of crime and delinquency is many-sided and the church, like any other social institution, can be effective if its leadership can accept the challenge.

6. Governmental Action

On the National Level

Several agencies and bureaus of the federal government are concerned with problems of the prevention of crime and the treatment of criminals. Aside from the law-enforcement units, such as the Federal Bureau of Investigation, the Treasury Department, and Post Office Inspectors, we find for example, the Children's Bureau and the Public Health Service. The purpose of the latter is to deal with emotionally disturbed delinquents and criminals sent to institutions. It works closely with the Federal Bureau of Prisons by supplying trained personnel in the medical and psychological fields for the various units and staffs at the Medical Unit at Springfield, Missouri, and the two narcotics centers at Lexington, Kentucky, and Fort Worth, Texas.

Perhaps the most important agency, dealing with problems of child life, one of which is obviously delinquency, is the United States Children's Bureau, operating within the Department of Health, Education, and Welfare. This organization was established in 1912 and from that date until 1946 it was a part of the Department of Labor. The purpose of this important bureau is to "investigate and report on all matters pertaining to the welfare of children and child life among all classes of the people." It is concerned with orphanages, juvenile courts, desertion, and employment of children, as well as raising standards in dealing with children. Obviously it is deeply interested in delinquency control as well as prevention.

For many years it compiled statistics from local juvenile courts and since 1946, when methods were changed, from state departments of welfare. It has also assisted the National Probation and Parole Association in drawing up the Standard Juvenile Court Act, the latest edition of which was published in 1949. The Bureau also publishes annual reports on children in public institutions, including institutions for delinquents. It may be seen that the activities of the Children's Bureau are advisory rather than in the field of social action.

The Children's Bureau actually stemmed from the first White House Conference on Child Welfare, which was called in 1909 by President Theodore Roosevelt. Since that date the Children's Bureau has been the sponsor and chief promoter of other such conferences every ten years. In 1919, designated as "Children's Year," the conference adopted a set of standards that had wide influence. In 1930, the third conference adopted a Children's Charter that embodied basic goals for the welfare of children. In 1940 the fourth conference on "Children in a Democracy" took as its slogan "Our Concern—Every Child." The midcentury conference —the fifth—was held in December 1950. Another is scheduled for 1960.

In an earlier chapter we mentioned the various studies made and conferences called for the purpose of probing into the causes, prevention, and control of delinquency and crime. The most famous of these, certainly, was the National Commission on Law Observance and Enforcement, called by President Hoover in 1929, and headed by George W. Wickersham, one-time Attorney General of the United States. In 1934 a Conference on Crime explored the many phases of violent crime, such as kidnaping, bank robbery, and gang warfare. It was attended by many law-enforcement agents as well as hundreds of persons interested professionally in the over-all problem of crime and its control.[27] A national conference on parole was called in 1939, and another in 1955.

The Attorney General's *Survey of Release Procedures,* with its five published volumes, should also be mentioned in connection with activity on the federal level. We have referred to this *Survey* on numerous occasions.

All of these conferences and surveys have had as their purpose the understanding of the intricate problems of prevention and control of delinquency and crime as well as the treatment of those who run afoul of the law. Certainly mention of these should be made in connection with prevention of this serious social problem.

On the State and Local Levels

We are not here concerned with the area in which states or towns and cities maintain and administer institutions for delinquents or adult offenders; nor are we interested in parole or probation units. Many states,

---

[27] See *Proceedings of the Attorney General's Conference on Crime,* December 10-13, 1934 (Washington, D.C.: U.S. Government Printing Office).

through their various agencies, also operate some sort of program in prevention and control of crime and delinquency. It is our purpose here to enumerate some of these projects.

Aside from the radical departure of caring for delinquents through state-wide control as well as by careful scientific diagnosis, the California Youth Authority has established an auxiliary service in prevention. This adjunct of the Authority makes surveys of local communities and counties, the results of which are turned over to local authorities for perusal and whatever action they may wish to take in crime control. The Authority maintains the "California Youth Committee," which consists of lay persons appointed by the governor. It is the function of this group to apprize local communities of the work of the Authority, and through periodic meetings to discuss ways and means of coping with problems of delinquency.

Illinois has organized a special division on delinquency prevention, now called "Division of Youth and Community Service" functioning within the state Department of Welfare. This division has set up clinics in various school districts throughout the state and, through a staff of experts, seeks to stimulate local action for youth programs, primarily through community councils.

New Jersey, also, attempts to work out state-local cooperation in juvenile welfare. This is done through the Division of Community Services for Delinquency Prevention of the Department of Institutions and Agencies. The activities are as follows: to render advisory and consultative service to local communities on all matters pertaining to child welfare; to raise the level of county juvenile court services and to compile an annual census of children dealt with by the juvenile courts; to develop a closer relationship between schools and social agencies, particularly in matters dealing with mental hygiene; and to engage in research projects.[28] Legislation was also passed in 1947 in New Jersey authorizing municipalities to create agencies to be called "Municipal Youth Guidance Councils" for the purpose of coodinating local child and youth welfare services.[29]

The state of New York, in a spectacular manner, has set up a state "Youth Commission" whose purpose it has been, since 1943, to stimulate local communities to spend funds for the welfare of youth. During its first year, some $300,000 of state funds and $500,000 from local communities outside New York City were expended for the following: (1) the establishment of local Youth Bureaus for young people; (2) recreation projects; and (3) educational projects that include guidance clinics. By 1948 this vast program was operating through 11 Youth Bureaus and

---

[28] For details of this program, see Douglas H. MacNeil, "Two-and-One-Half Years of State-Local Collaboration in Delinquency Prevention," *Yearbook,* N.P.P.A., 1948, pp. 252-262.

[29] See "Delinquency Prevention Begins At Home," by Paul G. Cressey, *Focus,* Vol. 28, No. 3 (May, 1949), pp. 78-82.

three traveling child-guidance clinics that were added to the previously organized eight units. There were also 520 recreation projects under way.[30]

In several states mobile mental hygiene clinics, organized on a statewide basis and staffed by psychologists and social workers, cruise throughout the state and provide preventive services in small towns and rural areas where such facilities are not available.

There is almost no limit as to the aid state services can render local communities in coping with delinquency and crime. Since legislation in these areas is centered in the states, it is natural that many legislatures, prompted by the great amount of publicity on child delinquency, have created special committees devoted to the problem of delinquency. These sometimes consist of members of the legislature only or are augmented by practitioners in the field, or by interested lay citizens. Their objective usually is the preparation of needed legislation, including the appropriation of funds for state services in this field. Detailed statewide surveys are often undertaken to aid their programs. Governors have called crime conferences that are attended by workers in the prevention field and by laymen interested in child life. Summarizing the whole picture of state activity in this area, Professor Carr states:

> Out of 39 states on which data were available, 23 had set up special bodies of some kind—5 central planning bodies; 6 coordinating and educational bodies; and 12 commissions whose functions ranged from advice to the drafting of new children's codes. This seemed to indicate a widespread feeling that something needed to be done either to strengthen the juvenile court or to supplement it with additional services. Organizations carrying on social action in the delinquency field were reported from twelve states.[31]

Space does not permit a more thorough discussion of what is being done in the field of delinquency prevention on the state level. We have merely cited a few samples. However, we can state that there has been an awakening to the serious problem of crime prevention throughout the entire country.

We have integrated activity on the local level with that in operation on the state level. States generally work through local agencies and public bureaus. In other connections we have discussed the role of the school, the activity of child-guidance clinics, parental schools, the work of police-women—all of which are supported through taxes or community chests. We have also dealt earlier in this chapter with recreation and character-building agencies, as well as group work. In addition, we could refer to the work of the church, press, radio, motion pictures, and television as agencies having great potential in the preventive field. Crime prevention is a many-sided battle and calls for mobilization of all facilities of the

---

[30] For details of this, see Lowell J. Carr, "Organization for Delinquency Control," *The Annals*, Vol. 261 (January, 1949), pp. 64-76, especially page 66.

[31] *Ibid.*, pp. 75-76.

community, public and private alike, in coping with the serious problem. We noted that citizen participation is greatly needed in community councils and in sponsoring youth activities. But the battle will not be won until every child has equal opportunities to come into the world without social or economic handicap and to grow to maturity in a healthy fashion. That is the challenge to America. Only in this way can we materially cut down the delinquency rate.

7. Research in Delinquency and Crime

In a provocative little monograph published in 1956, Helen L. Witmer and Edith Tufts conclude, after surveying the maze of delinquency prevention programs that have been attempted throughout the years:

> What does all this add up to, in knowledge about how to reduce delinquency? The answer, unfortunately, is "with certainty, rather little.". . . delinquency prevention . . . as a body of scientific knowledge and practice is just emerging from its infancy.[32]

The value of research in the etiology of delinquency and crime has long been recognized. Its limitation is, of course, financial since it takes well-trained people to engage in this important work. While there has been, through the years, much satisfactory work, the results of which have had important bearing on action programs, there is much more to be done in the future. The reader must check bibliographies, book-lists and publications of many organizations and agencies in order to become acquainted with the results of hundreds of studies being currently carried on in this field.

One of the most ambitious studies dealing with the problem of delinquency among children was begun in the Boston area in 1935 at the suggestion and with the support of Dr. Richard Cabot who, during his lifetime, was vitally interested in character formation. The basis of this study was, from the outset, the belief that whatever can be done to favor the growth of character is an effective prophylactic against delinquency. Dr. Cabot was also interested in using techniques found more often in the physical sciences than seems possible in the social sciences. In the beginning of the study, known as the Cambridge-Somerville Youth Study, two groups of boys were selected. There were 325 in each group. One of these, known as the "T" or Treatment group (no older than 12 when treatment began), received counseling services throughout the duration of the study. The other, the "C" or Control group, comprised boys whose behavior was noted throughout the study but who received no treatment.

Each of these two groups was divided into three additional groups on the basis of their behavior: 1. The difficult or "D" group whose tendencies

---

[32] Helen L. Witmer and Edith Tufts, *The Effectiveness of Delinquency Programs* (Washington, D.C.: U.S. Government Printing Office, 1956).

pointed toward delinquency; 2. The average or "A" group whose prognosis was considered good; and 3. The zero or "Z" group about whom there was considerable diagnostic and prognostic doubt.

From 1935 to 1939 six stages in the program were followed: These were (1) Pre-selection of the boys, which included the preparation of school authorities who cooperated with teachers of both public and private schools in selecting the types of boys needed for the study. Each boy selected was interviewed and tested, medically, psychologically, and socially, the latter through the various social agencies. Camping trips formed one part of sizing up the boys, as well as their families, before they were finally selected. (2) Selection: Each member of the committee reviewed the folder of each boy independently and, after much discussion, 782 boys were selected: 360 were diagnosed as difficult, 334 as average, and 88 as belonging to the "zero" group; (3) Matching: This was considered one of the most difficult steps. An attempt was made to match each boy of each of the two groups based on such attributes as chronological age, dominant stock, birthplace of each parent, religion, locality, school, grade placement, physical health, intelligence, attainment, and educational quotients, mental health, social adjustment, judgments of the selecting committee, home ratings, standards of living, occupational status of father, neighborhood, and school occupational level; (4) Assignment: This dealt with the allocation of each boy to a counselor on the basis of certain of the attributes mentioned above; (5) Treatment: Treatment was begun on June 1, 1939. Aside from the services furnished by the staff of the study, 88 organizations and agencies rendered specialized services. These included clinics, hospitals, child welfare organizations, state departments, etc.; (6) "Control" program: This dealt with the collection, integration and interpretation of data concerning the boys in the Control group and involved such items as school program of each boy, actual apprehended delinquency, community relationships, personality development, agency contacts, psychological tests, and mental examinations. These data were secured through interviews with teachers, probation officers, church officials, agency representatives, and summer camp directors.

The study covered a period of ten years. Several of the boys in both groups moved away, went into the armed services, and a few died, so the number observed or receiving treatment throughout the entire period diminished somewhat from that which began in 1935. The results therefore, cover only those who finished the whole ten-year period. The original plan called for the inclusion of boys of 6 and 7 years of age but this had to be modified later so that boys of 11 and 12 were included. At the end of the period most of the boys were 18 to 21 years of age.[33]

---

[33] The following information is digested from an article by the director of the study, Edwin Powers, "An Experiment in Prevention of Delinquency," *The Annals*, Vol. 261 (January, 1949), pp. 77-88. See also Edwin Powers and Helen Witmer, *The Prevention of Delinquency* (New York: Columbia University Press, 1951).

The conclusions of this significant study are listed as "First Conclusion" and "Second Conclusion." The first conclusion indicates "that the special work of the counselors was no more effective than the usual forces in the community in preventing boys from committing delinquent acts."[34]

The second conclusion, based on facts and not wishful thinking, shows (a) "that though the first stages of delinquency are not wholly averted when starting treatment at the 8-to-11-year level, the later and more serious stages are to some degree curtailed"; and (b) since delinquency is not the entire story "the making of good citizens or social adjustment" of the boys seemed to be more successful. "An examination of the records and interviews with the boys themselves offer evidence that in many cases, even in the lives of many of the delinquent boys, emotional conflicts were alleviated, practical problems were dealt with successfully and boys were given greater confidence to face life's problems."[35]

Summing up the results of the Cambridge-Somerville experiment, Helen L. Witmer and Edith Tufts have this to say:

> This experiment seems to indicate that the provision of friendly guidance and other services . . . afforded will not reduce delinquent acts or keep chronic delinquency from developing. This is not to say such services are not useful to certain children, especially to nondelinquents. The services were especially ineffectual, however, with the kinds of boys who became chronic delinquents: slum boys with indifferent, neglectful parents; seriously neurotic boys, from various kinds of neighborhoods, who had even more emotionally unfavorable homes; and feebleminded or neurologically handicapped boys whose homes, too, were poor.[36]

Yet it is just these types of boys that will have to be "reached" somehow if delinquency is to be controlled. The study, with allowances made for its many handicaps, at least points up the fact that delinquency will be controlled if at all only by deeper therapy or more penetrating analysis, or, as suggested earlier in this chapter, by beginning observation and treatment of troubled children much earlier than is now done, with children starting school earlier than the age of six.

The ten-year study on delinquency made by Drs. Eleanor and Sheldon Glueck, is also significant. It embraced two groups of boys; 500 delinquent and 500 non-delinquent. Their findings point up the fact that if a child's family life was adequate, the chances were only three in 100 that he would turn out to be a delinquent. On the other hand, if his family relationships were strained, the chances were 98 out of 100 that he would begin a delinquent career.

These authors found from their studies of family life that little progress can be made in delinquency control until the home relationships can be

---

34 *Ibid.*, p. 87.
35 *Ibid.*, p. 88.
36 Witmer and Tufts, *The Effectiveness of Delinquency Prevention Programs*, p. 30.

strengthened "by a large-scale, continuous, pervasive program designed to bring to bear all the resources of mental hygiene, social work, education, and religious and ethical instruction upon the central issue."

Summarizing their findings into what they call a dynamic causal pattern, the Gluecks held that delinquents as a group are distinguishable from non-delinquents in the following manner:

1. Physically: in being solid, closely knit, muscular in constitution;
2. Temperamentally: in being restlessly energetic, impulsive, extroverted, aggressive, destructive, often sadistic;
3. In attitude: by being hostile, defiant, resentful, suspicious, stubborn, socially assertive, adventurous, unconventional, non-submissive to authority;
4. Intellectually: in tending to direct and concrete, rather than symbolic, intellectual expression, and in being less methodical in their approach to problems;
5. Socio-culturally: in having been reared to a far greater extent than the non-delinquent boys in homes in which the "under the roof culture" was "bad," that is, homes of little understanding, affection, stability or moral fiber.[37]

8. Conclusion

Studies in prevention and control of delinquency and crime have suffered more from lack of integration than from the paucity of funds. There is a great need for some over-all national organization that can correlate and synthesize the work that has been done, is underway, and is being contemplated. The attack on delinquency and crime must be relentless and sustained. There can be no moratorium on the gathering of adequate statistics, the amelioration of the conditions that breed frustration, maladjustment, and bitterness, the strengthening of law-enforcement, or the improvement of our prisons and reform schools.

Any significant plan for the reduction of law-breaking, either by children or adults, must combine two characteristics: first, it must be comprehensive enough to achieve permanent results; and second, it must have certain elements that can be put into immediate operation. Permanency and immediacy are both essential. It is obvious that a large proportion of our laws could be repealed without too great a strain on the body politic, and thus we could reduce crime by that much.

Briefly we might state that any thoroughgoing system of crime prevention must include the accurate collection of criminal statistics and a more adequate body of data on delinquency. There can be no intelligent attack upon the crime problem that is not founded on a complete knowledge

---

[37] See their work, *Unraveling Juvenile Delinquency* (New York: Commonwealth Fund, 1950). This work has caused considerable discussion. For evaluations see "Symposium on the Gluecks' Latest Research," *Federal Probation,* Vol. 15, No. 1 (March, 1951), pp. 52-58.

of the relevant facts about the number, types, and frequency of the crimes committed, as well as of such accessory facts as the number of convictions and recommitments. Some comprehensive plan of gathering data regarding our prisons and jails must be developed. There is a shocking lack of information in this area. The ambitious *Survey of Release Procedures,* to which we have referred throughout this book, is an example of the type that should be available at five-year intervals, at least. At the moment it is impossible to secure over-all information concerning what is being done in our prisons so far as treatment is concerned. We have no way of knowing how many recidivists are in our institutions at any moment. We have few studies of a national character showing the efficacy of parole. We know almost nothing definite as to the status of productive labor in our prisons. The last over-all study in this area was done by the Bureau of Labor in 1940.

Whether one looks upon the criminal law chiefly from the standpoint of deterrence or of reformation, there can be no doubt that certainty of apprehension and conviction for criminal behavior is the first and most indispensable item in securing the immediate reduction in the volume and variety of crime. An honest and expert police system is the only answer to efficient apprehension. Defects in our court system should be remedied. This calls for a sweeping reconstruction of court procedure with the eventual elimination of the jury, the modification of the sporting theory of justice, and the adoption of pre-sentence clinics and diagnostic boards of professionally trained experts. The emphasis must be on keeping men out of prison rather than on committing them.

Index of Names

A

Abbott, Edith, 150, 181
Abels, Jules, 39
Abrahams, Joseph, 480
Abrahamsen, David, 103, 181
Acton, Lord John Emerich, 38
Addition, Henrietta, 414
Adler, Mortimer J., 279, 280
Ahern, Danny, 55
Aichhorn, August, 200
Alexander, Franz, 158, 200
Alexander, Myrl, 388, 390, 392, 393, 398
Allen, Charles R., Jr., 39
Allen, D. D., 33
Allen, Robert M., 581
Andenaes, Jos, 121
Anderson, Henry W., 3
Andrews, William, 293, 315, 316
Aptegar, Herbert H., 472
Arluke, Nat R., 576
Asbury, Herbert, 51
Aschaffenburg, Gustav, 143, 614
Ash, Ellis, 362
Ashley-Montagu, M. F., 129, 130, 131, 132
Atiyah, E. S., 15
Aubert, Henri, 142
Augustus, John, 553
Axelrad, Sidney, 170

B

Baan, Pieter A. H., 111, 590
Bacon, Selden, 91
Baily, William F., 381
Baker, Harry J., 138
Baldwin, Simeon, 593
Balistrieri, James J., 70
Ball, John C., 318
Banay, Ralph S., 67, 91, 138, 204
Barber, W. Charles, 513
Barber, Roland, 68
Barker, Gordon H., 178
Barnes, Harry Elmer, 197, 208, 325, 354, 385, 497, 536
Barnett, Elliott B., 212
Barron, Alfred J., 579

Barrows, Isabel C., 519
Barry, John Vincent, 418, 422, 425
Bates, Jerome E., 46
Bates, Sanford, 394, 397, 511
Battaglia, Bruno, 142
Bayle, Edmond, 228
Bazelon, David L., 256
Beattie, Ronald H., 76
Beaumont, Gustav de, 328, 345
Beccaria, Cesare, 246, 253, 262, 322, 323, 324, 327
Bechdalt, Frederick R., 383
Becker, Howard, 197, 208
Beier, Ernst G., 70
Belbenoit, René, 302, 303
Bender, Lauretta, 67
Bennett, James V., 6, 61, 383, 403, 461, 493
Bennett, Margaret E., 612
Bentham, Jeremy, 246, 262, 324, 325, 327, 335, 594
Berger, Meyer, 27
Berkman, Alexander, 364, 373, 517, 529
Berman, Louis, 129
Bernard, William, 66
Bertillon, Alfonse, 227, 228, 230
Biggs, John, Jr., 260
Bixby, F. Lovell, 104, 480
Blackstone, Sir William, 246, 335
Blanchard, Jean Pierre, 337
Blanche, Ernest E., 30, 31, 34
Bligh, Capt. William, 300
Bloch, Herbert A., 145, 193, 613
Bloom, Murray Teigh, 30
Blumer, Herbert S., 190, 191
Bonger, William A., 56, 142, 147, 148, 495, 614
Borchard, Edwin M., 266, 267, 268, 278
Botein, Bernard, 243, 249
Bowlby, John, 200
Bowler, Alida C., 165
Brace, Charles Loring, 95, 148
Bradford, John, 82
Bragg, Jonny, 518
Brancale, Ralph, 104, 477
Branch-Thompson, W., 298
Branham, Vernon C., 41, 100
Brean, Herbert, 52

General Index

I

Identical twins and crime, 130
Identification techniques in police work, 226-233
Idleness in prisons, 523, 536-537
Illegal operations as "white-collar" crime, 46
Illinois state penitentiary, Joliet-Stateville, 357 n., 358
Immigration:
  and the crime rate, 164-166
  and the Quota Law, 164
Immorality and crime, 71
Immunity and the professional staff in prisons, 474-475
Imprisonment:
  and loss of civil rights, 544-545
  and stigma attached thereto, 355, 544
  as substitute for corporal punishments, 325
  cost of, compared to probation, 557
  development of, 328-337, 584
  dubious value of, as a corrective device, 371, 584-587
  substitutes for, 592-594
Inadequate homes and delinquency, 180-185
Inadequate schooling and delinquency, 602-609
Incommunicado, 135-136
Indefinite sentence, see Indeterminate sentence
Indemnity:
  as a substitute for imprisonment, 593
  for those wrongfully accused, 266-268
Indeterminate sentence:
  and the Reformatory, 432
  history of, 422, 570
  types of, 570
Indians and crime, 176
Indictment:
  and the grand jury, 239
  defects in, 240-241
Individualized treatment of prisoners, 598; see also, Diagnostic clinic
Industrial Revolution and prison labor, 528
Informants, use of by police, 224
"Information" in lieu of grand jury indictment, 239
Information regarding crime problems, 79-81
Inmate-guard complex, 361-363, 463
Inmates:
  and cliques, 498
  and religious affiliations, 614-615
  and religious views, 495
  and routine of, 328
  councils, 504

Inmates: (cont.)
  number of in prisons today, 328
Inmates and participation in:
  community services, 514-517
  cultural pursuits, 517-519
  medical research, 515
Inspection of jails, 398
Insanity as a defense plea, 253-260
Insanity, tests of in courtroom, 254-256
In-service training of staff officers in prisons, 452
Inside cell blocks, 369
Installment payment of fines, 400
Intangible costs of crime, 9
Intelligence of criminals, 7, 135-136
Intermediate sentence, development in Ireland, 423
International Ass'n of Chiefs of Police, and crime statistics, 75
International Penal and Penitentiary Congresses, 81, 593
Interstate compacts, on detainers, 61
Irish system of penal treatment, 423-425
Iron yoke as a punishment, 351
"Irresistible impulse," as a test of insanity, 255
Islas Marias, prison colony in Mexico, 310
Isolation cells, use of in prisons, 351-352
Italian school of criminology, 124-128

J

Jail farms:
  and misdemeanants, 402-403
  for vagrants, 84
Jails:
  abolition of, 406
  and John Howard, 331-333
  and the fee system, 389-390
  and the kangaroo court, 392-396
  architecture of, 393
  as anachronisms, 388
  colonial, 389-390
  history of, 387-390
  inspection of, 398
  minors in, 392
  overcrowding in, 391
  reform of, 398-406
  women in, 396
Japanese and crime, 172
Jews, amount of crime among, 614
John Howard Ass'n:
  study of babies in women's prisons, 414-415
  study of lockups, 397
Joliet, Ill., prison, 357
Jones-Miller Narcotics Act of 1922, 87
Judges:
  ability of, 250

"Whirligig," a Revolutionary War Punishment for soldiers, 591
"White-collar" crime:
  analysis of, 41-49
  and abortions, 46
  and bond salesmen, 44
  and fraudulent advertising, 45
  and holding companies, 19
  and legal profession, 47-48
  and medical profession, 46-47
  and patent medicines, 45
  demoralizing effects of, 49
White House Conferences, 619
Wickersham Commission, 64, 79, 619 see National Commission on Law Observance and Enforcement
Wild beast test of insanity, 254
Willful intent and crime, 597-598
Wire service and bookmaking, 32
Wiretapping, 226
Witches, in Massachusetts, 119
Witnesses:
  expert, 227-230, 248, 277
  in the court room, 275-277

Women criminals, see Females
Women in jails, 396
Women Reformatories, see Reformatories for women
Workhouse:
  forerunner of the penitentiary, 330-331
  today, 406
Work for prisoners, see Labor for prisoners
World War II, and prison labor, 536

Y

Yale clinic, for studies of alcoholism, 91
"Yard-out":
  in Cherry Hill, Philadelphia, 460-461
  in prisons, 455
Young adults:
  as offenders, 64-70
  specialized treatment of at Highfields, N.J., 438
Youth Authority of California, see, chart, 436
Youth gangs, 67-68, 70

C